W9-BOZ-232

RADIATION BIOCHEMISTRY

Volume II: Tissues and Body Fluids

Radiation Biochemistry

By Kurt I. Altman
UNIVERSITY OF ROCHESTER
SCHOOL OF MEDICINE AND DENTISTRY
ROCHESTER, NEW YORK

Georg B. Gerber
EURATOM
CENTRE D'ETUDE DE L'ENERGIE NUCLEAIRE
MOL, BELGIUM

Shigefumi Okada
UNIVERSITY OF TOKYO
FACULTY OF MEDICINE
TOKYO, JAPAN

Volume II: Tissues and Body Fluids
By GEORG B. GERBER and KURT I. ALTMAN

1970

ACADEMIC PRESS New York and London

ACADEMIC PRESS, INC.
111 Fifth Avenue, New York, New York 10003

United Kingdom Edition published by
ACADEMIC PRESS, INC. (LONDON) LTD.
Berkeley Square House, London W1X 6BA

LIBRARY OF CONGRESS CATALOG CARD NUMBER: 69-18344

PRINTED IN THE UNITED STATES OF AMERICA

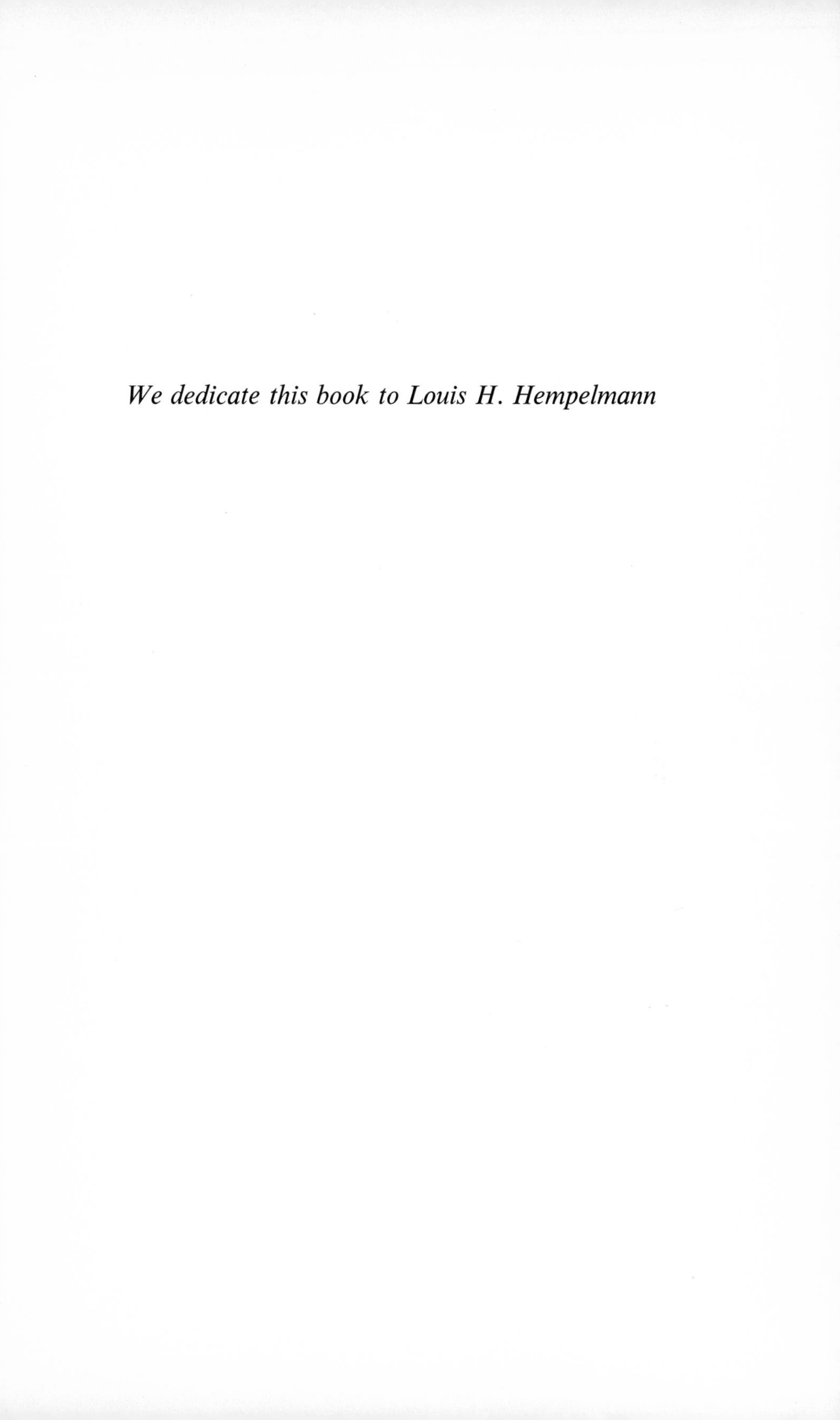

We dedicate this book to Louis H. Hempelmann

FOREWORD

A comprehensive review of the broad field of radiation biochemistry such as is presented in this treatise is long overdue. Studies of radiation biochemistry on the molecular and cellular level are now in vogue, due, in part, to new techniques, such as mammalian cell cultures and gradient centrifugation, now available for this type of experimentation. The intensive pursuit of studies in this field has led to discoveries that had only been dreamed of fifteen years ago. Although biochemical changes in tissues and body fluids of irradiated individuals are the sum of the molecular and cellular events throughout the body, there has been little effort to integrate these two aspects of radiation biochemistry or to explain the former changes in terms of the latter. As a result, interest in radiation biochemistry on the body and tissue level has declined in favor of the more basic cellular and molecular radiation biochemistry. This is illustrated by the relatively few biochemists now working on the gross metabolic phenomena caused by radiation exposure.

The main purpose of this work is to review critically the field of radiation biochemistry at all levels and to explain the observations at the body and tissue levels in terms of what is happening at the molecular and cellular levels. It is hoped that such an integrated approach will encourage interest in gross biochemical changes following irradiation. It is hoped also that the discussions in the text and the simplified schematic representation of gross biochemical changes may clarify the complex and sometimes contradictory reports in the literature. And, finally, it is my personal hope that repeated reference to possible biochemical indicators of radiation injury may stimulate work in this field and lead to the discovery of a simple, yet accurate, practical measure of acute radiation injury in man.

Since the biochemistry of the metabolic changes induced by radiation exposure is a complex subject fraught with many pitfalls, this treatise should be of enormous help to life scientists who are just embarking in the field of radiation biology. In particular, the discussions of the complications introduced by body changes secondary to radiation damage, such as partial starvation and changes in cell populations of a given tissue, should aid in avoiding errors in interpretation that have been committed in the past. Hopefully this will encourage more scientists to enter the field of radiation biochemistry.

This book represents an enormous investment of time and energy on the part of the authors. I trust that other readers will find it as useful as I have.

LOUIS H. HEMPELMANN

Department of Radiology
School of Medicine and Dentistry
University of Rochester
Rochester, New York

PREFACE

In our opinion, radiation biochemistry has three major aims: to describe radiobiological effects in biochemical terms; to elucidate the mechanisms underlying these effects; and to shed light on general biological principles in living organisms. In this two-volume treatise, we have tried to fulfill these aims to the best of our ability.

Because of space limitation, however, we have focused on the radiation biochemistry of mammalian systems, and cite only selected publications, a practice which is certain to result in the unjust omission of many important investigations. We have made a special effort to avoid mere descriptions of radiation-induced biochemical changes and have sought to seek correlations between the various biochemical changes and to interpret radiobiological effects at all levels in biochemical terms.

Volume I, written by Shigefumi Okada, shows the interplay among radiation chemistry, radiation biochemistry, radiation biophysics, and radiobiology which has brought us to our present understanding of cellular radiobiological effects in terms of molecules. The important feature of the interplay is the eradication of the demarcation lines between the disciplines of physics, chemistry, and biology. What excites many radiobiologists is the fact that the concepts developed through intensive radiobiological studies have had great bearing on the life sciences in general. These concepts involve hydrated electrons, DNA repair, artificially induced mutation, cell cycle, cell proliferation kinetics, and organ transplantation.

Volume II, written by Georg B. Gerber and Kurt I. Altman, deals with the radiation biochemistry of mammalian organs and body fluids. Emphasis is placed on descriptions of overall biochemical changes in irradiated tissues and animals, on the dependency of these changes on cellular responses, and on the interactions among different organ systems. Consideration is also given to a practical application of radiation biochemistry to the problem of assessing the nature, tissue localization, and extent of radiation injury in man and animals.

It is hoped that this work will be useful to radiobiologists at all stages of sophistication as well as to other scientists who are interested in radiobiology. It is also hoped that it will illustrate a unique facet of radiobiology in which one type of environmental insult to mankind has been studied at all levels, individually and integrated, i.e., molecular, cellular, tissue, and whole body. This may serve as a guideline for future studies of many other types of environmental insults besides ionizing radiation. We sincerely hope that readers will sense our feeling of excitement about radiation biochemistry.

S. Okada would like to thank Dr. L. H. Hempelmann for his encouragement and for the many hours he spent in reviewing the manuscript; Mrs. T. Brady for her cheerfulness and tremendous effort in typing, assembling, and checking of the manuscript; and Mrs. Y. Doida for drawing the charts in Volume I. He also thanks Drs. J. N. Stannard, F. Hutchinson, P. Reisz, S. Person, D. Billen, L. J. Tolmach, W. Szybalski, W. C. Dewey, D. K. Myers, M. W. Miller, and J. S. Wiberg for reviewing various parts of the manuscript and for their useful suggestions. Dr. Okada is grateful to the University of Rochester School of Medicine and Dentistry and to the U.S. Atomic Energy Commission [AT-(30-1)-1286 and W-7401-eng-49] for supporting his research through which the concept of this work was conceived and developed.

G. B. Gerber and K. I. Altman are indebted to many, in the United States and Europe, for their assistance at various stages in the preparation of Volume II, particularly to L. H. Hempelmann for his tireless effort in reading and criticizing all parts of this manuscript and for his often-sought advice. They are also grateful to Mrs. I. S. Ytreberg, Mr. Van Gerven, Mrs. J. Remy-Defraigne, and Mrs. K. I. Altman for typing various parts of the manuscript and to the latter also for invaluable help in editorial and stylistic matters; to Mrs. Ytreberg, Mrs. Remy-Defraigne, and Mrs. M. J. Soffer for rendering assistance in searching and collecting literature, a task in which one of us (G. B. Gerber) received much help from CID-Euratom (computerized information storage and retrieval system); to L. Schwartz, J. P. Cooper, and F. Geussens for the preparation of illustrations; to many colleagues for reading and critically discussing various parts of the manuscript among whom are Drs. Z. M. Bacq, J. F. Scaife, C. Streffer, M. F. Sullivan, R. Goutier, A. Sassen, A. M. Reuter, H. Hilz, A. Rothstein, L. L. Miller, J. N. Stannard, W. Staib, J. Hugon, J. R. Maisin, and L. G. Lajtha; and to Christine Mertz and Catherine

Robinson for help in proofreading. G. B. Gerber wishes to acknowledge support by the Division of Biology and Medicine of Euratom, an agency of the European Economic Community, as well as the use of facilities at the Centre d'Etude de l'Energie Nucleaire at Mol, Belgium. K. I. Altman is indebted to the Department of Radiation Biology and Biophysics (contract No. W-7401-eng-49, U.S. Atomic Energy Commission) and to the Department of Radiology at the University of Rochester, School of Medicine and Dentistry, Rochester, New York for support during all stages of this treatise.

We are also grateful to the staff of Academic Press for their aid and cooperation in the production of this work.

June, 1970

KURT I. ALTMAN
GEORG B. GERBER
SHIGEFUMI OKADA

CONTENTS

FOREWORD vii

PREFACE ix

CONTENTS OF VOLUME I xvi

Chapter I. **General Aspects of Radiation Biochemistry**

A. Introduction 1
B. Whole-Body and Local Irradiation 2
C. Conditions of Irradiation and Experimental Controls 4
D. Experimental Design in Radiation Biochemistry 7
E. Expression and Interpretation of Biochemical Data in Radiation Biology 9
References 10

Chapter II. **Bone Marrow and Red Blood Cells**

A. Bone Marrow 12
B. Red Blood Cells 30
References 41

Chapter III. **Lymphoid Organs**

A. Introduction 49
B. Immediate and Early Effects of Radiation on DNA Synthesis in Lymphoid Tissue 52

C. Intermediate and Late Effects on DNA Metabolism 55
D. Metabolism of RNA Proteins and Lipids 56
E. Chemical Changes in Nucleoproteins of Irradiated Lymphocytes 59
F. The Metabolic Fate of Macromolecules Present in Lymphoid Tissue at the Time of Irradiation 61
G. Energy Metabolism in Irradiated Lymphoid Organs 62
H. Lysosomal Enzymes in Irradiated Lymphoid Organs 68
I. Conclusions 71
References 73

Chapter IV. **Gastrointestinal Tract**

A. Introduction 80
B. Biochemical Changes in the Irradiated Intestine 84
C. Physiological Functions of the Gastrointestinal Tract after Irradiation 91
References 99

Chapter V. **The Liver**

A. Introduction 103
B. Lipid and Carbohydrate Metabolism 104
C. Oxidative Metabolism 141
D. Oxidative Phosphorylation 142
E. Lipid Peroxides and Lysosomal Damage 144
F. Amino Acid Metabolism 145
G. Aminotransferases 147
H. Protein Metabolism 153
I. Effect of Irradiation on the Formation and Activity of Adaptive ("Induced") Enzymes 161
J. Nucleic Acids 162
References 185

Chapter VI. **Radiation Biochemistry of Miscellaneous Organs**

A. Introduction 195
B. Connective Tissue 197
C. Blood Vessels and Adipose Tissue 199
D. Cartilage and Bone 199
E. Skin 203
F. Muscle 208
G. Kidney 214
H. Eye 217
I. Brain 219
References 225

Chapter VII. **Radiation Biochemistry of Tumors**

A. Introduction 235
B. Radiosensitivity and Composition of Tumor Cells 235
C. Metabolism of Macromolecules 237
D. Glycolysis, NAD Metabolism, and Other Biochemical Changes in Irradiated Tumor Cells 239
References 244

Chapter VIII. **Changes in the Biochemistry of Body Fluids after Irradiation**

A. Introduction 248
B. Fluid and Electrolyte Metabolism 250
C. Compounds of Low Molecular Weight in Blood and Urine after Irradiation 253
D. Enzymes in Body Fluids 268
E. Serum Proteins after Irradiation 268
F. Aids in the Biochemical Diagnosis of Radiation Injury 276
References 277

Chapter IX. **Hormones and Systemic Effects**

A. Introduction 287
B. Diencephalon and Central Nervous System Regulation after Irradiation 288
C. Pituitary Gland 289
D. Adrenal Cortex 291
E. Reproductive Organs 299
F. Thyroid Gland 303
G. Biogenic Amines 307
H. Systemic Effects 311
I. Irradiation, Stress, and Systemic Responses 314
References 323

AUTHOR INDEX 331

SUBJECT INDEX 367

CONTENTS OF VOLUME I

CELLS

I. Radiation Effects on Molecules
II. Actions of Radiation on Living Cells
III. DNA as Target Molecule Responsible for Cell Killing
IV. Are There Other Target Molecules?
V. Molecular Events in Irradiated Bacteria
VI. Radiation Effects on Cell Progress through the Life Cycle
VII. Radiation-Induced Death
VIII. Chromosome Aberrations and Mutations
Appendix 1. G Value (G_0)
Appendix 2. Survival Curves and Target Theory (Definitions of D_{34}, D_0, D_q, and n)
Appendix 3. Life cycle of Cultured Mammalian Cells in Log Growth Phase

Chapter I

GENERAL ASPECTS OF RADIATION BIOCHEMISTRY

A. Introduction

A delineation of the biochemical response of mammalian organisms to radiation must touch upon several levels: (a) it must define the biological events in the cell at the molecular level, as was attempted in Volume I of this treatise; (b) it must integrate the cellular events with the changes occurring in individual organs; (c) it must take into account the interaction between the different cells and organs and the humoral and nervous regulators of metabolism. From the beginning, radiation biology emphasized the morphological sequelae of irradiation as seen grossly or with the microscope. Recently the dynamic concept of the renewal of cells and their constituent molecules has been added to these static aspects of radiation damage. A quantitative account of the behavior of metabolic activity and its regulation is still a distant goal in radiation biology. Progress toward this goal can be achieved only to a limited extent because structure and biological function are not separate entities, but two aspects of the same picture.

This chapter introduces several basic notions of mammalian radiation biology and then points out the pitfalls and opportunities inherent in design and interpretation of experiments in radiation biochemistry.

B. Whole-Body and Local Irradiation

1. Whole-Body Irradiation

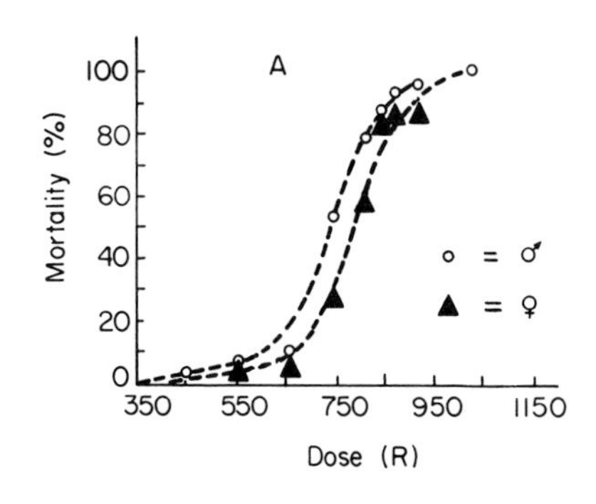

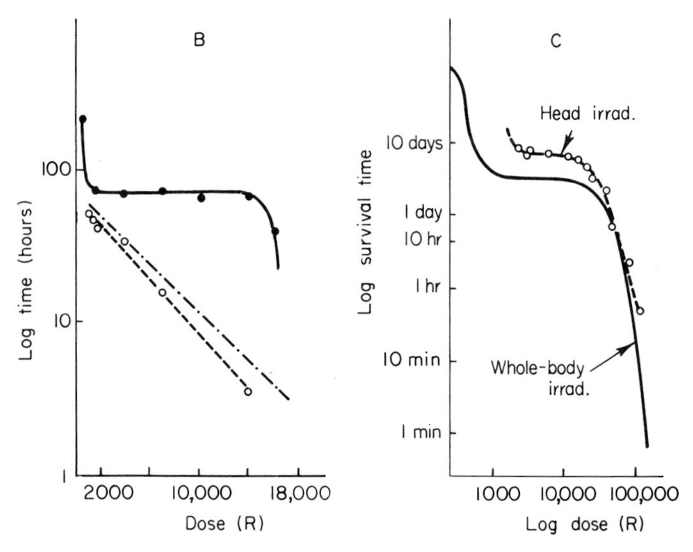

FIG. I-1. (A) Mortality of male and female mice after whole-body irradiation. [From W. H. Chapman, *Radiation Res.* **2**, 502 (1955).] (B) Mean survival time of normal, adrenalectomized, and hypophysectomized rats after whole-body irradiation: solid line—control; dash line—hypophysectomized; dot-dashed line—adrenalectomized. [From G. Gerber, "Wissenschaftliche Grundlagen des Strahlenschutzes," pp. 149–162, 1957.] (see also *10*). (C) Mean survival time of male mice after whole-body or head irradiation. [From B. Rajewski, K. Aurand, and O. Heuse, *Z. Naturforsch.* **8b**, 524 (1953).] Female mice are more radioresistant than male ones. The dose-independent survival time after doses causing gastrointestinal death becomes dependent on the dose in adrenalectomized or hypophysectomized rats. The survival time after head irradiation is independent of the dose but longer (about 9 days) than that after whole-body irradiation (3.5 days). Above 30 kR the curves for head and whole-body exposure coincide.

Exposure to ionizing radiation of the whole body or of large parts of the mammalian organism produces a characteristic sequence of acute symptoms which depend on the radiation dose (*3*). (i) Relatively low doses of radiation (up to about 1000 R) damage primarily the blood-forming tissues (see Chapter II). The renewal of stem cells in the bone marrow is impaired and widespread cell death occurs in lymphoid tissue. This injury to the cellular and immunological defense mechanisms of the organism may cause death as a result of infection, bleeding, or anemia from 1 to several weeks after exposure. Exposure to radiation which does not produce death within a few weeks may shorten the life-span by accelerating the unspecific processes of aging or by inducing cancer and leukemia. (ii) In the dose range from 1 kR to about 10 kR, cell renewal in the gastrointestinal tract is severely disturbed and the animals die shortly after the epithelial lining of the intestinal villi has been lost (3–7 days after exposure), presumably as a consequence of the loss of electrolytes and water. The survival time for gastrointestinal death is dose-independent (Fig. I-1). (iii) After still higher doses, injury to the central nervous system (see Chapter IX, I) plays the predominant role. Again, the survival time depends on the dose and, in mice, varies from almost 3 days after a dose of 20 kR to a few minutes after a dose of 100 kR (Fig. I-1). Death usually intervenes following convulsions and coma.

2. Local Irradiation

Local irradiation can produce a variety of symptoms, depending on the radiation dose and on the tissues exposed. If an organ responsible for death after whole-body irradiation (e.g., intestine or central nervous system) is submitted to local exposure, the symptoms will be similar to those seen after whole-body exposure (Fig. I-1). Uncommon modes of acute death can occur after radiation injury of other vital organs or areas (e.g., skin, kidney, and lungs). Most frequently, local symptoms predominate after local irradiation, and the outcome depends on the extent of the local tissue destruction, vascular damage, and fibrosis, and on possible late malignant changes.

The decision whether whole-body or local irradiation should be used in a biochemical investigation, depends, of course, on the particular problem being studied. Whole-body exposure is a simple and reproducible procedure, and is best suited for studies related to radiation illness. Local irradiation can be used in the examination of biochemical changes in a given organ. Local irradiation alleviates or eliminates the intervention by other organs and systemic reactions and, except at very high dose levels, circumvents the difficult problem of dietary controls (see p. 6). Thus, biochemical effects of radiation in organs can be studied for long periods or at relatively high dose levels. On the other hand, local irradiation usually requires restraining or anesthetizing the animals, and these factors must be considered in the interpretation of the data.

C. Conditions of Irradiation and Experimental Controls

Biochemical alterations due to radiation exposure have to be distinguished from those due to factors secondary to the exposure procedure. The radiation insult should therefore be administered in a reproducible manner and controls should be provided to eliminate extraneous factors. The conditions of radiation

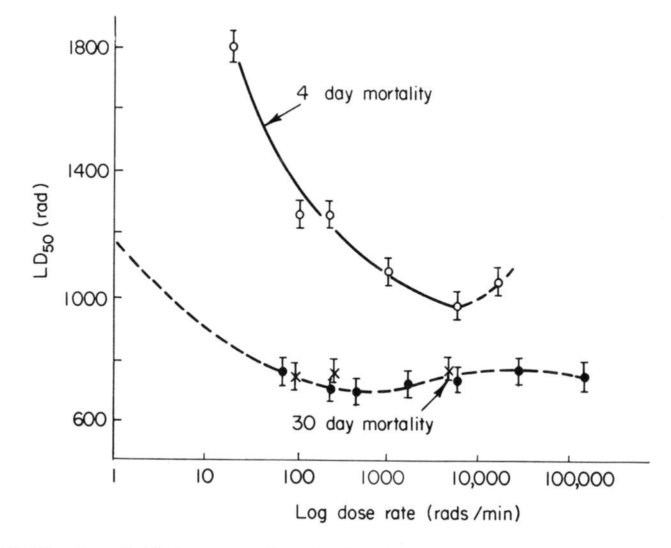

FIG. I-2. The 4- and 30-day mortality of mice (LD_{50}) after irradiation at different dose rates. Note that at high dose rates, mice become more sensitive to the gastrointestinal syndrome but not to the bone marrow death. [From S. Hornsey and T. Alper, *Nature* **210**, 212 (1966)] (see also *10a*, *26*).

should be chosen so as to ensure a homogeneous exposure of the volume desired and to reduce the stray radiation outside this volume to a minimum. In publications, one should present information not only about the dose but also about the dose rate (see Fig. I-2), the kilovoltage of x-rays, the filtering (or half-value layer), and the source to target distance.

Irradiation involves handling and restraining the animals, factors which, as such, may elicit a biochemical stress response. Control animals should, therefore, be sham-irradiated under the same conditions and at the same time as the animals are x-irradiated. Preconditioning the animals to the exposure conditions can substantially reduce the stress response to sham irradiation

(*6b*). Moreover, certain types of investigations, especially irradiation of exteriorized organs, require treatment with anesthetics or muscle relaxants. Again, the effects of these drugs must be controlled. It is quite obvious that only animals of the same strain, age, sex, and weight should be used as controls. Moreover, in certain critical investigations, such as the excretion of hormones or biogenic amines, it is often not appropriate to compare the results with those of previous experiments. Instead, the experimental groups should consist only of animals x-irradiated and sham-irradiated at the same time.

Radiosensitivity is not an immutable property of an animal strain, but

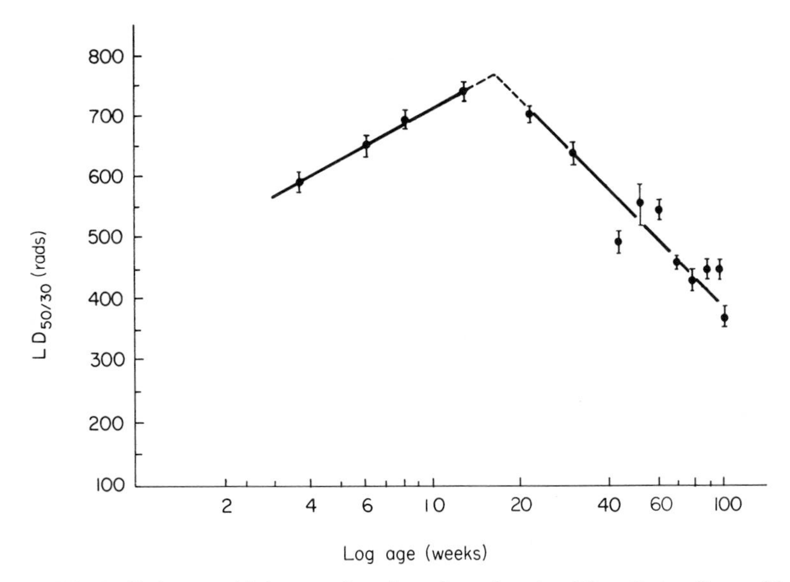

FIG. I-3. Radiation sensitivity as a function of age in mice. Note that radiosensitivity diminishes until young adulthood then rises again. [From J. F. Spalding, O. S. Johnson, and R. F. Archuleta, *Nature* **208**, 905 (1965).] Other authors report a period of high radioresistance immediately after birth.

depends on many factors. The LD_{50} should, therefore, be specified in publications and determination of the LD_{50} should be repeated at reasonable intervals or whenever a change has taken place in the conditions of animal care, e.g., housing, cleaning, air conditioning, temperature control, feeding, and sterilization of food and cages. A few of the factors influencing radiosensitivity need to be specified in more detail. Radiosensitivity of mice diminishes with age until young adulthood and then increases again (Fig. I-3) (*1*, *8a*, *20*, *27*). Newborn mice are, however, more radioresistant than those a few days old. (*6a*, *12*). The radiosensitivity of rats varies with age in a somewhat different way (*18a*, *18b*). Sex also plays a role in the response to radiation. Females are

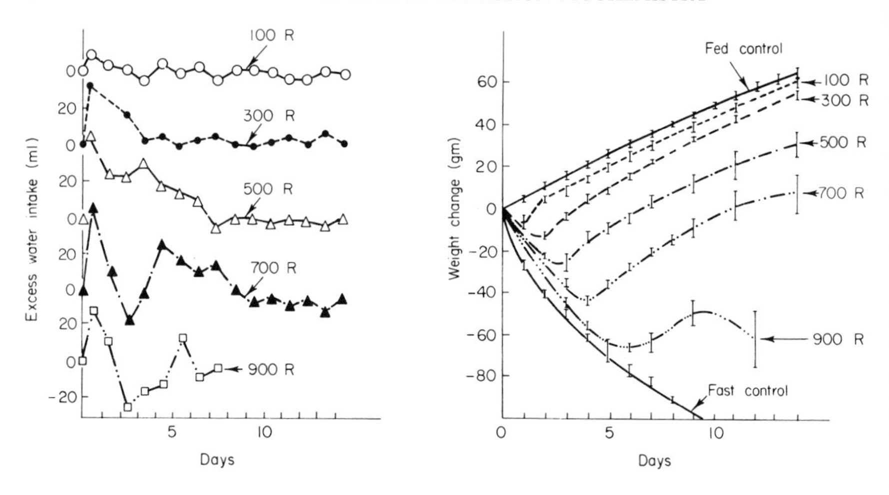

FIG. I-4. Water intake and weight changes in fed rats after whole-body irradiation. The values are expressed in percent of the controls on day 0. Note that water intake during days 1 and 2 increases with the irradiation dose, as does also the length of time during which the body weight diminishes. No recovery in body weight takes place after irradiation with 900 R. [From L. F. Nims and E. Sutton, *Am. J. Physiol.* **195**, 271 (1958); **198**, 762 (1960).]

usually more radioresistant than males (*1*, *6*, *12*, *13*, *19*) and display cyclical changes in their LD_{50} related to the phase of estrus. During estrus, sensitivity is low and in diestrus it is higher (*19a*). Less well appreciated are the diurnal and seasonal fluctuations in the response to irradiation. Rats are more resistant if irradiated in the morning than in the evening (*7*, *8*, *16*, *17*, *19*), or in the spring rather than the fall (*16*), but circadian rhythms, which may be related to adrenal activity (*7*), have not been found by other investigators (*18*). Mice live longer under conditions of continuous irradiation if kept at lower temperatures (*4*).

Infection may complicate any animal experiment but can be a particular problem in radiation biology, as resistance to infection is low during radiation illness. Many radiobiological experiments have been spoiled because an unexpected infection intervened. Pathogen-free or even germ-free animals are desirable but not easy to obtain. Addition of chlorine or antibiotics to the drinking water is a useful precaution to prevent infections, but its advantages must be balanced against its biochemical effects in certain studies. Individual housing of the animals after irradiation is preferable since it restricts the spread of infections during radiation illness.

A most difficult problem is controlling the dietary intake after irradiation. After whole-body irradiation, animals usually eat little or no food (*14*, *15*, *21*, *22*, *23*, *24*), have delayed gastric emptying, and may vomit (see p. 311); therefore, they lose weight for a period of a few days (Fig. I-4). Anorexia depends, however, on the species, and is much more pronounced in rats than in guinea pigs or hamsters (*14a*). Moreover, more liquid is ingested (Fig. I–4) and excreted in the urine from the first to the third day after irradiation (see p. 312). The state of partial starvation created by these factors must be taken into account when considering these effects. One can pair feed a control group of animals, giving them as much food as the irradiated animals ate the day before (*5*). Pair feeding is not the ideal control (*2*), and does not always give satisfactory results. Aside from the fact that the control group is delayed 1 day compared with the irradiated one, pair feeding does not take into account nonabsorbed food retained in the stomach.

Complete starvation, with water *ad libitum*, for both sham- and x-irradiated animals is, in the opinion of the authors, the best solution for short-term experiments, especially if an additional fed control group is investigated simultaneously. The data on the extent and direction of the changes due to starvation alone permit a better evaluation and interpretation of the radiation-induced alterations. Thus, the glycogen level in liver increases after irradiation compared with that of starved animals but decreases if compared with that of fed animals. This finding suggests that the normal adaption to starvation is disturbed in irradiated animals (see p. 140). In long-term experiments (more than 3–4 days in rats) one must resort to pair feeding. One should, however, never compare irradiated animals only with fed ones.

D. Experimental Design in Radiation Biochemistry

Metabolic investigations of irradiated organisms frequently involve three types of experimental designs. The opportunities and limitations of each design will be discussed briefly in this section.

1. Determination of Enzymatic Activity in Tissues

The enzymatic activity in tissues is determined after breaking up the structure of the organ and often of the cell. Homogenization involves, however, a destruction of the intracellular structures and it may thus alter the activity of cellular enzymes and obscure or accentuate the changes after irradiation. Moreover, excess substrate is added for the enzymatic assay and the pH adjusted to optimal conditions. In this way, the total activity of an enzyme *in vitro* is measured, but not its activity *in vivo*. An extrapolation from the amount of enzyme measured *in vitro* to the amount of substrate converted *in vivo* is legitimate only if it can be demonstrated that the concentration of the enzyme, and not that of the substrate at the enzymatic sites, limits the rate of the reaction *in vivo*. Since such a proof is not often presented, one should be reluctant to draw important conclusions solely on the basis of data on enzymatic activity *in vitro*.

2. Determination of Substances in Tissues and Body Fluids

The determination of substances in tissues and body fluids, like the determination of enzymatic activity, yields, essentially, a static picture of metabolism and does not reveal its dynamic aspects, such as the movement of metabolites in and out of organs. Nevertheless, both techniques are relatively simple to execute and useful as indicators for radiation-induced alterations of metabolism, but they should be combined with isotope experiments if a comprehensive interpretation of the changes is desired.

3. Experiments with Isotopically Labeled Precursors

Experiments with isotopically labeled precursors could, in principle, provide extensive information concerning the kinetics of metabolic reactions *in vivo*. For such an exhaustive interpretation, the plots of activity versus time, as well as the concentrations of metabolites in different organs of the body should be known. Yet, it is seldom possible to take enough sequential samples and to isolate and determine all metabolites in question in order to obtain such extensive data. Often, data on incorporation of a precursor at a single point in time are thought to be sufficient for an assessment of the synthesis of macromolecules in normal and x-irradiated animals. This procedure is, however, correct only if the precursor enters the cell at the same rate, and the time course of its incorporation and catabolism as well as its intracellular concentration are the same in sham- and x-irradiated animals. On the contrary, pool sizes change markedly after irradiation, and nonsteady conditions often prevail during the course of the experiment. Under these conditions, the interpretation of such simplified isotope experiments in the whole animal becomes uncertain. To some extent, one can discriminate between radiation-induced changes of precursor metabolism and those in the metabolic replacement of a macromolecule if one injects the labeled precursor some time prior to exposure. The changes in specific and total radioactivity of the macromolecules then permit an estimate of its replacement and loss (see pp. 61; 212). This method, however, does not allow the experimenter to recognize transient changes of short duration, even if of considerable magnitude. In conclusion, an interpretation of metabolic alterations in the irradiated organism should not rest solely on data from one type of experimental design; rather, all information available from different types of experiments should be integrated into a composite evaluation.

E. Expression and Interpretation of Biochemical Data in Radiation Biology

Experimental data should be expressed in a way that supports the interpretation proposed. Biochemical changes in the irradiated animal can be of significance for the organism as a whole and/or for a certain organ only. When pointing out changes in biochemical functions which affect the total organism, one could best present the data per organism or, if only one organ is involved in a certain function, per organ. Radiation-induced alterations in the biochemistry of an organ may represent changes at the level of single cells but, more frequently, reflect changes in the cell population of the organ. It should always be remembered that no organ has a uniform cell population; even in the simplest organs, connective tissue and macrophages are omnipresent.

1. Changes in Cell Population

Changes in cell population due to selective killing, failure to reproduce, or translocation of certain types of cells are common in many organs after irradiation e.g., bone marrow, testis, intestine, and lymphoid organs. Since the enzymatic endowment can vary from one cell type to another, a shift in the cell population in favor of one cell type may result in a change in the biochemical activity of an organ with a corresponding change in activity per cell.

2. Changes in Single Cells

Biochemical changes in single cells can be due to altered synthesis or catabolism of cellular constituents, or to the loss from their normal intracellular sites in the cell. Moreover, cells can be arrested by irradiation at a certain stage in their cell cycle, e.g., cannot progress from the resting stage, G_0 to S (DNA synthesis), and, therefore, will not produce the enzymes needed to synthesize DNA. Finally, cell death may manifest itself in a variety of biochemical alterations. Only by establishing the time sequence of the different biochemical changes can one avoid confusing cause and result in such changes in dying cells.

Biochemical changes important for the organ and its constituent cells should be expressed per average number of cells. In many, but not all (see below) cases, it is permissible to express the data per unit of DNA, e.g., as specific activity of an enzyme per milligram DNA. If one desires to point out to what extent the composition of the organ as a whole has changed after irradiation, data on the total activity per organ should also be presented.

Three examples, which are discussed in more detail elsewhere in this book, follow to illustrate these thoughts.

(i) The specific activity, i.e., the activity per average number of cells, of DNase and other lysosomal enzymes, increases in lymphoid organs after irradiation (see p. 68). Simultaneously, the number of cells and the weight of these organs diminish. Therefore, the total activity of the enzyme per organ remains constant or increases much less than the specific activity. The specific activity of the enzymes increases since more enzyme-rich, radioresistant reticulum cells are preserved in the organ as the radiosensitive lymphocytes die and are phagocytozed by reticulum cells and macrophages. On the other hand, when the total activity of an enzyme is elevated, an increase in the ratio of *de novo* synthesis to catabolism or a translocation of unusual cellular elements, e.g., macrophages rich in lysosomes, into the organ must have occurred. In this case one would be interested in the increase in lysosomal enzymes per average number of cells as well as in the *de novo* synthesis or translocation of the enzyme. Data on both specific as well as total enzymatic activity should, therefore, be presented.

(ii) The DNA content in the gastrointestinal tract diminishes after irradiation as cells are lost and cell renewal is impaired (see p. 85). In a certain dose range, the destruction of the intestinal epithelium causes the death of the irradiated animals (p. 81). The DNA content of the intestine is a function of the number of epithelial cells, and, therefore, constitutes a useful measure of the damage to the whole organism.

(iii) Liver regenerates after partial hepatectomy by division of its parenchymal cells (see page 161). Several enzymes, e.g., thymidine kinase, DCMP deaminase, and DNA polymerase, are not present in a normal liver cell and must be produced prior to DNA synthesis. Irradiation prior to the operation prevents the transition of cells from the resting stage G_0 to the S phase, and these enzymes are, therefore, not synthesized. The metabolism of a given cell type is altered in this case, and the change is best expressed per cell. Yet, the concentration of DNA per cell increases during the experimental period, and is thus not a suitable point of reference. Enzymatic activity may then be related to the number of nuclei in the organ.

REFERENCES

1. Abrams, H. L., *Proc. Soc. Exptl. Biol. Med.* **76**, 729 (1951).
2. Baker, D. G., and Hunter, C. G., *Radiation Res.* **15**, 573 (1961).
3. Bond, V., Fliedner, T. M., and Archambeau, J. O., "Mammalian Radiation Lethality." Academic Press, New York, 1965.
4. Carlson, L. D., Scheyer, W. J., and Jackson, B. H., *Radiation Res.* **7**, 190 (1957).
5. Caster, W. O., and Armstrong, W. D., *J. Nutr.* **59**, 57 (1956).
6. Chapman, W. H., *Radiation Res.* **2**, 502 (1955).

6a. Crosfill, M. L., Lindop, P.J., and Rotblat, J., *Nature* **183**, 1729 (1959).
6b. Flemming, K., *Strahlentherapie* **134**, 565 (1967).
7. Fochem, K., Michalika, W., and Picha, E., *Strahlentherapie* **128**, 441 (1965); **133**, 256 (1967); **135**, 223 (1968).
8. Fochem, K., Michalica, W., and Picha, E., *Strahlentherapie* **130**, 590 (1966).
8a. Fowler, J. F., *Proc. 2nd Internat. Congr. Radiation Res., Harrogate, Engl., 1962*, p. 310. North-Holland Publ., Amsterdam, 1963.
9. Gerber, G., "Wissenschaftliche Grundlagen des Strahlenschutzes," pp. 149–162. 1957.
10. Gerber, G. B., *Strahlentherapie* **110**, 510 (1959).
10a. Heuss, K., *Biophysik* **4**, 146 (1967).
11. Hornsey, S., and Alper, T., *Nature* **210**, 212 (1966).
12. Ingbar, S. H., and Freinkel, N., *Federation Proc.* **11**, 77 (1952).
13. Langendorff, H., and Koch, R., *Strahlentherapie* **94**, 250 (1954).
14. Ledney, G. D., and Wilson, R., *Radiation Res.* **22**, 207 (1964).
14a. Newson, B. D., and Kimeldorf, D. J., *Am. J. Physiol.* **195**, 271 (1958); **198**, 762 (1960).
15. Nims, L. F., and Sutton, E., *Am. J. Physiol.* **171**, 17 (1952).
16. Peters, K., *Strahlentherapie* **122**, 554 (1963).
17. Pizzarello, D. J., Witcotski, A. L., and Lyons, E. A., *Science* **139**, 349 (1962).
17a. Rajewsky, B., Aurand, K., and Heuse, O., *Z. Naturforsch.* **8b**, 524 (1953).
18. Reincke, U., *Z. Naturforsch.* **23b**, 393 (1968).
18a. Reincke, U., Mellmann, T., and Goldmann, E., *Strahlentherapie* **129**, 622 (1966); *Intern. J. Radiation Biol.* **13**, 137 (1967).
18b. Reincke, U., Goldman, E., Mellmann, J., and Jesdinsky, H. J., *Strahlentherapie* **136**, 349 (1968).
19. Rugh, R., *in* "Handbuch der Med. Radiol. Strahlenbiologie," Vol. II, p. 438. Sprinter, Berlin, 1966.
19a. Rugh, R., and Clugston, H., *Radiation Res.* **2**, 227 (1955).
20. Sacher, G. A., *Science* **125**, 1039 (1957).
21. Smith, D. E., Tyree, E. B., Patt, H. M., and Jackson, E., *Proc. Soc. Exptl. Biol. Med.* **78**, 774 (1951).
22. Smith, D. E., and Tyree, E. B., *Am. J. Physiol.* **177**, 251 (1954).
23. Smith, D. E., and Tyree, E. B., *Radiation Res.* **4**, 435 (1956).
24. Smith, W. W., Ackermann, I. B., and Smith, F., *Am. J. Physiol.* **168**, 382 (1952).
25. Spalding, J. F., Johnson, O. S., and Archuleta, R. F., *Nature* **208**, 905 (1965).
26. Woodward, K. T., Michaelson, S. M., Noonan, T. R., and Howland, J. W., *Intern. J. Radiation Biol.* **12**, 265 (1967).
27. Yuhas, J. M., and Storer, J. B., *Radiation Res.* **32**, 596 (1967).

Chapter II

BONE MARROW AND RED BLOOD CELLS

A. Bone Marrow

1. Introduction

Bone marrow is composed of a heterogeneous cell population, but investigators often fail to appreciate this important feature when interpreting radiation-induced biochemical changes in this organ. Nevertheless, unless one uses autoradiography (which permits identification of individual cell types) or investigates a substance (e.g., hemoglobin) which is present in only one cell type, all experiments concerned with bone marrow irradiated *in vivo* must deal with a heterogeneous cell population which contains several functionally distinct categories of rapidly dividing cells, namely, the erythropoietic, the myeolopoietic, and the thrombopoietic cells.

It is, therefore, unrealistic to interpret changes in bone marrow after irradiation solely in terms of biochemical processes; rather, one must also take into consideration changes in the composition of the prevailing cell population. This limitation of *in vivo* studies of bone marrow could be eliminated to a certain extent by appropriate tissue culture techniques which are capable of selecting cell types to obtain a nearly homogeneous cell population; a discussion of this approach, however, is beyond the scope of this book.

TABLE II-1

SYNOPSIS OF IMPORTANT BIOCHEMICAL AND MORPHOLOGICAL CHANGES IN BONE MARROW AFTER IRRADIATION

Time after exposure	Morphology	Nucleic acids	Nucleid acid precursors	Hemoglobin synthesis	Selected enzymatic reactions
Immediate (0–2 hr)	Arrest of mitosis	Decreased incorporation of precursors	No change	Stimulation: bone marrow, reticulocytes	Stimulation of respiration
Early (2–24 hr)	Decrease in number of bone marrow stem cells because of interphase death	Amount and synthesis diminish, breakdown from cell destruction	Increase in pyrimidine nucleotide (dCMP, dUMP, UMP, CMP); decrease in ATP, etc., content at 4 hr	Normal in bone marrow; decreased in circulating blood	Decreased glycolysis, increased synthesis and esterification of fatty acids; activation of lysosomal enzymes
	Lipocytes increase Reticulocytes disappear from bloodstream				
Intermediate (1 day to 1–2 weeks)	Number of stem cells and erythroblasts diminish progressively because of reproductive death; later, abortive recovery may occur	Progressive decrease in synthesis and content corresponding to cellularity		Progressive decrease	
Late (from 2 weeks on)	(a) Recovery	Increased incorporation; return to normal concentrations	dTMP, dTDP, and dTTP content increases	Return to normal	Return to normal
	(b) Continuing depletion of stem cells resulting in death	Decreased synthesis and content		Decrease	

2. Lipid Metabolism

In addition to the rapidly dividing cell lines mentioned above, bone marrow also contains lipocytes, which contain mainly triglycerides and thus resemble the cellular elements of adipose tissue found in other parts of the mammalian body (*224*). In rat bone marrow (*224*), triglycerides represent about 85% of the esterified fatty acids, and of these fatty acids, about 80% are palmitic and oleic acid. In addition, small quantities of phospholipids and cholesterol, as well as traces of glyceryl ethers and free fatty acids of varying chain lengths are found. Lipocytes can be easily separated from other cells because they float when centrifuged in a suitable medium (*224*).

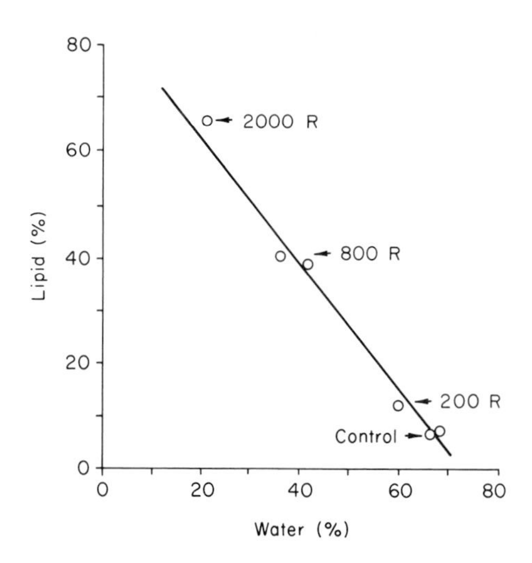

Fig. II-1. The effect of total body irradiation on the relationship of water and lipid content of rat bone marrow to radiation dose. All rats were sacrificed 8 days after total body exposure to radiation from ^{137}Cs sources at a dose rate of 4.08 R/min average tissue dose. Marrow was obtained from femurs. Nutritional state of the animals is not mentioned. [From F. Snyder and E. A. Cress, *Radiation Res.* **19**, 129 (1963).]

When hematopoietic cells disappear from the marrow cavity of irradiated animals, lipocytes increase in number and fill part of the free space remaining after cell loss (*224*, *226*). The lipid content of the marrow increases correspondingly (Fig. II-1), and the water content diminishes (*40*, *224*). Such an increase in lipid content has been observed in many mammals, e.g., rabbit (*11*, *40*, *47*) and rat (*18*, *181*, *223–226*), and takes place only in the marrow directly exposed to radiation (*226*). Only the triglyceride content (*40*, *47*, *223*, *224*) increases in irradiated bone marrow; the phospholipid and cholesterol content is moderately diminished (Fig II-6) (*134*, *149*, *224*).

The radiation-induced increase in triglycerides could result from: (i) increased esterification and synthesis of fatty acids via the acyl CoA transferase pathway, (ii) transport from other parts of the body to the bone marrow, and (iii) decreased catabolism of triglycerides.

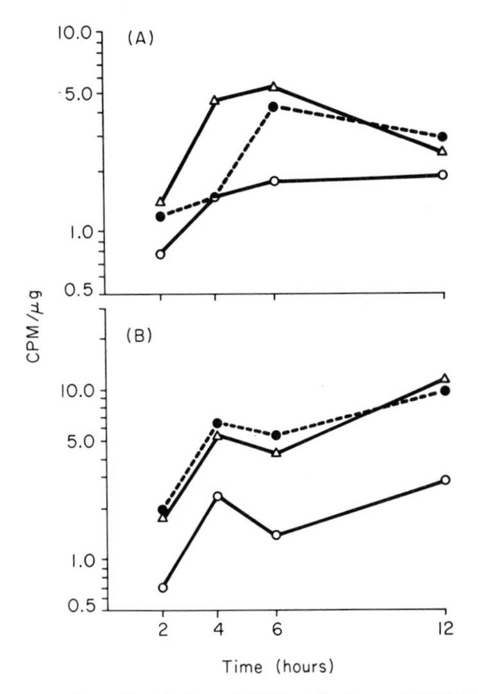

FIG. II-2. Incorporation of palmitate-1-^{14}C and oleate-1-^{14}C into lecithin and cephalin of rat bone marrow after total body irradiation. Rats were exposed to 800 R from 8 ^{137}Cs sources at a dose rate of 4.43 R/min air dose and sacrificed 4 days after irradiation. Irradiated and control animals were killed 2, 4, 6, and 12 hr after administration of labeled fatty acid. (A) Incorporation of oleate-1-^{14}C into lecithin and cephalin. (B) Incorporation of palmitate-1-^{14}C into lecithin and cephalin. (△) Irradiated rat, lecithin; (○) nonirradiated rat, lecithin; and (●) irradiated rat, cephalin. [From J. K. Lim, C. Piantadosi, and F. Snyder, *Proc. Soc. Exptl. Biol. Med.* **124**, 1158 (1967).]

The first alternative seems the more plausible one because the rate of incorporation of either ^{14}C-palmitate or ^{14}C-stearate into bone marrow triglycerides *in vitro* as well as *in vivo* is more rapid in irradiated than in nonirradiated bone marrow (Fig. II-2) (*1–5*, *224*, *225*); the accelerated incorporation of these precursors into triglycerides occurs in lipocytes and erythropoietic cells (*214*). By contrast, irradiation fails to stimulate the incorporation of oleic acid into triglycerides (*226*); perhaps the formation of phosphatidic-acid-derived precursors requires esterification of α-hydroxyl groups of glycerol by a saturated long-chain fatty acid (*226*). The triglycerides synthesized in irradiated bone marrow differ from those of normal marrow in that they contain a greater quantity of monoethenoic and diethenoic fatty acids of longer than C_{18} chain length (up to C_{22}), more stearic acid, but less arachidonic acid (*18*). Phospholipid synthesis, however, is inhibited after irradiation (*149*), and the pools of lecithin and cephalin (*134*, *181*) diminish as the marrow is depleted of hematopoietic constituents.

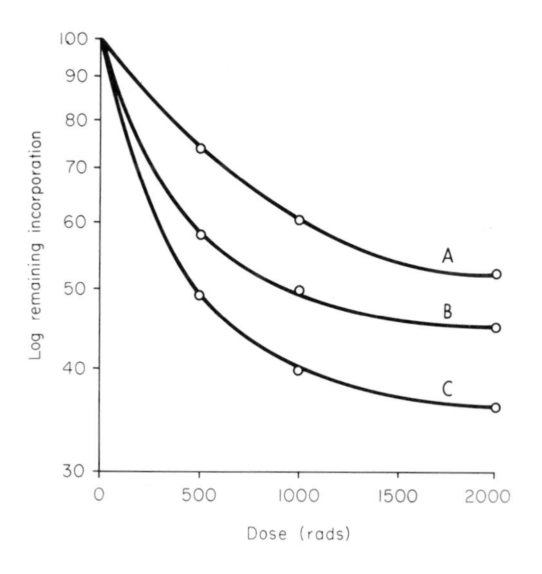

FIG. II-3. Incorporation of acetate-1-^{14}C into total lipids of human bone marrow cell cultures irradiated *in vitro*. Curve A: Acetate-1-^{14}C added 20 hr after irradiation (^{60}Co γ-irradiation); curve B: Acetate-1-^{14}C added 44 hr after irradiation; curve C: Acetate-1-^{14}C added 68 hr after irradiation. [From C. J. Miras, J. D. Mantzos, and G. M. Levis, *Radiation Res.* **22**, 682 (1964).]

Most probably, the fatty acids utilized for esterification have been synthesized *de novo* in irradiated bone marrow and are not derived from extramedullary sites for the following reasons: (i) The plasma level of fatty acids and triglyceride do not increase as much as the levels in the bone marrow (*47*); (ii) incorporation of acetate into saturated and unsaturated long-chain fatty acids is enhanced after irradiation (*5*).

The synthesis of oleic and other long-chain unsaturated fatty acids is maximal 1 week after exposure and may account for the accumulation of these unsaturated fatty acids in the irradiated bone marrow at this time. It appears conceivable that this increase in unsaturated fatty acids is not specific for radiation injury but rather reflects the anemic state of the bone marrow since it is seen also in anemic cats (*104*).

Contrary to events *in vivo*, acetate incorporation into lipids of irradiated human bone marrow culture (*158*) is as depressed as that when bone marrow homogenates are irradiated *in vitro* (Fig. II-3) (*180*). Lipid synthesis in marrow cultures, however, is not comparable with that occurring under *in vivo* conditions since, in the former, lipocytes are usually absent.

Oxidation of fatty acids diminishes in rat bone marrow after exposure to 100 R for a period of 55 days (*222*). As a consequence, more fatty acids become available for esterification and may therefore contribute to the increase in triglyceride content. The inhibition of fatty acid catabolism cannot be solely responsible for this increase, however, since it can be detected even after the postirradiation lipid levels in the bone marrow have returned to normal (*224*). Finally, it should be mentioned that homogenates of rabbit bone marrow prepared 48 hours after 1400 R of whole body exposure form peroxides on incubation (*11*), whereas homogenates of normal bone marrow do not. The significance of this observation remains uncertain except that it may be related to the excessive amounts of unsaturated fatty acids accumulating in irradiated bone marrow.

3. Hemoglobin Biosynthesis

Biosynthesis of hemoglobin, one of the major functions of nucleated red cells, involves two separate metabolic processes, namely, the synthesis of globin and that of heme. Only in the final steps do these processes converge to form hemoglobin. The following factors must be considered when interpreting radiation-induced changes: (i) the number of bone marrow cells capable of synthesizing hemoglobin; (ii) the availability of enzymes required for synthesis; (iii) the availability of precursors of hemoglobin, some of which originate outside the bone marrow; and (iv) the transport of these precursors in the bloodstream and across the membrane of hemoglobin-forming cells.

4. Biosynthesis of Heme

Exposure to ionizing radiation rapidly depletes the bone marrow of hematopoietic cells (*6*, *116*, *116a*, *119a*, *143–145*), including cells capable of synthesizing hemoglobin. One must ask, therefore, whether or not all the changes in hemoglobin synthesis can be explained satisfactorily on the basis of reduction in the number of heme-synthesizing cells (*121*, *234*). Heme synthesis after irradiation is often equated with the rate or ^{59}Fe uptake by bone marrow cells

(*8*, *9*, *93*, *95*, *121*, *234*, *247*) or reticulocytes (*7*), using mainly autoradiographic and chemical techniques. Autoradiographs reveal the presence of radioactivity in a cell but fail to disclose which intracellular constituent has incorporated the radioactive tracer. Except for a few studies in which the isolation of ^{59}Fe-labeled heme was actually carried out, adequate data relating to ^{59}Fe uptake to ^{59}Fe incorporation into heme (*20*, *53*, *81*) in a quantitative way appear lacking.

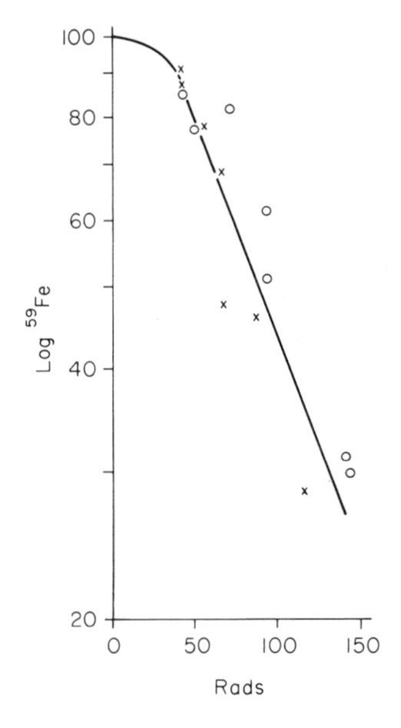

FIG. II-4. Uptake of ^{59}Fe by bone marrow. Semilogarithmic plot of total dose and ^{59}Fe uptake by rat bone marrow, calculated as percent of controls for each irradiation. Ordinate: ^{59}Fe uptake in bone marrow as percent of controls. (○) ^{60}Co γ-rays and (×) fission neutrons plus γ-ray contaminants. [From W. A. Rambach *et al.*, *Radiation Res.* **10**, 148 (1959).]

In the experiments to be described, one must distinguish between the uptake of precursors, e.g., ^{59}Fe (*7–9*, *20*, *200*) by cells of the bone marrow and by cells of circulating blood. The uptake by circulating blood cells is depressed as early as 6 hours after exposure of rats to 600 R when measured *in vitro* (*99*) and only slightly later after doses of 50–100 R. Uptake of ^{59}Fe is reduced on the first or second day after exposure (*6*, *8*, *9*, *72*, *78*, *186*, *256*), reflecting a decrease in the number of erythroblasts which stop dividing and, subsequently, diminish in number (Fig. II-4) (*49*, *72*, *86*, *121*, *174*, *186*). This does not impair the ability of surviving erythroblasts to take up ^{59}Fe at a normal rate (*69*), as shown by the observation that the grain counts in individual erythroblasts from rabbits 2 days after exposure to 220 R and from comparable control animals are similar (*234*).

The rate of incorporation of 5-aminolevulinate and porphobilinogen into heme does not change up to 21 days after exposure (rabbits) if the data are expressed on the basis of DNA content to approximate cellularity (*35*, *206*). Reticulocytes in the circulating blood take up only 3–4% of injected ^{59}Fe, the younger reticulocytes being most active in this respect (*4*, *6–9*, *15*). The number of circulating reticulocytes declines a few hours after irradiation (*51*) and so does the uptake of ^{59}Fe by these cells (Fig. II-5) (*4*, *5a,b–9*, *15*, *20*, *21*, *49*, *59*,

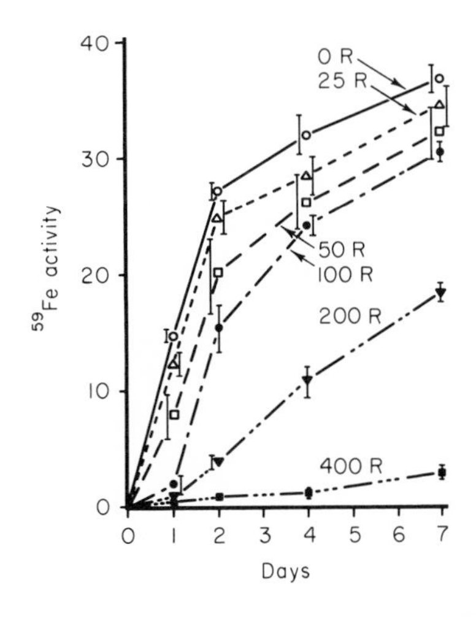

FIG. II-5. Uptake of ^{59}Fe by rat erythrocytes after exposure of rats to varying doses of x-rays. ^{59}Fe in citrate buffer was injected 48 hr after x-irradiation. ^{59}Fe activity represents percent of injected activity appearing in circulating blood. [From E. H. Belcher, I. G. F. Gilbert, and L. F. Lamerton, *Brit. J. Radiol.* **27**, 387 (1954).]

63–65, *70*, *72*, *73*, *77*, *78*, *80–83*, *85*, *86*, *88*, *93–95*, *97*, *98*, *101*, *102*, *110*, *116*, *121*, *123*, *138*, *156*, *183*, *186*, *205*, *232*, *234*, *247*). At this time the erythroblastic population in the bone marrow is not yet greatly altered and one must assume therefore that the release of reticulocytes from the irradiated marrow has been impaired (*8*). Other factors such as changes in permeability or rate of maturation of reticulocytes (*169*) may also contribute to the early depression of iron uptake by the circulating blood (*99*).

Avian erythrocytes retain their nuclei throughout their life-span. Nevertheless, duck erythrocytes take up 75% less ^{59}Fe on the first and 90% less on the second day after total-body irradiation with 600 R than nonirradiated red blood cells (*95*).

Whereas all data considered above are compatible with the idea that changes in cell population can account for apparent changes in hemoglobin synthesis (*7*, *88*, *121*, *174*, *234*), certain other observations demand a different explanation. These observations are as follows: (i) incorporation of glycine-2-^{14}C (*191*), aminolevulinate (*35*, *206*), or porphobilinogen (*35*) is stimulated in bone marrow homogenates isolated immediately after total body x-irradiation of rabbits with 800 R; (ii) the uptake of glycine-2-^{14}C by dog reticulocytes (*170*) or chicken erythrocytes (*239*) is enhanced 30 minutes after exposure; (iii) 3 or 38 hours after exposure, the incorporation of glycine into hemoglobin is lower than would be expected on the basis of erythroblast counts (*200*).

Early stimulation of precursor uptake may reflect permeability changes in cells and/or subcellular particulates, but could also result from utilizing heme precursors in metabolic pathways which do not lead to heme formation. It is noteworthy that other metabolic reactions, especially oxygen consumption (*192*), are also stimulated in bone marrow immediately after exposure to x-irradiation. The enzyme 5-aminolevulinate synthase, which forms 5-aminolevulinate from succinate and glycine, is one of the key enzymes regulating hemoglobin synthesis (*89*). Its activity is elevated in nucleated chicken red blood cells immediately after and for 2 days following total body irradiation with 750 R (*153*). Although increased activity of this enzyme could account for increased incorporation of glycine-2-^{14}C into heme (*191*), this does not apply to the contribution by 5-aminolevulinate (*35*, *206*) or porphobilinogen (*35*) to heme biosynthesis. Other enzymes involved in heme synthesis, namely, ferrochelatase (E.C. 4.99.1.1) (which catalyzes the incorporation of iron into protoporphyrin 9) and porphobilinogen synthase (E.C. 4.2.1.24) (which participates in the synthesis of porphobilinogen from 5-aminolevulinate) exhibit the following changes as compared to AL synthase.

PBG Synthase. There is an immediate rise in activity after total body x-irradiation of rats with 50 R but no rise with 400 and 650 R (*37*), although activity is elevated at 18 and 21 days after exposure.

Ferrochelatase. The activity does not change immediately after exposure, but decreases markedly on the second and third day after 1000 R in rabbit bone marrow (*36*, *155*).

AL Synthase. This enzyme exhibits increased activity immediately and 1 and 2 days after exposure when it is assayed *in vitro* in preparations of subcellular particles of red blood cells obtained from total body x-irradiated (750 R) hens (*153*). In this species, heme synthesis is also stimulated immediately after irradiation (*239*).

Transport of iron in the plasma by the β-globulin siderophilin and the release of iron to hemoglobin-synthesizing cells are additional factors which may affect hemoglobin synthesis after irradiation. Indeed, the release of iron from its protein carrier is significantly retarded in rabbits exposed total-body to 800 R although lower radiation doses are of little importance in this respect (*80*, *82*). The minimal radiation dose that retards iron release varies considerably from species to species (*137*, *138*).

It is conceivable that irradiation damages the synthetic processes of a cell in such a way that abnormal types of hemoglobin are formed. Such an effect, if demonstrable, would have considerable theoretical interest. Consequently, some effort has been devoted to the search for an abnormal hemoglobin. One such radiation-induced structural modification of hemoglobin is characterized by an altered O_2-saturation curve (*235*). More interesting, although this was not investigated further, is the occurrence of a hemoglobin fraction, present in trace amounts in irradiated chick embryos, but absent in nonirradiated controls (*182*, *201*).

5. Biosynthesis of Globin and Other Proteins

The synthesis of globin in irradiated bone marrow has been assessed on the basis of the extent of incorporation of labeled amino acid precursors. Immediately after *in vitro* or *in vivo* exposure to 800 R, dog reticulocytes incubated with phenylalanine-3-^{14}C *in vitro* after exposure, incorporate more ^{14}C into globin (*171*) than nonirradiated cells, an effect resembling that seen in the synthesis of heme. However, immediately and from 1 hour to 21 days after total body exposure of rabbits to 800 R, bone marrow homogenates (*191*) incorporate glycine-2-^{14}C into globin at a normal rate.

The synthesis of proteins other than globin has also been studied. Incorporation of glycine-1-^{14}C into proteins (*1*) is altered only slightly, but incorporation into nucleic acids is much depressed in bone marrow of rabbits 5 hours after exposure to 800 R. Increased incorporation of methionine-^{35}S into proteins, followed later by a decrease, is also found (*101*, *189*, *190*). In the latter case, the depression of protein synthesis is most marked at 7–10 days after exposure to 100 R, i.e., at a time coincident with the lowest levels of nucleic acid and protein content in bone marrow. The transfer of 5-methyl groups from methionine to intermediates of amino acid and nucleic acid metabolism is not impaired in bone marrow of rats x-irradiated with 750 R (total body). This is shown in experiments in which methionine-2-^{14}C and -methyl-^{14}C are used; only the level of incorporation of methionine-2-^{14}C is lowered slightly with time after irradiation (*46*). The apparent labeling of proteins is undoubtedly influenced by changes in cell population, but definitive studies correlating biochemical with autoradiographic data are still lacking in this area.

Changes in the protein composition of bone marrow also occur after irradiation, as demonstrated by electrophoretic (*172*) and immunoelectrophoretic means (*249*, *264*) (Table II-1). It is difficult, however, to distinguish changes in the behavior of bone marrow proteins from changes in the electrophoretic properties of proteins in blood. Electrophoresis on cellulose acetate strips reveals a decrease in prealbumin and globulin concentration in rat bone marrow either after a single exposure to 1500 R or after multiple exposures to 250 R (*172*). Transient changes in protein composition of bone marrow are detectable by means of immunoelectrophoresis in x-irradiated mice of several different strains and suggest the presence of precipitins in irradiated marrow (*249, 264*); these immunochemical changes may, however, not result entirely from irradiation. Nucleoproteins isolated from bone marrow of rabbits 1–6 days after exposure to 200 or 2000 R contain a larger percentage of the "residual protein" fraction than their nonirradiated counterparts (*106, 220, 251*).

6. Metabolism of Nucleic Acid Precursors

Changes in the concentration of nucleic acid precursors develop in the bone marrow beginning 4 hours after exposure (Fig. II-6). They include changes in pyrimidine deoxyribo- and ribonucleotides (*26, 108, 131, 195, 196, 228, 230*) and in ATP content (*148*). Pools of nucleic acid precursors change after irradiation if (i) nucleic acids liberated from dying cells are degraded, (ii) synthesis of nucleic acids decreases (*87*) while synthesis of precursors remains unchanged, (iii) changes in cell permeability affect the distribution of precursors between extra- and intracellular spaces, (iv) the amount of precursors supplied by other organs is altered (*69, 122*), and (v) the activities of key enzymes converting one precursor to another are modified.

As discussed below, the destruction of nucleic acids and the inhibition of their synthesis (i and ii) probably determine the early changes in precursor pool sizes, whereas at a somewhat later time only inhibition of synthesis is important. Changes in enzymatic activity (v) decide mainly which types of precursors are found after irradiation. Nothing seems to be known about the influence of the remaining factors (iii and iv). In summary, two biochemical characteristics of bone marrow must be considered when interpreting radiation-induced changes in the metabolism of nucleic acid precursors: First, bone marrow depends on liver for its supply of purines (*122, 178*); second, bone marrow reutilizes pyrimidines extensively (*54*) via the salvage pathway.

Free deoxyribopolynucleotides (soluble in 0.14 *M* NaCl and insoluble in 0.2 *M* $HClO_4$) accumulate in bone marrow of rats (*218, 219*) and rabbits (*196*) within a few hours after exposure to doses in the range of LD 50–LD 100/30 days, but as later observed a significant increase occurs in rats even after irradiation with only 30–50 or 100 R (*217, 250*). This effect in rats is a linear function of the dose up to about 300 R (*218*), so that one could consider utilizing this phenomenon to assess radiation damage on the basis of bone marrow specimens obtained from persons accidentally exposed to ionizing radiation.

Deoxyuridylic (dUMP) and deoxycytidylic acid (dCMP) comprise the major portion of free deoxyribonucleotide in the bone marrow of rabbits exposed to 1000 R (*228*). This may reflect altered activities of dCMP aminohydrolase (*233*) and thymidylate (dTMP) kinase (*19*). These are the key enzymes of nucleic acid metabolism which fail to appear after irradiation of regenerating liver. Later (days after 100 R), an increase in thymidylate and its di- and triphosphates can be detected (*228*), and this suggests that cells capable of forming thymidine (dTR) have reappeared in the course of recovery of irradiated bone marrow.

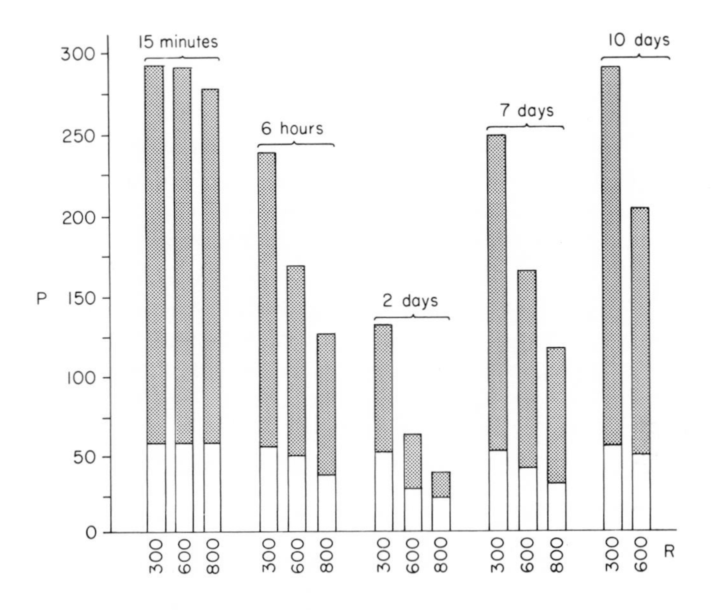

FIG. II-6. Changes in nucleic acid and lipid phosphorus content in rat bone marrow after total body x-irradiation. The unshaded areas represent lipid phosphorus and the shaded areas nucleic acid phosphorus, expressed as mg P/100 gm of tissue. Three different dose levels were employed: 300, 600, and 800 R. [From C. Lutwak-Mann, *Biochem. J.* **52**, 356 (1952).]

The ribonucleotides, particularly those of the pyrimidine bases, increase appreciably about 1–4 hours after exposure (*48*, *105*, *107*, *108*, *230*) of one hind leg (rabbit) to 2000 R. It is conceivable that the free nucleotides originate from degraded RNA in dying cells. Two days after local irradiation under these conditions the content of AMP, IMP, GMP, and UMP decreases precipitously (*105*, *107*), whereas ribonucleotide content of rat bone marrow remains elevated for 24 hours after total body x-irradiation with 600 R. Since the activity of the enzyme which specifically dephosphorylates UMP decreases after irradiation (*195*), UMP degradation may occur to only a limited extent. Little is known concerning the biosynthesis of ribonucleotides in bone marrow after irradiation.

Incorporation of ^{32}P phosphate into nucleotides is unaltered or even enhanced immediately after exposure but begins to diminish 4 hours thereafter (*44*, *137*). A slower rate of phosphorylation or a greater dilution of the nucleotides by inactive material derived from degradation of DNA might be responsible for such a decrease in incorporation.

In contrast to the increase in pyrimidine nucleotides, the concentration of ATP is reduced (*70*) in bone marrow soon after irradiation (500–1000 R). The level of ATP is determined mainly by the energy metabolism of the cells and only to a minor extent by RNA metabolism. Therefore, the content of ATP may be diminished when less AMP is phosphorylated or when more ATP is degraded. Indeed, considerable ATP might be degraded before or during isolation procedures, since, in dying cells, phosphatases are liberated from their subcellular structures and thereby become active even though their total activity does not change during the first day after exposure.

Finally, it should be mentioned that not all investigators agree that pyrimidine nucleotides increase after irradiation (*69*). Some investigators report a decrease in deoxyribo- and ribonucleotides in irradiated bone marrow. This discrepancy cannot be explained satisfactorily at the present time except by presuming that the procedures of isolation and determination may have differed sufficiently to yield divergent results.

7. Biosynthesis of Nucleic Acids

Incorporation of precursors into nucleic acids, especially into DNA, is reduced in bone marrow immediately after irradiation as it is also in other tissues (*1*, *10*, *12*, *44*, *45*, *50*, *87*, *101*, *102*, *117–120*, *128*, *130*, *132*, *152*, *177*, *218*, *221*, *229*, *246*, *250*). This has been demonstrated for all species studied using a variety of labeled precursors (see exception below). The degree of reduction in DNA and RNA synthesis depends mainly on the radiation dose and is usually detectable within 1 hour after exposure. Mitotic arrest is not responsible for inhibition of DNA synthesis immediately after exposure, but later on, because of the reduction in the number of DNA synthesizing cells, it is the apparent reason for impairment of DNA synthesis.

When nucleic synthesis is depressed and cells are lost either because of early death in interphase or release into the bloodstream, the nucleic acid content of the bone marrow diminishes (*26*, *107*, *120*, *148–150*, *157*, *193*, *197*, *240*) (Fig. II-7). One day after exposure to 1000 R, the content of DNA decreased to 50% of the controls, and that of RNA to 40% (*193*). The concentration of DNA per nucleus does not change under these conditions (*251*), whereas RNA is lost mainly from extranuclear sites, i.e., from the cytoplasm, microsomes, and mitochondria.

The early and intermediate effects of radiation on nucleic acid metabolism are influenced by several factors, e.g., by the type of precursor administered, by the conditions of irradiation, and by compounds stimulating nucleic acid synthesis.

Incorporation into DNA and RNA of ^{32}P (*26*, *44*, *101*, *103*, *118*, *154*, *221*, *239*) and of one- and two-carbon precursors (*1*, *10*, *12*, *119*, *152*, *178*, *228*, *246*) [formate (*10*, *12*, *50*, *152*, *178*, *228*, *246*) and glycine (*1*, *102*)], as well as incorporation of thymidine into DNA (*10*, *87*, *101*, *173*, *194*, *218*, *250*), are all inhibited to about the same extent shortly after irradiation. However, incorporation of adenine (*118*) into DNA and RNA of bone marrow is only slightly affected by irradiation. Differences in the utilization of purines and

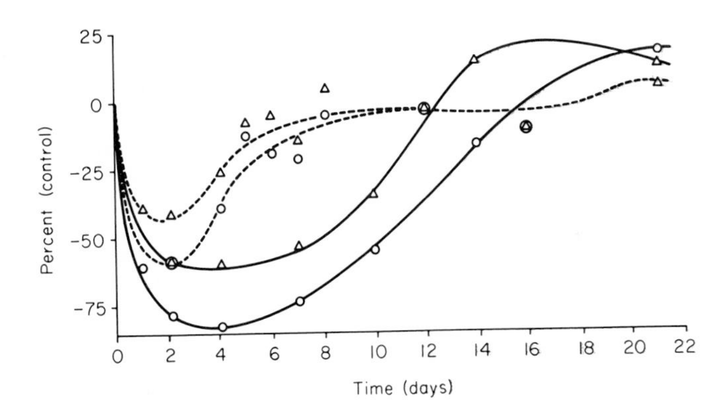

FIG. II-7. The effect of local and total body x-irradiation on the content of nucleic acids in rat bone marrow as a function of time after exposure. Total dose, 700 R. (△) RNA and (○) DNA. Broken lines indicate local x-irradiation given unilaterally to one hind leg, whereas solid lines signify total body irradiation. In all instances the rats were exposed only once. [From P. Mandel *et al.*, in "Advances in Radiobiology" (G. C. de Hevesy, A. G. Forssberg, and J. D. Abbatt, eds.), p. 59. Oliver & Boyd, Edinburgh and London, 1957.]

pyrimidine precursors for nucleic acid synthesis may be attributable to the following factors acting in concert: (i) inhibition of pyrimidine synthesis (ii) inadequate supply of purine nucleotide precursors either because of impaired ability to store precursors or because of limited availability of the latter from extramedullary sources, chiefly liver (*122*) (see also Chapter V), and (iii) change in the size of the intramedullary adenine pool because of radiation-induced changes in ATP metabolism paralleling changes in nucleic acid metabolism.

Since few data on the specific activity of various precursors in the cell are available, it cannot be decided which of the above interpretations is valid.

Extramedullary factors, such as starvation, can influence the degree of damage to irradiated marrow. Total body exposure of rats to 750 R reduces incorporation of ^{32}P into DNA to 45%, and into RNA to 65% of the controls. Local irradiation of the femur results in a depression to 60 and 75% respectively (*154*).

The addition of deoxyribonucleotides, such as dCMP, dAMP, or dTMP, to marrow suspensions (guinea pig) 6 hours after exposure to 600 R stimulates (*228*) incorporation of formate-^{14}C or ^{32}P into the DNA of reticulum cells. Injection of dCMP into irradiated animals restores the base composition of marrow DNA to normal and promotes mitotic division. No unifying interpretation entirely consistent with our current knowledge of DNA metabolism can be offered.

Depending on dose and species, recovery of bone marrow may commence by the end of the first week after exposure (*6*, *51*, *241*, *242*). The recovery is characterized by a return to normal of the nucleic acid content of the bone marrow. In rats exposed to neutrons (140 R) (*39*), recovery starts about one day later than in rats exposed to an equivalent dose of x-rays, although damage estimated on the basis of the decrease in DNA content is about the same in both cases. In the course of recovery, DNA and RNA concentrations in the bone marrow may increase above normal levels about 3 weeks after the irradiation and then gradually decline to normal (*154*). This "overshoot" reaction is probably related to the regulation of cell replacement in bone marrow and is more pronounced after total body than after local irradiation.

8. Other Metabolic Alterations in Irradiated Bone Marrow

Bone marrow possesses a high glycolytic activity, effectively utilizes acetate and various Krebs cycle intermediates (*2*), forms relatively large amounts of ammonia since its amino hydrolases are very active, and contains abundant quantities of hydrolytic enzymes such as phosphatases and nucleases. Phosphatases seem to be located mainly in myelocytic elements and cholinesterase chiefly in megakaryocytes, but little is known about the distribution of other enzymes among the different cell types. As has been stated, biochemical information available on irradiated bone marrow often ignores the heterogeneity of the cell population involved and, at best, provides only an indication of the over-all metabolism of this organ.

Unlike circulating blood cells, irradiated bone marrow cells exhibit a marked and progressive decrease in activity of several glycolytic enzymes and modifications in concentration of related metabolites (Table II-2). These

TABLE II-2

METABOLIC CHANGES IN BONE MARROW AFTER IRRADIATION

Enzymatic activity	Effects on metabolic activity
Oxygen consumption	Increase in homogenates immediately after 800 R X-irradiation of rabbits (*192*); 5% decrease on day 2 in rabbits after 800 R, 30% on day 14, and return to normal after 4 weeks; similar decrease in chicken (300 R) (*41*)
Succinate: (acceptor); oxidoreductase, (E.C. 1.3.99.1) succinate dehydrogenase	Decrease a few days after irradiation of chickens (*41*) but not rabbits (*240*)
Choline: (acceptor): oxidoreductase, (E.C. 1.1.99.1) choline dehydrogenase	Decrease in chicken a few days after 300 R (*41*)
Glycolysis	
ATP: D-hexose-6-phosphotransferase, (E.C. 2.7.1.1) hexokinase	
D-glyceraldehydephosphate: NAD oxidoreductase (E.C. 1.2.1.12) glyceraldehyde phosphate dehydrogenase	Progressive decrease in rat bone marrow immediately after 750 or 1500 R, falling to minimal levels on day 6 (*141*)
D-glucose-6-phosphate: NADP oxidoreductase (E.C.1.1.1.49) glucose-6-phosphate dehydrogenase	Decrease (*140*)
Various metabolic intermediates of glycolysis	Immediately after 750 R (rat), decrease of glucose-6-phosphate, fructose-6-phosphate, and fructose-1, 6-diphosphate content; increase dihydroxyacetone phosphate content; lactate-pyruvate ratio unchanged Two days after irradiation, lactate and almost all phosphorylated intermediates decrease; fructose-6-phosphate and pyruvate increase (*141*)
ATP	Progressive decrease until day 6 after 750 R (*141*)
Reduced-NADP: oxidized glutathione oxidoreductase (E.C. 1.6.4.2) glutathione reductase	Decrease early after exposure (*140*)

TABLE II-2 (contd)

Enzymatic activity	Effects on metabolic activity
Hydrolytic enzymes:	
ATP phosphohydrolase (E.C. 3.6.1.3) ATPase	Decrease after first day proportional to hematopoietic cell loss (*130*, *140*)
5′-ribonucleotide phosphohydrolase (E.C.3.1.3.5)	
Acetylcholinesterase (E.C.3.1.1.7) acetyl choline hydrolase	Diminishes to about 70% 2 days after 400 R; determined histochemically, activity/megakaryocytes unaltered after 5 kR (*74*, *146*, *147*)
Cholinesterase, (E.C. 3.1.1.8) Acylcholine acyl hydrolase	Increases to 150% in guinea pigs 24 hr after 400 R, reduced in plasma and liver (*146*, *147*)
Deoxyribonucleate oligo-nucleotidohydrolase (E.C. 3.1.4.5) DNase I	
Deoxyribonucleate 3′-nucleotidohydrolase (E.C. 3.1.4.6) DNase II	Activity increases progressively, maximal levels at 7 days after 780 R (23-fold normal), later return to normal; earlier recovery after treatment with bone marrow suspensions or when spleen is shielded (*114*)
	Effect on DNases possible result of inhibitor destruction; elicited in part also by an abscopal exposure (*115*, *129*)

changes begin immediately after *in vivo* exposure and are most marked when the functional cells of the bone marrow are depleted. Early after exposure to radiation electrolyte concentrations in bone marrow change subsequent to loss of K^+ from bone marrow cells and influx of Ca^{2+} from the bloodstream leaking through damaged capillary sinuses. Loss of NAD, an important feature in irradiated tumor cells (see Chapter VII, 239), and leakage of enzymes and metabolites from the cells could also impair glycolysis in bone marrow soon after irradiation.

Oxygen uptake increases markedly in bone marrow homogenates immediately after exposure (Table II-2) but diminishes later (*192*). The effect of radiation on dehydrogenases depends on the species in question (Table II-2) (*41*, *240*).

Increased nucleolytic activity (*108*, *114*, *115*, *128*, *129*) and decreased ATPase and 5′-nucleotidase (*195*) activity are the most significant changes in the activity of rabbit bone marrow hydrolases. Bone marrow homogenates (rabbit) allowed to autolyze, liberate RNA breakdown products on the first day after exposure (*109*). Later, this autolytic activity decreases, but since activity of the DNases increases progressively until recovery sets in (*129*) this may be only a result of the reduction in nucleic acid substrate occurring with time. It has been suggested that destruction of inhibitors is responsible for the increase in DNases (*114*), but, in the opinion of the authors, the presence of such inhibitors has not yet been established unequivocally and the methods employed in the assay of DNase activity, i.e., binding of methyl green to DNA, are open to criticism.

9. Erythropoietin

Since erythropoietic activity is greatly depressed by partial- or total-body exposure to ionizing radiations, the response of irradiated bone marrow to erythropoietin (ESF) and the effect of irradiation on ESF production in the kidney are of considerable interest because of the possible relation to mechanisms for erythropoietic recovery and radiation-induced inhibition of erythropoietic activity. The mechanism of action of ESF is still a matter of controversy and different investigators claim that ESF acts at various stages of red blood cell formation by promoting the following: (i) the transition from primitive "stem" cell precursors to ESF-responsive cells thus initiating differentiation into erythroblasts (*3a*, *72*, *73*, *83*, *116a*, *119a*, *185*, *205*), (ii) the proliferation and maturation of normoblasts (*6*), (iii) the release of reticulocytes from bone marrow into the circulating blood stream (*112*, *154a*), and (iv) the synthesis of DNA-dependent RNA (messenger RNA?) by stimulating, for example, the incorporation of uridine-2-^{14}C into RNA without affecting the incorporation of ^{14}C-dTR into DNA (*103*). The stimulating effect of ESF is detectable within 15 min, i.e., much earlier than the ESF-induced enhancement of hemoglobin synthesis. Since actinomycin D inhibits ESF-induced RNA synthesis, RNA of the messenger RNA type is probably involved (*103*). In this context it is of interest to note that actinomycin D and X-rays act on the same type of ESF-sensitive cells (*71*).

At low doses of total body irradiation (about 200 R) ESF is capable of restoring to normal levels the ability of erythrocytes to take up ^{59}Fe (*3a*, *72*, *112*), but efforts directed toward modifying the mortality rate by ESF administration have failed (*33a*). However, ESF enhances the efficacy of isologous bone marrow transfusions after exposure to 840 R (*266*, *267*). In a test animal such as the polycythemic mouse responsiveness to a challenging dose of ESF declines immediately after total body irradiation, but returns to normal or above normal levels 10 days later (*3a*, *70a*, *72*, *184*, *185*). The magnitude of the decrease in ESF responsiveness and the degree of recovery from radiation damage depends primarily on (i) the radiation dose received and (ii) the challenging dose of ESF administered (*3a*, *70a*, *116a*, *119a*, *184*, *205*). In the mouse irradiated with 200 R, low doses of ESF (about 1 unit) are most effective whereas in the nonirradiated mouse high doses (12 units) elicit a response (*205*).

Biosynthesis of ESF by the kidney and release from the site of biosynthesis into the bloodstream are not depressed after total body irradiation with 300, 500, or 1000 R (*55*, *84a*), but exposure of the exteriorized kidney to 3600 R causes significant reduction of ESF synthesis. After irradiation ESF levels in plasma (*96*, *97*) and in bone marrow (*70a*, *73*) are elevated. Radiation-induced destruction of radiation-sensitive erythroid cells may constitute a stimulus for increased production of ESF postirradiation.

B. Red Blood Cells

1. Introduction

The mature mammalian red blood cell or erythrocyte utilizes anaerobic glycolysis as its principal source of energy, but it is limited with respect to many other biochemical functions, especially synthetic functions of which most mammalian cells are capable. The erythrocyte has no nucleus and its life-span is limited once it enters the circulation; one might, therefore, consider the erythrocyte a dying cell and not representative of other mammalian cells. Despite these shortcomings, research in radiation biology has, from its beginning (*84*), employed the erythrocyte as a convenient object of study since it (i) is easily accessible and readily obtained in large quantities, (ii) is uniform in size and has a characteristic metabolism, (iii) has already been widely studied with respect to permeability and transport, and (iv) lends itself to facile isolation of cell membranes (red cell ghosts) which retain most of the metabolic, structural, and transport characteristics of the red blood cell.

Investigations of erythrocytes irradiated *in vivo* or *in vitro* have been concerned mainly with changes in permeability and cation transport (*23, 24, 28, 32, 34, 38, 61 ,67, 99, 100, 124, 159, 160, 162–164, 167, 168, 175, 179, 180, 211–213, 215, 216, 237, 238, 243, 245, 248, 252*), osmotic fragility (*38, 60, 66, 68, 90, 162, 204, 207, 208, 210, 244, 254*), hemolysis (*27, 28, 31, 38, 53, 61, 66, 75, 84, 100, 133, 135, 162, 180, 202, 203, 214–216, 231, 261*), and life-span of the cell (*33, 42, 63, 65, 75, 84, 88, 113, 136, 176, 199, 202, 203, 209, 210, 214–216, 231, 248, 255, 261*) as well as changes in various metabolic functions.

2. Changes in Ion Transport of Irradiated Erythrocytes

As with many other cell types, red cells possess a "pump-leak" system involving Na^+ and K^+. The system regulates the ion flux across the cell membrane at which site a Na^+-, K^+-activated, ouabain-inhibited ATPase is located. The latter enzyme appears to be an integral part of this transport system.

Exposure of red cells to 20–40 kR damages the cell membrane in such a way that K^+ leaks out and Na^+ enters the cell. After some delay, these prolytic

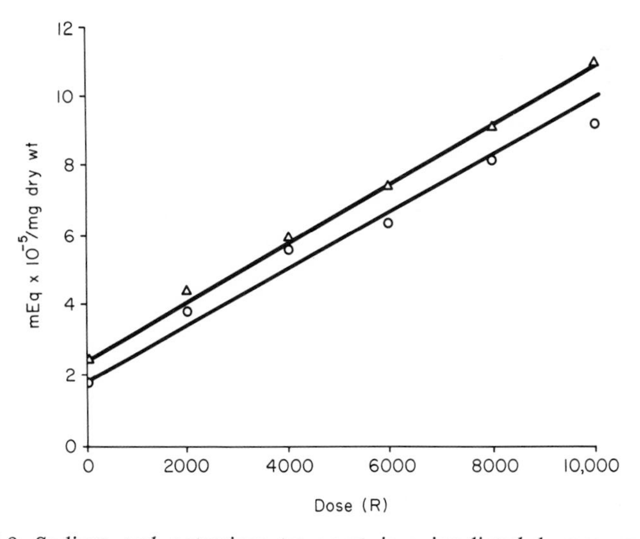

FIG. II-8. Sodium and potassium transport in x-irradiated human erythrocytes. Cation exchange 20 hr after exposure and at 4°C is plotted as a function of radiation dose. (△) Sodium uptake and (○) potassium loss. [From B. Shapiro, G. Kollmann, and J. Asnen, *Radiation Res.* **27**, 139 (1966).]

changes resulting in increased osmotic pressure may cause the cell to burst and lose its hemoglobin; i.e., hemolysis sets in. In a medium of 0.9% NaCl the quantity of Na^+ entering exceeds slightly that of K^+ leaving the cell. To maintain an osmotic equilibrium, water flows into the cell and the cell volume thus increases. Radiation-induced changes in cation flux in irradiated erythrocytes could be altered if (i) active transport is impaired, (ii) passive transport at certain sites of the cell membrane is facilitated, and (iii) gross structural defects in the membrane are induced.

Each of these factors influences the cation flux in irradiated erythrocytes to an extent which depends upon the radiation dose. Gross structural defects of the membrane probably develop only after intermediate doses of radiation or, with considerable delay, after extremely high doses. Composition and temperature of the medium in which the cells are kept after exposure largely determine the extent and the mechanism of the changes in permeability.

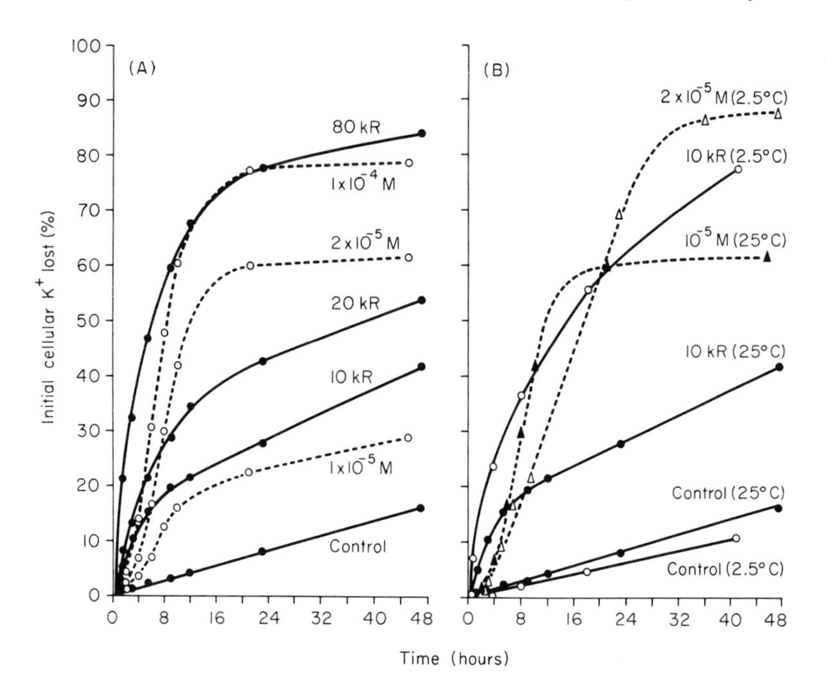

FIG. II-9. Leakage of K^+ from x-irradiated human erythrocytes. (A) K^+ loss as a function of radiation dose or concentration of PCMBS; (B) K^+ loss as a function of temperature. Incubation temperatures in (A) 25°C and (B) 25° or 2.5°C, as indicated. The solid lines represent x-irradiated red cells, whereas the broken lines describe the behavior of cells treated with *p*-chloromercuribenzenesulfonate. In (A) incubation commences at t_0. [From R. M. Sutherland, J. N. Stannard, and R. I. Weed, *Intern. J. Radiation Biol.* **12**, 551 (1967).]

Active transport of sodium has been studied in irradiated red blood cells by comparing the flux of ^{22}Na in the presence and absence of extracellular K^+. The dose–effect curve for active transport consists of two components, one radiosensitive (D_0 740 R) representing about 60% of the active transport and the other radioresistant (D_0 about 10,000 R) (*23*). A similar dose–effect curve is obtained for inactivation of ouabain-sensitive ATPase (*24*). The view that impairment of active Na^+ transport is responsible for loss of K^+ has been challenged (*162*, *165*, *237*), since radiosensitivity of ATPase could not be confirmed by others (*165*) and, also, since active transport of glucose and K^+ was found more radioresistant than that of Na^+ (*164*).

Passive flux of cation, i.e., a loss of K^+ and an uptake of Na^+, is accelerated by *in vivo* exposure to several kR (Fig. II-8). This acceleration is most pronounced when the cells are stored at low temperature after exposure (Fig. II-9).

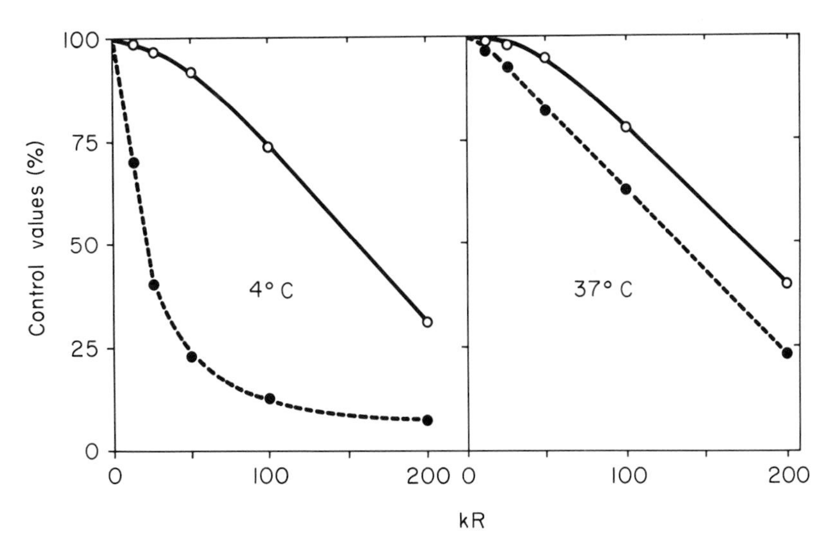

FIG. II-10. Changes of K^+ and hemoglobin content of rat erythrocytes at two different temperatures at various times after x-irradiation. Red blood cells suspended in 20 vol of Krebs–Ringer phosphate solution containing glucose and 20% of calf serum were incubated at 4°, or 37°C for 20 hr; (○–○) is hemoglobin and (●---●) is potassium. [From D. K. Myers and R. W. Bide, *Radiation Res.* **27**, 250 (1966).]

At low temperatures, Na^+ uptake exceeds K^+ loss to a greater extent than at normal temperatures, and the changes in volume are maximal. Incubation of the irradiated cell at 37°C before storing in the cold reduces the K^+ loss, as does bringing the cells to 37°C after cold storage (Fig. II-10). Both of these effects are abolished by ouabain treatment and may be considered to be repair of some sort (*162*). The development of hemolysis and the peroxidation of lipids, however, continue at the same rate at 5°C as at 37°C (Fig. II-11). Addition of glucose to the incubation media also diminishes K^+ loss at 37°C (*162*).

The changes in cation flux after doses from 10–100 kR are related to structural damage at certain sites of the cell membrane and are associated with oxidation of SH groups for the following reasons:

(i) Electron microscopy reveals localized damage to the cell membrane of irradiated RBC (*125*, *126*, *135*, *236*, *253*, *262*, *263*).

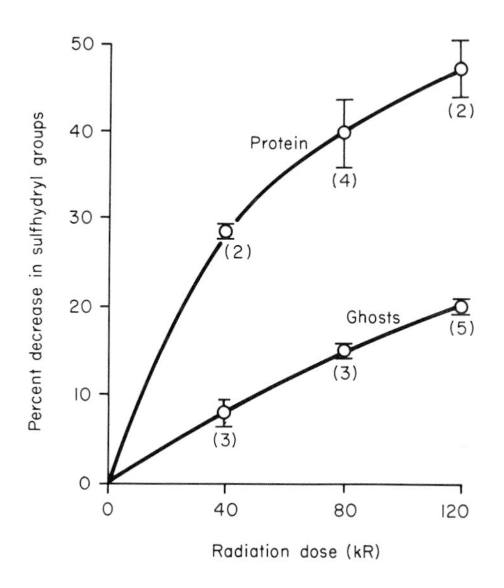

FIG. II-11. Decrease in the number of SH groups in whole human red cell ghosts and solubilized membrane protein. The SH group decrease is plotted as a function of radiation dose. Vertical lines at each point of the curve show the standard deviation; numbers in parentheses refer to the number of experiments. [From R. M. Sutherland, J. N. Stannard, and R. I. Weed, *Intern. J. Radiation Biol.* **12**, 551 (1967).]

(ii) No gross defects are produced in the membrane structure by irradiation since erythrocytes which have lost most of their K^+ can still exclude choline chloride and retain hemoglobin (*237*).

(iii) Proteins of erythrocytes exposed to 60 kR are more susceptible to degradation by proteolytic enzymes, and membranes already degraded in this way can be further broken down by phospholipase (*163*). These experiments thus indicate that damage to the membrane involves proteins linked to lipid structures (*163*).

(iv) The number of SH groups which can be titrated with parachloromercuribenzene sulfonate (PCMBS) (Fig. II-11) decreases in the irradiated membrane (*212*, *213*, *237*). The effect of radiation is more pronounced in defatted ghosts than in intact red blood cells, probably because the cells contain more proteins to protect the SH groups (*237*). The decrease in glutathione content in the interior of irradiated red cell may be related to a diminished ability to restore the S–S groups to their original state.

(v) Compounds which reduce S–S groups, such as mercaptoethyl guanidine (MEG) can repair radiation damage to the membrane even if added after exposure (*213*).

(vi) However, compounds such as PCMS, iodoacetate (*13*, *24*, *168*), iodacetamide (*166*), or *N*-ethylmaleimide (*13*, *14*, *213*) sensitize red cells to ionizing radiation. The sensitization by iodoacetamide probably involves not only reaction with SH groups in the membrane but may also interfere with repair of radiation damage or react with OH radicals to form a compound specifically capable of inactivating SH groups (*168*).

(vii) Damage to the cell membrane caused by the products of water radiolysis seems to occur near the surface of the membrane, since the extent of damage is diminished not only by protective agents that enter the cell (aromatic amino acids, sucrose, etc.) but also by those remaining outside the cell (glutathione, choline, and chloride). The distance which radiolytic products can travel in the surrounding medium to reach the cell is small. This is illustrated by the fact that changes in cell concentration do not influence radiation-induced K^+ loss, although they alter the time lag of the occurrence of hemolysis. The latter effect, however, may result from factors operative after exposure. In addition, some of the membrane sites are also accessible to radicals formed inside the cell, since MEG still protects the cell when the medium containing MEG has been replaced; MEG has been retained, in part, inside the cell (*213*).

(viii) It is of interest to note the additivity of the effect of PCMBS and x-irradiation on K^+ transport, on the one hand, and the lack of a radiation effect on glucose transport, on the other hand (*252*), whereas in nonirradiated red cells PCMBS markedly inhibits glucose transport. This apparent paradox remains to be resolved.

The problems discussed with reference to (vii) pose the question as to which products of radiolysis of water cause damage to the membrane. Presumably, the OH radical produces more than 40% of the damage to cellular permeability (*168*). This is suggested by the fact that *p*-aminobenzoate, a scavenger specific for OH radicals, protects against radiation-induced K^+ loss and reduces sensitization by iodoacetate. Only a small oxygen effect exists (OER 1.2) and catalase does not protect the cell (*168*). Hydrogen peroxide, therefore, is not the effective agent (*166*).

Although most evidence suggests that damage to SH groups is responsible for changes in permeability of the irradiated red cell (*162, 165, 205, 211, 237*), the following factors have also been suggested: peroxidation of lipids (*163*), decrease in antioxidant activity (*162*), and damage to stromal enzymes (*164*). Indeed, there exists a peroxidation of lipids, beginning after irradiation, but this effect is irreparable and independent of the incubation temperature. Therefore, it is not directly related to K^+ loss, but perhaps it is related to hemolysis.

3. Hemolysis and Fragility of Irradiated Erythrocytes

Hemolysis and oxidation of hemoglobin to methemoglobin in red blood cells exposed to irradiation were observed by Henri and Mayer (*79*) and by Chambers (*29*) as early as 1904 and 1911, respectively. We know that these authors used radium as the radiation source, but we do not have data about the dosage they used; we know that a high dose, of the order of 50 kR, is required to bring about hemolysis of mammalian or avian erythrocytes and, moreover, that this result does not occur until 1 or 2 days after exposure.

Hemolysis can be accelerated by irradiation or by depriving the medium of glucose; when erythrocytes are irradiated and then incubated in a medium devoid of glucose, hemolysis is considerably accelerated (*32*). This indicates, at first glance, that the damage produced by irradiation cannot be repaired when glucose is absent.

The structual integrity of erythrocytes can also be assessed on the basis of their ability to withstand osmotic, thermal, or mechanical stress. Irradiation *in vitro* with 20–40 kR increases osmotic (*56–58, 60, 90, 244*), mechanical (*38*), or thermal fragility (*68, 207, 208*), but, as in the case of hemolysis or shortening of life-span, the red cell must be incubated for some time for the effect to become manifest.

Whole body irradiation of rats with 150–250 R also increases osmotic fragility of red cell from the fourth to the eighteenth day after exposure, that is, at a time when glycolysis in red cells is depressed (*204*). Increased red cell fragility is also noticed in rabbits after exposure to 2 kR (*223*), but no such change is seen in dogs after irradiation with lower doses (150–250 R) (*38*).

It is still unknown how the different sequelae of radiation damage to the membrane (K^+ loss, prolytic swelling, increased fragility, and hemolysis) are related one to the other. Initially, the same type of radiation damage probably causes all these changes, which are modified, as they develop, by the experimental conditions; loss of K^+ depends to some extent on the temperature and is partly reparable, whereas hemolysis is not. Probably K^+ loss alone does not cause hemolysis (or vice versa), but, rather, the difference between K^+ loss and Na^+ influx makes erythrocytes swell, increasing osmotic pressure and thus preparing conditions for subsequent hemolysis. Changes in volume are detected after exposure to only 400 R and may, but need not, lead to hemolysis. Formation of lipid peroxides and loss of antioxidant activity are also factors related to hemolysis but not to loss of K^+ (*163*).

4. Life-span of Irradiated Erythrocytes

The ability to survive a normal length of time in circulation or when transfused into a host is a valuable criterion for the cellular integrity of irradiated red blood cells. Irradiation *in vitro* requires doses of about 10 kR (*254*) to shorten significantly the life-span of mammalian or avian erythrocytes, and this effect is more pronounced when the cells are incubated *in vitro* for some time before they are transfused. For example, the life-span of ^{51}Cr-labeled mouse erythrocytes irradiated *in vitro* with 20 or 40 kR and subsequently transfused into an isologous host is reduced by 25 or 33%, respectively (*33*). When the cells are incubated *in vitro* for up to 25 days after exposure, their life-span is reduced even more, especially after irradiation with higher doses (*33*). This observation indicates that even severe radiation damage which would prove fatal to the red cell *in vitro* can be repaired when the cells are transfused into an appropriate recipient. The recovery thus resembles that of red cells used for blood transfusion: Erythrocytes stored for 1 week before transfusion and only moderately depleted of ATP can recover their ATP loss rapidly and have a normal life-span, whereas those stored for more than 3 weeks cannot.

The average life-span of red cells in whole body irradiated animals is found to be shortened in several species: rat (*113*), mouse (*22, 33, 136*), and monkey (*64, 65*). After whole body irradiation of monkeys with 400–600 R, the red blood cell volume in these animals diminishes to one-half of that before exposure (*64, 65*), and about 2.5% of the circulating red cells are randomly destroyed each day. No difference in life-span is observed after partial body irradiation with single or multiple doses (*42, 136*).

An effect of radiation on the life-span of red cells in an animal may result from (i) direct damage to circulating erythrocytes, some of which might be reparable in a nonirradiated host; (ii) increased loss of erythrocytes through capillaries damaged by irradiation (*88*); and (iii) constitutional damage of erythrocytes which originates at the level of the irradiated stem cell. Probably all these factors are operative in irradiated animals, but loss through capillaries appears most important, as illustrated by the following observations: Erythrocytes transfused from an irradiated dog into a nonirradiated host showed no shortening in life-span unless they had been in the circulation of the irradiated animal for some time; conversely, nonirradiated erythrocytes transfused into an irradiated host had a shorter life-span (*231*).

TABLE II-3

BIOCHEMICAL CHANGES AND ALTERED TRANSPORT PHENOMENA IN ERYTHROCYTES IRRADIATED *in vitro* OR *in vivo*

Biochemical or physiological property	Effect
Transport properties (cation flux, hemolysis)	*In vitro* irradiation (several kR) damages cell membranes mainly by SH group oxidation. K^+ is lost, Na^+ enters the cell; Na^+ active transport may be impaired after doses of 1 kR or more; as more Na^+ enters and K^+ leaves, the cell swells; this may cause hemolysis (upon exposure to about 50 kR) after a lag period of a few days; osmotic fragility is affected at 20–40 kR, the extent depending on time of incubation after irradiation
Life-span	*In vitro* irradiation (about 10 kR) shortens survival after transfusion into normal recipients depending on conditions of postirradiation incubation prior to transfusion
	In vivo exposure to lethal doses shortens the average life-span only when red cells remain for some time in the irradiated donor; transfusion of red cells from a nonirradiated into an irradiated recipient shortens life-span during the early postirradiation period
Stromal enzymes	
Orthophosphoric monoester phosphohydrolase (E.C. 3.1.3.2) acid phosphatase	*In vitro* (ghosts): Enzymatic activity decreases after 4 kR (rat) (*164*)
Acetyl cholinesterase (see Table II-2 for nomenclature)	*In vitro* (ghosts): decrease to 60% (rat) after 64 kR (*164*). No effect of 42 kR on enzymatic activity in intact human red cell (*159*)
	In vivo: increase during the first week, postirradiation and decrease thereafter in mice (25 R or more) (*198*); no change in guinea pigs (400 R) (*147*)
ATPase, ouabain-sensitive, Na^+-K^+ activated (see Table II-2 for nomenclature)	*In vitro*: decrease in red cells parallels that of active transport (*24*); no change in ghosts after 200 kR (*162*, *165*); Na^+ required for inhibition (*211*); inhibited in pigeon, human, and dog red cells after 760 R (*100*)
ATPase, ouabain-resistant, Ca^{2+} activated (see Table II-2 for nomenclature)	*In vitro*: no change after inactivation with radiation doses exceeding 200 kR (*164*)
Glycolysis	*In vitro*: no change in glucose utilization (40 kR) (*229*), after 50–150 kR of γ-irradiation increased glycolysis, gradual inhibition (*61*), glycolytic enzymes activity (*16*), or oxidation of glucose-1-^{14}C in presence of electron acceptors (100 R); slight increase in glycolysis (5 kR) and glycolytic enzyme activity (*17*)
	In vivo: no change in glucose utilization or lactate formation after total body irradiation (*159*); decreased glycolysis after 100 R total body irradiation (rat) (*30*), decreased glycolysis in patients upon total body irradiation up to 10 days after 200 R (*127*, *157*, *204*) after (6×) 100 R/day phenylhydrazine—treated rabbits showed increased hexokinase and glucose-6-phosphate dehydrogenase activity 7 and 14 days postirradiation (*52*)

NAD, NADP	*In vivo*: increase on days 1–4 after radiation therapy (*157*) in patients or after total body irradiation in rabbits (200–400 R) (*157*)
ATP	*In vitro*: no immediate effect (50 kR) (*216*), later decrease of ATP and increase of AMP, ADP contents (*25*), if values for adenine nucleotides exceed those for nucleic acids (*161*) *In vivo*: decrease during and after radiation therapy as well as 7 days after total body irradiation of rats (700 R) (*30*), decreased ATP formation (rat) 7 days after 700 R (*30*), increase immediately after 780 R (rat) (*139*)
Nucleosides and nucleotides	*In vivo*: decrease in guanine nucleosides in rats (700 R) (*30*); decreased utilization of purine nucleosides 3 days after 500 R (*188*). Increased uric acid content in patients given radiation therapy (*157*)
Purine nucleoside : orthophosphate ribosyltransferase (E.C. 2.4.2.1) purine nucleoside phosphorylase	*In vivo*: decrease of enzymatic activity is dose-dependent (rats) 1–5 days after irradiation (*187*) (Fig. II-12)
Incorporation of $^{32}P_i$ in stromal lipids	*In vivo*: decreased (*180*) incorporation (rabbit)
Glutathione	*In vitro*: oxidation to disulfide (50% after 75 kR), then rapid return to normal (*162*) at room temperature; at 4°C return to normal is less rapid *In vivo*: increase in rats at 2 days after 400–200 R (*265*); no change in patients given radiation therapy (*62*, *256*). Decrease 48 hr after 1400 R total body x-irradiation (*265*), decrease of 40% after 700 R (guinea pig) during 5 days postirradiation (*227*)
S-Lactoyl glutathione : methylglyoxal lyase (isomerizing) (E.C. 4.4.1.5) lactoyl glutathione lyase	*In vitro*: decrease of enzymatic activity (50%) after 5–10 kR (*92*)
Hydrogen peroxide : hydrogen peroxide oxidoreductase (E.C. 1.11.1.6) catalase	*In vitro*: decrease of enzymatic activity after 10 kR. This effect depends on the O_2 concentration and diminishes at high dose rates (*151*)
Methemoglobin formation	*In vitro*: increases after exposure to about 50 kR (*76*) *In vivo*: total body irradiation (rat) with 3–6 kR produces methemoglobinemia as a function of dose (*43*, *142*)

5. Metabolic Changes in Irradiated Erythrocytes

Various alterations related to the metabolism of red cells irradiated *in vivo* or *in vitro* are compiled in Table II-3, concomitant with changes in permeability. Data on glycolysis in irradiated erythrocytes are extensive but contradictory. No decision as to whether or not glycolysis is affected after irradiation can be made on the basis of these data, but most investigators seem to favor the view that there is no inhibition.

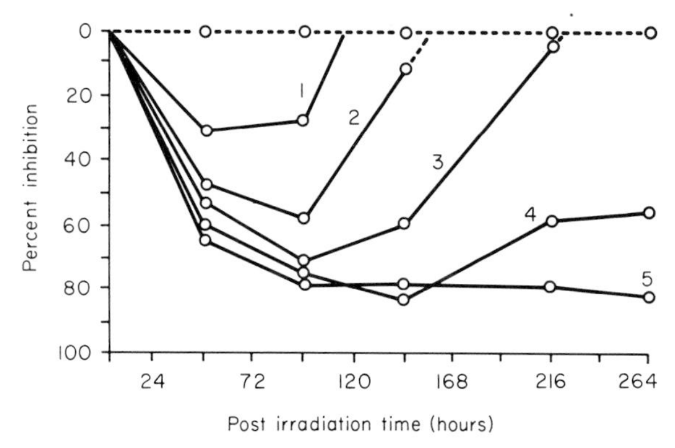

Fig. II-12. Radiation-induced change in the activity of nucleoside phosphorylase of rat erythrocytes. Adult rats were exposed to various doses of total body x-irradiation at a dose rate of 267 R/min. Enzymatic activities were measured at varying times after exposure. The dietary status of the rats is not mentioned. [From D. A. Rappoport, and R. R. Fritz, *Radiation Res.* **21**, 5 (1964).]

In order to attribute general significance to the decrease in nucleoside phosphorylase activity (*188*) (Fig. II-12) and to the increase in acetylcholine esterase activity (*198*), these findings should be confirmed in additional species. Nevertheless, these observations may eventually be useful as diagnostic aids in assessing radiation damage in cases of accidental exposure of human subjects to ionizing radiations.

Reactions related to the metabolism of glutathione and to the effect of hydrogen peroxide in irradiated erythrocytes are particularly interesting. Glutathione in rat red cells is oxidized immediately after *in vitro* exposure to more than 10 kR, whereas its concentration increases after whole body irradiation with 400–2000 R (*260*). Human erythrocytes with a hereditary defect in glutathione synthesis or glucose-6-phosphate dehydrogenase contain 50% less glutathione after irradiation with 100–880 R, as compared with normal irradiated cells, form Heinz bodies and exhibit a reduced glutathione stability in the presence of acetylphenylhydrazine. Higher radiation doses cause no further reduction in glutathione stability. Erythrocytes deficient in glutathione reductase have normal glutathione levels and behave as normal erythrocytes when irradiated, that is, they do not form Heinz bodies and their glutathione stability is normal (*256*, *257*).

Glyoxalase activity in red cells decreases after *in vitro* irradiation with 4 kR (*91*, *92*). Although the inhibition of glyoxalase activity probably is not caused by inactivation of the apoprotein or of glutathione, it is conceivable that irradiation interferes with the formation of the activated complex between the apoprotein and glutathione–methylglyoxal.

Human erythrocytes deficient in catalase form larger amounts of methemoglobin than normal cells after *in vitro* irradiation with 4 kR. Catalase, however, protects, in part, against radiation-induced methemoglobin formation. These findings suggest that hydrogen peroxide derived from radiolysis of water participates in methemoglobin formation (*197*) in irradiated erythrocytes. However, information gained by irradiating erythrocytes under anoxic conditions or in the presence of azide or thiol compounds (*3*, *76*, *151*, *258*) indicates that hydrogen peroxide is not the only factor involved.

REFERENCES

1. Abrams, R., *Arch. Biochem.* **30**, 90 (1951).
2. Abrams, R., Goldinger, J. M., and Barron, E. S. G., *Biochim. Biophys. Acta* **5**, 74 (1950).
3. Aebi, H., Heiniger, J. P., and Suter, H., *Experientia* **18**, 129 (1962).

3a. Alexanian, R., Porteous, D. D., and Lajtha, L. G., *Intern. J. Radiation Biol.* **7**, 87 (1963).

4. Altman, K. I., Noonan, T. R., Spraragen, S. C., and Salomon, K., *Federation Proc.* **11**, 4 (1952).
5. Altman, K. I., Richmond, J. E., and Salomon, K., *Biochim. Biophys. Acta* **7**, 460 (1951).

5a. Baum, S. J., *Am. J. Physiol.* **200**, 155 (1961).

5b. Baum, S. J., Davis, A. K., and Alpen, E. L., *Radiation Res.* **15**, 97 (1961).

6. Baxter, C. F., Belcher, E. H., Harriss, E. B., and Lamerton, L. F., *Brit. J. Haematol.* **1**, 86 (1955).
7. Belcher, E. H., and Courtenay, V. D., *Brit. J. Haematol.* **5**, 268 (1959).
8. Belcher, E. H., Gilbert, I. G. F., and Lamerton, L. F., *Brit. J. Radiol.* **27**, 387 (1954).
9. Belcher, E. H., Harriss, E. B., and Lamerton, L. F., *Brit. J. Haematol.* **4**, 390 (1958).
10. Beneš, L., Rotreklová, E., and Tačev, T., *Strahlentherapie* **135**, 443 (1968).
11. Bernheim, F., Ottolenghi, A., and Wilbur, K. M., *Radiation Res.* **4**, 132 (1956).
12. Berry, R. J., Hell, E. A., and Lajtha, L. G., *Intern. J. Radiation Biol.* **4**, 61 (1961).
13. Bianchi, M. R., Bocqacci, M., Misiti-Dorello, P., and Quintiliani, M., *Intern. J. Radiation Biol.* **8**, 329 (1964).
14. Bianchi, M. R., Quintiliani, M. Bocqacci, M., and Strom, E., *Minerva Radiol. Fis. Radiobiol.* **12**, 56 (1967).
15. Blackett, N. M., Roylance, P. J., and Adams, K., *Brit. J. Haematol.* **10**, 453 (1964).
16. Blandino, G., Caruso, P., Conti, F., and Caponetti, R., *Radiobiol. Radiotherap. Fis. Med.* **21**, 377 (1966).

17. Blum, K. U., and Voigt, E.D., *Verhandl. Deut. Ges. Inn. Med.* **69**, 200 (1963).
18. Bollinger, J. N., *Radiation Res.* **28**, 121 (1966).
19. Bollum, F. J., Anderegg, J. W., McElya, A. B., and Potter, V. R., *Cancer Res.* **20**, 138 (1960).
20. Bonnichsen, R., and Hevesy, G., *Acta Chem. Scand.* **9**, 509 (1955).
21. Bose, A., *Intern. J. Radiation Biol.* **1**, 383 (1959).
22. Brecher, G., Endicott, K. M., Gump, H., and Brawner, H. P., *Blood* **3**, 1259 (1948).
23. Bresciani, F., Auricchio, F., and Fiore, C., *Radiation Res.* **21**, 394 (1964).
24. Bresciani, F., Auricchio, F., and Fiore, C., *Radiation Res.* **22**, 463 (1964).
25. Breuer, H., and Parchwitz, H. K., *Strahlentherapie, Sonderbände* **46**, 234 (1960).
26. Brodskii, V. Y., and Suetina, I. A., *Biofizika* **3**, 92 (1958).
27. Buchsbaum, R., and Zirkle, R. E., *Proc. Soc. Exptl. Biol. Med.* **72**, 27 (1949).
28. Buhlmann, A., Liechti, A., and Wilbrandt, W., *Strahlentherapie* **71**, 285 (1942).
29. Chambers, H., and Russ, S., *Proc. Roy. Soc.* **84B**, 124 (1911).
30. Chambon, P., Karon, H., and Mandel, P., *Bull. Soc. Chim. Biol.* **43**, 127 (1961).
31. Chanutin, A., and Word, E. L., *Am. J. Physiol.* **179**, 221 (1954).
32. Cividalli, G., *Radiation Res.* **20**, 564 (1963).
33. Cividalli, G., and Horn, Y., *Israel J. Med. Sci.* **2**, 560 (1966).
33a. Cumming, J. D., and Khoyi, M. A., *Nature* **193**, 1196 (1962).
34. D'Amore, A., Marotta, U., Marsella, A., and Monacelli, R., *Rend. Ist. Super. Sanita* **26**, 1007 (1963).
35. Dancewicz, A. M., Altman, K. I., and Salomon, K., *Intern. J. Radiation Biol.* **5**, 85 (1962).
36. Dancewicz, A. M., Gutniak, O., Mazanowska, A., Malinowska, T., and Kowalski, E., *Proc. 9th Congr. European Soc. Haematol., Lisbon, 1963* p. 631. Karger, Basel, 1963.
37. Dancewicz, A. M., and Lipiński, B., *Nucleonika* **3**, 79 (1958).
38. Davis, R. W., Dole, N., Izzo, M. J., and Young, L. E., *J. Lab. Clin. Med.* **35**, 528 (1950).
39. Davis, W. E., and Cole, L. J., *Radiation Res.* **14**, 104 (1961).
40. Dietz, A. A., and Steinberg, B., *Arch. Biochem.* **23**, 222 (1949).
41. Dinning, J. S. Meschan, I., Keith, C. K., and Day, P. L., *Proc. Soc. Exptl. Biol. Med.* **74**, 776 (1950).
42. Di Pietrantorij, F., *Minerva Nucl.* **9**, 288 (1965).
43. Dowben, R. M., and Walker, J. K., *Proc. Soc. Exptl. Biol. Med.* **90**, 398 (1955).
44. Drašíl, V., and Soška, J., *Biochim. Biophys. Acta* **28**, 667 (1958).
45. Drašíl, V., Soška, J., and Beneš, L., *Folia Biol. (Prague)* **5**, 334 (1959).
46. Edwards, C. H., Gadsen, E. L., and Edwards, G. A., *Radiation Res.* **22**, 116 (1964).
47. Elko, E. E., and Di Luzio, N. R., *Radiation Res.* **11**, 1 (1959).
48. Ellis, M. E., and Cole, L. J., *Federation Proc.* **17**, 366 (1958).
49. Eltgen, D., Koch, R., and Langendorff, H., *Strahlentherapie* **114**, 118 (1961).
50. Erickson, C. A., Main, R. K., and Cole, L. J., *Radiation Res.* **5**, 332 (1956).
51. Fantoni, A., and Botteghi, F., *J. Nucl. Biol. Med.* **10**, 104 (1966).
52. Fantoni, A., Segni, P., and Leoncini, G., *J. Nucl. Biol. Med.* **10**, 110 (1966).
53. Fedorov, N. A., *in* "Progress in Radiobiology" (J. S. Mitchell, B. E. Holmes, and C. L. Smith, eds.), p. 310. Oliver & Boyd, Edinburgh and London, 1956.
54. Feinendegen, L. E., Bond, V. P., and Hughes, W. L., *Exptl. Cell Res.* **43**, 107 (1966).
55. Fisher, J. W., Roh, B. L., and Shenep, K., *Sangre* **9**, 112 (1964).
56. Flemming, K., *Folia Haematol.* **77**, 147 (1960).

57. Flemming, K., *Strahlentherapie* **116**, 404 (1961).
58. Flemming, K., *Strahlentherapie* **120**, 456 (1963).
59. Fujii, M., *Naika Hokan* **11**, 43 (1964).
60. Galvao, J. P., Magalhaes, E. M. Azevedo, N. L. C., and Teixeira, F. P., *Lab. Fis. Engenharia Nucl.* (*Portoguese*) p. 15 (1966).
61. Gerasimova, G. K., *Radiobiologiya* **7**, 392 (1967).
62. Gerhardt, P., *Strahlentherapie, Sonderbaende* **64**, 136 (1967).
63. Gilbert, C., Paterson, E., and Haigh, M., *in* "Advances in Radiobiology" (G. C. de Hevesy, A. G. Forssberg, and J. D. Abbatt, eds.), p. 357. Oliver & Boyd, Edinburgh and London, 1957.
64. Gilbert, C., Paterson, E., and Haigh, M., *Intern. J. Radiation Biol.* **5**, 1 (1963).
65. Gilbert, C., Paterson, E., Haigh, M., and Schofield, R., *Intern. J. Radiation Biol.* **5** 9 (1963).
66. Gitelzon, I. I., and Terskov, I. A., *Biofizika* **5**, 180 (1960).
67. Glynn, I. M., *J. Physiol.* (*London*) **134**, 278 (1956).
68. Goldschmidt, L., Rosenthal, R. L., Bond, V. P., and Fishler, M. C., *Am. J. Physiol.* **164**, 202 (1951).
69. Graevskii, E. Ya., *Proc. 2nd U.N. Intern. Conf. Peaceful Uses At. Energy, Geneva, 1958*, **22**, 441. United Nations, Geneva, 1958.
70. Griffin, E. D., and Alpen, E. L., *Radiation Res.*, **3**, 230 (1955).
70a. Gurney, C. W., Lajtha, L. G., and Oliver, R., *Brit. J. Haematol.* **8**, 461 (1962).
71. Gurney, C. W., and Hofstra, D., *Radiation Res.* **19**, 599 (1963).
72. Gurney, C. W., and Wackman, N., *Nature* **190**, 1017 (1961).
73. Gutnitsky, A., Nohr, M. L., and Van Dyke, D., *J. Nucl. Med.* **5**, 595 (1964).
74. Hajduković, S., *Experientia* **11**, 356 (1955).
75. Halberstaedter, L., and Leibowitz, J., *Biochem. J.* **41**, 235 (1947).
76. Heiniger, J. P., and Aebi, H., *Helv. Chim. Acta* **46**, 225 (1963).
77. Hennessy, T. G., Herrick, S. E., and Okunewick, J. P., *Health Phys.* **10**, 49 (1964).
78. Hennessy, T. G., and Huff, R. L., *Proc. Soc. Exptl. Biol. Med.* **73**, 436 (1950).
79. Henri, V., and Mayer, A., *Compt. Rend. Soc. Biol.* **138**, 521 (1904).
80. Hevesy, G., *Strahlentherapie* **102**, 341 (1957).
81. Hevesy, G., and Bonnichsen, R., *Ann. Acad. Sci. Fennicae: Ser. A II* **60**, 295 (1955).
82. Hevesy, G., and Dal Santo, G., *Acta Physiol. Scand.* **32**, 339 (1954).
83. Hodgson, G. S., *Blood* **19**, 460 (1962).
84. Holthusen, H., *Strahlentherapie* **14**, 561 (1922).
84a. Høst, H., and Skjaelaaen, P., *Scand. J. Haematol.* **3**, 154 (1966).
85. Huff, R. L., Bethard, W. F., Garcia, J. F., Roberts, B. M., Jacobson, L. O., and Lawrence, J. H., *J. Lab. Clin. Med.* **36**, 40 (1950).
86. Ingram, M., and Mason, W. B., *U.S. At. Energy Comm., Rept.* **UR-122** (1950).
87. Jasinska, J., and Michalowski, A., *Nature* **196**, 1326 (1962).
88. Kahn, J. B., and Furth, J., *Blood* **7**, 404 (1952).
89. Kappas, A., and Granick, S., *J. Biol. Chem.* **243**, 346 (1968).
90. Karanovic, J., Meardi, G., and Robbiani, M. G., *Boll. Soc. Ital. Biol. Sper.* **33**, 577 (1957).
91. Klebanoff, S. J., *J. Gen. Physiol.* **41**, 725 (1958).
92. Klebanoff, S. J., *J. Gen. Physiol.* **41**, 737 (1958).
93. Klein, J. R., *Arch. Biochem. Biophys.* **54**, 556 (1955).
94. Klein, J. R., *Proc. Soc. Exptl. Biol. Med.* **91**, 630 (1956).
95. Klein, J. R., and Cavalieri, R., *Proc. Soc. Exptl. Biol. Med.* **89**, 28 (1955).
96. Klener, V., *Physiol. Bohemoslov.* **15**, 378 (1966).

97. Klener, V., *Physiol. Bohemoslov.* **15**, 385 (1966).
98. Koch, R., and Seiter, I., *Strahlentherapie* **129**, 608 (1966).
99. Koloušek, J., and Györgyi, S., *Neoplasma* **3**, 289 (1966).
100. Köver, G., and Schoffenfiels, E., *Intern. J. Radiation Biol.* **9**, 461 (1965).
101. Kozinets, G. I., and Tsessarskaya, T. P., *Med. Radiobiol.* **7**, 43 (1962).
102. Kozinets, G. I., Tsessarskaya, T. P., and Bogotavlenskaya. M. P., *Med. Radiobiol.* **7**, 50 (1962).
103. Krantz, S. B., and Goldwasser, E., *Biochim. Biophys. Acta* **103**, 325 (1965).
104. Krause, R. F., *J. Biol. Chem.* **149**, 395 (1943).
105. Kritskii, G. A., *Biokhimiya* **24**, 73 (1959).
106. Kritskii, G. A., *Radiobiologiya* **7**, 193 (1967).
107. Kritskii, G. A., and Kopylov, V. A., *Biokhimiya* **25**, 34 (1960).
108. Kritskii, G. A., and Roitman, F. I., *Biokhimiya* **26**, 148 (1961).
109. Kritskii, G. A., Arends, I. A., and Nicolaev, J. V., *Biokhimiya* **32**, 452 (1967).
110. Künkel, H. A., and Maass, H., *Strahlentherapie* **111**, 52 (1960).
111. Künkel, H. A., and Mauss, H.-J., *Strahlentherapie, Sonderbände* **49**, 234 (1962).
112. Künkel, H. A., Mauss, H.-J., and Klapheck, U., *Strahlentherapie* **129**, 263 (1966).
113. Künkel, H. A., Mauss, H.-J., and Mihm, K., *Strahlentherapie* **131**, 137 (1966).
114. Kurnick, N. B., Massey, B. W., and Sandeen, G., *Ann. N. Y. Acad. Sci.* **75**, 61 (1958).
115. Kurnick, N. B., and Nokay, N., *Radiation Res.* **23**, 239 (1964).
116. Lajtha, L. G., *Physiol. Rev.* **37**, 50 (1957).
116a. Lajtha, L. G., *in* "Current Topics in Radiation Research" (M. Ebert and A. Howard, eds.), Vol. 1, p. 139. North-Holland Publishing Co., Amsterdam, 1965.
117. Lajtha, L. G., and Oliver, R., *Lab. Invest.* **8**, 214 (1959).
118. Lajtha, L. G., Oliver, R., and Ellis, F., *Brit. J. Cancer* **8**, 367 (1954).
119. Lajtha, L. G., Oliver, R., and Ellis, F., *in* "Advances in Radiobiology" (G. C. de Hevesy, A. G. Forssberg, and J. D. Abbatt, eds.), p. 54. Oliver & Boyd, Edinburgh and London, 1959.
119a. Lajtha, L. G., Oliver, R., and Gurney, C. W., *Brit. J. Haematol.* **8**, 442 (1962).
120. Lajtha, L. G.. Oliver, R., Kumatori, T., and Ellis, F., *Radiation Res.* **8**, 1 (1958).
121. Lajtha, L. G., and Suit, H. D., *Brit. J. Haematol.* **1**, 55 (1955).
122. Lajtha, L. G., and Vane, J. R., *Nature* **182**, 191 (1958).
123. Lamerton, L. F., Pontifex, A. H., Blackett, N. M., and Adams, K., *Brit. J. Radiol.* **33**, 287 (1960).
124. Lehmann, F., and Wells, P., *Arch. Ges. Physiol.* **213**, 628 (1926).
125. Lessler, M. A., *Science* **129**, 1551 (1959).
126. Lessler, M. A., and Herrera, F. M., *Radiation Res.* **17**, 111 (1962).
127. Levin, W. C., Schneider, M., and Bresnick, E., *J. Lab. Clin. Med.* **52**, 921 (1958).
128. Libinzon, R. E., *Byull. Radiats. Med.* **3**, 57 (1956).
129. Libinzon, R. E., *Biokhimya* **24**, 625 (1959).
130. Libinzon, R. E., and Konstantinova, V. V., *Biokhimiya* **25**, 1018 (1960).
131. Libinzon, R. E., and Tseveleva, I. A., *Radiobiologiya* **4**, 503 (1964).
132. Libinzon, R. E., and Tseveleva, I. A., *Radiobiologiya* **6**, 9 (1966).
133. Liechti, A., and Wilbrandt, W., *Strahlentherapie* **70**, 541 (1941).
134. Lim, J. K., Piantadosi, C., and Snyder, F., *Proc. Soc. Exptl. Biol. Med.* **124**, 1158 (1967).
135. Lindemann, B., *Fortschr. Gebiete Röntgenstrahlen Nuklearmed.* **75**, 523 (1951).
136. Lockner, D., Ericson, U., and Hevesy, G., *Intern. J. Radiation Biol.* **9**, 143 (1965).
137. Loeffler, R. K., *Am. J. Roentgenol., Radium Therapy Nucl. Med.* **77**, 836 (1957).
138. Loeffler, R. K., Collins, V. P., and Hyman, G. A., *Science* **118**, 161 (1953).

139. Löhr, E., *Strahlentherapie* **130**, 245 (1966).
140. Löhr, E., *Strahlentherapie* **130**, 528 (1966).
141. Löhr, E., *Strahlentherapie* **130**, 361 (1966).
142. Loiseleur, J., and Petit, M., *Compt. Rend.* **251**, 2800 (1960).
143. Lord, B. I., *Brit. J. Haematol.* **11**, 525 (1965).
144. Lord, B. I., *Brit. J. Haematol.* **10**, 496 (1964).
145. Ludwig, F. C., and Kohn, H. I., *Radiation Res.* **21**, 233 (1964).
146. Lundin, J., and Clemendson, C. J., *Biochim. Biophys. Acta* **42**, 528 (1960).
147. Lundin, J., Clemendson, C. J., and Nelson, A., *Acta Radiol.* **48**, 52 (1957).
148. Lutwak-Mann, C., *Biochem. J.* **49**, 300 (1951).
149. Lutwak-Mann, C., *Biochem. J.* **52**, 356 (1952).
150. McCulloch, E. A., and Till, J. E., *Radiation Res.* **13**, 115 (1960).
151. Magdon, E., *Strahlentherapie* **126**, 258 (1965).
152. Main, R. K., Cole, L. J., and Bond, V. P., *Arch. Biochem. Biophys.* **56**, 143 (1955).
153. Malinowska, T., and Dancewicz, A. M., *Bull. Acad. Polon. Sci.* **13**, 13 (1965).
154. Mandel, P., Gros, C. M., Rodesch, J., Jaudel, C., and Chambon, P., *in* "Advances in Radiobiology" (G. C. de Hevesy, A. G. Forssberg, and J. D. Abbatt, eds.), p. 59. Oliver & Boyd, Edinburgh and London, 1957.
154a. Mauss, H.-J., Künkel, H. A., and Mayr, H.-J., *Strahlentherapie, Sonderbände* **64**, 155 (1967).
155. Mazanowska, A., and Dancewicz, A. M., *Bull. Acad. Polon. Sci.* **12**, 293 (1964).
156. Mechali, D., Taranger, M., and Godiniaux, F., *Rev. Franc. Etudes Clin. Biol.* **8**, 829 (1964).
157. Mills, G. C., Burger, D. O., Schneider, M., and Levin, W. C., *Radiation Res.* **20**, 341 (1963).
158. Miras, C. J., Mantzos, J. D., and Levis, G. M., *Radiation Res.* **22**, 682 (1964).
159. Morczek, A., *Strahlentherapie* **96**, 618 (1955).
160. Morczek, A., *Radiobiol. Radiotherap.* **1**, 339 (1960).
161. Myers, D. K., *Can. J. Biochem. Physiol.* **40**, 619 (1962).
162. Myers, D. K., and Bide, R. W., *Radiation Res.* **27**, 250 (1966).
163. Myers, D. K., Bide, R. W., and Tibe, T. A., *Can. J. Biochem.* **45**, 1973 (1967).
164. Myers, D. K., and Church, M. L., *Nature* **213**, 636 (1967).
165. Myers, D. K., and Levy, L., *Nature* **204**, 1324 (1964).
166. Myers, D. K., and Slade, D. E., *Radiation Res.* **30**, 186 (1967).
167. Nakhilnitskaya, Z. N., *Radiobiologiya* **5**, 25 (1965).
168. Nakken, K. F., *Strahlentherapie* **129**, 586 (1966).
169. Nizet, A., Bacq, Z. M., and Herve, A., *Arch. Intern. Physiol.* **60**, 449 (1952).
170. Nizet, A., Lambert, S., Herve, A., and Bacq, Z. M., *Arch. Intern. Physiol.* **62**, 129 (1954).
171. Nizet, A., Lambert, S., Herve, A., and Bacq, Z. M., *Arch. Intern. Physiol.* **62**, 429 (1954).
172. Notario, A., Slaviero, G. E., and Zampaglione, G., *Haematologica (Pavia)* **51**, 509 (1966).
173. Nygaard, O. F., and Potter, R. L., *Radiation Res.* **12**, 131 (1960).
174. Odartchenko, N., Cottier, H., Feinendegen, L. E., and Bond, V. P., *Radiation Res.* **21**, 413 (1964).
175. Orlova, L. V., and Trincher, K. S., *Radiobiologiya* **6**, 915 (1966).
176. Pabst, H. W., Hahn, B., and Klemm, J., *Strahlentherapie, Sonderbände* **64**, 147 (1967)
177. Pany, J., *Z. Biol.* **111**, 401 (1960).

178. Perretta, M., Rudolph, W., Aguirre, G., and Hodgson, G., *Biochim. Biophys. Acta* **87**, 157 (1964).
179. Perris, A. D., and Myers, D. K., *Nature* **207**, 986 (1965).
180. Petropoulos, M. C., Altman, K. I., and Salomon, K., *Biochim. Biophys. Acta* **22**, 471 (1956).
181. Pfleger, R. C., and Snyder, F., *Radiation Res.* **30**, 325 (1967).
182. Philipps, G. R., Hildinger, M., Schall, H., and Turba, F., *Biochem. Z.* **335**, 223 (1961).
183. Pontifex, A. H., and Lamerton, L. F., *Brit. J. Radiol.* **33**, 736 (1960).
184. Porteous, D. D., and Lajtha, L. G., *Brit. J. Haematol.* **12**, 177 (1966).
185. Porteous, D. D., Tso, S. C., Hirashima, K., and Lajtha, L. G., *Nature* **206**, 204 (1965).
186. Rambach, W. A., Cooper, J. A. D., Alt, H. L., Vogel, H. H., Clark, J. W., and Jordan, D. L., *Radiation Res.* **10**, 148 (1959).
187. Rappoport, D. A., and Fritz, R. R., *Radiation Res.* **21**, 5 (1964).
188. Rappoport, D. A., and Sewell, B. W., *Nature* **184**, 846 (1959).
189. Revis, V. A., *Vopr. Med. Khim.* **10**, 513 (1963).
190. Revis, V. A., *Radiobiologiya* **2**, 919 (1962).
191. Richmond, J. E., Altman, K. I., and Salomon, K., *J. Biol. Chem.* **190**, 817 (1951).
192. Richmond, J. E., Altman, K. I., and Salomon, K., *Science* **113**, 404 (1951).
193. Rodesch. J., Jaudel, C., Chirpaz, B., and Mandel, P., *Experientia* **11**, 437 (1955).
194. Ruhenstroth-Bauer, G., Gostmzyk, J. G., and Arnold, R., *Strahlentherapie* **122**, 56 (1963).
195. Rysina, T. N., *Biokhimiya* **24**, 556 (1959).
196. Rysina, T. N., "Biological Effect of Radiation and Questions on the Distribution on Radioactive Isotopes," p. 9. Moscow, 1961.
197. Rysina, T. N., and Libinzon, R. E., *Biokhimiya* **25**, 825 (1961).
198. Sabine, J. C., *Am. J. Phsyiol.* **187**, 275 (1956).
199. Sauerbier, W., *Strahlentherapie* **107**, 468 (1958).
200. Savoie, J. C., *2nd Colloq. Intern. Biol. Saclay, 1962* p. 269. *Presse Univ. France*, Paris, 1962.
201. Schall, H., and Turba, F., *Biochem. Z.* **339**, 224 (1963).
202. Schiffer, L. M., Atkins, H. L., Chanana, A. D., Cronkite, E. P., Greenberg, M. L., Johnson, H. A., Robertson, J. S., and Stryckmans, P. A., *Blood* **27**, 831 (1966).
203. Schnappauf, H., Joel, D. D., Chanana, A. D., DiGiacomo, R., and Cronkite, E. P., *Radiation Res.* **25**, 646 (1965).
204. Schneider, M., Levin, W. C., and Bresnick, E., *Radiation Res.* **9**, 178 (1958).
205. Schooley, J. C., Hayes, J. M., Cantor, L. N., and Havens, V. W., *Radiation Res.* **32**, 875 (1967).
206. Schulman, M., Altman, K. I., and Salomon, K., *Intern. J. Radiation Biol.* **2**, 28 (1960).
207. Šebestík, V., *Physiol. Bohemoslov.* **14**, 343 (1965).
208. Šebestík, V., *Physiol. Strahlenwirk.* p. 389 (1964).
209. Šebestík, V., Jelinek, J., and Dienstbier, Z., *Strahlentherapie* **115**, 609 (1961).
210. Šebestík, V., Jelinek, J., Dienstbier, Z., and Viktora, L., *Physiol. Bohemoslov.* **11**, 510 (1962).
211. Shapiro, B., and Kollmann, G., *Radiation Res.* **22**, 70 (1964).
212. Shapiro, B., and Kollmann, G., *Radiation Res.* **34**, 335 (1968).
213. Shapiro, B., Kollmann, G., and Asnen, J., *Radiation Res.* **27**, 139 (1966).
214. Sheets, R. F., Hamilton, H. E., DeGowin, E. L., and Janney, C. D., *J. Clin. Invest.* **33**, 179 (1954).

215. Sheppard, C. W., and Beyl, G. E., *J. Gen. Physiol.* **34**, 691 (1951).
216. Sheppard, C. W., and Stewart, M., *J. Cellular Comp. Physiol.* **39**, Suppl. 2, 189 (1952).
217. Skalka, M., and Matyášová, J., *Intern. J. Radiation Biol.* **7**, 41 (1963).
218. Skalka, M., and Matyášová, J., *Folia Biol.* (*Prague*) **11**, 378 (1965).
219. Skalka, M., Matyášová, J., and Chlumeckà, V., *Folia Biol.* (*Prague*) **11**, 113 (1965).
220. Skalka, M., Matyášová, J., and Soška, J., *Proc. Symp. Biol. Effects Ionizing Radiation Mol. Level, Brno, 1962* p. 323. I.A.E.A., Vienna, 1962.
221. Smellie, R. M. S., Humphrey, G. F., Kay, E. R. M., and Davidson, J. N., *Biochem. J.* **60**, 177 (1955).
222. Snyder, F., *Nature* **206**, 733 (1965).
223. Snyder, F., *Radiation Res.* **27**, 375 (1966).
224. Snyder, F., and Cress, E. A., *Radiation Res.* **19**, 129 (1963).
225. Snyder, F., and Wood, R., *Radiation Res.* **33**, 142 (1968).
226. Snyder, F., and Wright, R., *Radiation Res.* **25**, 417 (1965).
227. Sonka, J., Malina, L., Pospisil, J., and Dienstbier, Z., *Rev. Agressol.* **6**, 205 (1965).
228. Soška, J., Beneš, L., Drašíl, V., Karpfel, Z., Paleček, E., and Skalka, M., *in* "The Initial Effects of Ionizing Radiations on Cells" (R. J. C. Harris, ed.), p. 153. Academic Press, New York, 1961.
229. Soška, J., Karpfel, Z., and Drašíl, V., *Folia Biol.* **5**, 190 (1959).
230. Soška, J., and Soškova, L., *Folia Biol.* (*Prague*) **6**, 425 (1959).
231. Stohlman, F., Jr., Brecher, G., Scheiderman, M., and Cronkite, E. P., *Blood* **12**, 1061 (1957).
232. Storer, J. B., Harris, P. S., Furchner, J. E., and Langham, W. H., *Radiation Res.* **6**, 188 (1957).
233. Sugino, Y., and Potter, R. L., *Radiation Res.* **12**, 477 (1960).
234. Suit, H. D., Lajtha, L. G., Oliver, R., and Ellis, F., *Brit. J. Haematol.* **3**, 165 (1957).
235. Sukhomlonov, B. F., and Merenov, V. V., *Dopovidi Akad. Nauk USSR* **5**, 624 (1962).
236. Sundharagiati, B., and Wright, C. S., *J. Clin. Invest.* **32**, 979 (1953).
237. Sutherland, R. M., and Pihl, A., *Radiation Res.* **34**, 300 (1968).
238. Sutherland, R. M., Stannard, J. N., and Weed, R. I., *Intern. J. Radiation Biol.* **12**, 551 (1967).
239. Szot, Z., Malinowska, T., and Kowalski, E., *Acta Physiol. Polon.* **12**, 559 (1961).
240. Thomson, J. F., Tourtellotte, W. W., Carttar, M. S., and Storer, J. B., *Arch. Biochem. Biophys.* **42**, 185 (1953).
241. Till, J. E., and McCulloch, E. A., *Radiation Res.* **14**, 213 (1961).
242. Till, J. E., and McCulloch, E. A., *Radiation Res.* **18**, 96 (1963).
243. Ting, T. P., and Zirkle, R. E., *J. Cellular Comp. Physiol.* **16**, 189 (1940).
244. Ting, T. P., and Zirkle, R. E., *J. Cellular Comp. Physiol.* **16**, 197 (1940).
245. Ting, T. P., and Zirkle, R. E., *J. Cellular Comp. Physiol.* **16**, 269 (1940).
246. Totter, J. R., *J. Am. Chem. Soc.* **76**, 2196, (1954).
247. Travníček, T., Neuwirt, J., Táborsky, J., and Stoklasová, H., *Strahlentherapie* **130**, 376 (1966).
248. Trincher, K. S., and Orlova, L. V., *Radiobiologiya* **5**, 797 (1965).
249. Tsuchiya, T., and Tamanoi, I., *Intern. J. Radiation Biol.* **10**, 203 (1966).
250. Tsuya, A., Bond, V. P., Fliedner, T. M., and Feinendegen, L. E., *Radiation Res.* **14**, 618 (1961).
251. Uyeki, E. M., Leuchtenberger, C., and Salerno, P. R., *Exptl. Cell. Res.* **17**, 405 (1959).
252. Van Steveninck, J., Weed, R. I., and Rothstein, A., *J. Gen. Physiol.* **48**, 617 (1965).

253. Vinetski, Y. P., *Radiobiologiya* **2**, 370 (1960).
254. Vorobyev, V. N., Sheremet, Z. I., and Raushenbakh, M. O., *Med. Radiol.* **6**, 65 (1959).
255. Waggener, R. E., and Hunt, H. B., *Am. J. Roentgenol., Radium Therapy Nucl. Med.* **79**, 1050 (1958).
256. Waller, H. D., and Schoen, H. D., *Klin. Wochschr.* **44**, 186 (1966).
257. Waller, H. D., and Schoen, H. D., *Klin. Wochschr.* **44**, 189 (1966).
258. Warburg, O., Schröder, W., and Gattung, H. W., *Z. Naturforsch.* **15b**, 163 (1960).
259. Watras, J., *Radiobiol. Radiotherap.* **3**, 451 (1962).
260. Wernze, H., Braun, H., and Koch, W., *Strahlentherapie* **126**, 291 (1965).
261. Wright, C. S., Dodd, M. C., Bouronele, B. A., and Morton, T. L., *J. Lab. Clin. Med.* **40**, 962 (1952).
262. Začek, J., and Rosenberg, R., *Compt. Rend.* **144**, 462 (1950).
263. Začek, J., and Rosenberg, M., *Biochim. Biophys. Acta* **5**, 315 (1950).
264. Zaorska, B., Deckers, C., and Dunjic, A., *Rev. Franc. Etudes Clin. Biol.* **7**, 169 (1967).
265. Zicha, B., Dienstbier, Z., Borova, J., Beneš, J., and Neuwirt, J., *Strahlentherapie* **126**, 299 (1965).
266. Zographov, D. G., and Baev, I. A., *Strahlentherapie* **123**, 254 (1964).
267. Zographov, D. G., and Baev, I. A., *Strahlentherapie* **126**, 1595 (1965).

Chapter III

LYMPHOID ORGANS

A. Introduction

As early as 1903, Heineke (*71*) recognized that all lymphoid organs are extremely radiosensitive. Failure to reproduce, which is generally considered the most important effect of radiation on cell populations undergoing constant renewal, is only of minor, although not negligible, importance (*77a*, *130*) in lymphoid tissues. In these cells, immediate death in interphase is the salient feature of the radiation response. Therefore, lymphocytes are convenient for the study of the morphological and biochemical aspects of interphase death in mammalian cells (see also Volume I, p. 247).

When evaluating biochemical changes in irradiated lymphoid organs, it is mandatory to establish relationships of these changes to morphological alterations (see Table III-1). Lymphocytes, including thymocytes, react similarly in all irradiated lymphoid organs, i.e., thymus, spleen, and lymph nodes, but the presence of nonlymphoid tissue in different organs can modify the individual pattern of destruction, phagocytosis, and regeneration. Thus, in the spleen, besides lymphoid white pulp, we encounter the more radioresistant red pulp. Furthermore, formation of ectopic bone marrow can take place in the spleen of certain species 1–2 weeks after irradiation.

The weight of the irradiated lymphoid organs diminishes as lymphocytes are destroyed and their fragments are removed (Fig. III-1). The dose-response curves of isolated lymphocytes and particularly of spleen and lymph nodes exhibit biphasic shapes, the second, more radioresistant components probably being represented by more resistant types of lymphocytes (*4*, *15a*, *138a*, *190*)

TABLE III-1

TIME SEQUENCE OF MORPHOLOGICAL AND BIOCHEMICAL ALTERATIONS IN IRRADIATED LYMPHOID TISSUE

Time	Morphological alteration	Biochemical alteration
Immediate changes: 1st hr (latent period)	No evident histological damage	↓ 1. Decreased incorporation of precursors in DNA 2. Partial degradation of DNA 3. Release of histones and nuclear globulins 4. Inhibition of nuclear phosphorylation 5. Loss of glycolytic enzymes and ions from nuclei 6. Oxidation of SH to SS, loss of glutathione
Early changes: 1–3 hr (interphase death)	Aggregation of chromatin near nuclear membrane. Changes in nuclear membrane and in mitochondria. "Homogenization" (loss of structure) of nuclei. Release of histones into cytoplasm. Pycnosis of nuclei. Formation of pseudopodia by reticulum cells preparing for phagocytosis	7. Changes in synthesis of RNA 8. Decreased mitochondrial phosphorylation 9. Defect in electron transport of mitochondria 10. Increased lactate formation, loss of NAD (These changes develop progressively during the immediate and early periods after exposure. Probably the earliest alterations are 1–4 and occur in the order indicated by the arrow.)
Intermediate changes: 2–24 hr (phagocytosis)	Phagocytosis of lymphocytes and cellular debris by macrophages and reticulum cells. Removal of dead lymphocytes from organ	Increased specific and total activity of lysosomal enzymes (DNase, RNase, etc.) Decreased O_2 consumption. Enzymatic changes related to the altered cell composition of the organ.
Late changes: 1–10 days (repair)	Low doses: regeneration; large monocytoid lymphoblasts reappear first, mature lymphocytes 1–2 days later	Resumption of DNA synthesis
	High doses: growth of connective tissue	Return to normal enzymatic composition

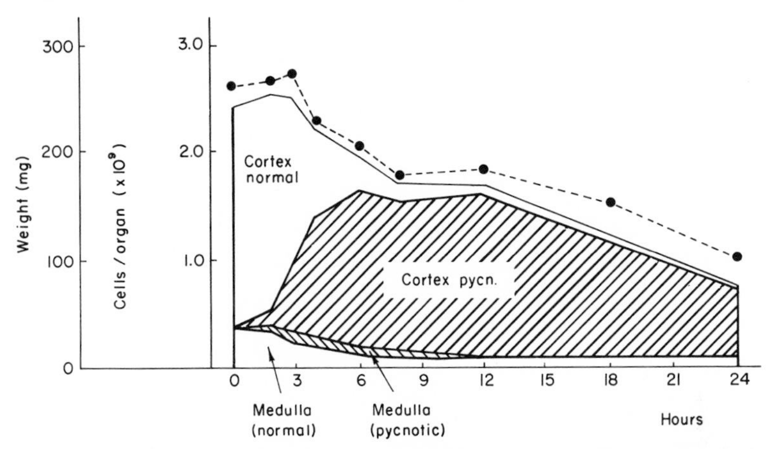

FIG. III-1. Pycnotic nuclei in cortex and medulla of thymic lobules (solid line), and weight of thymus (dashed line) at different times after exposure of rats to 1000 R. Note the increase in pycnotic nuclei, particularly in the cortical areas and the decrease in organ weight. Pycnoses are more numerous in the cortex than in the medulla, and a migration of thymocytes from the medulla takes place during the first hours after irradiation. [From H. Braun, *Strahlentherapie* **121**, 567 (1963); **122**, 248 (1963); **126**, 236 (1965).]

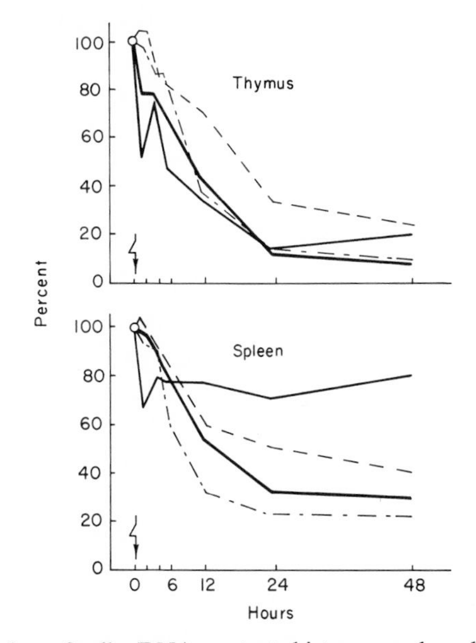

FIG. III-2. Number of cells, DNA content, histones, and nuclear globulins in spleen and thymus of rats after whole-body irradiation (1000 R). Note that decrease in DNA content is preceded by loss of nuclear globulins and, in the thymus, of histones (*39*). Dashed line—number of cells; dot-dash line—DNA; heavy solid line—histone; thin solid line—globulin. [From H. Z. Ernst, *Naturforsch.* **17b**, 300 (1962).]

and by reticulum cells and connective tissue, respectively. The DNA content of lymphoid organs decreases proportionately to destruction of lymphoid cell nuclei (Figs. III-2 and III-8). The average content of DNA per cell remains constant until 12 hr after exposure, and decreases thereafter. Less RNA is lost than DNA, since the reticulum cells remaining after irradiation contain more RNA than lymphocytes.

Studies *in vitro* with irradiated lymphocytes bear out the great radiosensitivity of this cell type, since the *in vitro* survival is shortened by interphase death after exposure to as little as 10–30 R (*156*). Moreover, lymphocytes and related cells are also radiosensitive with respect to their reproduction and their differentiation to antibody-producing plasma cells (*99a*, *105a*). The ability of irradiated lymph-node cells to form colonies *in vivo* has a D_0 of 74 R and an extrapolation number n of 2.7 (*164*). It is possible, however, that not all lymphocytes but only a few stem cells are capable of forming colonies if assayed by colony formation *in vivo*. Primary antibody synthesis in lymphocytes is about as radiosensitive as the capacity to reproduce (D_0 = 30–70 R) (*14*), and the ability to reject homografts is only slightly more resistant (*22*). On the other hand, lymphocytes differentiated to plasma cells (*99a*), dedifferentiated to lymphoblasts (*156a*), or cultured after addition of phytohemagglutinin (*46*) are considerably more radioresistant. For comparison it should be mentioned that cells from radiation-induced lymphosarcoma are about as sensitive as lymphocytes (*69*), but certain lymphoma cells are more resistant (D_0 = 110–130 R (*20*).

B. Immediate and Early Effects of Radiation on DNA Synthesis in Lymphoid Tissue

Within 30 min after exposure there is reduced incorporation into DNA of many metabolic precursors, such as phosphate (*2*, *157*), thymidine (*24*, *113*, *115*, *118*) (Fig. III-3), adenine and formate (*70*), and glycine (*138*). Radioautographs show that this effect results from a decrease in incorporation into DNA within each cell synthesizing DNA (as demonstrated by the diminished grain number per cell in radioautographs), and not from a decrease in the number of cells in the S phase. The dose-response curve for depression of the *in vivo* incorporation as measured autoradiographically by grain counting (*76*) or from the specific activity of DNA (*125*) consists of two components, a radiosensitive one, designated S_0, and a more radioresistant one, designated S_1 (Fig. III-4). The radiosensitive fraction is not detectable after irradiation *in vitro*, but, instead, a still more radioresistant component S_2 emerges (*4*, *76*) (Fig. III-4). The slopes and number of components after *in vitro* irradiation depend, however, on the temperature at which irradiation and incubation with

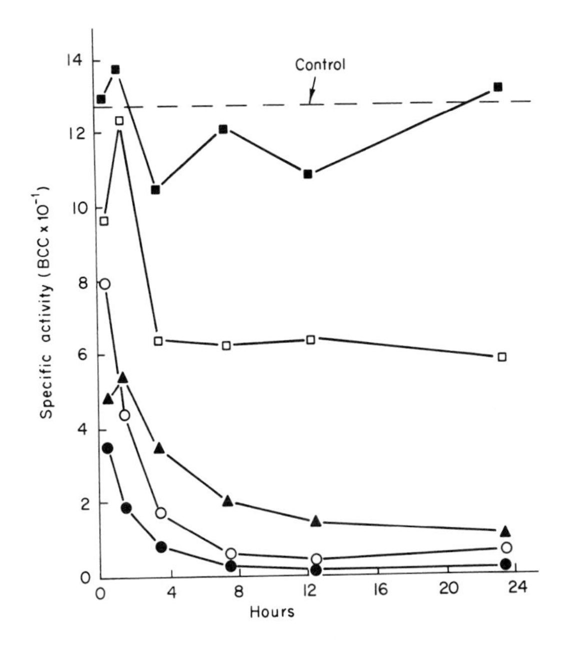

FIG. III-3. Specific activity of DNA isolated from rat thymus after different doses of whole-body irradiation. The animals were sacrificed 1 hr after injection of thymidine-2-^{14}C and the values are plotted at the midpoints of this 1-hr period. Note that incorporation diminishes immediately after exposure to doses of 100 R or more, and remains low during the subsequent hours when the lymphocyte population is depleted. [From O. F. Nygaard and R. L. Potter, *Radiation Res.* **10**, 462 (1959).] Solid squares—50 R; open squares—100 R; solid triangles—200 R; open circles—400 R; solid circles—800 R.

precursors are carried out. Thus, lymphocytes irradiated at 4°C and incubated at 37°C yield a biphasic dose-response curve (D_0 = 4 krad and 28 krad), whereas the curve is monophasic (D_0 = 7 krad) if the cells are irradiated and incubated at 37°C (*100, 113*).

It should be noted, however, that cells *in vitro* are living under nonphysiological conditions, and other pathways which bypass the radiosensitive process may be operative in such "unhappy" cells. Another factor complicating the interpretation of the changes in incorporation into DNA is the presence of repair replication (see Volume I, p. 301). Thus, the immediate decrease in incorporation of thymidine into DNA from spleen or thymus is followed 15–30 min later by an increase (*128h*) which, although best detectable after lower doses (*128g*), has been observed in blood lymphocytes even after 1 krad (*40a*). This newly formed labeled DNA is readily soluble in phenol/trichloroacetate (*128h*) and may not be bound to proteins.

The cause of the immediate depression of incorporation of precursors into

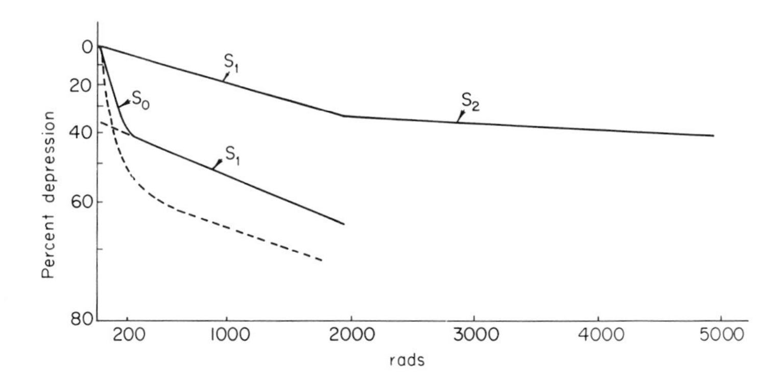

FIG. III-4. Incorporation of thymidine into DNA of lymphocytes as a function of the radiation dose. Upper curve—irradiation *in vitro*; lower curve—irradiation *in vivo*; continuous line—incorporation determined from grain counts after injection of thymidine; dashed line—incorporation determined from specific activity of DNA after injection of ^{32}P (*125*). Differences exist between different precursors with respect to the dose effect curve (*1a*). Note that the radiosensitive component S_0 is absent after an *in vitro* irradiation. [From G. H. Huntley and L. G. Lajtha, *Intern. J. Radiation Biol.* **5**, 447 (1962).]

DNA is still obscure. Changes in the uptake of precursors by the cells, in the conversion to nucleotide triphosphates, and in the size of the precursor pools must be considered in addition to direct effects on the rate of DNA synthesis. For example, products of DNA breakdown could dilute the radioactivity in the precursor pool and result in a lower specific activity of the newly synthesized DNA. Indeed, free deoxyribonucleotides and nucleosides accumulate in irradiated spleen and thymus (*12*, *16*, *124*), but such an increase might as easily result from an excess of precursors due to an inhibition of DNA synthesis as from radiation-induced breakdown of DNA. Effects on incorporation of DNA due to changes in the pools of precursors have been described for cultured cells (*160*), but data on lymphocytes suggest no changes in the size of the thymidine pool after irradiation (*1a*). If incorporation had been depressed due to dilution by breakdown products of DNA, one would expect the incorporation into DNA to decrease progressively with the time after exposure. Indeed, such a decrease has been found only within the initial 3 min (*128b*) or 30 min (*113*) after exposure. The radiosensitive component of DNA synthesis (S_0) has also been related to a defect in ATP production in the nuclei of irradiated lymphocytes (see p. 63). Damage to the DNA template represents another factor which most likely is involved in the more resistant components S_1 and S_2. Loss of ions from the nucleus might be still another possible factor which depresses DNA synthesis after irradiation (*26*, *28*), and this could be more effective *in vivo* than in the *in vitro* medium (physiological saline). Indeed, loss of K^+ from lymphocytes is observed *in vivo* (*18*, *128c*) and *in vitro* (*26*), but only little (*109*) or no (*77*) changes in ionic composition are reported for nuclei isolated from irradiated lymphocytes. Nuclei isolated from irradiated thymus take up inorganic phosphate from the surrounding medium more readily than nonirradiated ones (*193*), but the significance of this observation is not clear, and phosphate uptake is unaltered in intact thymocytes (*155a*). More investigations on the pools of DNA precursors and their relation to the inhibition of incorporation of various precursors are needed to decide which mechanisms are responsible for the immediate effects of radiation on DNA synthesis.

Enzymes involved in synthesis of DNA and its precursors are not significantly altered in activity immediately after irradiation, and do not appear to be responsible for depressing DNA synthesis. Thymidine kinase and DNA polymerase remain at normal levels up to several hours after exposure, and only decrease later on (*42*, *78*, *177*). The activity of deoxycytidylate aminohydrolase (but not that of other deoxynucleotide aminohydrolases) diminishes in lymphoid organs 18 or 24 hr after exposure (*143*, *166*), but no data are available for the early time period after exposure. On the other hand, deoxycytidylate aminohydrolase in thymocyte suspensions is inactivated rapidly following irradiation (*109*).

C. Intermediate and Late Effects on DNA Metabolism

The incorporation of precursors into DNA remains low or falls further for a period of several hours to a few days after exposure (Fig. III-3) (*115*, *116*, *145*) as lymphocytes are destroyed. After whole-body exposure of rats to 400 R, incorporation of thymidine into the DNA of thymus and spleen increases again on the second day (Fig. III-5) and then markedly overshoots the normal incorporation into DNA of nonirradiated organs before returning to normal. Simultaneously, the amount of DNA begins to rise again, and continues to increase until about the tenth day after exposure when the lymphoid organ has almost regained its original size. Incorporation of deoxycytidine into DNA resumes only one day after the peak incorporation of thymidine (*47*, *166*) (Fig. III-5). The different patterns of incorporation of the two precursors probably reflect the changes in cell population during regeneration. In the thymus, the large monocytic lymphoblasts reappear on the fourth day after exposure to a dose of 400 R. These cells might be unable to synthetize thymidine *de novo* and would, therefore, require exogenous thymidine. More mature thymocytes, which emerge about the fifth day, can utilize deoxycytidine to form thymidine. Indeed, deoxycytidylate aminohydrolase and thymidylate synthetase, which had disappeared from the thymus within 1 day after exposure to 400 R, reappear again on the third day, but do not reach normal levels until the fifth day after irradiation (*166*).

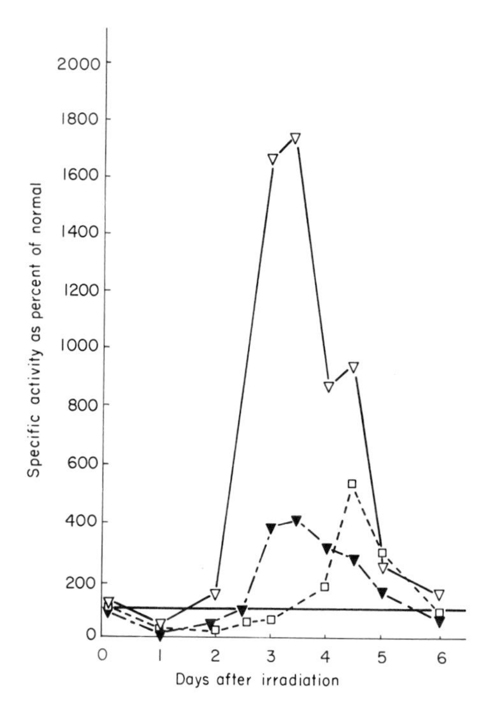

FIG. III-5. Incorporation of deoxycytidine, thymidine, and ^{32}P into DNA of the thymus of whole-body irradiated rats (400 R). Note that the incorporation of thymidine and ^{32}P into DNA is resumed 2 to 3 days after exposure, i.e., 1 day before that of deoxycytidine. [From Y. Sugino, E. P. Frenkel, and R. L. Potter, *Radiation Res.* **19**, 682 (1963).] Open triangles—thymidine-^{3}H; open squares—deoxycytidine-^{3}H; solid squares—^{32}P.

D. Metabolism of RNA, Proteins, and Lipids

Immediately after irradiation, the *in vivo* incorporation of precursors into RNA and proteins of lymphoid organs (*97*, *137*, *138*) is less severely affected than incorporation into DNA. This fact was noted even by early investigators using isotopes (*2*). The radiosensitivity of the biosynthetic process may differ for various types of RNA. Thus, incorporation of ^{32}P injected 30 min after exposure to 300 R is diminished markedly in nuclear RNA I of thymus but less so in nuclear RNA II (*82*). Incorporation of orotate into nuclear RNA 3 hr after 800 R shows similar changes (*153*; see also *106*). In thymocyte suspensions, as contrasted to *in vivo* thymocytes, the difference between radiosensitivity of incorporation of precursors into RNA and DNA is much less pronounced (*113*), due to the greater radioresistance of DNA incorporation in these "unhappy" cells. Synthesis of nuclear proteins, as determined by the incorporation of valine, is enhanced immediately after 1000 R, but decreases later (*162*; see also *73*). This increase could be an indication of stimulation via the adrenals, since it is not observed in adrenalectomized animals (see also p. 316). Nuclear proteins soluble in HCl also incorporate less leucine from 4 hr after exposure to 650 R (*73a*). At later times after irradiation, when the cell population of the lymphoid organ is altered, *in vivo* incorporation of precursors into RNA (*101*) and proteins (*137*) diminishes markedly.

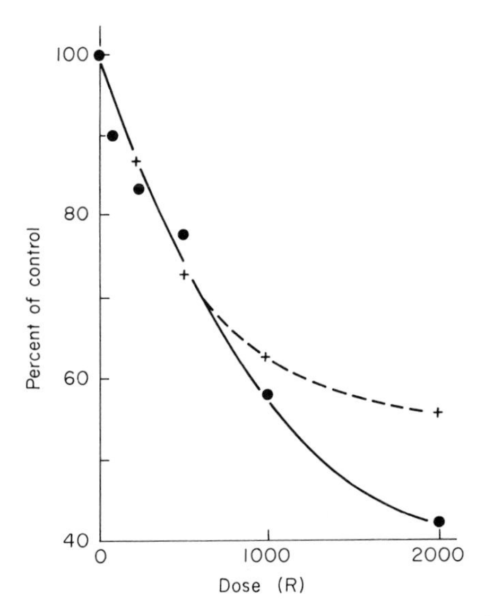

FIG. III-6. Incorporation of valine into proteins (circles), and activation of amino acids (crosses), by irradiated thymus nuclei (in sucrose medium). Note the decreased incorporation of valine and the activation of amino acids dependent on the dose. Both reactions are about equally radiosensitive up to a dose of 1000 R. [From J. A. Smit and L. A. Stocken, *Biochem. J.* **89**, 37 (1963); **91**, 155 (1964).]

Isolated nuclei irradiated *in vitro* with 300 R incorporate less uracil, orotate, or adenine into RNA than do controls and show a similar depression in uptake of phenylalanine into nuclear proteins (about 60% of the controls) (*95, 96, 169*; see also *73*). The decrease in incorporation of thymidine is even more pronounced, amounting to about 30% of the controls. It is of interest to note that addition of denatured DNA partially restores protein synthesis in irradiated nuclei (*41, 169*). Irradiation of the nuclei with higher doses (900 rads) does not depress further the synthesis of RNA and proteins (*41*). Other investigators studying dose levels up to 2000 R, however, claim a dose-dependent decrease of incorporation of valine into nuclear proteins *in vitro* (*162*) (Fig. III-6). The activation of amino acids in isolated nuclei decreases as a function of dose (*161*), although, after higher doses, this inhibition is somewhat less marked than the depression in the incorporation into proteins (Fig. III-6). Since isolated activating enzymes are radioresistant, it is likely that damage to the nuclear structure rather than damage to the enzymes is responsible for the effects of radiation on amino acid activation in isolated nuclei.

Information concerning the metabolism of low molecular weight precursors is scanty. Utilization of formate (one precursor of nucleic acid bases) is depressed after 500 R (*135*), and the folic acid content of the spleen is slightly diminished after whole-body exposure to 400–600 R (*62*, *114*) (folic acid is a cofactor in formate–nucleic acid base conversion).

Irradiation depresses the *in vivo* incorporation of RNA and protein precursors to the same extent; from this it can be concluded that irradiation acts on a common synthetic pathway rather than on specific precursors. Two possible mechanisms can be postulated: defective formation of messenger RNA at the DNA template or depression of nucleotide phosphate formation.

The specific or total activity of several enzymes related to the metabolism of nucleic acids and proteins is altered in irradiated lymphoid organs. These changes occur only during the early and intermediate postirradiation period, and, thus, may reflect alterations in cell population or death of the lymphocytes. Aside from the lysosomal enzymes, which are discussed in more detail later (p. 68), the specific activity of ATPase (*7*, *19*, *32*, *171*; see also p. 62), 5-nucleotidase (*32*), alkaline phosphatase (*7*), nucleoside phosphorylase (*33*, *163*), guanase, and adenosine aminohydrolase increases (*35*). A decrease in specific activity of splenic aminotransferases has been described 1–5 days after exposure (*165*).

The concentration of histidine, tyrosine, tryptophan (*87*, *136*), valine, taurine, and cysteic acid (*105*) in the spleen increases from 3 to 6 hr after exposure, and remains elevated for several days (*94*). These changes are perhaps related to altered permeability of lymphoid tissue to amino acids (*11*). The concentration of taurine decreases on the second day, whereas that of cysteic acid remains elevated until the tenth day after irradiation (*105*). In contrast, the concentrations of cysteine and reduced glutathione (*94*, *152*) diminish. In the case of reduced glutathione in the spleen a decrease begins immediately after an exposure to 200 rads (*8*).

Changes in the composition and metabolism of lipid parallel the changes seen in liver (p. 104) and blood (p. 14). Triglycerides and ^{14}C-acetate incorporation increase in rabbit thymus 4 hr after exposure to 500 rad, whereas ^{14}C-palmitate utilization decreases (*105b*). Incorporation of ^{32}P into phospholipids depends upon the nutritional state of the animals studied (*24a*); it increases in mouse spleen 24–48 hr after 800 R (*94a*).

E. Chemical Changes in Nucleoproteins of Irradiated Lymphocytes

Damage to the molecular structure of nucleoproteins in lymphocytes begins within the first hour after irradiation, and increases with time. More nucleic acid can be extracted from the nucleoproteins of irradiated lymphoid tissues than from controls using strong salt solutions (*23*, *103*, *126*, *128d*, *128e*, *146*, *155*, *155a*, *158*, *159*), trichloroacetate (*57*, *58*, *66*), or acids (*120*) (see review *168a*) (Figs. III-7 and Fig. III-8). The rate at which DNA in the gel form dissolves and relaxation of DNA filaments after loading (*61a*) are altered almost immediately after irradiation. DNA may be liberated from the nucleus in living cells and is detectable in the "microsomal" fractions of irradiated spleen

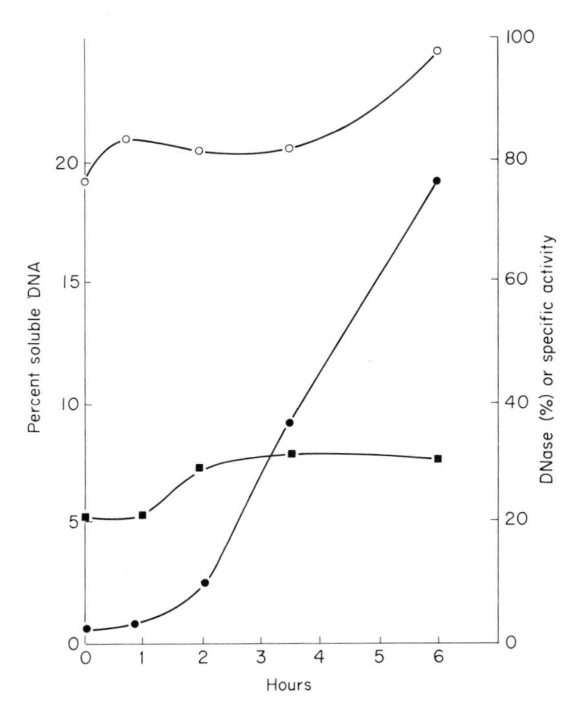

FIG. III-7. Specific activity of DNase II (U/mg protein) of thymus homogenates (open circles); nonsedimentable activity (in the supernatant fraction) as percent of total DNase II activity (squares); and concentration of DNA, soluble in 1% trichloroacetate, at different times after whole-body irradiation (solid circles) (700 R). Note that the nonsedimentable DNase and soluble DNA increase simultaneously beginning 1–2 hr after exposure. The specific activity of total DNase is enhanced 4–6 hr after exposure. [From R. Goutier, M. Goutier-Pirotte, and A. Raffi, *Intern. J. Radiation Biol.* **8**, 51 (1964), and R. Goutier, Contribution à l'étude du mode d'action des radiations ionisantes et des substances radioprotectrices sulfhydrylées sur la biosynthèse de l'ADN chez le rat. Thèse d'Aggregation Liège Arscia S. C., Bruxelles, 1966.]

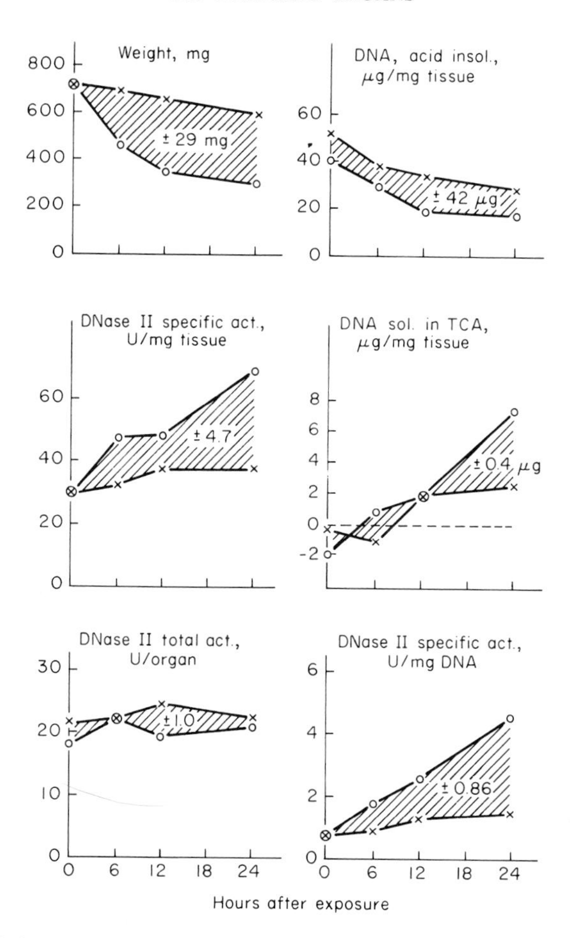

FIG. III-8. Spleen weight, DNA, DNase content during the first day after total-body exposure of rats to 756 R. The variations given represent the pooled standard deviation for the difference between nonirradiated and irradiated rats. Note that the weight and DNA concentration decrease, whereas the amount of DNA, soluble in trichloroacetic acid (TCA), and the specific and total activities of DNase II increase. [From S. Okada, *et al.*, *Arch. Biochem. Biophys.* **70**, 469 (1957).] Circles: irradiated; crosses: nonirradiated.

(*72*). By 3 hr postexposure [earlier if denatured DNA is assayed (*70a*)] the viscosity of nucleoprotein solutions isolated from irradiated lymphoid organs is diminished (*65*, *92*, *146*). Changes in the chromatographic profile reveal a partial degradation and a liberation of proteins from nucleoproteins (*10*, *79a*, *92*, *99*). Determination of the endgroups of the DNA liberated after irradiation suggest that the degradation occurs in a manner similar to that initiated by DNAse I, since the free phosphate group is found at the C-5 positions of the deoxyribose (*168*, *168a*).

Histones and globulins are liberated from nuclear structures and released into the cytoplasm of lymphocytes 30 min after irradiation (*37*, *38*, *40*) (Fig. III-2), and the composition of specific histones is altered 24 hr after exposure to 600 R (*166a*). However, an early postirradiation decrease in the histone to DNA ratio in spleen, thymus, and liver (Fig. III-2) (*39*) has not been confirmed in recent experiments (*169a*). Moreover, phosphorylation of histones is depressed after irradiation (*128a*, *128b*). Protein-bound sulfhydryl compounds are more reactive in nuclei of irradiated than in those of normal thymocytes. Since about 70% of the protein-bound SH groups in the nuclei are associated with arginine-rich histones (*28*), the increase in SH activity also suggests changes in the structure of the histones (*28*, *112*, *128*, *152*). Not only are deoxyribonucleoproteins degraded in lymphocytes after irradiation, but also ribonucleic acid and, as a result, the amount of acid-soluble ribonucleotides increases after irradiation (*109*, *110*, *151*). Although cell death usually follows the labilization of the nucleoprotein molecules in the irradiated lymphocyte, these two effects are not necessarily causally related. Liver cells, for example, display similar, although smaller, changes in the structure of nucleoproteins (*40*) but do not die in interphase. Furthermore, in lymphocytes, the nuclear structure can be partially restored by postirradiation treatment with Ca salts (*66*, *108*), whereas addition of phosphate accelerates loss of nuclear structure and histones (*128f*, *155a*). On the other hand, nuclear changes are prevented when respiration (*155a*) (and phosphate uptake?) is blocked (*128f*) immediately after irradiation. Apparently, the latent reactions leading to pycnosis can be divided into two phases. The first, irreversible, proceeds in the absence of phosphate; the second, occurring after about 40 min, requires extracellular phosphate, at least part of the time (*188a*). Histological observations confirm the chemical changes in the nuclei of irradiated lymphocytes; while nuclei lose their normal structure (homogeneous nuclei) and become pycnotic (*181*, *183*, *186*) histones are released into the cytoplasm and can be visualized by staining with Sakaguchi reagent or with methyl green (*182*, *183*, *186*).

F. The Metabolic Fate of Macromolecules Present in Lymphoid Tissue at the Time of Irradiation

If macromolecules are labeled with an isotopic precursor prior to exposure, their fate can be followed by observing subsequent changes with time in total and specific radioactivity. Metabolic replacement of molecules (relative to the amount present) or unequal loss from different metabolic or cellular pools are estimated by means of the alterations in specific activity, whereas the total loss of labeled molecules from an organ is assessed on the basis of the decrease in total activity. If DNA is labeled with radioactive thymidine 3 hr or 1 to 3 days prior to irradiation, the total radioactivity of DNA is found to decrease markedly after exposure as DNA is lost from the lymphocytes. On the other hand, the specific activity time curve of DNA in spleen and thymus is not influenced significantly by x-irradiation (*51*, *117*, *144*). This observation implies that labeled and nonlabeled dividing and nondividing cells are lost to about the same extent (*117*, *144*). According to one investigator the specific activity of DNA 3 days after exposure decreases (*51*) but others disagree with this conclusion (*117*, *144*). This fall in specific activity (*51*) could be the result of a resumption of DNA synthesis, as suggested also by the increase in the mitotic index at this time. Concurrent autoradiographic studies (*52*) demonstrate that labeled DNA originating in lymphocytes destroyed by irradiation is phagocytosed by the macrophages during the first day after exposure. The specific activity of RNA (*48*) and proteins labeled prior to exposure decreases markedly (*49*, *50*), a finding which suggests that metabolically active components of the RNA and protein pools are lost rapidly after irradiation.

G. Energy Metabolism in Irradiated Lymphoid Organs

1. Mitochondrial Oxidative Phosphorylation

One of the earliest and most sensitive signs of radiation damage to lymphocytes is a defect in the ability to produce energy-rich compounds. Lymphocytes produce ATP either by oxidative phosphorylation in mitochondria, using energy provided by electron transport, or by phosphorylation in the nucleus of endogeneous AMP, employing still unknown sources of energy. The significance of "nuclear" phosphorylation for ATP synthesis in the intact lymphocyte is, however, uncertain.

Oxidative phosphorylation decreases markedly in mitochondria of spleen or thymus 2–4 hr after exposure of rats to 400 R of whole-body irradiation (*104*, *123*, *147*, *173*). It is difficult, however, to decide to what extent the energy production is impaired *in vivo* and which biochemical reactions are involved in the impairment, since mitochondria carefully isolated and provided with excess cytochrome c fail to show a reduction in oxidative phosphorylation at an early time after irradiation (*171*). Differences in phosphorylation at the various sites along the electron transport chain for enzyme reactions are seen only from 24 hr after irradiation (*191*, *192*).

The formation (*175*) and the levels of ATP (*15*, *98*, *172*) are diminished in lymphocytes irradiated *in vivo*, but these data must be evaluated with care, since reactions occurring as rapidly as the formation of ATP are not easy to measure *in vivo*. Moreover, this decrease in ATP content develops only at a time when cell degeneration becomes manifest (*154*). The activation of ATPase in irradiated lymphoid organs may result in breakdown of some ATP *in vivo* or during the isolation procedures, since it is difficult to obtain mitochondrial preparations free of lysosomes from irradiated lymphoid organs. On the other hand, oxidative phosphorylation has been claimed to be depressed before the activation of ATPase (*174*), and even if an inhibitor of ATPase was added to the mitochondria (*104*).

After irradiation, the respiration of lymphoid organs remains unaltered until marked changes in cellular composition have occurred (*40b*, *104*, *167*, *184*). Cytochrome oxidase (*45*, *170*) and NADPH cytochrome reductase (*34*) show little changes in irradiated lymphoid organs. Succinate dehydrogenase was reported to decrease (*178*), or to remain within normal limits (*7*, *123*, *170*). Moreover, the activity of various other Krebs cycle dehydrogenases (*80*), acyl CoA dehydrogenases, L-amino acid oxidase (E.C. 1.4.3.2) (*178*), and of dehydrogenases pertaining to the pentose cycle (*81*) decreases very early after irradiation. Citric acid synthesis is depressed in irradiated spleen after total-body exposure of rats to 800 R. (*31*). Apparently, radiation damage to the mitochondrial dehydrogenases becomes manifest only when certain electron acceptors are used, and, therefore, most likely does not represent a direct effect of radiation on the dehydrogenases (*149*). A defect in the electron-transport chain is postulated (perhaps the loss of a soluble factor) which causes a reduction in as well as a diversion of the flow of electrons to other than normal electron acceptors (*150*). The nature of this soluble factor remains to be defined but several substances, e.g., cytochrome c, can restore the function of the electron-transport system to (near) normal (*154*). Moreover the terminal step of oxidative phosphorylation seems to be the one most easily damaged by irradiation (*191*, *192*), although this reaction becomes manifest at a relatively late time after irradiation.

In conclusion, it seems probable that the structure of the mitochondria is damaged after irradiation in such a manner that, under certain conditions, the systems for electron transport and mitochondrial phosphorylation are affected. Mitochondrial enzymes, however, are not affected by direct action of radiation, since exposure of isolated mitochondria to more than 100 kR is needed to depress oxidative phosphorylation and respiration. The same dose is required to liberate cytochrome c from mitochondria irradiated in concentrated suspensions or in tissue slices (*134*; see also p. 144).

2. "Nuclear Phosphorylation"

Phosphorylation in nuclei isolated from lymphoid organs after whole-body irradiation is depressed more than that in mitochondria. Whole-body exposure to as little as 25 R was found to decrease "nuclear phosphorylation" to 20% of the control values within 1 hr (*25*; see also *83*, *127*) (Fig. III-9). Irradiation of isolated nuclei was also reported to depress "nuclear phosphorylation" (*25*), but this has not been confirmed (*83*). The erratic results of the early studies on the effect of irradiation on "nuclear phosphorylation" (*83*,

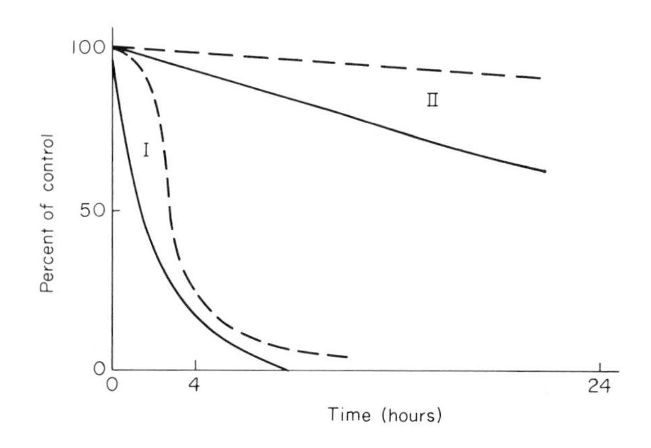

FIG. III-9. Oxidative phosphorylation in cells from thymus, myeloid leukemia, and C57 Bl lymphosarcoma (curves I) as well as in cells from F1 and CBA lymphosarcoma (curves II) after whole-body irradiation (900 R). Continuous lines represent nuclear phosphorylation, dashed lines represent mitochondrial phosphorylation. Note that nuclear and mitochondrial phosphorylation are considerably more radiosensitive in the cells of group I (thymus, myeloid leukemia, and C57 Bl lymphosarcoma) than in those of group II (F1 and CBA lymphosarcoma). [From H. M. Klouwen, A. W. M. Appelman, and M. J. de Vries, *Nature* **209**, 1149 (1966).]

193) can now be attributed to failure to deplete the ATP content of the nuclei by aging through incubation (*83*, *84*). Regeneration of ATP depends, however, on the original ATP concentration. Decreases in "nuclear phosphorylation" and ATP content are, therefore, interrelated, and an assay of "nuclear phosphorylation" will reflect ATP breakdown but will not represent the total energy production by the nucleus (*14a*). Experiments *in vivo* suggest that less ATP is formed in the nuclei after irradiation (*83*, *127*), although this reaction may also depend on ATP levels. It is of interest that certain types of lymphosarcoma cells show a defect in "nuclear phosphorylation" (i.e., perhaps ATP loss) after irradiation and are particularly susceptible to interphase death whereas others do not show these characteristics (*85*, *86*) (Fig. III-9).

3. Glycolysis

Anaerobic lactate production increases markedly in thymocytes after irradiation (*6*, *118a*, *185*, *189a,b*) (Fig. III-10, but compare *74*) and fructose-1,6-diphosphate accumulates as ATP diminishes. Activation of phosphofructokinase (*189a*) might divert energy-rich phosphate from ATP and trap it as fructose-1,6-diphosphate, a reaction that would stimulate lactate formation. Moreover, glycolytic enzymes leak from the nucleus into the cytoplasm (*68*)

(Fig. III-11). Irradiated lymphoid cells also leak pyridine nucleotide coenzymes (NAD, NADP); the loss of these coenzymes plays an important role in the production of radiation damage in tumor cells (see p. 239) as well as in

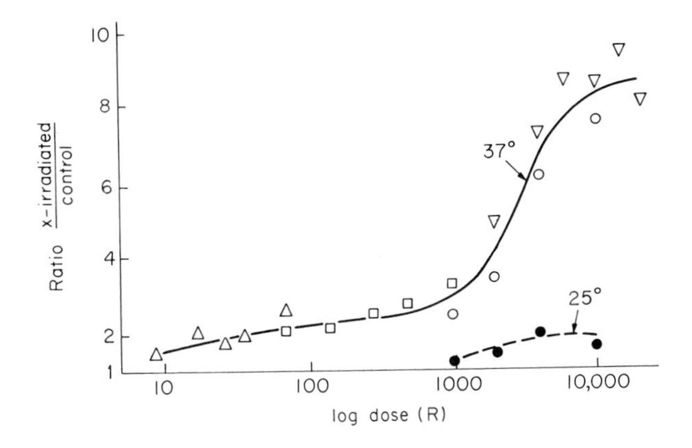

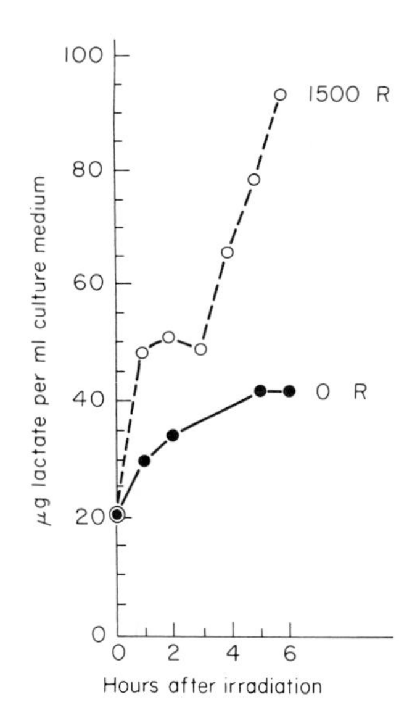

FIG. III-10. Lactate production by thymocytes (in the ratio x-irradiated to controls) during a 2-hr period (at 37°C or 25°C) after exposure to varying doses (upper figure). [From K. Araki and D. K. Myers, *Can. J. Biochem. Physiol.* **41**, 21 (1963).] and lactate production by thymocytes (in μg/ml medium) as a function of time after exposure (lower figure). [From J. F. Whitfield, H. Brohée, and T. Youdale, *Exptl. Cell Res.* **37**, 637 (1965).] Note that lactate production increases soon after exposure, even to low doses of x-rays.

thymocytes. Whereas the amount of NAD in irradiated thymus (*148*) and isolated thymocytes (*21*, *110*) (Fig. III-12) is diminished, the ratio of oxidized to reduced coenzymes remains unchanged until 1 hr after exposure to 5 kR (*79*). NAD(P) nucleosidase activity decreases temporarily in irradiated spleen and returns to normal levels 3–5 hr after exposure (*75*). Enzymes of NAD catabolism are liberated at intracellular sites in irradiated thymus (*149*), and catabolism of NAD is increased in irradiated lymphosarcoma cells (*5*) after doses of 500 R. Addition of nicotinamide to irradiated thymocytes protects

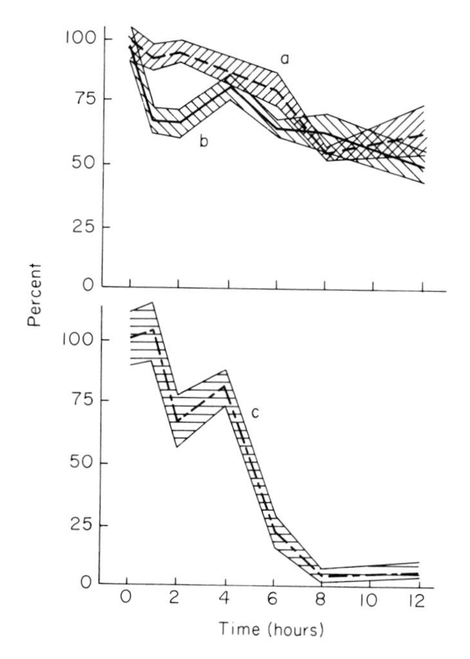

FIG. III-11. Glycolytic enzymes in the nuclei of thymus cells after a whole-body exposure of rats to 800 R of x-rays. (a) Aldolase; (b) lactate dehydrogenase; (c) triose phosphate dehydrogenase. Note that glycolytic enzymes are lost from nuclei after irradiation, especially triose phosphate dehydrogenase. [From U. Hagen, H. Ernst, and I. Cepicka, *Biochim. Biophys. Acta* **74**, 598 (1963).]

NAD from enzymatic breakdown and, at the same time, reduces the damage in nuclei (*184*). However, synthesis of NAD is unaltered shortly after exposure (*107*) but decreases 2–3 hr postirradiation in the thymus (*148*). Enzymes related to NAD synthesis (NAD pyrophosphorylase) appear unaltered early after irradiation (*1*). Therefore, the decrease in NAD content in irradiated lymphocytes is probably the result of leakage from the cells as well as increased catabolism. The role, if any, of pyridine nucleotide coenzymes in the interphase death of lymphocytes remains to be elucidated.

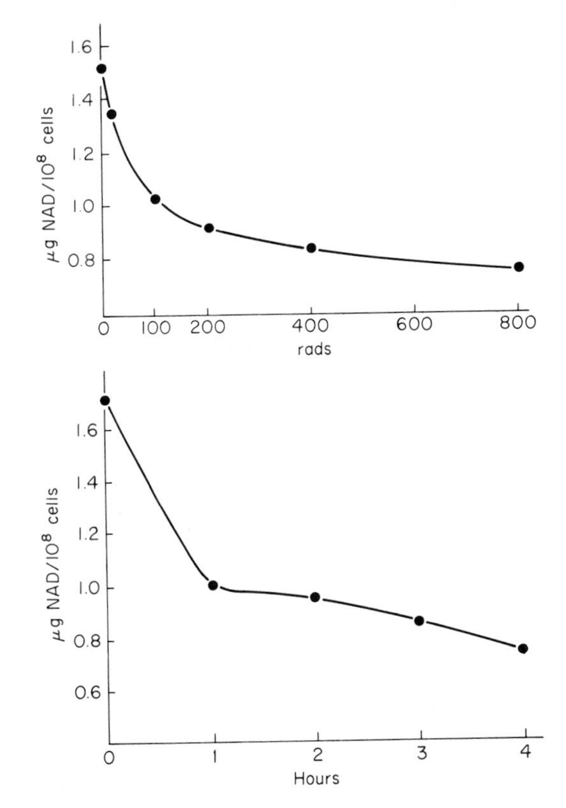

FIG. III-12. Nicotinamide adenine dinucleotide (NAD) content in thymocytes as a function of dose (4 hr) (upper curve) and time after irradiation (800 rads) (lower curve). The thymocytes were irradiated *in vitro*. Note that the decrease in NAD content parallels homogenization of the nuclear structures. [From J. F. Scaife, *Can. J. Biochem. Physiol.* **41**, 1469 (1963).]

4. SH COMPOUNDS

Oxidation of SH groups to S—S groups has long been considered a factor in certain radiation effects (see p. 34). As mentioned above, nonprotein SH compounds (e.g., glutathione) diminish progressively in the irradiated lymphocyte (see p. 58), perhaps in relation to changes in triosephosphate dehydrogenase (*8*, *152*), whereas protein-bound SH compounds in the nucleus become more reactive (*28*) (see p. 61). Changes in the ratio of SH to S—S compounds in the cell can be detected after inhibiting GSSG reductase (*128*). The ability to restore thiol compounds is probably not altered after irradiation (*152*). Yet, it is conceivable that this reduction cannot fully restore the original configuration of the macromolecule. Glutathione which diminishes after exposure is involved in a pathway detoxifying H_2O_2 via glutathione peroxidase. Catalase, which also destroys H_2O_2, is lost from nuclei of irradiated lymphocytes (*27*), but it is uncertain whether these two effects influence the H_2O_2 level after irradiation.

H. Lysosomal Enzymes in Irradiated Lymphoid Organs

Bacq and Alexander (*9*) postulated in 1955 that irradiation releases lytic enzymes from their compartments within the cell and thereby permits them to attack vital cellular structures. In rudimentary form, this idea can be traced back to the very early days of radiation biology and has stimulated research on the "activation" of enzymes in many organs. Indeed, specific or total enzyme activity most frequently increases with respect to lysosomal enzymes (DNase II, RNase, cathepsin, acid phosphatase, β-glucuronidase, and arylsulfatase) but also with respect to related lytic enzymes, e.g., phosphatase, 5-nucleotidase, and ATPase (see p. 58). These changes are not found immediately after exposure, but develop progressively after 1–3 hr, and thus represent early and intermediate effects of the irradiation.

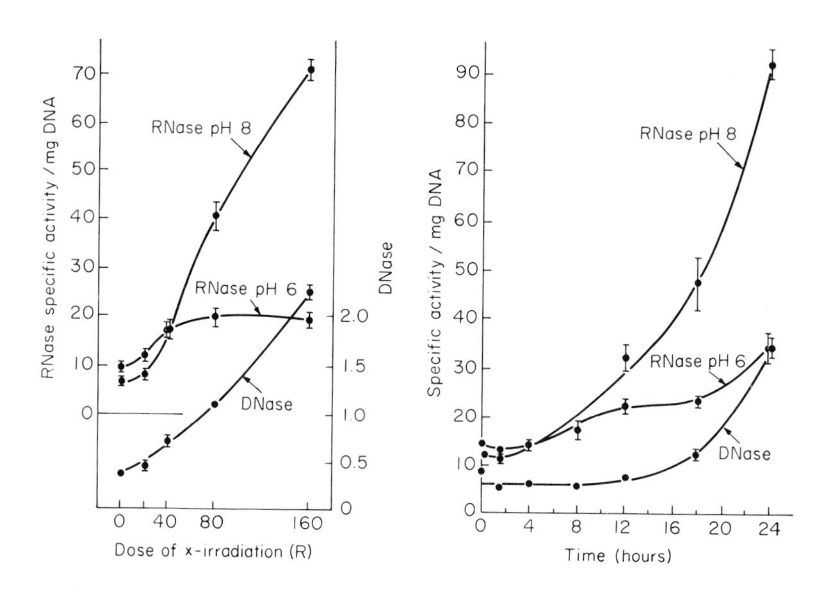

FIG. III-13. Specific activity of DNase II, RNase pH 6, and RNase pH 8 in thymus of C57 Bl mice as a function of the dose (at 24 hr) (left figure) and the time (after 160 R) (right figure) after whole-body exposure. Note that differences in specific activities between the three lysosomal enzymes exist. [From P. P. Weymouth, *Radiation Res.* **8**, 307 (1958).]

The specific activity of DNase II increases in lymphoid organs beginning 2 hr after an exposure to 50 R or more, attains maximal values from 12 to 24 hr after irradiation, and declines thereafter (*30*, *53–55*, *56*, *57*, *59*, *88*, *89*, *120*, *121*, *135a*) (Fig. III-7, III-8 and III-13). (The specific activity refers to the enzymatic activity per gram organ, per milligram DNA, per milligram protein, etc.) Rat spleen shows a maximal response after a whole-body dose of 300 R; higher doses produce no additional effect (*90*, *179*).

The weight and the DNA content of lymphoid organs diminish, whereas the specific activity of DNase II increases. The total enzymatic activity, however, remains constant or increases only slightly in lymphoid organs of the rat, whereas in the mouse a marked increase in total activity of spleen and thymus DNase II is reported, even after low doses of radiation (*129*). In irradiated lymphoid organs, a significant part of the DNase II is already active without treatment with water, detergents, or ultrasound due to increased fragility of the lysosomes (*180*). More enzymatic activity is present in the supernatant fraction of irradiated lymphoid tissue than in that of normal tissue (*42*, *58*, *59*, *121*) (Fig. III-8). RNase (*139*, *140*) and β-galactosidase (*142*) exhibit similar changes.

In evaluating data on lysosomal enzymes after irradiation, attention should be paid to the methods employed in determining enzymatic activity and to the way the results are presented. Data on specific activity of the enzymes are insufficient and should be complemented by obtaining data on the total activity, weight, and DNA content of the organ (*36*). Care must be exercised when judging data on subcellular fractionation, since autolysis in the irradiated cells may change the size (*131*) and the distribution of subcellular particles in an unpredictable manner. If possible, the activity in the total organ should approximate the sum of that in the various fractions isolated. DNase II is not the only lysosomal enzyme which increases in specific activity after irradiation. RNase (*88*, *139*, *140*, *141*, *179*) proteolytic enzymes (*63*, *64*, *67*; but see *47a*, *47b*), glycylglycine peptidase, alanyl and leucyl aminases (*60*), β-galactosidase (*142*), β-glucuronidase (*61*, *129*, *141*), acid phosphatase (*132*, *139*, *141*), and phosphoprotein phosphatase (*44*, *141*) display similar changes in specific activity, but the extent and time course of these activations differ from one enzyme to the other (*179*) (Fig. III-13).

The increase in enzymatic activity in irradiated lymphoid organs may be explained by one or more of several mechanisms; the following among them are most likely: (a) change in cell population, (b) change in the rates of synthesis or catabolism of enzymes, (c) translocation of enzymes from one organ to another. In addition, the following other mechanisms have been proposed but, at the present time, seem of minor importance, (d) changes in inhibitors or activators. Such inhibitors, however, have not been demonstrated for DNase II and most lysosomal enzymes, although changes in inhibitors have been found for DNase I (*88*) and cathepsin (*43*). (e) No evidence exists for changes in lysosomal enzymes related to a shift in the distribution of cells between different stages of their life cycles. (f) Immediate release of enzymes (arylsulfatase) (*29*, *189*) and formation of peroxides (*189*) in lysosomes have been observed only after doses of 1 kR and more. In the case of DNase II, release requires doses of more than 100 kR (*119*).

As we proceed to discuss the various mechanisms likely to be responsible for the activation of lysosomal enzymes, we should remember that the increase in specific as well as total enzymatic activity must be explained. It should also be realized that not only one but several mechanisms may operate simultaneously. The cell population in lymphoid organs changes drastically during the intermediate period after irradiation. The radiosensitive lymphocytes containing few lysosomes are destroyed; the more resistant reticulum cells and macrophages with many lysosomes persist and start to phagocytose the cellular debris. The average number of lysosomes per cell is thereby increased. Simultaneously, lysosomes increase in number, cytolysomes are formed, and the distribution of enzymes in the subcellular fractions changes as a result of liberation of enzymes from subcellular particles. Histochemical techniques have demonstrated (*3*) that DNase II and acid phosphatase are located in the macrophages engaged in active phagocytosis after irradiation. The increase in specific activity of lysosomal enzymes reflects the progress of auto- and heterophagocytosis. The difference between the various lysosomal enzymes with respect to the extent and the time course of their activation (*179*) could result from a different distribution of enzymes among the various cell types, but might, of course, also reflect different rates of *de novo* synthesis or of catabolism of enzymes after irradiation.

When the total activity of an enzyme increases, as reported for DNase II (*122*, *129*), RNase, and acid phosphatase (*132*), a change in cell population obviously cannot be the sole cause, but increased *de novo* synthesis, decreased catabolism, and/or translocation from outside the organ must also have taken place. About 15% of thymic RNase (*142*) and about 10% of splenic DNase II (*122*) found in the rat after irradiation represents enzyme not previously present in the organs. Although it seems unlikely that such a large number of DNase-rich macrophages were transported into these organs via the circulating blood, the possibility of a translocation of cells cannot be excluded entirely until quantitative histochemical studies have been carried out. Treatment with inhibitors of protein synthesis, e.g., actinomycin and ethionine, may serve to determine whether *de novo* synthesis of enzyme takes place after irradiation. Unfortunately, these studies have not yielded consistent results in that an inhibition (*122*), as well as no effect (*129*), have been encountered. This may be due to the fact that the doses which inhibit protein synthesis are very close to toxic levels.

An enhanced activity of lysosomal enzymes in lymphatic tissue is a reaction commonly observed after exposure to many agents which damage this tissue. It also occurs after treatment with cortisone (*53, 54, 129*), thyroxine (*133*), or after starvation. Irradiation of the head also produces an increase in specific activity of splenic RNase (*102*; see also *56, 89, 91*), probably by stimulating phagocytes via humoral pathways. On the other hand, head irradiation has no influence on weight, cytology, RNase activity, or DNA content in the thymus (*151*) or on nuclear phosphorylation in the spleen (*176*). A decrease in mitochondrial phosphorylation after exposure of the head (*13*) may have been due to the contamination of the isolated mitochondria with lysosomes of high ATPase activity (*133*).

To recapitulate, the increase in the specific activity of lysosomal enzymes in the rat spleen appears to be the result of changes in cell population. For example about 90% of DNase II present in the spleen at the time of irradiation (756 R) is still retained there 24 hr later. *De novo* synthesis or a translocation of enzyme account for only about 20% of the enzyme present after 24 hr, as the increase in total activity demonstrates (*122*). During the late period after irradiation, when regeneration of the different types of cells takes place, the activity of lysosomal enzymes return to normal levels. It is of interest that in thymus the activity of certain enzymes is restored only when the immunologically competent cells have recovered (*60, 61*).

I. Conclusions

The immediate and early effects of irradiation on lymphoid organs are associated with the interphase death of lymphocytes. Interphase death is a fascinating phenomenon. If we understood this mode of cell death, we could follow the path of radiation injury from the absorption of radiation energy in the various molecular species of the cell to the immediate biochemical lesions and, eventually, to the death of the cell. As our present knowledge is still inadequate, we cannot fuse all biochemical and morphological data into a coherent picture. We must depend on working hypotheses to coordinate the experimental findings and to guide us in future investigations, but we must recognize clearly the limitations inherent in such hypotheses.

Structural alterations in the nucleoprotein molecules (*111*), release of histones and other substances from the nucleus, decrease in the incorporation of precursors into DNA, and changes in the energy metabolism of the cell are the biochemical events detectable in the lymphocytes within 1 hr after irradiation. The cause and temporal sequence of these changes are still unknown. The different immediate biochemical lesions may be unrelated, but may also be mutually interdependent and originate via similar mechanisms. For example, it has been suggested that the biochemical lesions caused by irradiation become manifest at the nuclear and mitochondrial membranes and alter permeability and metabolism of these structures. Alternatively or concurrently, damage to nucleoprotein molecules may weaken the bond between DNA and the proteins, disorganizing the nuclear structure and metabolism and affecting the synthesis of DNA and perhaps that of messenger RNA. Furthermore, irradiation may alter the proportion between SH and S—S groups (*128b*), and thereby affect several biochemical systems simultaneously, e.g., delay mitosis, which is thought (*104a*) to be controlled by the ratio between protein and nonprotein SH groups, disarrange the nucleoprotein molecule, and affect carbohydrate metabolism, nuclear phosphorylation, and detoxication of H_2O_2. The formation of "quinones," postulated by Kuzin (*93*), should also be mentioned as another possible common mechanism for the immediate effects of irradiation.

Once established, all biochemical lesions in the irradiated lymphocytes become, to some extent, interconnected, and form a vicious cycle which disorganizes cellular structure and function and brings about cell death. Whitfield (*188*) suggested that the concentration of phosphate in the cell increases, since less phosphate is used for the production of ATP. This could result in the removal of bivalent ions (e.g., Ca) thought to be essential for preserving the organized nuclear structure. The bonds between nucleic acids and proteins may thus be weakened, so that histones diffuse into the cytoplasm. Phosphate concentration is, however, found to be unaltered (*155a*), and it is doubtful whether less ATP is formed in the cell at such an early time (*14a*), so that this hypothesis must be reevaluated. Changes in mitochondrial metabolism, loss of glycolytic enzymes, and NAD catabolism could be other areas in which the action of radiation becomes manifest. The activation of lysosomal enzymes in reticulum cells and, perhaps, in lymphocytes finally contributes to the death and autolysis of the cells.

Several agents protect lymphocytes from this destructive cycle, even when added after irradiation. Just how these protective agents act is still being debated. The nuclear structure is preserved and interphase death delayed by treatment with spermidine or high salt concentrations. Protection is afforded by substances which act on respiration (*155a*) and glycolysis (*155*, *188*) or prevent the accumulation of phosphate in the cell, e.g., Ca ions (*108*), nicotinamide, cyanide, and dinitrophenol (*128*, *180*, *187*). It is of interest that high salt concentrations and Ca ions also protect against mitotic delay in irradiated lymphocyte cultures (*128d*, *128g*, *180*). These findings suggest that the initial biochemical lesions may be similar for the different effects of irradiation on cells.

Nevertheless, it should be emphasized that some of the most crucial questions related to the mechanism of interphase death still await an answer. Does an irradiated lymphocyte indeed produce less ATP than it needs, or is the diminished *in vitro* phosphorylation of nuclei and mitochondria only a sign, but not a cause, of the interphase death? Is the immediate depression in incorporation of precursors caused by a decrease in the rate of DNA synthesis, by changes in the metabolism of precursors, or by breakdown of DNA diluting the precursor pools? Are the lesions that impair DNA synthesis present in non-S stage cells as well as in the S stage cells? What could be the significance of these lesions in the many small lymphocytes which are not in the S phase and represent the bulk of the cells sustaining interphase death in lymphoid organs? How does repair of DNA interact with interphase death? Why are lymphocytes much more sensitive to interphase death than other nondividing cells in the body, e.g., liver cells? Is this difference due to fewer mitochondria and less cytoplasm in the lymphocytes?

Intermediate and late effects follow the death of the lymphocytes. The lysosomal enzymes in reticulum cells and macrophages remain activated and degrade the nucleic acids and proteins of lymphocytes. The dead lymphocytes are phagocytosed and then disappear entirely. In this way, the cellular composition of the lymphoid organ changes profoundly, and with it, the various enzymatic activities as well as the concentration of nucleic acids, proteins, and low-molecular compounds.

As regeneration begins, DNA synthesis is resumed, but the newly formed cells do not possess, at first, the full enzymatic endowment of mature lymphocytes. Gradually the cellular composition of the organ returns to normal, and as the more complex functions of the lymphocytes and the immunological competence recover, the biochemical functions of the lymphoid organ are fully restored.

REFERENCES

1. Adelstein, S. J., and Biggs, S. L., *Radiation Res.* **16**, 422 (1962).

1a. Adelstein, S. J., and Manasek, G. B., *Intern. J. Radiation Biol.* **12**, 587 and 593 (1967).

2. Ahlström, L. H., von Euler, H., and von Hevesy, G., *Arkiv. Kemi, Mineral. Geol.* **A19**, 16 (1944).

3. Aldridge, W. G., Hempelmann, L. H., and Emmel, V. M., *Radiation Res.* **12**, 49 (1960).

4. Alpen, E. L., Cooper, E. H., and Barkley, H., *Intern. J. Radiation Biol.* **2**, 425 (1960).

5. Altenbrunn, H-J., Steenbeck, L., Ehrhardt, E., and Schmaglowski, S., *Intern. J. Radiation Biol.* **9**, 43 (1965).

6. Araki, K., and Myers, D. K., *Can. J. Biochem. Physiol.* **41**, 21 (1963).

7. Ashwell, G., and Hickman, J., *J. Biol. Chem.* **201**, 651 (1953).

8. Ashwood-Smith, M. J., *Intern. J. Radiation Biol.* **3**, 125 (1961).

9. Bacq, Z. M., and Alexander, P., "Fundamentals of Radiobiology." Academic Press, New York, 1955.

10. Bauer, R. D., Loring, W., and Kurnick, N. B., *Radiation Res.* **26**, 507 (1965).

11. Bekhor, I. J., Shah, P., Thomas, D. W., and Bavetta, L. A., *Radiation Res.* **21**, 223 (1964).

12. Beneš, L., Soška, J., and Lukášová, E., *Folia Biol. (Prague)* **11**, 123 (1965).
13. Benjamin, T. L., and Yost, H. T., Jr., *Radiation Res.* **12**, 613 (1960).
14. Berlin, B. S., *Radiation Res.* **26**, 554 (1965).
14a. Betel, I., *Intern. J. Radiation Biol.* **12**, 459 (1967).
15. Bettendorf, G., Maass, H., Kirsten, E., and Künkel, H. A., *Strahlentherapie* **112**, 74 (1960).
15a. Blomgren, H., and Révész, L., *Exptl. Cell Res.* **51**, 92 (1968).
16. Bishop, C. W., and Davidson, J. N., *Brit. J. Radiol.* **30**, 367 (1957).
17. Braun, H., *Strahlentherapie* **121**, 567 (1963); **122**, 248 (1963); **126**, 236 (1965).
18. Breuer, H., Knuppen, R., and Parchwitz, H. K., *Z. Naturforsch.* **13b**, 741 (1958).
19. Breuer, H., and Parchwitz, H. K., *Strahlentherapie* **108**, 93 (1959).
20. Bush, R. S., and Bruce, W. R., *Radiation Res.* **25**, 503 (1965).
21. Campagnari, F., Whitfield, J. F., and Bertazzoni, U., *Exptl. Cell Res.* **42**, 646 (1966).
22. Celada, F., and Carter, R. R., *J. Immunol.* **89**, 161 (1962).
23. Cole, L. J., and Ellis, M. E., *Radiation Res.* **7**, 508 (1957).
24. Cooper, E. H., and Alpen, E. L., *Intern. J. Radiation Biol.* **1**, 344 (1959).
24a. Cornatzer, W. E., Davison, J. P., Engelstad, O. D., and Simonson, C., *Radiation Res.* **1**, 544 (1954).
25. Creasey, W. A., and Stocken, L. A., *Biochem. J.* **72**, 519 (1959).
26. Creasey, W. A., *Biochim. Biophys. Acta* **38**, 181 (1960).
27. Creasey, W. A., *Biochem. J.* **77**, 5 (1960).
28. Deakin, H., Ord, M. G., and Stocken, L. A., *Biochem. J.* **89**, 296 (1963).
29. Desai, I. D., Sawant, P. L., and Tappel, A. L., *Biochim. Biophys. Acta* **86**, 277 (1964).
30. Douglass, C. D., and Day, P. L., *Proc. Soc. Exptl. Biol. Med.* **89**, 616 (1955).
31. Dubois, K. P., Cochran, K. W., and Doull, J., *Proc. Soc. Exptl. Biol. Med.* **76**, 422 (1951).
32. Dubois, K. P., and Petersen, D. F., *Am. J. Physiol.* **176**, 282 (1954).
33. Eichel, H. J., *Proc. Soc. Exptl. Biol. Med.* **88**, 155 (1955).
34. Eichel, H. J., *Proc. Soc. Exptl. Biol. Med.* **95**, 38 (1957).
35. Eichel, H. J., *Proc. Soc. Exptl. Biol. Med.* **105**, 103 (1960).
36. Eichel, H. J., and Roth, J. S., *Radiation Res.* **12**, 258 (1960).
37. Ernst, H., *Naturwissenschaften* **48**, 575 (1961); **50**, 333 (1963).
38. Ernst, H., *Z. Naturforsch.* **16b**, 329 (1961).
39. Ernst, H., *Z. Naturforsch.* **17b**, 300 (1962).
40. Ernst, H., *Intern. J. Radiation Biol.* **6**, 395 (1963).
40a. Evans, R. G., and Norman, A., *Nature* **217**, 455 (1968).
40b. Evans, N. T. S., *Radiation Res.* **35**, 465 (1968).
41. Faures, A., and Errera, M., *Intern. J. Radiation Biol.* **4**, 477 (1961).
42. Fausto, N., Smoot, A. O., and Van Lancker, J. L., *Radiation Res.* **22**, 288 (1964).
43. Feinstein, R. N., and Ballin, J. C., *Proc. Soc. Exptl. Biol. Med.* **83**, 6 (1953).
44. Feinstein, R. N., *Radiation Res.* **4**, 217 (1956).
45. Fischer, M. A., Coulter, E. P., and Costello, M., *Proc. Soc. Exptl. Biol. Med.* **83**, 266 (1953).
46. Fliedner, T. M., "The Lymphocyte in Immunology and Haemopoiesis" (J. M. Yoffey, ed.), p. 198. Arnold, London, 1966.
47. Frenkel, E. P., Sugino, Y., Bishop, R. C., and Potter, R. L., *Radiation Res.* **19**, 701 (1963).
47a. Furlan, M., *Strahlentherapie* **133**, 68 (1967).
47b. Furlan, M., and Jericijo, M., *Strahlentherapie* **133**, 262 (1967).

48. Gerber, G. B., Gerber, G., and Altman, K. I., *Intern. J. Radiation Biol.* **4**, 67 (1961).
49. Gerber, G. B., Gerber, G., Altman, K. I., and Hempelmann, L. H., *Intern. J. Radiation Biol.* **4**, 609 (1962).
50. Gerber, G. B., Gerber, G., Altman, K. I., and Hempelmann, L. H., *Intern. J. Radiation Biol.* **4**, 615 (1962).
51. Gerber, G. B., Gerber, G., Altman, K. I., and Hempelmann, L. H., *Intern. J. Radiation Biol.* **6**, 17 (1963).
52. Gerber, G. B., Aldridge, W. G., Gerber, G., Altman, K. I., and Hempelmann, L. H., *Intern. J. Radiation Biol.* **6**, 23 (1963).
53. Gordon, E. R., Okada, S., and Hempelmann, L. H., *Radiation Res.* **5**, 479 (1956).
54. Gordon, E. R., Gassner, E., Okada, S., and Hempelmann, L. H., *Radiation Res.* **10**, 545 (1959).
55. Goutier, R., *Arch. Intern. Physiol. Biochim.* **67**, 15 (1959).
56. Goutier, R., *Compt. Rend. Soc. Biol.* **159**, 1001 (1965).
57. Goutier, R., Goutier-Pirotte, M., and Raffi, A., *Intern. J. Radiation Biol.* **8**, 51 (1964).
58. Goutier, R., Contribution à l'étude du mode d'action des radiations ionisantes et des substances radioprotectrices sulfhydrylées sur la biosynthèse de l'ADN chez le rat. Thèse d'Aggregation Liège Arscia S.C., Bruxelles, 1966.
59. Goutier-Pirotte, M., and Thonnard, A., *Biochim. Biophys. Acta* **22**, 396 (1956).
60. Greenberg, L. J., Cole, L. J., and Martin, R. L., *Radiation Res.* **26**, 413 (1965).
61. Greenberg, L. J., and Cole, L. J., *Proc. Soc. Exptl. Biol. Med.* **121**, 1120 (1966).
61a. Grigorovitch, N. A., *Dokl. Akad. Nauk SSSR* **167**, 1171 (1966).
62. Guggenheim, K., Halevy, S., Hochman, A., and Ben-Porath, M., *Radiation Res.* **12**, 567 (1960).
63. Hagen, U., *Naturwissenschaften* **44**, 402 (1957).
64. Hagen, U., *Strahlentherapie* **106**, 277 (1958).
65. Hagen, U., *Strahlentherapie* **116**, 385 (1961).
66. Hagen, U., *Strahlentherapie* **117**, 119 (1962).
67. Hagen, U., *Strahlentherapie* **117**, 201 (1962).
68. Hagen, U., Ernst, H., and Cepicka, I., *Biochim. Biophys. Acta* **74**, 598 (1963).
69. Hanks, G. E., and Kaplan, H. S., *Radiation Res.* **26**, 84 (1965).
70. Harrington, H., and Lavik, P. S., *Arch. Biochem. Biophys.* **54**, 6 (1955).
70a. Harrington, H., *Arch. Biochem. Biophys.* **117**, 615 (1966).
71. Heineke, H., *Muench. Med. Wochschr.* **50**, 2090 (1903).
72. Herranen, A., *Nature* **203**, 891 (1964).
73. Herranen, A., *Arch. Biochem. Biophys.* **107**, 158 (1964).
73a. Herranen, A., and Brunkhorst, W. K., *Arch. Biochem. Biophys.* **119**, 353 (1967).
74. Hickman, J., and Ashwell, G., *J. Biol. Chem.* **205**, 651 (1953).
75. Hilgertovà, J., Šonka, J., Pospíšil, J., and Dienstbier, Z., *Radiation Res.* **27**, 211 (1966).
76. Huntley, G. H., and Lajtha, L. G., *Intern. J. Radiation Biol.* **5**, 447 (1962).
77. Jackson, K. L., and Christensen, G. M., *Radiation Res.* **27**, 434 (1966).
77a. Jackson, K. L., Christensen, G. M., and Forkey, D. J., *Radiation Res.* **34**, 366 (1968).
78. Jacob, M., Chambon, P., and Mandel, P., *J. Physiol.* (*Paris*) **53**, 373 (1961).
79. Jamieson, D., *Nature* **209**, 361 (1966).
79a. Janík, B. *Folia Biol.* (*Prague*) **12**, 189 (1966).
80. Kivy-Rosenberg, E., Cascarano, J., and Zweifach, B. W., *Radiation Res.* **20**, 668 (1963).

81. Kivy-Rosenberg, E., Cascarano, J., and Zweifach, B. W., *Radiation Res.* **23**, 310 (1964).
82. Klouwen, H. M., *Biochim. Biophys. Acta* **42**, 366 (1960).
83. Klouwen, H. M., Betel, I., Appelman, A. W. M., and Reuvers, D. C., *Intern. J. Radiation Biol.* **6**, 441 (1963).
84. Klouwen, H. M., *in* "Cellular Radiation Biology," p. 142. Williams & Wilkins, Baltimore, Maryland, 1965.
85. Klouwen, H. M., Appelman, A. W. M., and de Vries, M. J., *Nature* **209**, 1149 (1966).
86. Klouwen, H. M., Appelman, A. W. M., and Betel, I., *in* "The Cell Nucleus: Metabolism and Radiosensitivity " p. 295. Taylor & Francis, London, 1966.
87. Konstantinova, M. S., *Tashkentsk. Konf. po Mirnomu Ispol'z At Energii Akad. Nauk Uz. SSSR* **3**, 133 (1961).
88. Kurnick, N. B., Massey, B. W., and Sandeen, G., *Radiation Res.* **11**, 101 (1959).
89. Kurnick, N. B., Massey, B. W., and Montano, A., *Radiation Res.* **13**, 263 (1960).
90. Kurnick, N. B., and Nokay, N., *Radiation Res.* **17**, 140 (1962).
91. Kurnick, N. B., and Nokay, N., *Ann. N. Y. Acad. Sci.* **114**, 528 (1964).
92. Kuzin, A. M., *Intern. J. Radiation Biol.* **6**, 201 (1963).
93. Kuzin, A. M., "Radiation Biochemistry." Israel Progr. Sci. Transl., Jerusalem, 1964.
94. Langendorff, H., Streffer, C., and Melching, H. J., *Strahlentherapie* **124**, 445 (1964).
94a. Lee, T. C., Salmon, R. J., Loken, M. K., and Mosser, D. G., *Radiation Res.* **17**, 903 (1962).
95. Logan, R., Errera, M., and Ficq, A., *Biochim. Biophys. Acta* **32**, 147 (1959).
96. Logan, R., *Biochim. Biophys. Acta* **35**, 251 (1959).
97. Maass, H., and Timm, M., *Biophysik* **1**, 334 (1964).
98. Maass, H., and Timm, M., *Strahlentherapie* **123**, 64 (1964).
99. Main, R. K., Cole, L. J., and Ellis, M. E., *Nature* **180**, 1285 (1957).
99a. Makinodan, T., Nettesheim, P., Morita, T., and Chadwick, C., *J. Cellular Physiol.* **69**, 355 (1967).
100. Manasek, F. J., Adelstein, S. J., and Lyman, C. P., *Intern. J. Radiation Biol.* **10**, 189 (1966).
101. Mandel, P., and Chambon, P., *Intern. J. Radiation Biol.* Suppl. 1, 71 (1960).
102. Maor, D., and Alexander, P., *Intern. J. Radiation Biol.* **6**, 93 (1963).
103. Matyášová, J., Skalka, M., and Soška, J., *Compt. Rend. Soc. Biol.* **157**, 451 (1963).
104. Maxwell, E., and Ashwell, G., *Arch. Biochem. Biophys.* **43**, 389 (1953).
104a. Mazia, D., *in* "The Cell" (J. Brachet and A. C. Mirsky, eds.), Vol. III, p. 77. Academic Press, New York, 1961.
105. Melching, H. J., Streffer, C., and Mavely, D., *Strahlentherapie* **127**, 368 (1965).
105a. Miller, J., and Cole, L., *J. Immunol.* **98**, 982 (1967).
105b. Miras, C. J., Mantzos, J. D., Legakis, N. J., and Levis, G. M., *Radiation Res.* **36**, 119 and 208 (1968).
106. Mori, K. J., and Morita, T., *Nature* **200**, 1323 (1963).
107. Myers, D. K., *Nature* **187**, 1124 (1960).
108. Myers, D. K., and De Wolfe, D. E., *Intern. J. Radiation Biol.* **5**, 195 (1962).
109. Myers, D. K., De Wolfe, D. E., Araki, K., Arkinstall, W. W., *Can. J. Biochem. Physiol.* **41**, 1181 (1963).
110. Myers, D. K., and De Wolfe-Slade, D., *Can. J. Biochem.* **42**, 529 (1964).
111. Myers, D. K., *Nature* **205**, 365 (1965).
112. Myers, D. K., and Bide, R. W., *Radiation Res.* **25**, 221 (1965).

113. Myers, D. K., and Skov, K., *Can. J. Biochem.* **44**, 839 (1966).
114. Noronha, J. M., and Shah, N. S., *Radiation Res.* **15**, 604 (1961).
115. Nygaard, O. F., and Potter, R. L., *Radiation Res.* **10**, 462 (1959).
116. Nygaard, O. F., and Potter, R. L., *Radiation Res.* **12**, 120 (1960).
117. Nygaard, O. F., and Potter, R. L., *Radiation Res.* **12**, 131 (1960).
118. Nygaard, O. F., and Potter, R. L., *Radiation Res.* **16**, 243 (1962).
118a. Ohyama, H., Yamada, T., and Kumatori, T., *Intern. J. Radiation Biol.* **13**, 457 (1967).
119. Okada, S., and Kallee, F., *Exptl. Cell Res.* **11**, 212 (1956).
120. Okada, S., Gordon, E. R., King, R., Hempelmann, L. H., and Schlegel, B., *Arch. Biochem. Biophys.* **70**, 469 (1957).
121. Okada, S., Schlegel, B., and Hempelmann, L. H., *Biochim. Biophys. Acta* **28**, 209 (1958).
122. Okada, S., and Hempelmann, L. H., Personal Communication (1967).
123. Ord, M. G., and Stocken, L. A., *Brit. J. Radiol.* **28**, 279 (1955).
124. Ord, M. G., and Stocken, L. A., *Biochim. Biophys. Acta* **29**, 201 (1958).
125. Ord, M. G., and Stocken, L. A., *Nature* **182**, 1787 (1958).
126. Ord, M. G., and Stocken, L. A., *Biochim. Biophys. Acta* **37**, 352 (1960).
127. Ord, M. G., and Stocken, L. A., *Biochem. J.* **84**, 600 (1962).
128. Ord, M. G., and Stocken, L. A., *Nature* **200**, 136 (1963).
128a. Ord, M. G., and Stocken, L. A., *Biochem. J.* **101**, 34 (1966).
128b. Ord, M. G., and Stocken, L. A., *Proc. Roy. Soc. Edinburgh* **B70**, 117 (1968).
128c. Parchwitz, H. K., Wittekindt, E., and Wittekindt, W., *Strahlentherapie*, **126**, 94 (1965).
128d. Perris, A. D., and Whitfield, J. F., *Intern. J. Radiation Biol.* **11**, 399 (1966).
128e. Perris, A. D., and Whitfield, J. F., *Nature* **214**, 302 (1967).
128f. Perris, A. D., Youdale, T., and Whitfield, J. F., *Exptl. Cell Res.* **45**, 48 (1967).
128g. Perris, A. D., Whitfield, J. F., and Rixon, R. H., *Radiation Res.* **32**, 550 (1967).
128h. Pierucci, O., *Radiation Res.* **32**, 760 (1967).
129. Pierucci, O., and Regelson, W., *Radiation Res.* **24**, 619 (1965).
130. Puck, T. T., *Radiation Res.* **27**, 272 (1966).
131. Rahman, Y. E., *J. Cell Biol.* **13**, 253 (1962).
132. Rahman, Y. E., *Proc. Soc. Exptl. Biol. Med.* **109**, 378 (1962).
133. Rahman, Y. E., *Radiation Res.* **20**, 741 (1963).
134. Rajewsky, B., Gerber, G., and Pauly, H., *Strahlentherapie* **102**, 517 (1957).
135. Rappoport, D. A., Seibert, R. A., and Collins, V. P., *Radiation Res.* **6**, 148 (1957).
135a. Rathgen, G., *Strahlentherapie* **130**, 305 (1966).
136. Reuter, A. M., Sassen, A., and Kennes, F., *Radiation Res.* **30**, 445 (1967).
137. Revis, V. A., *Vopr. Med. Khim.* **10**, 513 (1964).
138. Richmond, J. E., Ord, M. G., and Stocken, L. A., *Biochem. J.* **66**, 123 (1957).
138a. Rixon, R. H., *Radiation Res.* **32**, 42 (1967).
139. Roth, J. S., and Eichel, H. J., *Radiation Res.* **9**, 173 (1958).
140. Roth, J. S., and Eichel, H. J., *Radiation Res.* **11**, 572 (1959).
141. Roth, J. S., Bukovsky, J., and Eichel, H. J., *Radiation Res.* **16**, 27 (1962).
142. Roth, J. S., and Hilton, S., *Radiation Res.* **19**, 42 (1963).
143. Roth, J. S., Wagner, M., and Koths, M., *Radiation Res.* **22**, 722 (1964).
144. Sanders, J. L., Dalrymple, G. V., and Robinette, C. D., *Nature* **201**, 206 (1964).
145. Sanders, J. L., Dalrymple, G. V., and Robinette, C. D., *Nature* **202**, 919 (1964).
146. Scaife, J. F., and Alexander, P., *Intern. J. Radiation Biol.* **3**, 389 (1961).
147. Scaife, J. F., and Hill, B., *Can. J. Biochem. Physiol.* **40**, 1025 (1962).

148. Scaife, J. F., *Can. J. Biochem. Physiol.* **41**, 1469 (1963).
149. Scaife, J. F., *Can. J. Biochem. Physiol.* **41**, 1486 (1963).
150. Scaife, J. F., *Can. J. Biochem.* **42**, 431 (1964).
151. Scaife, J. F., *Can. J. Biochem.* **42**, 623 (1964).
152. Scaife, J. F., *Can. J. Biochem.* **42**, 1717 (1964).
153. Scaife, J. F., *Can. J. Biochem.* **43**, 119 (1965).
154. Scaife, J. F., *Can. J. Biochem.* **44**, 433 (1966).
155. Scaife, J. F., *in* "The Cell Nucleus: Metabolism and Radiosensitivity." Taylor & Francis, London, 1966.
155a. Scaife, J. F., and Brohée, H., *Intern. J. Radiation Biol.* **13**, 111 (1967).
156. Schrek, R., *Ann. B. Y. Acad. Sci.* **95**, 839 (1961).
156a. Schrek, R., *Arch. Pathol.* **85**, 31 (1968).
157. Sibatani, A., *Exptl. Cell Res.* **17**, 131 (1959).
158. Skalka, M., Matyášová, J., and Chlumecká, V., *Folia Biol. (Prague)* **11**, 113 (1965).
159. Skalka, M., and Matyášová, J., *Intern. J. Radiation Biol.* **11**, 193 (1966); *Nature* **212**, 1253 (1966); *Folia Biol. (Prague)* **11**, 378 (1965), **13**, 457 (1967).
160. Smets, L. A., *Nature* **211**, 527 (1966); *Intern. J. Radiation Biol.* **11**, 197 (1966).
161. Smit, J. A., and Stocken, L. A., *Biochem. J.* **89**, 37 (1963).
162. Smit, J. A., and Stocken, L. A., *Biochem. J.* **91**, 155 (1964).
163. Smith, K. C., and Low-Beer, B. V. A., *Radiation Res.* **6**, 521 (1957).
164. Smith, L. H., and Vos, O., *Radiation Res.* **19**, 485 (1963).
165. Streffer, C., and Melching, H. J., *Strahlentherapie* **129**, 282 (1966).
166. Sugino, Y., Frenkel, E. P., and Potter, R. L., *Radiation Res.* **19**, 682 (1963).
166a. Suhar, A., and Furlan, M., *Strahlentherapie* **132**, 604 (1967); **134**, 457 (1967).
167. Sullivan, M. F., and Dubois, K. P., *Radiation Res.* **3**, 202 (1955).
168. Swingle, K. F., and Cole, L. J., *Radiation Res.* **30**, 81 (1967).
168a. Swingle, K. F., and Cole, L. J., "Current Topics in Radiation Research," Vol. IV, p. 191. North Holland Publ. Co. Amsterdam (1968).
169. Tchoe, Y. T., and Sibatani, A., *Nature* **183**, 1334 (1959).
169a. Telia, M., Reuter, A., Gerber, G. B., and Kennes, F., *Intern. J. Radiation Biol.* **16**, 389 (1969).
170. Thomson, J. F., Tourtelotte, W. W., and Carttar, M. S., *Proc. Soc. Exptl. Biol. Med.* **80**, 268 (1952).
171. Thomson, J. F., *Radiation Res.* **21**, 46 (1964).
172. Uyeki, E. M., *Radiation Res.* **11**, 617 (1959).
173. van Bekkum, D. W., and Vos, O., *Brit. J. Exptl. Pathol.* **36**, 432 (1955).
174. van Bekkum, D. W., *Biochim. Biophys. Acta* **16**, 437 (1955).
175. van Bekkum, D. W., *Biochim. Biophys. Acta* **25**, 487 (1957).
176. van Bekkum, D. W., de Vries, M. J., and Klouwen, H. M., *Intern. J. Radiation Biol.* **9**, 449 (1965).
177. Van Lancker, J. L., *Biochim. Biophys. Acta* **33**, 587 (1959).
178. Waldschmidt, M., *Strahlentherapie* **128**, 586 (1965); **132**, 463 (1967).
179. Weymouth, P. P., *Radiation Res.* **8**, 307 (1958).
180. Whitfield, J. F., Brohée, H., and Youdale, T., *Exptl. Cell Res.* **35**, 207 (1964).
181. Whitfield, J. F., Brohée, H., and Youdale, T., *Exptl. Cell Res.* **36**, 335 (1964).
182. Whitfield, J. F., Brohée, H., and Youdale, T., *Exptl. Cell Res.* **36**, 341 (1964).
183. Whitfield, J. F., Brohée, H., and Youdale, T., *Exptl. Cell Res.* **36**, 417 (1964).
184. Whitfield, J. F., Brohée, H., and Youdale, T., *Intern. J. Radiation Biol.* **9**, 421 (1965).
185. Whitfield, J. F., Brohée, H., and Youdale, T., *Exptl. Cell Res.* **37**, 637 (1965).

186. Whitfield, J. F., and Youdale, T., *Nature* **209**, 730 (1966).
187. Whitfield, J. F., and Brohée, H., *Nature* **211**, 775 (1966).
188. Whitfield, J. F., and Youdale, T., *Exptl. Cell Res.* **43**, 153 (1966).
188a. Whitfield, J. F., Youdale, T., and Perris, A. D., *Exptl. Cell Res.* **48**, 461 (1967).
189. Wills, E. D., and Wilkinson, A. E., *Biochem. J.* **99**, 657 (1966).
189a. Yamada, T., and Ohyama, H., *Intern. J. Radiation Biol.* **14**, 169 (1968).
189b. Yamada, T., Ohyama, H., Kumatori, T., and Minakami, S., *Intern. J. Radiation Biol.* **15**, 497 (1969).
190. Yamaguchi, T., *Intern. J. Radiation Biol.* **12**, 235 (1967).
191. Yost, H. T., Yost, M. T., and Robson, H. H., *Biol. Bull.* **33**, 697 (1967).
192. Yost, M. T., Robson, H. H., and Yost, H. T., *Radiation Res.* **32**, 187 (1967).
193. Zimmerman, D. H., and Cromray, H. L., *Life Sci.* **6**, 621 (1967).

Chapter IV

GASTROINTESTINAL TRACT

A. Introduction

Animals exposed to a single whole-body dose of radiation of from less than 1 kR to several kR die a few days after exposure once the epithelial lining of the small intestine has been lost. The survival time which is characteristic for the species (mouse and rat 3–4 days; guinea pig, hamster, and monkey about 5 days) (*7*) (Fig. I-1, p. 2) is independent of the dose over the whole range of doses causing gastrointestinal death. If early death can be prevented, a delayed gastrointestinal syndrome is sometimes observed. Under these conditions, mice and rats die from 5 to 8 days after exposure (*7*). Although other regions of the gastrointestinal tract are only slightly less radiosensitive than the small intestine, they assume a major role in death only under special circumstances, since their rate of cell replacement is slower than that of intestine. Thus, irradiation of the lip region or the head of mice with 2 kR or more results in so-called "oral death" after 9 days, apparently because the animals are unable to eat (see p. 2, Fig. I-1) (*78*). The high radiosensitivity of the intestine and the dose independence of the survival time were known to some of the earliest investigators. Thus, Krause (*47*), in 1906, observed that after "intense" irradiation, mice die 4 days after exposure, regardless of the dose. In 1922 Warren (*99*) recognized that early death after abdominal or whole-body irradiation resulted from destruction of the intestinal epithelium. He also observed diarrhea, dehydration, bacteriemia, and liberation of proteolytic enzymes as being important manifestations of this syndrome (*99*).

A few cells in the small intestine show morphological alterations after exposure to as little as 100 R. Changes in the cell population of the organ become prominent, however, only after doses of 500 R or more. The time sequence of the most important morphological changes in the small intestine and their relation to biochemical and physiological alterations are summarized in Table IV-1. Although some epithelial cells die in interphase during the first hours after exposure, this mode of death is not as important for the irradiated intestine as is the inability of the cells to reproduce as normal viable daughter cells, the so-called reproductive death. During recent years, an understanding

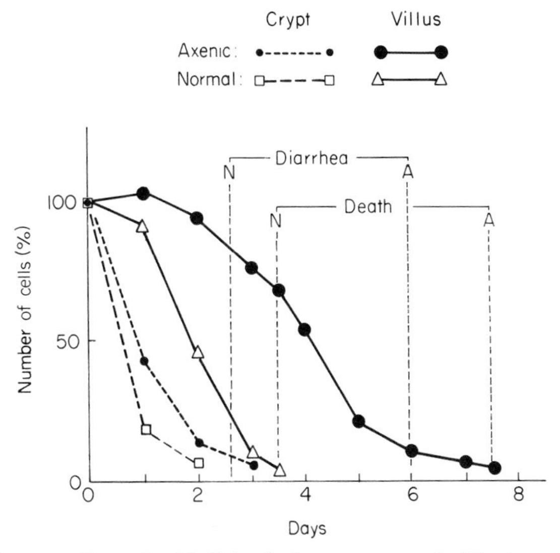

FIG. IV-1. The rate of loss of epithelial cells from crypts and villi of normal and axenic mice after exposure to 3000 R in per cent of cells of the nonirradiated intestine. Note that cells diminish first in the crypts then on the villi. Diarrhea commences when the villi are almost denuded of cells. In axenic mice (A), cells remain longer on the villi, therefore diarrhea and death ensue several days later than in normal mice (N). [From T. Matsuzawa and R. Wilson, *Radiation Res.* **25**, 15 (1965).]

of cellular kinetics in the irradiated intestine has allowed us to relate the time sequence of the histological, functional, and biochemical changes to cell renewal (*7, 73*). The generation time of stem cells in the crypts of the small intestine of the mouse is about 12 hr, of which 7–8 hr are needed for DNA synthesis. Some of the newly formed cells reproduce, others differentiate and move up to the villi. After a transit time to the tip of the villi of about 2 days, the cells are shed into the intestinal lumen (Fig. IV-1). Because no new cells are formed after irradiation, the villi become largely denuded 3 days later and the animal dies. As the shedding of the cells from the villi follows its inherent time schedule after irradiation, the survival time in the gastrointestinal syndrome is essentially independent of the dose. The threshold dose for the gastro-

TABLE IV-1

OUTLINE OF THE TIME SEQUENCE OF MORPHOLOGICAL, BIOCHEMICAL, AND PHYSIOLOGICAL ALTERATIONS IN THE IRRADIATED SMALL INTESTINE

Time	Morphological alterations	Biochemical alterations	Physiological alterations
0–3 hr	1. Cessation of mitotic activity plus chromosomal damage 2. Lesions in mitochondrial and nuclear membranes 3. Swelling of mitochondria 4. Formation of cytolysomes plus increased number of lysosomes	1. Decreased incorporation of precursors into DNA 2. Increase in lysosomal enzymes	1. Delay in gastric emptying 2. Changes in intestinal motility
3–24 hr	1–4 (above) 5. Occasional cell death in interphase 6. Increased nuclear volume 7. Relative increase in number of goblet cells 8. Vascular changes (occlusion)	Low doses: 1 + 2 (above) 3. Decreased mucosal weight, protein and DNA content, etc. 4. Gradual decrease in synthesis of RNA and proteins (enzymes) 5. Increase in phosphatase and proteolytic enzymes 6. Decreased serotonin, increased acetylcholine, and changes in enzymes degrading biogenic amines High doses: 7. Decrease in utilization of glucose and formation of citrate	1 + 2 (above) 3. Decreased active transport of electrolytes, fluids, and carbohydrates

1–3 days	Low doses: Gradual recovery, resumption of mitotic activity in crypts High doses (≥1000 R): 1–8 (above) 9. Progressive denudation of crypts, then of villi 10. Villi sometimes covered with abnormal cells	Low doses: Gradual recovery, increased utilization of glucose High doses: 1–7 (above) 8. Decreased K^+, increased Na^+ in mucosa	1–3 (above) 4. Changes in intestinal flora
After 3 days	Low doses: Recovery High doses (≥1000 R): Denudation of crypts and villi	Low doses: Recovery with "overshooting" of synthesis of DNA, proteins, and utilization of glucose High doses: 1–8 (above)	Low doses: Recovery High doses: 1–4 (above) 5. Decreased absorption of fats and other nutrients 6. Leakage of proteins from intestinal lumen 7. Loss of electrolytes and H_2O 8. Diarrhea 9. Death

intestinal syndrome may be estimated from the following considerations (*34*, *97a*, *101*). A crypt contains about 35 cells, and 7–8 crypts contribute cells to one villus. A dose of 800–1000 R would, therefore, be needed to render enough cells incapable of dividing so that the remainder could not maintain the epithelial cover of the villi needed to sustain life. The D_0 of intestinal crypt cells is not very large (100 R) (*101a*), equaling that of bone marrow stem cells, but split-dose experiments indicate that crypt cells have an excellent capacity for repair (recovery factor about 60).

The following two observations further underscore the key role of cellular kinetics in the gastrointestinal syndrome. (1) Axenic, i.e., germ-free, mice have a slower rate of cell replacement and a longer villus transit time (4.3 instead of 2.1 days) (*59*) than normal mice. Presumably, cells are shed more slowly from the villi into the lumen in the absence of bacteria. Therefore, the time of survival after doses producing the gastrointestinal syndrome is correspondingly longer (7.3 instead of 3.5 days) (Fig. IV-1), and the threshold dose is somewhat higher (*59*). (2) The normal rate of cellular proliferation is disturbed by continuous irradiation, but the intestinal epithelium can adapt itself readily to relatively high dose rates (400 R/day) (*49*). In such animals, the cell population in the crypts diminishes and the villi become shorter. After a few days, a new steady state is attained and the life cycle of surviving cells is shortened, an effect which may represent a return to a more youthful pattern of cell renewal. In this way, the epithelial lining of the villi is maintained. When the irradiation is terminated, cells regenerate rapidly and reestablish a population of normal epithelial cells within a few days. It seems likely that the cell cycle would also be shortened when the epithelium regenerates following a single radiation exposure.

B. Biochemical Changes in the Irradiated Intestine

1. Metabolism of Macromolecules

The weight as well as DNA and protein content of the intestine diminish after irradiation as a result of the loss of the intestinal epithelium (Fig. IV-2) (*11*, *21*, *43*, *56*). The DNA content is apparently a more sensitive indicator of radiation damage to the intestine (*63*) than is the weight.

The incorporation of radioactive precursors into DNA is reduced 10 min (*52*) after exposure (*1*, *56*, *67*, *68*, *70*, *84*) (Figs. IV-3 and IV-4). The number of cells synthesizing DNA remains largely the same at this time, but the cells in S phase incorporate less radioactive precursor into DNA (*52*). As a result, the average grain number per cell diminishes, as shown by autoradiographic studies. Later, the number of cells in the S phase is reduced. It is noteworthy that DNA synthesis after irradiation depends not only on dose, but also on dose rate (*52*) (Fig. IV-5). Stewart (*86*) has studied the incorporation of thymidine into DNA of normal and x-irradiated intestine after injection of different amounts of precursor. He evaluated the data so obtained with the aid of a computer and applied a model which takes into account the influence of the following four factors on thymidine metabolism (Fig. IV-6): (1) the rate

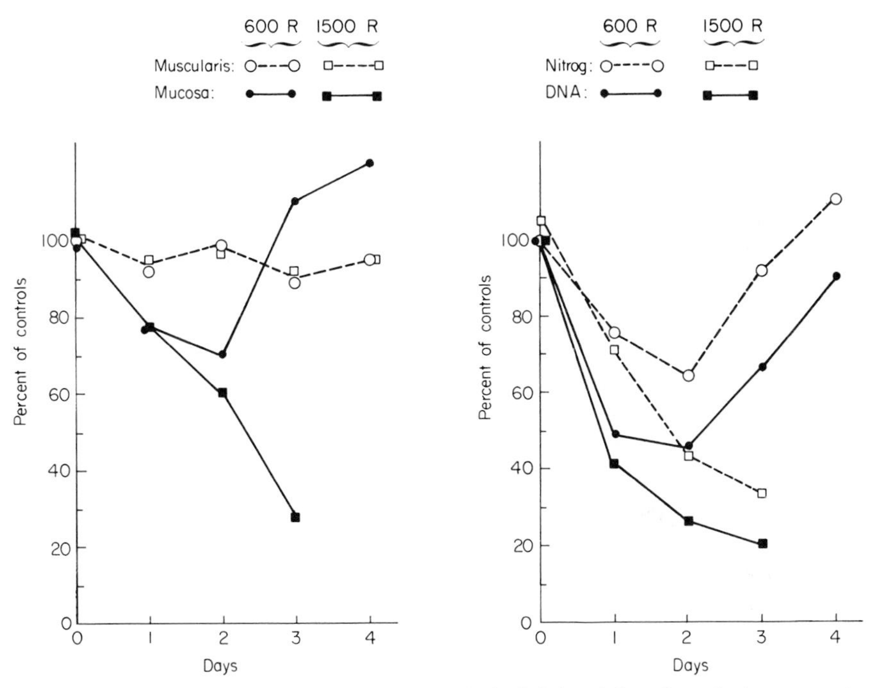

FIG. IV-2. Weight of the mucosa and muscularis (left-hand figure), and nitrogen and DNA contents of the mucosa (right-hand figure) in per cent of the controls at the given day of the small intestine of the rat after irradiation with 600 R or 1500 R. The weight and DNA and nitrogen contents of the starved controls decreased about 30% during the experimental period. Note that the DNA and nitrogen contents as well as the weight of the mucosa diminish rapidly after irradiation and return to normal after 600 R, but not after 1500 R. On the contrary, the weight of the muscle layer changes but slightly. [From R. E. Kay and C. E. Entenman, *Am. J. Physiol.* **197**, 13 (1959).]

of synthesis of DNA; (2) the rate of endogenous formation of thymidine phosphates; (3) the rate of catabolism of thymidine phosphates; (4) the amount of exogenous thymidine added or derived from breakdown of DNA.

The results of this mathematical analysis suggest that the effective pool of thymidine triphosphate increases in size after irradiation. Such an effect which, for example, could arise when the low-molecular compounds formed from DNA are degraded immediately after exposure, would dilute the radioactivity in the precursor pool. Indeed, degradation of DNA has been observed in the

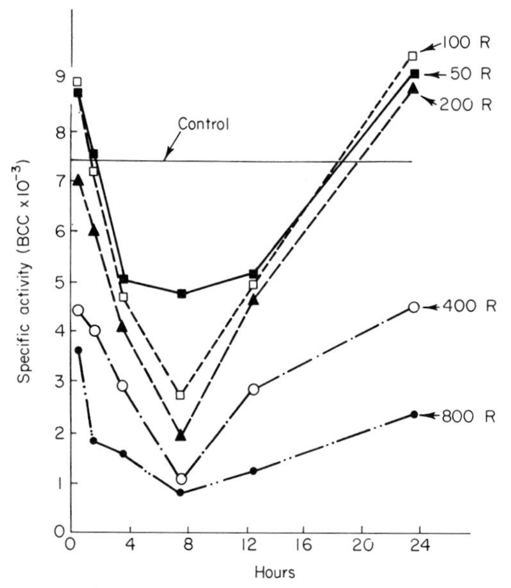

FIG. IV-3. Incorporation of thymidine-^{14}C into DNA of the small intestine of rats after different doses of whole-body x-irradiation. Note that the incorporation diminishes immediately after doses of more than 200 R and falls progressively until 8 hr after exposure. Later, DNA synthesis recovers after doses of 50–200 R, but remains low after doses of 400 R or more. [From O. F. Nygaard and R. L. Potter, *Radiation Res.* **10**, 462 (1959).]

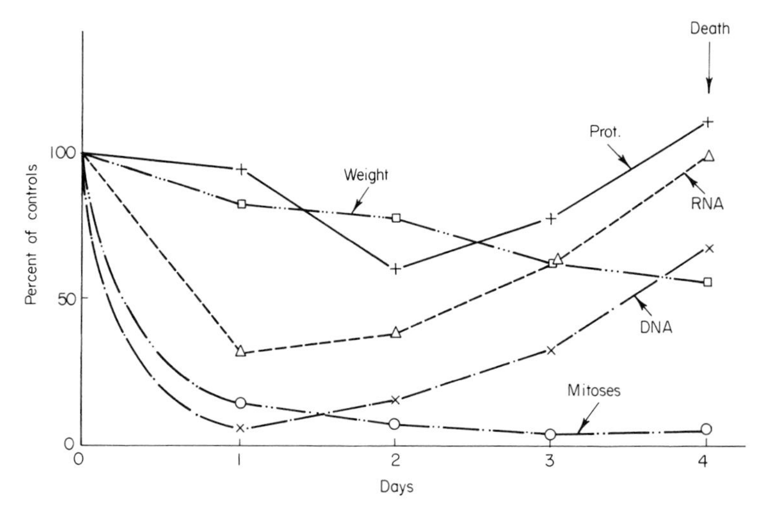

FIG. IV-4. Mitotic activity, weight of intestine, incorporation of ^{32}P into RNA or DNA, and incorporation of glycine-^{14}C into proteins of mice after whole-body irradiation with 1500 R. Note that the synthesis of DNA and RNA decreases markedly during the first day after irradiation, whereas synthesis of proteins decreases only on days 2 and 3. [From J. R. Maisin, "Radiation, radioprotecteur et syndrome gastro intestinal." G. Thone, Liège, 1966.]

intestine after irradiation (*21*), but data on DNA breakdown are less extensive for the intestine than for other organs, e.g., lymphoid tissue (see Chapter 3). On the other hand, studies of isolated perfused intestine (*28*) demonstrate that uptake, catabolism, and probably turnover in the pools of thymidine and its nucleotides are not greatly altered after irradiation. Until precise direct measurements of specific and total activities of different DNA precursors

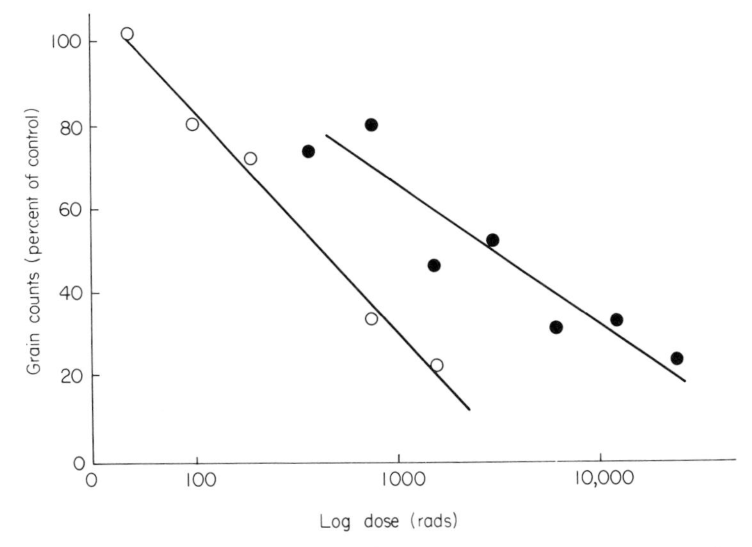

FIG. IV-5. Incorporation of thymidine-^{3}H into DNA (grain counts) as a function of the time of exposure at 12.5 R/min (open circles) and 300 R/min (closed circles). Thymidine-^{3}H was injected immediately after exposure. Note that the effect of radiation on DNA synthesis depends not only on the dose of radiation but also on the time during which it was delivered. [From W. B. Looney, *Intern. J. Radiation Biol.* **10**, 97 (1966).] It should be pointed out, however, that the interval from delivery of half the dose until measurement of DNA synthesis is quite different for the two dose rates.

become available, a decision as to the mechanism of the immediate effect of radiation on incorporation into DNA cannot be made. During the hours following irradiation, incorporation into DNA diminishes progressively, reaching a minimal value about 8 hr after exposure (Fig. IV-3). This is the time when all cells in S or late G_1 at the time of irradiation have completed DNA synthesis. The subsequent behavior of DNA synthesis is determined by the length of mitotic delay and the extent of regeneration in the crypt epithelium.

As in the case of lymphoid organs (see Chapter 3), the fate of macromolecules in the irradiated intestine has been studied by administering labeled precursors before exposure. DNA (*26*) and RNA (*25a*) present in the intestine at the time of exposure are lost to a certain extent during the post-irradiation period when cells die and are lost. Yet, the specific activity, time plot of gut DNA, is barely altered by irradiation (*69*). This observation suggests that the cells in the crypts and those on the villi are about equally susceptible to radiation-induced cell death.

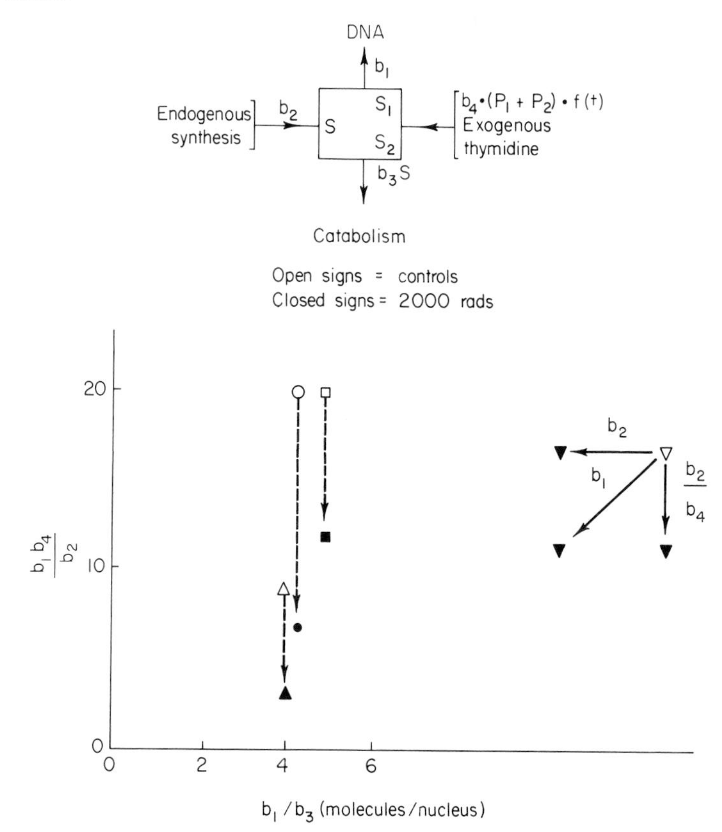

FIG. IV-6. Schematic of the four factors influencing DNA synthesis. The size of the thymidine triphosphate (TTP) pool (S) depends on the rates of DNA synthesis b_1, of thymidilate synthesis b_2, of catabolism b_3, and of the supply of exogenous thymidine. It is assumed that the rates of DNA and TTP synthesis (b_1 and b_2) are independent of the TTP pool, whereas the rate of catabolism is proportional to the TTP available. The thymidine injected consists of labeled (P_1) and nonlabeled (P_2) molecules which arrive in the intestinal cell from the blood according to the function $f(t)$. The values b_1/b_3 and $b_1\ b_4/b_2$ can be calculated from the data on incorporation of different amounts of labeled thymidine. A semilog plot of these values demonstrates that a change in b_1 would displace a point diagonally, a change in b_3 horizontally, and a change in the ratio of b_4 to b_2 vertically. Data on intestinal crypts (circles), forestomach epithelium (squares) and hair follicles (triangles) after irradiation show a vertical displacement, indicating a decrease in the ratio b_4/b_2. This observation suggests that the TTP pool increases after irradiation, possibly as the result of a dilution by precursors released from damaged DNA molecules. [From Stewart *et al. Radiation Res.* **24**, 521 (1965).]

Incorporation of precursors into RNA (*1, 56, 96*) is less sensitive to the early action of radiation than is incorporation into DNA (Fig. IV-4). Later, following irradiation, RNA synthesis diminishes and RNA precursors accumulate (*82*). The synthesis of proteins is not altered immediately after exposure (Fig. IV-4) (*1, 25, 52, 96*) but diminishes during the second and third day in the intestine (*38, 51*) and the stomach (*87*), when the epithelial cells have become abnormal and contain only a few ribosomes. During regeneration, protein synthesis increases above normal 6 days after irradiation with 700 R (*10a, 20*). An increase in chondroitin sulfate has also been reported 2 days after a sublethal exposure (*28a*).

2. Enzymatic Activities in the Intestine after Irradiation

Enzymes in the irradiated intestine have not been studied as extensively as those in liver or lymphoid tissue. Moreover, many studies have been carried out using only semiquantitative histochemical techniques. This is regrettable, since radiation damage of the intestine has great practical importance. Also, reproductive death and associated biochemical changes might be investigated conveniently in this organ. The changes reported in enzymatic activities of the

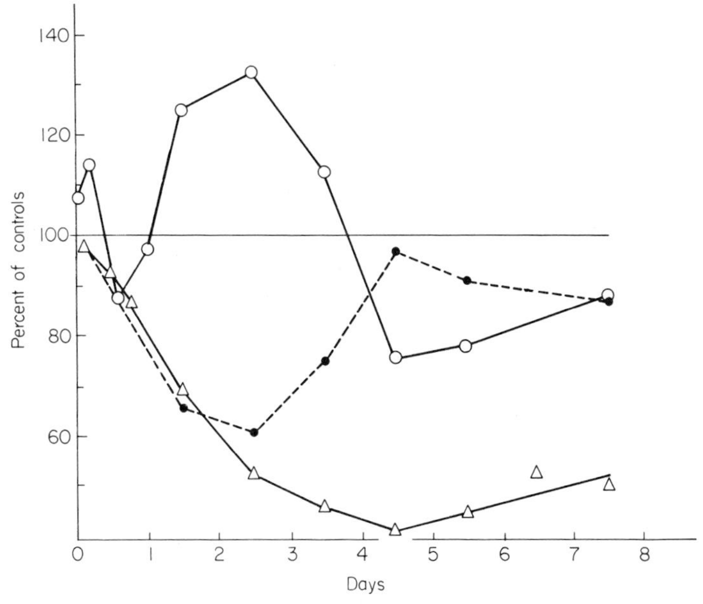

Fig. IV-7. Cholinesterase activity (triangles), acetyl choline (open circles), and weight (closed circles) of the small intestine of rats after 500 R of whole-body x-irradiation. Note that from days 2–4 cholinesterase decreases, whereas acetylcholine content increases. [From R. A. Conard, *Radiation Res.* **5**, 167 (1956).]

TABLE IV-2

ENZYMATIC ACTIVITIES IN THE INTESTINE AFTER IRRADIATION

Enzyme	Change	Assay	Time	Remarks
Arylsulfatase	Increase	Histochemical	1–3 hr	Mice: 1300 R (*36*, *37*); rats: chemical determination (*101b*)
β-Glucuronidase	Increase	Histochemical	1 hr	Rat: 1200 rads (*2*)
		Chemical	3–4 days	Rats: 400–1200 R (*31*, *60*); synthesis of glucuronides diminished (*31a*)
RNase (alakaline)	Increase	Chemical	1st day	Rats: 700 R; also following head irradiation. No change acid RNase, slightly decreased phosphodiesterase (*30a*)
Acid phosphatase	Increase in Golgi appar.	Histochemical	1 hr	Guinea pigs: remains elevated after 400 R (*41*); mice: increases 1–3 days after 1 kR (*36*); mice: increases 1–2 days after irradiation, dependent on dose (*9*, *60*, *85*, *101b*)
Alkaline phosphatase	Decrease in brush-border microvilli	Histochemical	1st day	Mice, rats, and guinea pigs (*2*, *36*, *46*): changes in distribution on gel electrophoresis (*5*); abnormal cells formed during abortive recovery contain no alkaline phosphatase (*29*)
ATPase	Decrease	Histochemical	1st day	Activity follows that of alkaline phosphatase (*2*, *36*, *37*, *46*)
Proteolytic enzymes	Increase	Histochemical	1–3 days	Dogs (*85*, *99*): perhaps via destruction of inhibitor (*22*)
Esterases (E600 resistant)	Increase	Histochemical	1–4 days	Guinea pigs: 400 R (*41*) probably related to carboxypeptidase
Choline esterase (nonspecific)	Decrease	Chemical	1–7 days	Rats, mice, and guinea pigs: 500–1500 rads (*2*, *8*, *11*, *23*) (Fig. IV-7). Particularly in upper parts of intestine (*5*, *56*); increase or no change in monkeys (*15*, *23*); specific acetylcholinesterase not altered in rats (*8*)
Monoaminooxidase	Decrease	Histochemical	1 hr	Rats: 1200 rads; later increase in crypts (*2*); 48 hr decrease in spermine and spermidine (*5a*)
Histidine decarboxylase	Decrease	Chemical	1 day	Rat stomach: 2000 rads (*45*)
Cytochrome oxidase	Decrease	Histochemical	1–2 days	Rats: 1200 rads (*2*); also decreased succinate dehydrogenase and glucose 6 phosphate dehydrogenase (*101b*)
NADH diaphorase	Decrease	Histochemical	4th day	Guinea pigs: 400 R (*46*)
Lactate dehydrogenase	Decrease	Chemical electrophoresis	3rd day	Rats: 800 R; changes in isoenzymes (*42*)
Phospholipase A and B	Decrease	Chemical	1st day	Rats: 800 R; also decreased antioxidant activity (*72a*)
Lecithinase	Decrease	Chemical	1st day	Rats: 800 R; production of inhibitor? (*72b*)

irradiated intestine are summarized in Table IV-2. Lysosomal and related enzymes increase in the intestine during the first hours after irradiation as they do in lymphoid tissues (see Chapter 3). The enhanced proteolytic activity in the intestines of irradiated dogs was reported by Warren (*99*) as early as 1922. These changes probably reflect the auto- and heterophagocytosis of damaged cells.

Other enzymatic alterations are related to the metabolism of biogenic amines and the gastrointestinal motility (see also p. 311). As complex biochemical pathways have rarely been studied in the irradiated intestine, we must again be satisfied to enumerate the changes reported rather than to present a coherent picture. The oxygen consumption of the gastric (*58*) or intestinal mucosas (*6*) diminish after irradiation. Aerobic lactate formation (*74*, *75a*), glucose oxidation (*100*), and perhaps citric acid formation (*19*, *75a*) are reduced from 1 day after exposure to 650–1200 R. The increase in glucose oxidation reported by one investigator (*44*) may have been due to the use of a nonphysiological preparation (*75a*). The synthesis of glucosamine increases (*10*), whereas that of glucuronides decreases after irradiation (*31*). The synthesis of lipids from acetate (*12*), the esterification of fatty acids, and the transformation of fatty acids to phospholipids (*56*) are also impaired 3 days after exposure.

C Physiological Functions of the Gastrointestinal Tract after Irradiation

1. Secretion of Enzymes

The gastrointestinal tract forms and secretes the enzymes needed for digestion, transports the food, and absorbs the digested food. It also provides a barrier which prevents bacteria and macromolecules from entering the body and vital substances from leaving the animal. Damage to the physiological functions of the gastrointestinal tract is characteristic of the terminal phase of the gastrointestinal syndrome, but is not prominent as long as the epithelial lining is intact.

Synthesis of enzymes by the gastrointestinal system and its accessory glands is not affected greatly by intermediate doses of radiation, but is impaired after doses which damage the morphological structure of the organs. Synthesis of amylase in the pancreas, for example, is unaltered 4 hr after a dose of 2000 R (*33*; see also *4*), but it is depressed after doses of 5000 R and more (*98*). Nevertheless, the secretory functions of the gastrointestinal tract may be modified by much lower doses of radiation via humoral or neural mechanisms (*48*). Production of saliva and of gastric, pancreatic, or intestinal juices undergoes cyclical variations after whole-body irradiation with only 200 R (*48*), but it is uncertain how these changes affect the animals during radiation illness. On the other hand, the bile plays an important role in the reactions of animals during the gastrointestinal syndrome, as is discussed in detail later (see p. 98). Except for a transient rise in bile acids (*27*) and bilirubin (Popkov, cited in *48*), the composition of bile is apparently not greatly altered after irradiation.

2. Physiological Changes in the Irradiated Intestine

Exposure to low doses of radiation produces marked changes in motility, i.e., a delay in gastric emptying and a reduction in intestinal motility, sometimes preceded by a period of excessive intestinal peristalsis and antiperistalsis (see p. 311). As a result of these changes, less food is available for absorption by the intestine, and the bacterial flora in the intestine may change. On the other hand, food present in the gastrointestinal tract at the time of exposure can affect the severity of the mucosal lesion (*32a*).

After exposure to doses causing bone marrow death, the resistance of the organism is lowered and bacteria can invade the body by way of the intestinal tract or the oral cavity. Nevertheless, it is doubtful whether bacteriemia results primarily from damage to the intestinal epithelium, since it occurs only at a time when regeneration of the intestines is largely complete. In animals dying of the gastrointesinal syndrome, bacteriemia is occasionally, but not always, found at the time of death (*95*). Treatment with antibiotics or bone marrow transplants, particularly if carried out in combination with electrolyte therapy, prolongs the survival time in rats exposed to doses of 1000 to 2000 R (*95*). This observation suggests that bacteria participate in the intestinal syndrome either by entering the blood stream or by forming toxic substances which are absorbed by the organism. Axenic mice survive somewhat longer beyond the time of destruction of the intestinal epithelial lining than do normal mice, but the considerable lengthening in total survival time results mainly from the longer villus transit time of the epithelial cells (see Fig. IV-1). Moreover, if axenic mice are reinfected with different types of bacteria, they display the same ratio (1.7) of survival time to villus transit time as normal mice (*100*).

The destruction of cells, altered gastrointestinal motility, and later diarrhea could affect the intestinal bacteria and their metabolism. Indeed, coliform bacteria and enterococci in the intestinal lumen increase in number at the expense of lactobacilli 3 days after whole-body exposure of rats to 1400R (*44a*). Alterations in the bacterial flora after irradiation should, therefore, be considered when explaining radiation-induced changes in metabolism. Since many metabolites, e.g., bile acids and bilirubin, undergo enterohepatic circulation, they may be subject to the action of intestinal bacteria. To mention one example, excess indoxyl sulfate excreted after irradiation (see p. 257) may be produced from tryptophan and its metabolites by the intestinal bacteria.

Degradation of urea diminishes in irradiated mice (2 kR) to an even greater extent than in starved controls (*28a*).

3. Absorption of Nutrients

The uptake of food from the intestine appears to be impaired significantly

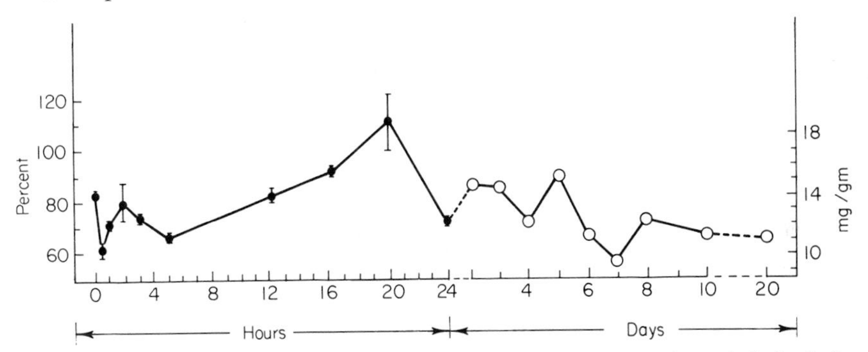

Fig. IV-8. Absorption of glucose by everted rat intestinal sacs after whole-body irradiation with 800 R. Note that the absorption decreases shortly after exposure; normal values are attained after 20 hr, and a second decrease follows later. [From H. Poppei and K. Erdman, *Naturwissenschaften* **53**, 20 (1966).]

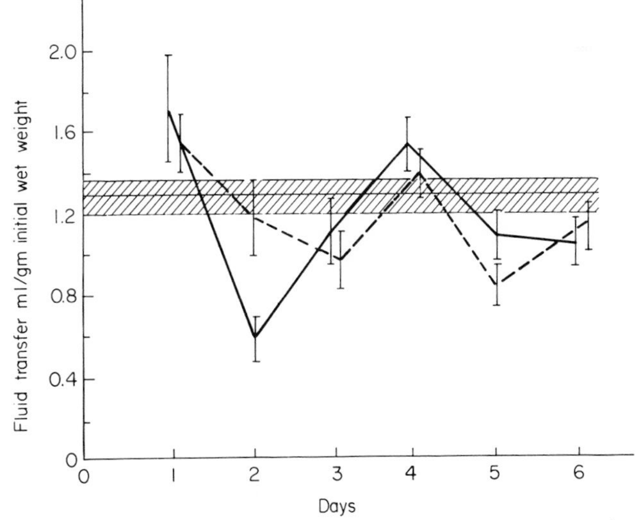

Fig. IV-9. Fluid transfer by everted sacs prepared from the distal portion of the lower small intestine of rats irradiated with 600 R (continuous line) or sham irradiated (dotted line). Shaded area is mean and standard deviation of nonirradiated controls. Note that fluid transfer increases on the first day and decreases on the second day after irradiation. Sacs obtained from the middle of the small intestine show no increased transfer on day 1 but a decrease on days 2 and 3. [From A. D. Perris, E. L. Jervis, and D. H. Smyth, *Radiation Res.* **28**, 13 (1966).]

at a time when histological damage to the intestine is severe, i.e., 2–3 days after doses in excess of 500 R. On the other hand, active transport of electrolytes, glucose, and liquid may be modified as early as the first day after exposure (Figs. IV-8 and IV-9) (see p. 96). Nevertheless, it seems unlikely that after doses causing gastrointestinal death the reduction in absorption of food per se could ever encroach upon the nutritional requirements of the animal and thereby influence the outcome of the radiation illness. Malabsorption may, however, become an important factor after local irradiation of the intestine.

The data published on intestinal absorption of food by irradiated animals appear contradictory at first sight, but it should be realized that the evaluation of *in vivo* experiments on absorption is difficult because of changes in motility and secretion of the gastrointestinal tract. Furthermore, starvation per se may alter the intestinal absorption of certain substances. Some, but not all, interfering factors can be eliminated if the material is administered directly into the lumen of the duodenum or if perfused intestinal loops or everted intestinal sacs are studied. The latter *in vitro* preparations are, however, not entirely satisfactory if changes after irradiation are to be assessed in a quantitative way

a. Carbohydrates

The absorption of carbohydrates is reduced in patients undergoing radiation therapy as well as in animals after whole-body exposure, but the changes do not become pronounced until 2–3 days after irradiation (*65*, *89*). Carbohydrates move across the intestinal wall either via active transport (glucose for the most part) or via passive diffusion (e.g., xylose). The active transport of glucose is linked with that of ions and requires energy, while producing an electrical potential across the intestinal wall. The transport of glucose is depressed during the initial period after irradiation (*74*, *75*) and then returns to normal or is enhanced after 24 hr (*75a*, *76*, *102*, *103*) (Fig. IV-8). During the following days, glucose transport diminishes markedly again (*32*, *74*, *96a*, *102*). Not only the active transport of glucose, but also the passive diffusion of xylose is impaired (*89*) at this time after irradiation.

b. Fats

A defect in the absorption of fats after irradiation was detected as early as 1922 by Mottram (*66*), and has since been noted by many investigators in patients (*14*, *80*) as well as in animals (*64*, *83*). In contrast, reduced absorption of fat introduced directly into the duodenum is not observed until the third day after irradiation (*19a*, *55*, *56*, *89*). Everted intestinal sacs absorb more fats on the first day and less on the second day after exposure to 1200 R, a behavior which corresponds to that of glucose absorption (*100*).

c. Other Substances

The intestinal absorption of many other substances has been studied after irradiation. Reduced absorption was observed for thiamine (*16*), pyridoxine (*18*), vitamin B_{12} (*93*), bile acids (*91*), iron (*12a, 77*), alanine (*62*), methionine (*20, 62, 66, 71*), and iodide (*1a*). No change in absorption was reported for proteins (*6b*) and vitamin A (*6a*). Absorption of Sr and Ca increased 2 hr after exposure to 400 R (*50*), but decreased 2–3 days after exposure to 1150 R (*57*).

4. EXCRETION OF FLUIDS AND ELECTROLYTES

Loss of substances from the organism via the intestine appears to be a more crucial factor in the gastrointestinal syndrome than the impaired absorption of food. Proteins leak through the intestinal wall on the third day after exposure and thereafter (*97*; see also p. 273), and this mechanism may contribute to the fall in concentration of serum proteins (see p. 269). The leakage of proteins, however, does not cause gastrointestinal death, since protein loss and survival have different responses to varying LET, oxygen tension, and fractionation of irradiation (*33a, 35*). Other macromolecules, e.g., polyvinylpyrrolidone (*88*), as well as low-molecular substances (*N*-methylnicotinamide) (*17*) may also be lost in increased amounts via the intestine after irradiation.

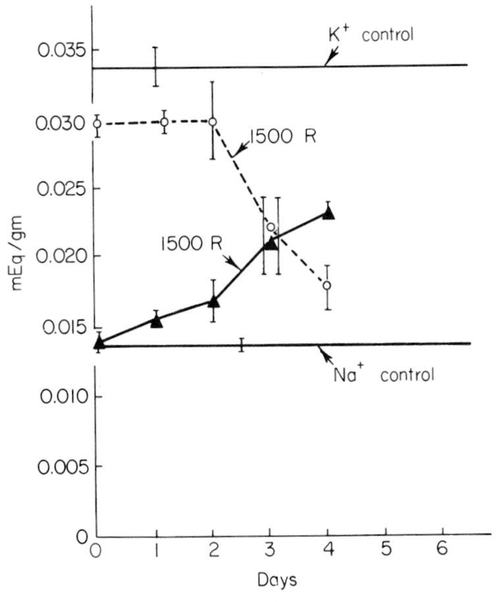

FIG. IV-10. Sodium and potassium concentration in the small intestine of mice after 1500 R of whole-body irradiation. Note that the concentration of Na^+ increases, whereas that of K^+ decreases compared with the controls, especially 3–4 days after exposure. [From J. R. Maisin and R. A. Popp, *Am. J. Physiol.* **199**, 251 (1960).]

Diminished uptake and loss of water and electrolytes during the terminal phase of the gastrointestinal syndrome precede and probably cause death. The active transport of sodium ions and water (Fig. IV-9) in the intestine is depressed within the first hours after irradiation (*74*), returns to normal after 24 hr (*103*), and progressively diminishes later (*13, 61, 97b*). The passive flux of Na^+ is only slightly altered; nevertheless, the changes in active transport result in a marked reduction in the net uptake of sodium ions on days 3 and 4 after exposure (*13*). After irradiation, more sodium ions are excreted into the intestine. The concentration of Na^+ in the wall and contents of the small intestine increases markedly after irradiation of rats with 600 to 1500 R (*5, 9a, 54, 56, 104*). On the other hand, the K^+ content in the intestinal wall diminishes 3 days after a dose of 1500 R (*54, 56*) (Fig. IV-10) when epithelial cells are shed,

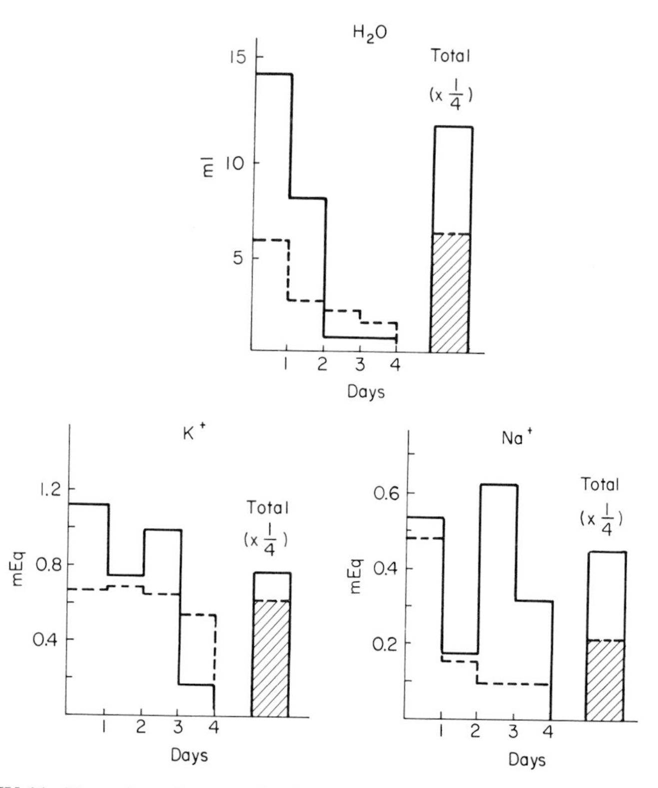

FIG. IV-11. Excretion of water, Na^+, and K^+ by starved normal (dotted lines and shaded areas) and x-irradiated (1500 R) (solid lines) rats. Note that water and K^+ are excreted in excess on days 1 and 2, whereas Na^+ is lost on days 3 and 4 after exposure. The total excretion of water and Na^+ over the 4-day period is about two times greater after irradiation than after starvation alone. [From K. L. Jackson, R. Rhodes, and C. Entenman, *Radiation Res.* **8**, 361 (1958).]

but the K^+ levels are normal or elevated 3 days after a dose of 600 R (*5*). Animals irradiated with doses causing gastrointestinal death excrete more Na^+ and K^+ into the urine and the feces than do starved controls (*39*) (Fig. IV-11). It should be pointed out, however, that starvation per se reduces the excretion of electrolytes; compared with fed controls, the fall in electrolyte excretion is less pronounced in irradiated rats than in starved ones (*39*, *79*). Nevertheless, the excess Na^+ lost by an irradiated animal appears incompatible with survival, since if such an amount is removed from a nonirradiated rat by intraperitoneal dialysis the animal will die (*39*).

On the other hand, several observations do not seem to fit the explanation that loss of electrolytes and water engenders gastrointestinal death. Radioactive $^{22}Na^+$ injected into rats prior to exposure is lost at the same rate in starved nonirradiated and starved irradiated animals, and this loss is markedly slower than in fed animals (*53*) (Fig. IV-12). If the irradiated rats had excreted a considerable amount of their body sodium, the $^{22}Na^+$ activity in the body should have decreased substantially after irradiation. Mice irradiated with 3 kR and starved thereafter lose less than 15% of their body sodium pool over

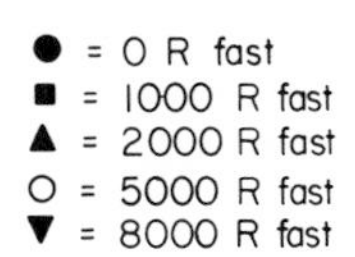

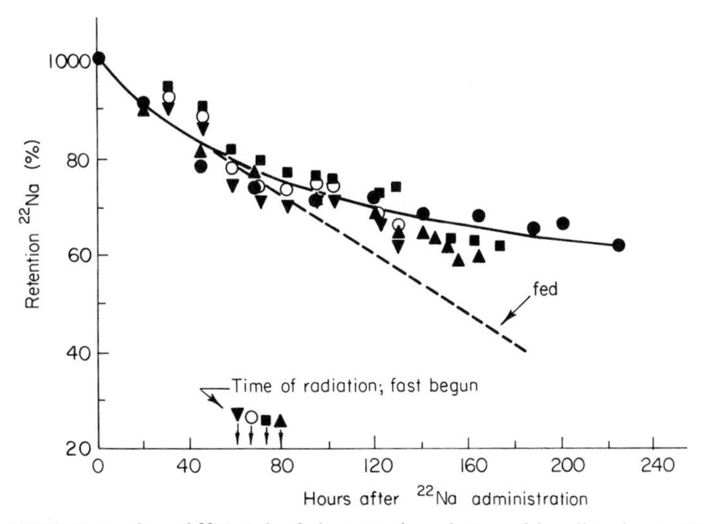

FIG. IV-12. Retention of $^{22}Na^+$ by fed, starved, and starved irradiated rats. Note that starvation causes $^{22}Na^+$ to be retained by the body. Starved irradiated and starved nonirradiated animals lose $^{22}Na^+$ at the same rate. [From C. C. Lushbaugh, J. Sutton, and C. R. Richmond, *Radiation Res.* **13**, 814 (1960).]

the 3 days until death. Although this is slightly more than is lost by starved nonirradiated mice, the sodium concentration (as measured from the ratio $^{22}Na^+$ to ^3H_2O) in organs remains unaltered (*58a*). Water is evidently lost as a consequence of diarrhea, and the plasma volume diminishes (see p. 250); yet water is not uniformly depleted during the terminal phase of the gastrointestinal syndrome, as many organs of irradiated rats contain as much water as those of nonirradiated rats (*81*).

Gastrointestinal death can be modified by several procedures. Thus, death is delayed, but not prevented when electrolytes and antibiotic solutions are administered to irradiated animals (*95*). Partial shielding of the intestine during irradiation also prevents diarrhea, presumably because the nonirradiated segments are capable of reabsorbing the sodium ions and water lost by the irradiated ones. A rat can also be protected from gastrointestinal death by removal of the irradiated part of the exteriorized intestine after exposure (*72*).

The action of bile in gastrointestinal death is of particular interest. If the bile duct is ligated after exposure of rats to 1000–2000 R (*3, 40, 90, 94*), diarrhea is prevented, and death is delayed until the fifth to seventh day. Ligation of the pancreatic duct also lessens intestinal damage after irradiation (*3, 60a*). Considerable amounts of water and sodium are normally excreted with the bile and are not reabsorbed during the gastrointestinal syndrome (*30a*); nevertheless, the loss of these substances is not the sole cause of this ligation effect, since extracorporeal division of bile also prevents diarrhea in irradiated rats. Moreover, replacement of bile or bile acids into the duodenum reestablishes the normal pattern of gastrointestinal death, whereas the intraduodenal administration of an electrolyte solution does not (*90*). On the other hand, survival time is lengthened in dogs with a gall bladder–colon fistula, but diarrhea is not prevented (*3*).

It has been postulated that bile acids remove the barrier of mucus from the intestinal wall (*30, 92*), and thus promote the loss of liquids and salts. Indeed, the movement of water and sodium ions from the intestinal lumen into the blood is increased, and that from the blood into the lumen diminished in the absence of bile (*93*). However, leakage of sodium ions by the irradiated intestine is not influenced by the presence of bile (*22a*). Other mechanisms of the action of bile in gastrointestinal death must be considered, such as its action on the intestinal flora, the replacement of cells, the absorption of toxic substances, the proteolytic reactions in the intestine, and the intestinal motility. Cell renewal is not affected by bile acids (*24*), but the reactivity of the intestine to serotonin and acetylcholine is decreased in animals with bile duct cannulas (*30*). Nevertheless, more studies are needed to elucidate the role of the bile in the gastrointestinal syndrome. It would be especially interesting to learn whether axenic animals are also protected by bile duct ligation.

In conclusion, loss of electrolytes and liquids appears to be an important factor during the terminal phase of the gastrointestinal syndrome, but it has not been established whether it is the principal cause of death. Also, the role played by toxic products, bacteria, malabsorption, proteolytic enzymes, and depression of leucocytes is not clearly understood.

REFERENCES

1. Abrams, R., *Arch. Biochem.* **30**, 90 (1951).
1a. Acland, J. D., *Intern. J. Radiation Biol.* **12**, 177 (1967).
2. Ansari, P. M., Eder, H., and Naegele, W., *Strahlentherapie* **120**, 275 (1963).
3. Archambeau, J. O., Maetz, M., Jesseph, J. E., and Bond, V. P., *Radiation Res.* **25**, 173 (1965).
4. Archambeau, J. O., Griem, N., and Harper, P., *Radiation Res.* **28**, 243 (1966).
5. Baker, D. G., Mitchell, R. I., Coles, D., and Christie, S., *Gastroenterology* **44**, 291 (1963).
5a. Barbiroli, B., Mozurri, M. S., and Rossoni, C., *Boll. Soc. Ital. Biol. Sper.* **43**, 1434 (1967).
6. Barron, E. S. G., Wolkowitz, W., and Muntz, J. A., ANL-MDDC 1241 (1947).
6a. Bennett, L. R., Bennett, V. C., Shaver, A., and Grachus, T., *Proc. Soc. Exptl. Biol. Med.* **74**, 439 (1950).
6b. Bennett, L. R., Chastain, S. M., Decker, A. B., and Mead, J. F., *Proc. Soc. Exptl. Biol. Med.* **77**, 715 (1951).
7. Bond, V. P., Fliedner, T. M., and Archambeau, J. O., "Mammalian Radiation Lethality." Academic Press, New York, 1965.
8. Burn, J. H., Kordik, P., and Mole, R. H., *Brit. J. Pharmacol.* **7**, 58 (1952).
9. Calzavara, F., and Pozza, F., *Minerva Radiol. Fisioterap. Radiobiol.* **10**, 274 (1965).
9a. Caster, W. O., and Armstrong, W. D., *Radiation Res.* **5**, 189 (1956).
10. Cescon, I., and Cavina Pratesi, A., *Arch. Sci. Biol.* (*Bologna*) **46**, 187 (1962).
10a. Chaika, Ya. P., *Biol. Deistvie Radiatsii, L'vovsk. Gos. Univ.* **3**, 10 (1965).
11. Conard, R. A., *Radiation Res.* **5**, 167 (1956).
12. Coniglio, J. G., McCormick, D. B., and Hudson, G. W., *Am. J. Physiol.* **185**, 577 (1956).
12a. Cultrera, G., Santagati, G., Robustelli della Cuna, G., and Zanolla, W., *Haematol. Latina* (*Milan*) **49**, 1057 (1964).
13. Curran, P. F., Webster, E. W., and Hovsepian, J. A., *Radiation Res.* **13**, 369 (1960).
14. dalla Palma, L., *Proc. Symp. Gastrointestinal Radiation Injury, 1966* p. 261. Excerpta Med. Found., Amsterdam, 1968.
15. Demin, N. N., Korneeva, N. V., and Shaternikov, V. A., *Biokhimiya* **26**, 494 (1961).
16. Detrick, L. E., Upham, H. C., Dunlap, A. K., and Haley, T. J., *Radiation Res.* **15**, 520 (1961).
17. Detrick, L. E., Upham, H. C., Springsteen, R. W., and Haley, T. J., *Intern. J. Radiation Biol.* **7**, 161 (1963).
18. Detrick, L. E., Upham, H. C., Springsteen, R. W., McCandless, R. G., and Haley, T. J., *Radiation Res.* **21**, 186 (1964).
19. Dubois, K. P., Cochran, K. W., and Doull, J., *Proc. Soc. Exptl. Biol. Med.* **76**, 422 (1951).
19a. Dulcino, J., Maisin, J. R., and Deroo, J., *Intern. J. Radiation Biol.* **12**, 93 (1967).
20. Edwards, C. H., Gadsden, E. L., and Edwards, G. A., *Radiation Res.* **22**, 116 (1964).
21. Feinstein, R. N., and Butler, C. L., *Proc. Soc. Exptl. Biol. Med.* **79**, 181 (1952).

22. Feinstein, R. N., and Ballin, J. C., *Proc. Soc. Exptl. Biol. Med.* **83**, 6 (1953).
22a. Forkay, D. J., Jackson, K. L., and Christensen, G. M., *Intern. J. Radiation Biol.* **14**, 49 (1968).
23. French, A. B., and Wall, P. E., *Am. J. Physiol.* **188**, 78 (1957).
24. Fry, R. J. M., Kisieleski, W. E., Sullivan, M. F., Brennan, P., and Fritz, T., *Proc. Symp. Gastrointestinal Radiation Injury, 1966* p. 142. Excerpta Med. Found., Amsterdam, 1968.
25. Gerbaulet, K., Maurer, W., and Brückner, J., *Biochim. Biophys. Acta* **68**, 462 (1963).
25a. Gerber, G. B., Gerber, G., and Altman, K. I., *Intern. J. Radiation Biol.* **4**, 67 (1961).
26. Gerber, G. B., Gerber, G., Altman, K. I., and Hempelmann, L. H., *Intern. J. Radiation Biol.* **6**, 17 (1963).
27. Gerber, G. B., and Remy-Defraigne, J., *Intern. J. Radiation Biol.* **10**, 141 (1966).
28. Gerber, G. B., and Remy-Defraigne, J., *Arch. Physiol. Biochem.* **74**, 785 (1966).
28a. Gerber, G. B., and Remy-Defraigne, J., *Ges. Exp. Med.* **150**, 33 (1969).
28b. Glaubitt, D., *Proc. Symp. Gastrointestinal Radiation Injury, 1966* p. 452. Excerpta Med. Found., Amsterdam, 1968.
29. Goessner, W., and Luz, A., *Proc. 3rd Symp. European Soc. Radiation Biol.*, p. 55. 1967.
30. Gorizontov, P. D., Fedorovskii, L. L., and Lebedeva, G. A., *Arkiv. Pathol.* **27**, 19 (1965).
30a. Gresham, P. A., and Pover, W. F. R., *Intern. J. Radiation Biol.* **13**, 275 (1967); *Radiation Res.* **34**, 256 (1968).
31. Hartiala, K., Näntö, V., and Rinne, U. K., *Acta Physiol. Scand.* **53**, 376 (1961).
32. Hehl, R., and Wilbrandt, W., *Helv. Physiol. Acta* **22**, 319 (1964).
32a. Hiatt, N., and Warner, N. E., *Proc. Soc. Exptl. Biol. Med.* **124**, 937 (1967).
33. Hokin, M. R., and Hokin, L. E., *J. Biol. Chem.* **219**, 85 (1956).
33a. Hornsey, S., and Hedges, M. J., *Brit. J. Radiol.* **42**, 278 (1969).
34. Hornsey, S., and Vatistas, S., *Brit. J. Radiol.* **36**, 795 (1963).
35. Hornsey, S., and Vatistas, S., *Proc. Symp. Gastrointestinal Radiation Injury, 1966* p. 396. Excerpta Med. Found., Amsterdam, 1968.
36. Hugon, J., and Borgers, M., *Histochemie* **6**, 209 (1966).
37. Hugon, J., Personal Communication (1967)
38. Il'Ina, L. I., and Petrov, R. V., *Tsitologiya* **1**, 289 (1959).
39. Jackson, K. L., Rhodes, R., and Entenman, C., *Radiation Res.* **8**, 361 (1958).
40. Jackson, K. L., and Entenman, C., *Radiation Res.* **10**, 67 (1959).
41. Jonek, J., Kosmider, S., and Kaiser, J., *Intern. J. Radiation Biol.* **7**, 411 (1963).
42. Kärcher, K. H., and Kato, H. T., *Strahlentherapie* **125**, 548 (1964).
43. Kay, R. E., and Entenman, C., *Am. J. Physiol.* **197**, 13 (1959).
44. Kay, R. E., and Entenman, C. E., *J. Biol. Chem.* **234**, 1634 (1959).
44a. Kent, T. H., Osborne, J. W., and Wende, C. M., *Radiation Res.* **35**, 635 (1968).
45. Kobayashi, Y., *Radiation Res.* **24**, 503 (1965).
46. Kosmider, S., Jonek, J., and Kaiser, J., *Strahlentherapie* **124**, 261 (1964).
47. Krause, P., and Ziegler, K., *Fortsch. Gebiete Roentgenstrahlen* **10**, 126 (1906–1907).
48. Kurtsin, I. T., "Effects of Ionizing Radiation on the Digestive System." Elsevier, Amsterdam, 1963.
49. Lamerton, L. F., *Radiation Res.* **27**, 119 (1966).
50. Lengemann, F. W., and Comar, C. L., *Radiation Res.* **14**, 662 (1961).
51. Lipkin, M., Quastler, H., and Muggia, F., *Radiation Res.* **19**, 277 (1963).
52. Looney, W. B., *Intern. J. Radiation Biol.* **10**, 97 (1966).

53. Lushbaugh, C. C., Sutton, J., and Richmond, C. R., *Radiation Res.* **13**, 814 (1960).
54. Maisin, J. R., and Popp, R. A., *Am. J. Physiol.* **199**, 251 (1960).
55. Maisin, J. R., Dulcino, J., Verly, W., Deroo, J., and Lambiet, M., *Compt. Rend. Soc. Biol.* **157**, 916 (1963).
56. Maisin, J. R., "Radiation, radioprotecteur et syndrome gastro intestinal." G. Thone, Liège, 1966.
57. Marcus, C. S., and Vos, O., *Proc. Soc. Exptl. Biol. Med.* **121**, 885 (1966).
58. Marsiti, G., and Paoletti, M., *Radiol. Med.* **37**, 948 (1951).
58a. Mathieu, O., and Gerber, G. B., *Atomkernenergie* **15**, 71 (1970).
59. Matsuzawa, T., and Wilson, R., *Radiation Res.* **25**, 15 (1965).
60. McFadyen, D. M., and Baker, D. G., *Proc. Symp. Gastrointestinal Radiation Injury, 1966* p. 401. Excerpta Med. Found., Amsterdam, 1968.
61. McKenney, J. R., *Proc. Symp. Gastrointestinal Radiation Injury, 1966* p. 206. Excerpta Med. Found., Amsterdam, 1968.
62. Mehran, A. R., and Blais, R., *J. Nutr.* **89**, 235 (1966); *Arch. Intern. Physiol. Biochim.* **75**, 27 (1967).
63. Mole, R. H., and Temple, D. M., *Intern. J. Radiation Biol.* **1**, 28 (1959).
64. Morehouse, M. G., and Searcy, R. L., *Radiation Res.* **4**, 175 (1956).
64a. Morgenstern, L., and Hiatt, N., *Gastroenterology* **53**, 923 (1967).
65. Moss, W. T., *Am. J. Roentgenol., Radiation Therapy Nucl. Med.* **78**, 851 (1957).
66. Mottram, J. C., Cramer, W., and Drew, A. H., *Brit. J. Exptl. Pathol.* **3**, 179 (1922).
67. Nygaard, O. F., and Potter, R. L., *Radiation Res.* **10**, 462 (1959).
68. Nygaard, O. F., and Potter, R. L., *Radiation Res.* **12**, 120 (1960).
69. Nygaard, O. F., and Potter, R. L., *Radiation Res.* **12**, 131 (1960).
70. Nygaard, O. F., and Potter, R. L., *Radiation Res.* **16**, 243 (1962).
71. Okulov, N. M., *Med. Radiol.* **1**, 41 (1956).
72. Osborne, J. W., *Radiation Res.* **17**, 22 (1962).
72a. Ottolenghi, A., and Bernheim, F., *Radiation Res.* **12**, 371 (1960).
72b. Ottolenghi, A., and Bernheim, F., *Radiation Res.* **15**, 609 (1961).
73. Patt, H. M., and Quastler, H., *Physiol. Rev.* **43**, 357 (1963).
74. Perris, A. D., Jervis, E. L., and Smyth, D. H., *Radiation Res.* **28**, 13 (1966).
75. Perris, A. D., *Radiation Res.* **29**, 597 (1966).
75a. Perris, A. D., *Radiation Res.* **34**, 523 (1968).
75b. Ponz, F., and Lluch, M., *Rev. Espan. Fisiol.* **23**, 117 (1967).
76. Poppei, H., and Erdman, K., *Naturwissenschaften* **53**, 20 (1966).
77. Prasad, K. N., and Osborne, J. W., *Intern. J. Radiation Biol.* **7**, 245 (1963).
78. Quastler, H., and Zucker, M., *Radiation Res.* **10**, 402 (1959).
79. Rakovich, M., and Gregora, V., *Med. Radiol.* **8**, 68 (1963).
80. Reeves, R. J., Sanders, A. P., Sharpe, K. W., Thorne, W. A., and Isley, J. K., *Am. J. Roentgenol., Radium Therapy Nucl. Med.* **89**, 122 (1963).
81. Rugh, R., *Nucleonics* **12**, 28 (1954).
82. Rysina, T. N., *Biokhimiya* **24**, 556 (1959).
83. Schwartz, E. E., and Shapiro, B., *Radiology* **77**, 83 (1961).
84. Sherman, F. G., and Quastler, H., *Exptl. Cell. Res.* **19**, 343 (1960).
84a. Sigdestad, C. P., Osborne, J. W., and Levine, D. L., *Intern. J. Radiation Biol.* **15**, 65 (1969).
85. Spiro, H. M., and Pearse, A. G. F., *J. Pathol. Bacteriol.* **88**, 55 (1964).
86. Stewart, P. A., Quastler, H., Skougaard, M. R., Wimber, D. R., Wolfsberg, M. F., Perrotta, C. A., Ferbel, B., and Carlough, M., *Radiation Res.* **24**, 521 (1965).
87. Sukhomlinov, B. F., and Chaika, Y. P., *Ukr. Biokhim. Zh.* **38**, 532 (1966).

88. Sullivan, M. F., *Intern. J. Radiation Biol.* **2**, 393 (1960).
89. Sullivan, M. F., *Am. J. Physiol.* **201**, 1013 (1961).
90. Sullivan, M. F., *Nature* **195**, 1217 (1962).
91. Sullivan, M. F., *Am. J. Physiol.* **209**, 158 (1965).
92. Sullivan, M. F., Hulse, E. V., and Mole, R. H., *Brit. J. Exptl. Pathol.* **46**, 235 (1965).
93. Sullivan, M. F., *Proc. Symp. Gastrointestinal Radiation Injury, 1966* p. 216. Excerpta Med. Found., Amsterdam, 1968.
94. Taketa, S. T., and Swift, M. N., *Radiation Res.* **14**, 509 (1961).
95. Taketa, S. T., *Radiation Res.*, **16**, 312 (1962).
96. Toal, J. N., Reid, J. C., Williams, R. B., and White, J., *J. Nat. Cancer Inst.* **21**, 63 (1958).
96a. Van Duyet, P., Nejtek, V., and Dienstbier, Z., *Strahlentherapie* **135**, 610 (1965).
97. Vatistas, S., and Hornsey, S., *Brit. J. Radiol.* **39**, 547 (1966).
97a. Vatistas, S., Herdan, A., and Ellis, R. F., *Proc. Symp. Gastrointestinal Radiation Injury, 1966* p. 433. Excerpta Med. Found., Amsterdam, 1968.
97b. Vaughan, B. E., and Alpen, E. L., *Intern. J. Radiation Biol.*, **3** 265 (1961).
98. Volk, B. W., and Wellmann, K. F., *Proc. Symp. Gastrointestinal Radiation Injury, 1966* p. 73. Excerpta Med. Found., Amsterdam, 1968.
99. Warren, S. L., and Whipple, G. H., *J. Exptl. Med.* **35**, 187, 205, and 213 (1922); **38**, 741 (1923).
100. Wesemann, I., Kuss, B., Breuer, H., and Parchwitz, H. K., *Strahlentherapie* **119**, 425 (1962).
100a. Wilson, R., Bealmear, P., and Matsuzawa, T., *in* "Proc. Symp. Gastrointestinal Radiation Injury," 1966 p. 148. Excerpta Med. Found., Amsterdam, 1968.
101. Wilson, C. W., *Radiology* **83**, 120 (1964).
101a. Withers, H. R., and Elkind, M. M., *Radiology* **91**, 998 (1968).
101b. Wrigglesworth, J. M., and Pover, W. F. R., *Intern. J. Radiation Biol.* **12**, 243 (1967).
102. Zehnder, H., *Radiol. Clin.* **29**, 127 (1960).
103. Zsebök, Z. B. Jánossy, G., Petrányi, G., and Kovacs, L., *Strahlentherapie* **131**, 122 (1966).
104. Zsebök, Z., and Petrányi, G., *Acta Radiol. Therapy Phys. Biol.* **2**, 377 (1964).

Chapter V

THE LIVER

A. Introduction

The statement made by Friedman (*81*) almost 30 years ago still holds true: "Without much conclusive evidence it has been assumed that the liver as an organ is relatively resistant against radiation injury." He then continued by pointing out that metabolic changes in irradiated liver are by far more important than morphological ones. Indeed, liver should be considered a "moderately radiosensitive organ" (*67*, *68*) with respect to some of its metabolic activities because it is involved in many metabolic reactions which are associated with radiosensitive organs in the body and because it is subject to various metabolic regulations with which radiation may interfere.

The morphology of the parenchymal cell is, in general, not altered immediately after irradiation (*218*), although an accumulation of glycogen and lipids may be seen on microscopic inspection. High doses of radiation cause necrosis of liver parenchyma, but only after a considerable period of time. Although parenchymal cells constitute the bulk of liver tissue (about 90%) and biochemists usually take it for granted that only these parenchymal cells are important in biochemical studies of the liver, a few other types of cells are also present in liver, namely, biliary epithelium, vascular endothelium, Kupffer cells, fibrocytes, and some lymphocytes. In general, changes in cell population which are so important in bone marrow, spleen, intestine, etc., present no serious problem of interpretation in liver. Dietary factors and hormonal regulation contribute more significantly to radiation-induced metabolic changes than changes in cell populations in irradiated liver, but the role of nutritional factors and the function of hormones in this context are difficult to evaluate at the present state of our knowledge. This consideration was neglected by many earlier investigators.

B. Lipid and Carbohydrate Metabolism

The interrelation between metabolism of lipids and carbohydrates in irradiated liver is of particular interest, and these metabolic activities are therefore discussed in the same section. Dietary conditions as well as the time elapsed after exposure must be considered carefully when evaluating lipid and carbohydrate metabolism in the liver of irradiated animals. Little is eaten by animals for 3–5 days after irradiation,* and much is retained in the stomach. A

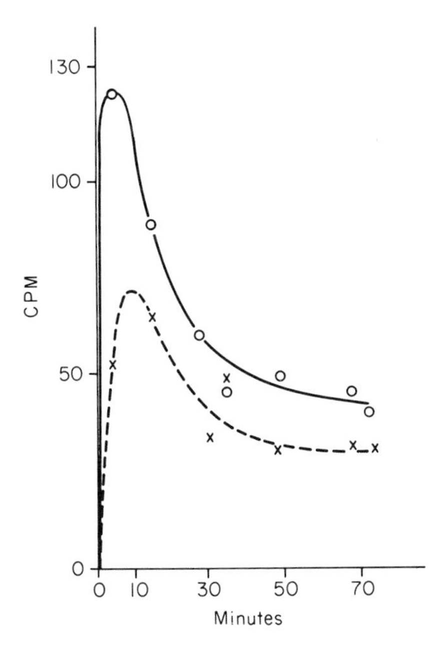

FIG. V-1. Lipogenesis from glucose in mouse liver after total body x-irradiation. (○) Specific ^{14}C activity (cpm/100 mg lipid) of hepatic lipids in mice irradiated with 2000 R x-rays (total body) after subcutaneous injection of glucose-UL-^{14}C. (×) Nonirradiated animals serving as controls for the above group. Irradiation began at t_0. All mice were fed until the time of exposure to x-rays and were subsequently starved. The ^{14}C activity (ordinate) is expressed in terms of cpm/100 mg lipid. [From M. Lourau-Pitres, *Arkiv. Kemi* 7, **211** (1954).] (Fig. 26.)

* The hamster represents an exception because this species does not reduce its food intake after irradiation.

valid comparison can, therefore, be made only between starved irradiated and starved nonirradiated animals, and, unless otherwise stated, the following discussion pertains only to the starved state. Food is again consumed later in the recovery period, but it is uncertain which controls are to be considered appropriate for animals just recovering from a prolonged state of starvation only to endure the early stages of radiation disease. Starvation of animals for 24 hours prior to sacrifice has been suggested as the appropriate control condition, but this alone seems inadequate.

In liver, the early postirradiation period is characterized by (i) increased lipogenesis and (ii) an accumulation of glycogen accompanied by gluconeogenesis. The shift toward lipogenesis is reflected in the intact irradiated animal by a fall in the respiratory quotient (*146*) as a result of reduction of carbon dioxide expiration without a change in oxygen consumption (*121a*, *146*, *225*).

Incorporation of various precursors into lipids is enhanced during the first days after an exposure to 300–5000 R, and pertinent data for different types of lipids and for various precursors are compiled in Table V-1 (see also Fig. V-1). Note that in addition to the dietary state several factors influence incorporation of precursors into lipids: (1) the type of precursor utilized, (2) the species, (3) the age, and (4) the dose and time of assay.

(1) The increased incorporation of labeled precursors into cholesterol after irradiation is most pronounced with acetate-^{14}C (*97*), less striking with mevalonate-^{14}C and squalene-^{14}C (*109*, *110*), and not detectable when mevalonate is used in the form of its lactone (*98*). This effect may reflect the loss of negative feedback control of cholesterol biosynthesis which normally operates at the level of hydroxymethylglutaryl-CoA reductase, the enzyme catalyzing the conversion of hydroxymethyl glutarate to mevalonate. Cholesterol, the end product of the metabolic reaction chain, is bound to the reductase at an allosteric site and causes a change in the conformation of the enzyme protein. The loss of this control mechanism may be responsible for the increased synthetic activity associated with cholesterol.

(2) Enhancement of incorporation of acetate-^{14}C into cholesterol after irradiation is greater in rats than in mice or guinea pigs (*97*), whereas no change in cholesterol synthesis can be detected in irradiated rabbits. It is interesting that the degree of inanition after irradiation follows the same order for the different species.

(3) Only old rats show an enhanced incorporation of butyrate-1-^{14}C (*109*) or glucose-UL-^{14}C into hepatic fatty acids after irradiation (*51*, *117*). This observation remains to be explained.

(4) The radiation dose largely determines the time course of enhanced lipogenesis. Thus, the rate of labeling of fatty acids and cholesterol is maximal at 48 hours after an exposure to 1200 R, at 24 hours after 800 R, and at 6 hours (for fatty acids) and 24 hours (for cholesterol) after 400 R (*282*).

TABLE V-1

The Effect of Irradiation on Hepatic Lipid Biosynthesis

Species	Isotopically labeled precursor	Nutritional status	Radiation dose (R)	Time of sacrifice after irradiation	Radiation-induced effects	References
			A. Total Fatty Acids			
Rat	Acetate-1-^{14}C injected 4 hr before sacrifice	Starved	300	2 days	Slight increase in ^{14}C incorporation rate	*97*[b], *158*
	Acetate-1-^{14}C injected 4 hr before sacrifice	?	400	3 days	Slight increase in ^{14}C incorporation	*282*[b]
	Acetate-2-^{14}C or butyrate-1-^{14}C	Starved	450	40 hr	Increased synthesis of palmitic acid; similar labeling pattern with either precursor	*117*[a]
	Acetate-2-^{14}C	Fed	450	6 hr–7 days	Decreased ^{14}C incorporation, minimal 3 days after irradiation; decreased iodine number, early postirradiation period	*116*[c]
	Acetate-1-^{14}C, injected 0.5, 6, 12, or 18 hr after irradiation	Starved	750	1, 6.5, 12.5, or 18 hr	Increased 6-hr ^{14}C incorporation; incorporation constant at 12 hr, rising at 18 hr postirradiation	*53*[b]
	Injected 20–22 hr after irradiation with acetate-1-^{14}C	Starved	750, 800	22.3, 22.7, 23, and 24 hr	Increased ^{14}C incorporation	*54*[b], *54a*[b] *276*[b], *282*[b]
	Acetate-1-^{14}C, injected 20 hr after irradiation	Starved	800	1 day	Increased ^{14}C incorporation in normal rats and adrenalectomized rats given a 0.9% NaCl solution	*274*[b]

Acetate-1-^{14}C	Fed	800	4 hr–2 days	Decreased ^{14}C incorporation into maternal and fetal liver on 19th and 22nd day of pregnancy	*95*[a]
Acetate-1-^{14}C injected 4 hr before sacrifice	Starved	900	2 days	Fourfold increase of ^{14}C incorporation	*97*[b]
Acetate-1-^{14}C injected 22 hr after irradiation	Force-fed glucose after irradiation	1000	1 day	No change	*206*[b]
Acetate-1-^{14}C added 30 min after the start of a 2-hr perfusion	Starved	1000	1 day	Increased ^{14}C incorporation	*12*[e]
Glucose-UL-^{14}C or fructose-UL-^{14}C	Force-fed glucose or fructose after irradiation	1000	4 hr; 1, 2, or 3 days	Decreased ^{14}C incorporation from glucose-^{14}C, but not fructose-^{14}C	*123*[a]
Acetate-1-^{14}C	?	1200	6–24 hr	Decreased ^{14}C incorporation	*282*[b]
Acetate-1-^{14}C injected 4 hr before sacrifice	Starved	1200	2 days	Eightfold increase in ^{14}C incorporation rate	*97*[b]
Acetate-1-^{14}C fed 24 or 48 hr after irradiation	Starved	1500	26 or 50 hr	Tenfold increase in ^{14}C incorporation	*182*[b]
Acetate-1-^{14}C	Starved	2000	1 day	Increased ^{14}C incorporation	*110*[a]
Acetate-1-^{14}C injected 4 hr before sacrifice	Starved	2400	1 or 2 days	Twofold increase in ^{14}C incorporation rate	*97*[b]

TABLE V-1 (contd.)

Species	Isotopically labeled precursor	Nutritional status	Radiation dose (R)	Time of sacrifice after irradiation	Radiation-induced effects	References
Rat	Acetate-1-^{14}C, butyrate-1-^{14}C, mevalonate-2-^{14}C, or glucose-UL-^{14}C	Starved	2500	1 day	Sixfold increase in ^{14}C incorporation from acetate-^{14}C; no change in incorporation with other 3 substrates	*109*[a]
	Acetate-1-^{14}C	Starved or fed	2500	2 or 3 days	Twofold increase in ^{14}C incorporation in starved as compared with fed rats postirradiation	*158*[a]
	Acetate-1-^{14}C	Starved	3500	1–4 days	Increased ^{14}C incorporation	*110*[b]
	Acetate-1-^{14}C or mevalonate-2-^{14}C	Starved	3500[f]	1 day	Moderately increased ^{14}C incorporation	*110*[a]
	Acetate-1-^{14}C	Starved	3500, 3750	1 day	Increased ^{14}C incorporation	*110*,[a,g] *207*
	Mevalonate-2-^{14}C, butyrate-1-^{14}C	Starved	3500, 3750	1 day	Decreased ^{14}C incorporation slight to moderate	*83*[a], *109*[a], *110*[a,g]
	Acetate-1-^{14}C	Starved	3500, 3750	1 day	Increased ^{14}C incorporation	*110*[e]
	Acetate-1-^{14}C	Starved	3500, 3750	1 day	Increased ^{14}C incorporation, glucose added to incubation mixture	*110*[a]
	Acetate-1-^{14}C	Starved	5000	1 day	Increased ^{14}C incorporation	*109*[a]
Mouse	Acetate-1-^{14}C	Starved	200	1 day	No change	*158*[a]
	Glucose-UL-^{14}C injected immediately or 24 hr after irradiation	Fed	500	1 or 25 hr	Increased ^{14}C incorporation	*164*[b]

	Glucose-UL-^{14}C injected immediately or 24 hr after irradiation	Fed	1000	1 or 24 hr	Increased ^{14}C incorporation	*164*[b]
	Glucose-UL-^{14}C injected immediately or 24 hr after irradiation	Fed	2500	1 or 24 hr	Increased ^{14}C incorporation (most marked at 2500 R)	*164*[b]
	Acetate-1-^{14}C injected 4 hr before sacrifice	Starved	2400	1 or 2 days	No change	*97*[b]
	Acetate-1-^{14}C	Starved and fed	2500	2 or 3 days	No change	*157*[a]
Rabbit	Acetate-1-^{14}C	Starved	2400	1 or 2 days	No change	*97*[b]
Guinea pig	Acetate-1-^{14}C	Starved	2400	1 or 2 days	No change	*97*[b]
			B. Phospholipids			
Rat	Dioleylpalmitin-^{14}C given by stomach tube 3 or 6 hr before sacrifice	Starved	500	4 hr	Increased rate of incorporation into phospholipid-contained palmitic acid	*184*[b]
	Palmitate-1-^{14}C or oleate-1-^{14}C; sacrificed 2, 4, 6, 12 hr after injection of precursor	?	800 (Cs-137)	4 days	No change in ^{14}C incorporation	*159*[a]

TABLE V-1 (contd.)

Species	Isotopically labeled precursor	Nutritional status	Radiation dose (R)	Time of sacrifice after irradiation	Radiation-induced effects	References
Rat	Stearate-1-^{14}C or acetate-1-^{14}C	Starved	500	2 days	Decreased incorporation of stearate-^{14}C, but no change with acetate-1-^{14}C	*181*[b]
	$^{32}P_i$ injected 3, 6, 9, or 12 days after irradiation	Fed	540–600	3.25, 6.25, 9.25, or 12.25 days	Increased ^{32}P incorporation on 3rd postirradiation day	*248*[b]
	$^{32}P_i$	Starved	700	18 or 120 hr	No change	*56*[a]
	$^{32}P_i$ injected immediately after irradiation	Starved	1000[h]	2 hr	Decreased rate of nuclear and cytoplasmic phospholipid renewal	*120*[b]
			950	8 hr	No change in ^{32}P uptake	*121*
	$^{32}P_i$ injected 24 hr after irradiation	Starved	1000	25 hr	No change	*288*[b]
	Dioleylpalmitin-^{14}C (by stomach tube) 3 or 6 hr before sacrifice	Starved	1000	4 hr	Increased ^{14}C incorporation into palmitic acid of phospholipids	*184*
	$^{32}P_i$ injected 1 hr before sacrifice	?	1000**r**	2, 4, 6, or 24 hr	Increased relative specific ^{32}P activity of total phospholipids, maximal at 6 hr	*130*[b]
	Stearate-1-^{14}C	Starved	1000	2 days	Decreased incorporation of stearate-^{14}C	*181*[b]
	Acetate-1-^{14}C	Starved	1000	1 day	Increased ^{32}P incorporation in nuclei, mitochondria, microsome, and supernatant	*288*[a]
	Acetate-1-^{14}C	Starved	1000	1 day	No change	*12*[e]
	Dioleylpalmitate-^{14}C	Starved	1500	2 days	Increased rate of ^{14}C incorporation into palmitic acid	*184*[b]

	Stearate-1-^{14}C	Starved	1500	2 days	No effect	*181*[b]
	Acetate-1-^{14}C	Starved	1500	50–52 hr	Twofold increase of ^{14}C incorporation	*183*[b]
	$^{32}P_i$	Starved	2000	1 day	Increased incorporation of ^{32}P	*288*[a]
	$^{32}P_i$	Starved	2500	1 day	No change	*288*[b]
	$^{32}P_i$	Starved	3000	20 hr	Increased ^{32}P incorporation in lysosomal and microsomal fractions; rate of ^{32}P turnover diminished in mitochondria	*240*[b]
	Dioleylpalimitate-^{14}C	Starved	1500–2000, 500 R alternate weeks		Increased ^{14}C incorporation into palmitic acid of phospholipid	*184*[b]
Mouse	$^{32}P_i$ injected immediately after irradiation	?	510	0.5 or 6 hr	Increased specific activity, highest in sphingomyelin	*244*[b]
	$^{32}P_i$ injected immediately after irradiation	Fed	800	2, 4, 8, or 24 hr	Increased ^{32}P incorporation, highest in sphingomyelin	*150*[b]
	$^{32}P_i$	Starved	800, 1850	Immediately	No change	*80*[a]
	$^{32}P_i$ injected immediately after irradiation	Fed(?)	2000	35 or 71 hr (partially hepatectomized 34 or 70 hr after irradiation)	No change	*129*[b]
	$^{32}P_i$	Starved	20,000	Immediately	No change	*80*[b]
			C. Triglycerides			
Rat	Dioleylpalmitin-^{14}C (by stomach tube) 3 or 6 hr before sacrifice	Starved	500	4 hr	Decreased specific ^{14}C activity	*184*[b]

TABLE V-1 (contd.)

Species	Isotopically labeled precursor	Nutritional status	Radiation dose (R)	Time of sacrifice after irradiation	Radiation-induced effects	References
Rat	Dioleylpalmitin-^{14}C (by stomach tube) 3 or 6 hr before sacrifice	Starved	1000	4 hr	Decreased specific ^{14}C activity	*184*[b]
	Dioleylpalmitin-^{14}C (by stomach tube) 3 or 6 hr before sacrifice	Starved	1500	4 hr	Decreased specific ^{14}C activity	*184*[b]
	Dioleylpalmitin-^{14}C (by stomach tube) 3 or 6 hr before sacrifice	Starved	1500–2000, single doses of 500 R on alternate weeks	4 hr	Decreased specific ^{14}C activity	*184*[b]
	Acetate-1-^{14}C injected 2–4 hr before sacrifice	Starved	1500	50–52 hr	Increased ^{14}C activity six times higher than controls	*183*[b]
	Acetate-1-^{14}C added to perfusate 30 min after start of perfusion	Starved	1000	1 day	Moderately increased specific ^{14}C activity	*12*[e]
Rabbit	Palmitate-1-^{14}C (albumin-bound) injected 14–18 hr after irradiation	Starved	1000	18–22 hr	Decreased specific ^{14}C activity concurrent with increased concentration	*66*[b] *208*[b]

	Palmitate-1-^{14}C (complexed with albumin) injected 2, 4, 16, and 24 hr after irradiation	Starved	1000	2, 4, 16, or 24 hr	Increased ^{14}C incorporation (73%, 263%, and 63%, respectively, 2, 4, and 16 hr after irradiation); no significant change at 24 hr	*41*[b]
	Palmitate-1-^{14}C (complexed to plasma protein) injected 0·5 hr before sacrifice	Starved	1000	2.5 and 4.5 hr	Increased specific activity	*295*[b]
				D. Cholesterol		
Rat	Acetate-1-^{14}C, ^{3}HOH	Starved	300	2 days	330% and 165% increase in ^{14}C and ^{3}H incorporation rate, respectively	*97*[b]
	Acetate-1-^{14}C	?	400, 800	1 day	Short-lived turnover rate increase maximal after 800 R	*282*[b]
	Acetate-2-^{14}C injected 6 hr before sacrifice	Fed	400	0, 1, 3, and 7 days	Initial increase followed by decrease in cholesterol synthesis	*115*[a]
	Acetate-1-^{14}C, ^{3}HOH	Starved	600	2 days	480% and 240% increase in ^{14}C and ^{3}H incorporation rate, respectively	*97*[b]
	Acetate-1-^{14}C injected 4 hr before sacrifice	Starved	800	1 day	Increased specific ^{14}C activity	*276*[b]
	Acetate-1-^{14}C injected 4 hr before sacrifice	Starved	800	1 day	Increased incorporation in normal rats and in adrenalectomized rats given a 0.9% NaCl solution; cortisone stimulates, deoxycorticosterone inhibits postirradiation cholesterol synthesis	***274***[b]**,** ***275***[b]
	Acetate-1-^{14}C, ^{3}HOH	Starved	900	2 days	750% and 375% increase in ^{14}C and ^{3}H incorporation rate, respectively	*97*[b]

TABLE V-1 (contd.)

Species	Isotopically labeled precursor	Nutritional status	Radiation dose (R)	Time of sacrifice after irradiation	Radiation-induced effects	References
Rat	Acetate-1-^{14}C injected 2 hr before sacrifice	Force-fed glucose	1000	1 day	No change	*206*[b]
	Acetate-1-^{14}C, ^{3}HOH	Starved	1200	2 days	1320% and 930% increase in ^{14}C and ^{3}H incorporation rate, respectively	*97*[b]
	Acetate-1-^{14}C	?	1200	6, 24, or 48 hr	Decreased ^{14}C incorporation at 6–24 hr, but increased incorporation at 48 hr after irradiation	*282*[b]
	Acetate-1-^{14}C, mevalonate-2-^{14}C	Starved	2000	1 day	Increased specific ^{14}C activity	*110*[a]
	Acetate-1-^{14}C, mevalonate-2-^{14}C, squalene-^{14}C (biosynthetic)	Starved	2400	2 days	300-fold increase of cholesterol synthesis from acetate-^{14}C, but smaller increase from other precursors	*40*[k]
	Acetate-1-^{14}C, mevalonolactone-2-^{14}C	Starved(?)	2400 or 3400	2 days	Increased ^{14}C incorporation from acetate-^{14}C, but not from mevalonolactone-^{14}C; cholesterol feeding depresses ^{14}C incorporation increase	*98*[c] *110*[a]
	Acetate-1-^{14}C, ^{3}HOH	Starved	2400	1 and 2 days	243% and 2320% increase, respectively, ^{14}C incorporation	*97*[b]
	Acetate-1-^{14}C	Starved	2500 and 3750	1 day	200 and 380% increase, respectively, ^{14}C incorporation	*51*[a]

	Acetate-1-^{14}C or mevalonate-2-^{14}C	Starved	3500, 3750, or 5000	1–4 days	Markedly increased ^{14}C incorporation with acetate-^{14}C; increases somewhat less with mevalonate substrate; increased incorporation not affected by hypophysectomy; butyrate-^{14}C and glucose-^{14}C incorporation not affected by x-irradiation	*110*[a], *207*[a,t] *83*[a], *109*[a]
	Acetate-1-^{14}C	Starved	1000 or 3750	1 day	Increased ^{14}C incorporation (livers from normal or irradiated animals and blood for perfusion from other normal or irradiated animals)	*12*[e], *110*[e]
Mouse	Acetate-2-^{14}C injected 6 hr before sacrifice	Fed	400	1.25, 3.25, or 7.25 days	Decreased ^{14}C incorporation 1st postirradiation day; increased incorporation thereafter	*115*[b]
	Acetate-1-^{14}C, ^{3}HOH	Starved	2400	2 days	Slightly increased incorporation rate	*97*[b]
Rabbit	Acetate-1-^{14}C, ^{3}HOH	Starved	2400	2 days	No change	*97*[b]
Guinea pig	Acetate-1-^{14}C, ^{3}HOH	Starved	2400	2 days	No change	*97*[b]
			E. Oxidative Conversion to CO_2			
Rat	Acetate-1-^{14}C	Fed	800	2 days	Decreased ^{14}C activity from maternal liver, 19th day of pregnancy; increased ^{14}C incorporation in fetal liver	*95*[a]
	Acetate-1-^{14}C	Starved	2000	1 day	Decreased ^{14}C incorporation	*110*[a], *109*[a]
	Acetate-1-^{14}C	Starved	2500	1 day	No change	*83*[a], *110*[a]

TABLE V-1 (contd.)

Species	Isotopically labeled precursor	Nutritional status	Radiation dose (R)	Time of sacrifice after irradiation	Radiation-induced effects	References
Rat	Acetate-1-^{14}C, butyrate-1-^{14}C, mevalonate-2-^{14}C	Starved	3500, 3750, or 5000	1 day	No change	*83*[a]
	Acetate-1-^{14}C	Starved	1000	1 day	Slightly increased ^{14}C incorporation	*12*[e]
Mouse	Acetate-1-^{14}C	Fed	800	5–6 days	Increased ^{14}C activity	*220*[a]

[a] Incubation of liver slices from total body irradiated animals with labeled precursor.
[b] Isolation of labeled lipid components from livers of irradiated animals injected with labeled precursor.
[c] Incubation of tissue minces or homogenates of livers obtained from irradiated animals.
[d] Adrenalectomy was performed 8 days before irradiation.
[e] Isolated, perfused liver.
[f] X- or γ-radiation.
[g] Animals hypophysectomized 15 days before irradiation.
[h] γ-irradiation.
[i] Hypophysectomy was performed before x-irradiation.
[k] Cell-free preparation.

In contrast to the striking changes in the rate of incorporation of precursors, there is a less marked change in the concentration of lipids in the liver after a single exposure (see Table V-2), but conditions of exposure, species, and age of animals determine whether or not a "fatty liver" will develop. Daily exposure of rats to 50 R over a period of 3 weeks or more causes "fatty liver" visible by microscopic inspection and an accumulation of triglycerides, whereas no such changes occur when the same cumulative dose is delivered as a single exposure (*3*, *47*). Moreover, the concentration of CoA in liver diminishes only when fatty liver develops (*3*), a pattern duplicated in fatty liver caused by agents other than radiation (*15*). Catabolism of lipids and their transport to other sites in the body seem to determine the extent of lipid accumulation after irradiation.

Glycogen levels in liver are higher in starved irradiated animals than in starved, nonirradiated ones, although, compared to fed animals, liver glycogen after irradiation is low (see Tables V-3 and V-4 and Fig. V-2). When irradiated and nonirradiated rats are pair-fed, the irradiated partner has a higher liver glycogen level than the nonirradiated control rat until the sixth day after exposure (*118*). Compared with the normal rat fed *ad libitum*, the fed irradiated rat has a lower hepatic glycogen level on the first day after exposure, the same glycogen content on the fourth day, and an elevated glycogen content on the sixth day (*118*).

Little is known so far about the catabolism of glycogen present in liver at the time of irradiation (*24*). Changes in enzymatic activity suggest a reduction of glycolytic activity (*23*, *24a*, *25a*, *255*), but this impairment is probably not very extensive, as indicated by the enhanced incorporation of glucose-^{14}C into hepatic lipids of irradiated rats (*164*) (Fig. V-1).

If one compares irradiated and nonirradiated animals fed a diet high in carbohydrates, one finds that incorporation of fructose into lipids is not altered after irradiation (*129*); the decreased incorporation of glucose under these conditions is explained by changes in glucokinase activity (as pointed out below). Glucose could also be degraded via the oxidative shunt, but degradation via this pathway apparently is not altered in relation to glycolysis as shown by the oxidation to CO_2 of glucose-1-^{14}C and glucose-6-^{14}C *in vivo* (*225*) or in rat liver slices and homogenates (*50*, *51*, *225*).

De novo synthesis of glycogen is enhanced in animals starved prior to irradiation. This is shown by the observation that the low level of glycogen in starved animals (*175*) increases subsequent to irradiation (compare row 3 in Table V-4). The substrate for increased glycogen synthesis is probably provided by increased gluconeogenesis, since glycogen deposition in liver is paralleled by an increase in blood glucose (see Fig. V-2).

One of the important metabolic routes to glycogen entails the enzymatic phosphorylation of glucose. An assessment of radiation-induced changes in the activity of a number of hepatic enzymes catalyzing various reactions of

TABLE V-2

LIPID CONTENT OF IRRADIATED LIVER[a]

Species	Nutritional status	Radiation dose (R)	Time of sacrifice after irradiation	Radiation-induced effects	References
			Total Lipids		
Rat	Pair-fed	500, 650, 800	1–42 days	No significant change	*1a, 52, 276*
	Starved	500		Increased, various constituent, polyunsaturated fatty acids increased; decreased oleic, palmitic, and behenic acid content	*195*
	Starved or fed	1000	1 day	No overall change	*1a, 195, 238*
	Fed	2800	30 hr	No change; effects reduced 50% in irradiated, adrenalectomized rat	*149*
	Fed or starved	3000	24 hr		*1a, 239*
	?	300[b]		Irradiated after partial hepatectomy; fatty livers only in hepatectomized rats	*174*
	Fed	630[c]	8–11 days	Left and anterior lobe have same increased content	*247*
	Fed	1400[d]	3, 6, and 12 hr	Increased in exteriorized lobe 12 hr after irradiation	*2*
Mouse	Fed	540–580	7–10 days	Increase in all but the bilateral stages of adrenalectomy; fatty livers develop only if at least one-half of an adrenal gland remains (cortisone can replace adrenal tissue)	*247*
	Starved	800	2, 4, and 6 days	Very slight increase	*58*
Rabbit	Fed	1000	1, 3, 5 and 6 days	Increased one day after irradiation	*27*

			Phospholipids		
Rat	Pair-fed	500, 650	2–42 days	No change in phosphatidyl content; hepatomegaly	*52*
	Pair-fed	800[e]	4 days	No change in total lipid-P	*159a*
	Starved	1000	9–40 min, 1 day	Increase in phosphatidyl-glyceride and -choline in mitochondria; slight increase in membrane fraction (also in linoleic and arachidonic acids in lecithin) in mitochondria	*126, 238, 239*
	Starved	1000[f]	1 day	No change	*130*
	Starved	1500	0.5, 3, 6, 12, and 24 hr	No change in total phospholipid content; decrease in arachidonic acid	*64*
	Starved	3000	1 day	Increase in phosphatidyl glycerol and phosphatidyl choline; decrease in cardiolipids and phosphatidyl ethanolamine	*239*
Mouse	Starved	800	2, 4, and 6 days	No change	*58*
Rabbit	Starved	800[f]	1 day	No change in mitochondrial phospholipids	*246*
	Fed	1000	1, 3, 5, and 6 days	Decrease in phospholipids	*27, 65*
		5000[f]	Immediately	Decrease in total phospholipids; lecithin, sphingomyelin, and cephalin redistribution	*130*
			Triglycerides		
Rat	Starved	1000	9–40 min, 24 and 48 hr	No loss from mitochondria; no change in total content	*126, 208*
	Fed	1150[g]	Immediately	Increase	*3*
	Fed	1150[h]	After last irradiation	Increased, microscopically visible globules	*3*
	Fed	1700–2350[i]	Immediately after last irradiation	Increased triglyceride content (not observed with single exposure)	*47*
Mouse	Starved	800	2, 4, and 6 days	Increased at 6 days after irradiation	*58*

TABLE V-2 (contd.)

Species	Nutritional status	Radiation dose (R)	Time of sacrifice after irradiation	Radiation-induced effects	References
Rabbit	Fed	1000	4 and 16 hr 1, 3, 5, and 6 days	Increased in microsomes; less of an increase in mitochondria	*27, 65*
	Starved	1000	22 hr	Decreased to 65% of controls	*66*
	Starved	1000	4.5 hr	Increased 86%	*295*
			16.5 hr	Increased 73%	*295*
			Cholesterol		
Rat	Pair-fed	500–650	2–45 days	No change	*52*
	Starved	800	1 day	Increase; increase in adrenalectomized rats	*274–276*
	Starved	1000	9–40 min	No loss from mitochondria; no change in total content	*126*
	?	2000	2 days		*282*
	Starved	3500	1 and 2 days	No change	*208*

[a] Unless otherwise stated, animals were exposed total body to x-rays.
[b] Irradiation over head and liver.
[c] Irradiation to the body, anterior lobe shielded.
[d] Exteriorized median lobe x-irradiated, remainder of the body shielded.
[e] Irradiation from eight ^{137}Cs sources.
[f] γ-irradiation.
[g] Single dose.
[h] Multiple doses: 50 R/day for 23 days.
[i] 50 R/day, fractionated.

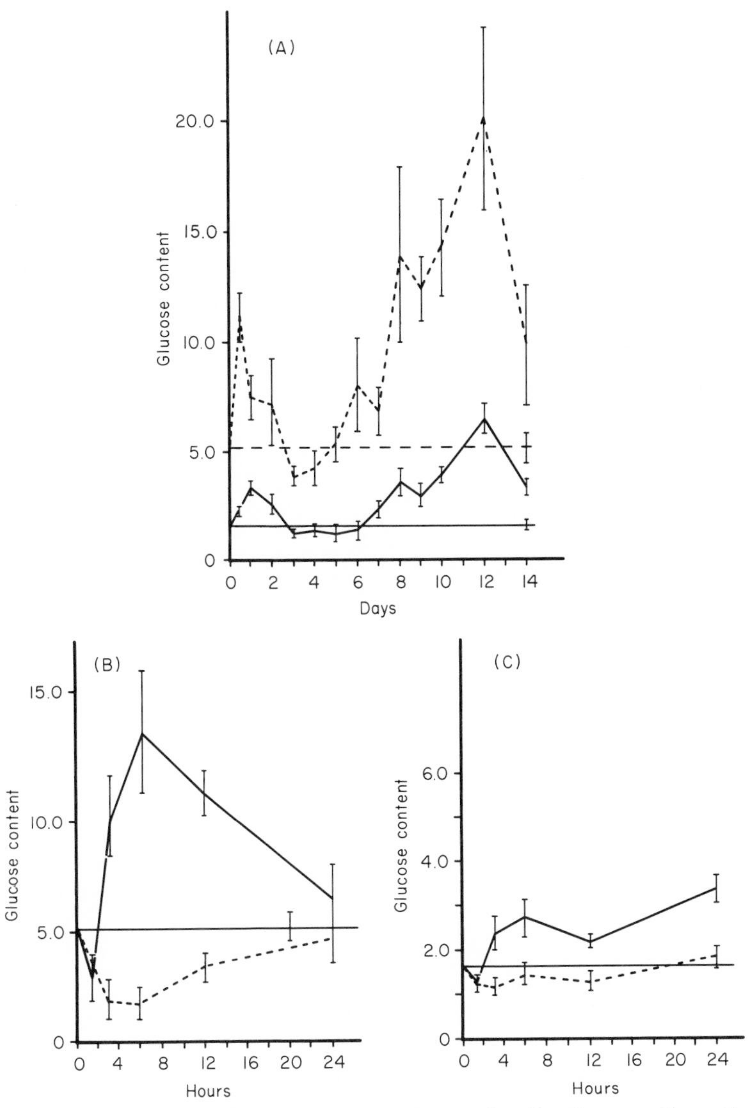

FIG. V-2. Glucose and glycogen contents in the x-irradiated mouse liver. Content expressed as μmoles glucose per gram of tissue. (A) Hepatic glycogen and glucose content after total body x-irradiation, examined for 2 weeks following x-irradiation; (——) glucose and (- - -) glycogen. (C) Hepatic glucose content investigated for 24 hr after exposure. (B) Hepatic glycogen content for the first 24 hr after exposure. The total body radiation dose was 690 R. All animals were starved 24 hr before sacrifice. In (B) and (C) the solid line represents the irradiated group, whereas the broken line represents the sham-irradiated group. [From C. Streffer, *Intern. J. Radiation Biol.* **11**, 179 (1966).] (Figs. 1 and 2.)

TABLE V-3
THE EFFECT OF IRRADIATION ON THE GLYCOGEN CONTENT OF LIVER[a]

Species	Nutritional status	Radiation dose (R)	Time of sacrifice after irradiation	Radiation-induced changes—hepatic glycogen content	References[b]
Rat	?	200	2 hr	Decreased glycogen content after local irradiation of liver	*77*
	Fasted before and after irradiation or pair-fed	300	3 hr	Increased glycogen accumulation (irradiated animals and irradiated partner of pair-fed pair); shielding of liver and exposure of rest of animal to 1000 R does not abolish accumulation of glycogen	*222*
	Fasted 20 hr before irradiation	250–1000	0–100 hr	Decreases during first few hours after irradiation, then increases reaching a peak at 50 hr and declines at 100 hr	***170***[c]
	Fasted before and after irradiation or fed *ad lib.*	500–7000	0–7 days	Decreased glycogen content, minimal 24 hr postirradiation; rising to significantly higher levels in starved and fed rats (fed rats attain higher glycogen levels than starved rats)	*190*, *191*, *194*, cf. also *30*
	Pair-fed or fed	450	24 hr	Irradiated partner of the pair exhibited increased hepatic glycogen content, exceeded by that of the fed, irradiated rat liver	*118*
	Starved 48 hr before sacrifice	200–1000	24 or 48 hr	24 hr postirradiation, a 6- to 10-fold hepatic glycogen content increase	*180*
	Fed *ad lib.*	500–3000	24 and 48 hr	Increased glycogen content after 500 and 1000 R at 24 and 48 hr postirradiation, but only at 48 hr after 3000 R	*1a*
	Fast started 24 hr before irradiation and continued thereafter	500–5000	24, 48 hr	Marked increase in glycogen content during 24 hr after irradiation; after 5000 R, glycogen levels at 48 hr exceed those at 24 hr	*133*

	Starved 18 hr before irradiation	600	0–8 days	Increased glycogen content, maximal 3 days postirradiation, on basis of glycogen content/hepatocyte	*287*
	Fasted before and after irradiation	750	18 hr	Increased glycogen content	*53, 54, 54a*
	Fasted 24 hr before irradiation	800	0–48 hr	Increased glycogen content, at peak 32 hr postirradiation	*78*
	Starved 24 hr before and/or after irradiation	880 or 1000	24 hr	20-fold increase of glycogen content	*31, 59, 123, 225, 270*[b]
	Fed high glucose diet before irradiation	1000	48 hr	Increased glycogen content as in starved animals	*31*
	Continuous infusion of glucose or fructose for 3 days postirradiation	1000	3 days	No increase in hepatic glycogen with glucose; marked rise of glycogen content with fructose	*123*
	Starved 48 hr before and 24 hr after irradiation	1500	24 hr	Increased glycogen content	*182*
	Fasted after irradiation	3 daily exposures to 300 rad each	48 hr	Glycogen content increased 20-fold over controls; 60 days postirradiation, glycogen content still four times higher than controls	*43*
	Starved	2000	3 hr	Glycogen content increased almost immediately after irradiation	*273*
	Starved 24 hr or fed	1000	24 hr	Slightly decreased glycogen	*76*
Mouse	Starved at time of irradiation; glucose *per os* before irradiation	500–2000	4 or 48 hr	Increased glycogen content after initial decrease, minimal 24 hr after 2000 R; after 1000 R the increase is less marked, no effect after 500 R	*162*
	Fed or fasted	600	15 min to 6 hr	Increase in "cold trichloroacetic acid soluble" fraction of glycogen	*245*
	Starved 24 hr before sacrifice	690	0–14 days	Increased glycogen content: 2 maxima, at 12–18 hr and 8–12 days after irradiation	*23, 25a, 255*
	Fed	690	0–14 days	Decreased glycogen content	*23, 25a, 255*

TABLE V-3 (contd.)

Species	Nutritional status	Radiation dose (R)	Time of sacrifice after irradiation	Radiation-induced changes—hepatic glycogen content	References[b]
Mouse	Fasted for 18 hr and refed for 5 hr before irradiation	10–60 kR (10 kR increments)	0–80 min	Marked decrease of glycogen content within 30 min, followed by further decrease	*156*
	Fed	60 kR	Immediately	Decreased glycogen content	*155*
	18-hr fast followed by 5-hr feeding before irradiation	50 kR	Immediately	Decreased glycogen level: decreased one-third of controls	*154*
	Fed	110 kR	Immediately	Decreased glycogen content	*159*
	Fed	3 daily doses of 350 R followed by 2 daily doses of 400 R	15 min to 6 hr	Glycogen content decreased, 27% of controls	*138*, *139*
Guinea Pig	Starved at time of irradiation	500–2000	4 or 48 hr	Same as for mouse (cf. reference *162*)	*162*, *163*
	Fed	500 or 1000	3 min to 24 hr	Increased deposition of glycogen, most marked after 1000 R	*202*
Hamster	Fed	110 kR	Immediately	Marked decrease of glycogen content	*159*
			Effect of Hypophysectomy on Glycogen Content		
Rat	Fasted 24 hr before irradiation	500	0–3 days	No increase in glycogen content	*191*
	Fasted 18 hr before death	600	0–8 days	No increase in glycogen content	*287*

	Fasted 24 hr before irradiation	1000	24 hr	No increase in glycogen content	*31, 270*
	Fasted 20 hr before irradiation	1500	22 or 41 hr	No increase in glycogen content	*170d*
	High glucose diet for 3 days before and 24 hr after irradiation	1000	24 hr	Marked decrease in glycogen content	*31*
			The Effect of Bilateral Adrenalectomy		
Rat	Fed	200–1000	24 or 48 hr	Increase in glycogen content abolished (surgery shortly before irradiation)	*180*
	Fasted before and after irradiation	880	24 hr	Glycogen level exceeds control level	*59*
	Fasted at time of irradiation and thereafter	1000	0–3 days	Increased glycogen content abolished (surgery 2 weeks before irradiation)	*192*
	Fasted 20 hr before irradiation	1500	22 or 41 hr	Increased glycogen content abolished	*170*
			Glycogen in Newborn Rats		
	Starved 18 hr before irradiation	1000	18–24 hr.	No increased glycogen content in 6-day-old rat livers	*269*

[a] Unless otherwise stated, all exposures were single, total body x-irradiations.
[b] For reviews of the pertinent literature until 1953 and 1960 consult references *203* and *205*, respectively.
[c] At 70 hr after 1500 R no increase in glycogen content was noted.
[d] γ-irradiation (^{60}Co source).

TABLE V-4

EFFECTS OF X-IRRADIATION AND VARIATIONS IN CARBOHYDRATE CONTENT OF THE DIET ON GLUCOKINASE ACTIVITY IN RAT LIVER[a,b]

Nutritional disposition		Time of sacrifice after irradiation[c] (hr)	Glucokinase (% control activity)	Glycogen content[d] (% control)
Before x-irradiation	After x-irradiation			
High glucose diet for 3 days	High glucose diet (stomach tube)	3	91	89
		24	74	155
		48	67	156
High glucose diet for 3 days	Fasted	3	91	
		24	126	
		48	168	
Fasted for 24 hr	Fasted	3	109	200
		24	99	732
		48	101	412
Fasted for 24 hr	High glucose diet (stomach tube)	3	109	
		24	120	
		48	67	
High glucose diet for 3 days	High glucose diet (stomach tube) at intervals[e] for 24 hr	24	66	29
High glucose diet for 3 days	Fasted for 24 hr[f]	24	65	100

[a] The values shown in this table were computed from Borreback *et al.* (*31*).
[b] The "controls" were nonirradiated rats treated like irradiated rats in all respects other than irradiation.
[c] Total body radiation dose, 1000 R.
[d] Glycogen content/weight of liver.
[e] Immediately after irradiation 2.5 ml of a solution containing 70% glucose, 3% Hawk–Oser salt mixture, and 0.5% vitamin B mixture was administered: this procedure was repeated thereafter every 6 hr.
[f] These rats were hypophysectomized 7 days before the experiment was started.

carbohydrate metabolism reveals that the activity of glucokinase in the liver of starved, irradiated rats decreases less than that of nonirradiated, starved rats (*123*) (Table V-4). Glucokinase is an "adaptive" enzyme and differs in its physiochemical properties from the nonspecific hexokinase in that its Michaelis constant (K_M) is larger, it is more labile *in vitro*, it is not subject to allosteric inhibition by glucose-6-phosphate (*31*), and its synthesis is claimed to be

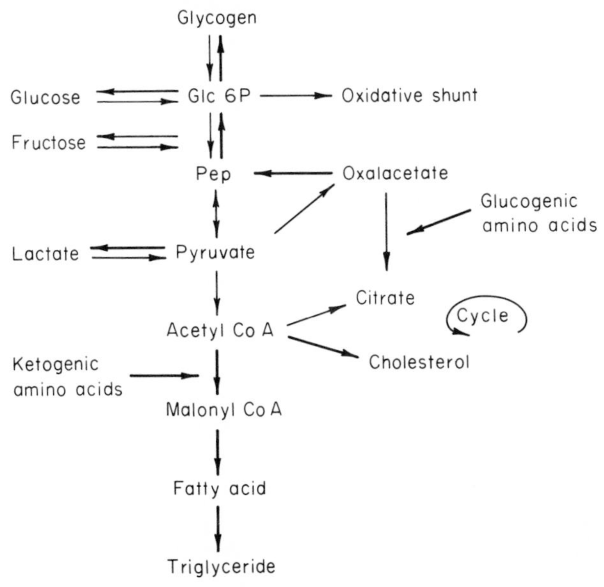

FIG. V-3. Interrelationships of carbohydrate, lipid, and protein metabolism in the irradiated mammalian liver. The size of the arrow indicates the approximate extent to which metabolism is thought to proceed in the direction of a given product, i.e., the heavier the arrow, the greater the metabolic flux in the direction of this arrow.

stimulated by insulin and depressed by glucagon (*250*). In rats fed a high carbohydrate diet, induction of glucokinase is retarded by irradiation. Accordingly, these irradiated animals catabolize glucose to CO_2, or utilize it for glycogen or lipid synthesis less efficiently than do nonirradiated ones, whereas irradiation has no effect on utilization of fructose, a carbohydrate phosphorylated by hexokinase but not by glucokinase (*123*). Control by the pituitary and the diencephalon play an important role in regulating changes in lipid and carbohydrate metabolism after irradiation. As pointed out elsewhere (*269*) (see Chapter IX), glycogen does not accumulate in newborn, fasted animals where control by the diencephalon is not yet established (Fig. IX-1) or in hypophysectomized animals (*31*, *281*). Removal of the pituitary, however, does not influence the radiation-induced changes in glucokinase activity (*31*).

TABLE V-5

EFFECTS OF IONIZING RADIATION ON ENZYMES OF CARBOHYDRATE AND LIPID METABOLISM IN MAMMALIAN LIVER[a]

Enzyme	Species	Nutritional status	Radiation dose (R)	Time after irradiation	Radiation-induced changes in enzymatic activity	References
1. D-Glucose-6-phosphate phosphohydrolase (E.C. 3.1.3.9) glucose-6-phosphatase	Rat	As	in	22(b)	Slightly lower specific activity 24–48 hr postirradiation; no change in total activity	*182*
	(b)	As	in	7(b)	Increased activity on the 3rd postirradiation day in hypophysectomized rats	*287*
	(c)	Fasted 18 hr before sacrifice	1150 (total) 50 R/day	24 hr after last irradiation	Microsomal fraction: activity unchanged	*3*
	(d)	Fed	750	0, 2, or 24 hr	Twofold increase in activity of nuclear fraction immediately after irradiation; 2 hr postirradiation, enzymatic activity in the nuclear, mitochondrial fraction elevated; 24 hr postirradiation only nuclear and mitochondrial fractions show increased activity	*57*
	(e)	Starved after irradiation	750 or 1000	0.5, 24, or 48 hr	No change	*53*
	(f)	As	in	22(c)	No change	*36*
	(g)	Fasted after irradiation	700	0–72 hr	No change	*30*
2. D-Fructose-1, 6-diphosphate 1-phosphohydrolase (E.C. 3.1.3.11) hexose diphosphatase	Mouse					
	(a)	Starved 24 hr before sacrifice	690	11–17 days	Slight increase in enzymatic activity; in fed mice, observed only on the 11th postirradiation day	*24a*
	(b)	Starved 24 hr before sacrifice	690	7–9 days	Slight increase as above	*255*

3. Lactate: NAD oxidoreductase (E.C. 1.1.1.28) D-lactate dehydrogenase	Rat	(a)	Fed diet of 60% glucose or fructose 1 week before, fasted after irradiation until death	900	24 hr	Specific enzymatic activity increased with either glucose or fructose	*79*
		(b)	Fed	500 or 1000	5 days	No change	*196a*
		(c)	Starved 24 hr before sacrifice	1400	0–24 hr	Moderate increase in activity at 12 hr; comparable decrease at 24 hr	*305*
4. L-Malate: NAD oxidoreductase (E.C. 1.1.1.37) malate dehydrogenase		(a)	Fed	500 or 1000	5 days	Moderate increase in activity	*196*
		(b)	?	650	5 min to 30 days	Moderate decrease in activity	*135d*
5. D-Glyceraldehyde-3-phosphate: NAD oxidoreductase (E.C. 1.2.1.12)		(a)	As	in	3(a)	No change	*79*
		(b)	Fed	500 or 1000	5 days	No change	*196a*
6. Glucose-6-phosphate: NADP oxidoreductase (E.C. 1.1.1.49) glucose-6-phosphate dehydrogenase		(a)	As	in	3(a)	No change	*79*
		(b)	Fed	1000	5 days	Slight increase in activity early after irradiation	*196a*
		(c)	?	625	5 min to 28 days	Decrease immediately, increase 4–17 days, decrease 21–23 days, and normal 28 days after irradiation	*136*
7. 6-Phosphogluconate: NADP oxidoreductase (E.C. 1.1.1.44) 6-phosphogluconate dehydrogenase		(a)	As	in	3(a)	No change	*79*
		(b)	Starved 18 hr before sacrifice	600	0–8 days	Increased activity/average cell in intact rat livers; decreased activity, hypophysectomized animals	*287*
		(c)	?	625	5 min to 28 days	Similar response as glucose-6-phosphate dehydrogenase except no rise in activity on days 21–23 postirradiation	*136*

TABLE V-5 (contd)

Enzyme	Species	Nutritional status	Radiation dose (R)	Time after irradiation	Radiation-induced changes in enzymatic activity	References
8. L-Glycerol-3-phosphate NAD oxidoreductase (E.C. 1.1.99.5) α-glycerolphosphate dehydrogenase	(a)	As	in	3(a)	Increased specific enzymatic activity after 60% fructose diet	*79*
	(b)	Starved 24 hr before sacrifice	1400	0–24 hr	Decreased activity, 1 hr; increased, 6 and 12 hr; no change, 24 hr	*305*
9. Butyryl-CoA: tetrazole purple oxidoreductase (E.C. 1.3.99.2) butyryl-CoA dehydrogenase	Mouse	Starved 24 hr before sacrifice	690	0–40 hr	No change; no change in palmityl CoA content	*285*
10. D-Glucose-6-phosphate ketol-isomerase (E.C. 5.3.1.9) glucose-6-phosphate isomerase	Rat	As	in	7(b)	Increased activity/cell, first (65%) and second (52%) postirradiation day; 39% increase, hypophysectomized rats on third postirradiation day only	*287*
11. D-Glyceradehyde-3-phosphate ketol-isomerase (E.C. 5.3.1.1)		Starved 24 hr before sacrifice	1400	0–72 hr	Decreased activity 12–24 hr; increased activity, 72 hr	*305*
12. ATP: D-Glucose-6-phosphate transferase, (E.C. 2.7.1.2) glucokinase	Rat					
	(a)	Starved after exposure	700	12–72 hr	No change	*30*
	(b)	Fasted before and after exposure	1000	3–48 hr	Enzymatic activity decreased 24 and 48 hr postirradiation	*31*
	(c)	Fed a high-glucose diet before and after irradiation	1000	3–48 hr	No change in enzymatic activity[b]	*31*

Enzyme	Species	Conditions	Dose	Time	Effect	Ref.
13. ATP: D-Hexose-6-phosphate transferase, (E.C. 2.7.1.1) hexokinase	(a)	Starved after exposure	700	0–72 hr	No change	
	(b)	Starved 24 hr before sacrifice	1400	0–72 hr	Increased at 1 hr; marked decrease 12 and 48 hr after exposure	*305*
14. ATP: D-Fructose-1-phosphate transferase (E.C. 2.7.1.3) ketohexokinase		Starved before and after exposure	1000	3–48 hr	No change	*31*
15. α-D-Glucose-1, 6-diphosphate: α-D-glucose-1-phosphate transferase, (E.C. 2.7.5.1) phosphoglucomutase	(a)	As	in	7(b)	Progressive increase in enzymatic activity, maximal at 3 days, decreasing 7–8 days after irradiation; hypophysectomy abolishes this response	*287*
	(b)	As	in	3(a)	Increased enzymatic activity with either glucose or fructose diets	*287*
	(c)	As	in	20(b)	No change in specific activity, but elevated total enzymatic activity 12 hr after exposure	*30*
	(d)	Starved 24 hr before irradiation or fed	1000	24 hr	No change in either starved or fed rats	*76*
	(e)	?	750	0, 2, or 24 hr	Activity decreases (60% of controls) 2 hr after exposure; a smaller decrease is noted at 24 hr, confined mainly to the mitochondrial fraction of this enzymatic activity	*57a*
16. ATP: D-Fructose-6-phosphate 1-phosphotransferase (E.C. 2.7.1.11) phosphofructokinase	Mouse	Starved 24 hr before sacrifice	690 or 810	0–17 days	After 3 days enzymatic activity increases progressively (150% of controls on day 12)	*24a*
		Fed			Increased activity 8–14 postirradiation day	

TABLE V-5 (contd.)

Enzyme	Species	Nutritional status	Radiation dose (R)	Time after irradiation	Radiation-induced changes in enzymatic activity	References
17. ATP: 3-Phospho-D-glycerate-1-phosphate transferase (E.E. 2.7.2.13) phosphoglycerate kinase	Rat	As	in	3(a)	No change in enzymatic activity on either glucose or fructose diet	*79*
18. ATP: pyruvate phosphotransferase, (E.C. 2.7.1.40) pyruvate kinase	Mouse (a)	Fed or starved 24 hr before sacrifice	690	0–12 days	Enzymatic activity normal until 9th postirradiation day; thereafter, a marked decrease; decreased activity, fed mice	*25a*
	Rat (b)	Fed	500 or 1000	5 days	Decreased activity after 1000 R	*196a*
	(c)	Starved 24 hr before sacrifice	1400	0–24 hr	Decreased activity (58% of control) at 1 hr, constant through 12 hr; decreased (31% of control) at 24 hr after irradiation	*305*
19. Sedoheptulose-7-phosphate: D-glyceraldehyde-3-phosphate-transferase (E.C. 2.2.1.1) transketolase		Fed	500 or 1000	5 days	Slight increase in activity	*196a*
20. UTP: α-D-glucose-1-phosphate uridylyltransferase (E.C. 2.7.7.9) UDPG pyrophosphorylase	(a)	Fed or starved 24 hr before irradiation	1000	24 hr	No change in fed or starved rats; in starved rats, hepatic UDPG concentration decreased	*76*
	(b)	Starved after irradiation	700	0–72 hr	Slight increase in activity at 72 hr	*30*

21. UDP glucose: glycogen α-4-glucosyltransferase (E.C. 2.4.1.11) UDPG glycogen glucosyltransferase		Starved after irradiation	700	0–72 hr	No change	*30*
22. α-1, 4- Glucan: orthophosphate glucosyltransferase (E.C. 2.4.1.1) glycogen phosphorylase	Mouse (a)	Fed or starved 24 hr before sacrifice	690	0–14 days	No change	*255*
	Rat (b)	Fasted after irradiation	700	0–72 hr	Decreased enzymatic (specific) activity	*30*
	(c)	Fed	10–100 kR	60–90 min incubation	Increased activity of "active" phosphorylase; incubation of slices for 120 min after irradiation with 35 kR decreased enzymatic activity 60%[c]	*37*, *36*
23. CTP: oxalacetate carboxy-lyase (E.C. 4.1.1.32) phosphopyruvate carboxylase	Mouse	Starved 24 hr before sacrifice	690	0–12 days	Decreased activity at 18 hr, 3–4 days, and 6–12 days	*25a*
24. Acetyl CoA: carbon dioxide ligase (ADP) (E.C. 6.4.1.2.) acetyl CoA carboxylase		Starved 24 hr before sacrifice	690	2–24 hr	X-irradiation does not overcome the starvation-induced reduced enzymatic activity, even though excess acetyl CoA is available, due to starvation; this rate-limiting enzyme, by controlling malonyl CoA formation, is probably responsible for most of the irradiation effects on fatty acid synthesis	*25*

TABLE V-5 (contd.)

Enzyme	Species	Nutritional status	Radiation dose (R)	Time after irradiation	Radiation-induced changes in enzymatic activity	References
25. Fructose-1, 6-diphosphate D-glyceraldehyde-3-phosphate-lyase (E.C. 4.1.2.13) fructose diphosphate aldolase	Rat					
	(a)	As	in	3(a)	No change	*79*
	(b)	As	in	19(b)	No change	*196a*
	(c)	Starved 24 hr before sacrifice	1400	0–72 hr	Significant decrease in activity at 3, 6, and 72 hr; marked increase at 12, 24, and 48 hr (maximal value at 24 hr)	*305*

[a] Unless otherwise stated, all exposures were single, total body x-irradiations.

[b] See also Table V-4.

[c] Liver slices were irradiated *in vitro*.

Acetyl CoA may be viewed as the point of convergence of carbohydrate and lipid metabolism in mammalian liver and is required in catalytic amounts as an activator of pyruvate carboxylase which itself catalyzes formation of oxalacetate from pyruvate (see Fig. V-3). The latter enzyme may also generate acetyl CoA by decarboxylation; the decarboxylation product may then serve as a substrate for acetoacetate and steroid or citric acid, fatty acid synthesis catalyzed by acetyl CoA carboxylase. The stimulation of pyruvate carboxylase by acetyl CoA favors the formation of oxalacetate from pyruvate; hence the formation of such reaction products of oxalacetate as phosphoenol pyruvate, citrate, or aspartate depends on the availability of, among other reactants, acetyl CoA, in the extra- as well as the intramitochondrial compartment. The complex regulatory machinery which directs the flow of metabolites in the direction of gluconeogenesis, lipogenesis, or oxidation via the Krebs tricarboxylic acid cycle can effect changes rapidly by means of "allosteric" effects on the enzyme protein or slowly by "inductive" means (Fig. V-3).

On the first day after irradiation PEP-carboxykinase increases in fed mice but decreases in starved ones (*23*, *24*), whereas the concentration of PEP and acetyl CoA increases slightly under both dietary regimens (*23*, *24*). Although earlier data on ATP content appear contradictory, the most recently published data on starved mice and rats indicate that ATP content decreases during the first and, in some cases, the second day after total body exposure to 690–1400 R (*23*, *165*, *167*, *285a*, *305*).

The ratio of lactate to pyruvate increases 4–14 hr after exposure but decreases later, whereas the ratio of β-hydroxybutyrate to acetoacetate diminishes beginning with the early postirradiation period (see Fig. V-4). In starved rats metabolic activity involving the citric acid cycle declines following irradiation (*49a*, *63*). Citrate accumulates in irradiated rats of both sexes after poisoning with fluoroacetate, but in nonirradiated rats only the livers of females accumulate citrate (*63*), suggesting a possible role of hormonal control by steroids of the female sex hormone type (Fig. V-5).

The available metabolic data, taken in their entirety (see Tables V-1 to V-6), still fail to provide a satisfactory basis for an understanding of the mechanism(s) underlying the radiation-induced changes in lipid and carbohydrate metabolism. Such an understanding would be required to formulate an integrated thesis which could serve to elucidate and interrelate all of the observed phenomena in this area of metabolism. Further studies on the effect of exposure to ionizing radiation on the metabolic interplay of glycogenesis, gluconeogenesis, lipogenesis, activity of the Krebs tricarboxylic acid cycle, and protein and amino acid metabolism seem necessary. It is clear that the assay of enzymatic activities and the determination of substrate concentration must be supplemented by suitable isotope experiments in order to evaluate rates of metabolic turnover and the kinetics of metabolic changes in man and animal.

TABLE V-6

CARBOHYDRATE METABOLISM IN IRRADIATED LIVER[a]

Species and nutritional state	Radiation dose (R)	Experimental conditions[b]	Time of death after irradiation	Radiation-induced changes[b]	References
Rat, fed	35,000 delivered in 5 min	Slices examined 90 min after *in vitro* irradiation	—	Decreased incorporation of glucose-UL-^{14}C into glycogen (28% of controls), no effect on oxidation to $^{14}CO_2$; oxidation ratio of glucose-6-^{14}C/glucose-1-^{14}C to $^{14}CO_2$ unaltered; glucose lost from irradiated slices	*36*, *37*
Rat, starved after irradiation	1000[c]	*In vivo* irradiated rat injected IP with $NaH^{14}CO_3$, glucose-UL-^{14}C or alanine-2-^{14}C	51–55 hr	Increased $^{14}CO_2$ (expired air); increased $^{14}CO_2$ fixation (glycogen) Increased incorporation of glucose-^{14}C and alanine-^{14}C into glycogen	*225*
	200, 400, or 800	*In vivo* irradiated rat, collection of $^{14}CO_2$ from oxidation of $H^{14}CO_3$-, $H^{14}COO^-$ acetate-1- or -2-^{14}C-glucose-UL-, -1-, -6-^{14}C-fructose-UL-^{14}C, ribose-1-^{14}C	51–55 hr	Decreased oxidation rate of acetate, glucose, fructose, and ribose; increased oxidation of formate; no change in $H^{14}CO_3^-$ metabolism	
Rat, starved until sacrifice and before irradiation; fed rats	1500	*In vivo* irradiated rat—injected IP with glycine-1-^{14}C, alanine-1-^{14}C or leucine-1-^{14}C, 2 hr before sacrifice	26 or 50 hr	Increased incorporation of ^{14}C activity into glycogen by irradiated, compared with control, rats; alanine showed the most marked change, leucine the least	*185*
Rat, fasted just before irradiation	1000	*In vivo* irradiated. Slices incubated *in vitro* with glucose-UL-^{14}C or fructose-UL-^{14}C	4 or 24 hr	Increased conversion of glucose-^{14}C and fructose-^{14}C to glycogen and lactate; for glucose, there is a decrease in labeled CO_2 and lactate formation at 24 hr and a decrease in conversion to amino acids	*282a*

				at 4 hr after irradiation; decreased labeling of glucose by fructose at 4 hr, increased at 24 hr after irradiation	
Rat, fed	500 or 1000	*In vivo* irradiation; *in vitro* incubation of homogenates	Several days	Decreased anaerobic glycolysis	*196*, *196a*
Rat, force-fed glucose and protein (or fructose and protein) by stomach tube at 4-hr intervals	1000	*In vivo* irradiation. Slices incubated *in vitro* with glucose- or fructose-UL-^{14}C	4 to 72 hr	Conversion of glucose to CO_2, fatty acid and glycogen markedly decreased; conversion of fructose to these products unaffected; defective glucose utilization by irradiated liver not corrected by feeding sufficient amounts of glucose	*123*
Rat, fasted 24 or 48 hr before and 24 hr after irradiation	1000	*In vivo* irradiation. Slices incubated with glucose-1-^{14}C or glucose-6-^{14}C	24 or 48 hr	Increased ^{14}C incorporation into glycogen 48 hr after starvation and irradiation; increased oxidation to CO_2; glucose-6-^{14}C/glucose-1-^{14}C ratio of conversion to CO_2 unaltered; increased incorporation from labeled glucose preparations into long-chain fatty acids	*50*, *51*
Mouse, starved after irradiation to sacrifice	2000 or 2500	*In vivo* irradiation. Injected subcutaneously with glucose-UL-^{14}C	0–180 min	Threefold increase in ^{14}C incorporation into glycogen, maximal 50 min after irradiation; peak levels in blood 10 min after irradiation (see also Fig. V-1); increased incorporation into hepatic lipids most marked at 2500R	*164*
Mouse, fed	2000	***In vivo*** **irradiation. Glucose-**UL-^{14}C injected subcutaneously immediately after irradiation	10–60 min	Decrease in expired $^{14}CO_2$ (80% of control)	*121a*
Mouse, starved 24 hr before sacrifice	690	*In vivo* irradiation	Until 17 days	No inhibition of glycolysis	*24a*

TABLE V-6 (contd.)

Appendix: The Concentration of Intermediary Metabolites of Glycolysis and of Cofactors in Irradiated Liver[a, b, d]

Species and nutritional state	Radiation dose (R)	Glucose	G6P	F6P	F1, 6P	DHAP	α-Gly-PA	3-PGal	3-PGA	PEP	Lactic acid	Pyruvic acid	ATP	NAD	References
Rat, starved 24 hr before x-irradiation	1400	Increase 1200% 24 hr	Increase 350% 72 hr	Increase twofold 48 hr	No change	Increase 24–72 hr	Decrease 1–72 hr	No change	Increase 1–72 hr	Increase 24 hr	Decrease 3–72 hr	Decrease 3–72 hr	Decrease 12, 24 hr	NAD^+ decrease 1, 12, 48 hr	*305*
Starved before irradiation	650–3000													Decrease several hr to 6 days NADH up to 30%	*1, 59, 64a, 167, 169, 172, 272*
Starved before irradiation	980														
Mouse, starved 24 hr before sacrifice	690	Increase 3–48 hr	Increase 12–24 hr, 4–14 days		Increase 12 hr–10 days	Increase 12 hr–10 days				Slight increase at 3 days	Decrease 2, 18 hr	Decrease 18 hr	Decrease 12–24 hr, 4–6 days		*23, 255, 285a*
	690						Increase 1–10 days								*22*
Mouse, starved 24 hr before sacrifice	690 or 810													NADH, NAD^+ decrease 6–14 days	*257*

[a] Unless otherwise stated, all exposures were single total body irradiations.

[b] Abbreviations: UL, uniformly labeled; G6P, glucose-6-phosphate; F6P, fructose-6-phosphate; F1, 6P, fructose-1, 6-phosphate; DHAP, dihydroxyacetone phosphate; α-Gly-PA, α-glycerophosphoric acid; 3-PGal, 3-phosphoglyceraldehyde; 3-PGA, 3-phosphoglyceric acid; PEP, phosphoenol pyruvate; and IP, intraperitoneal.

[c] γ-irradiation (^{60}Co).

[d] The time periods stated in the appendix refer to time elapsed after irradiation.

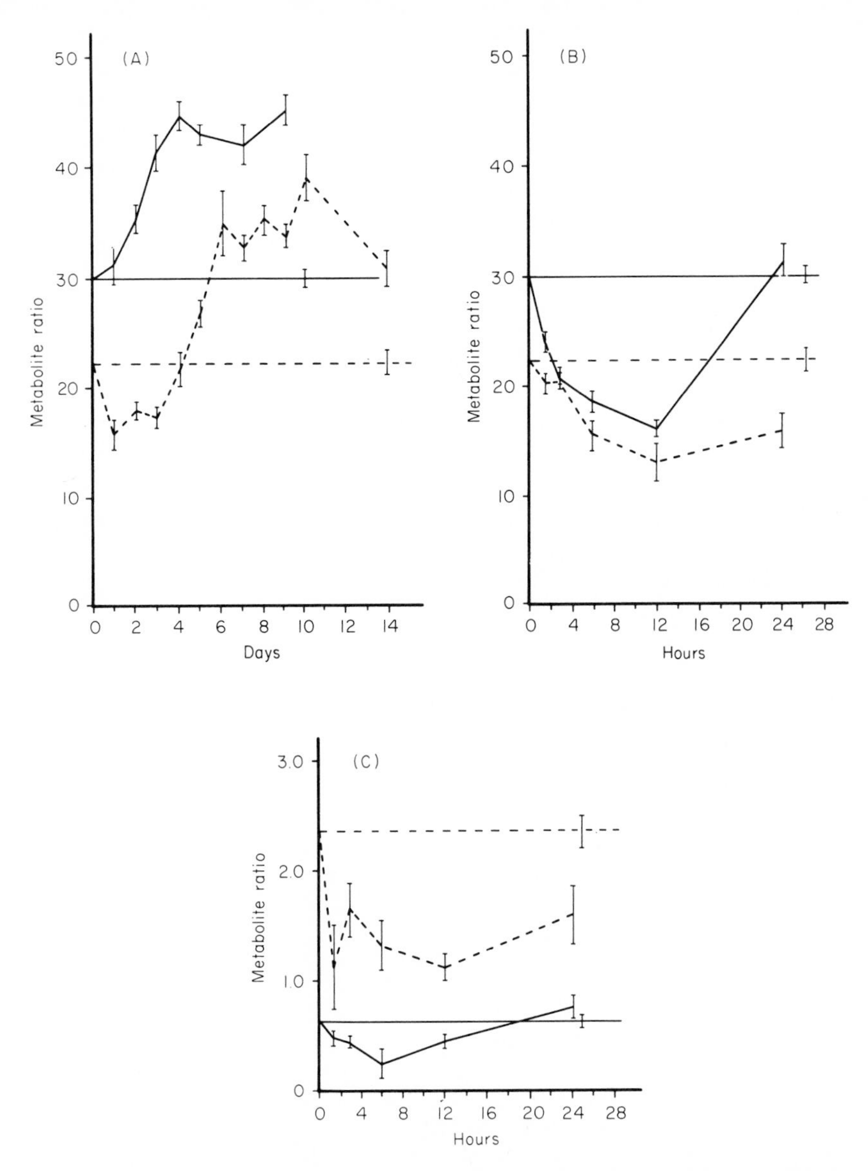

FIG. V-4. Lactate to pyruvate and β-hydrobutyrate to acetoacetate ratios. (A) Changes of the lactate to pyruvate ratio during 14 postirradiation hours. (B) Changes of the lactate to pyruvate ratio during 28 postirradiation days. (C) Changes of the β-hydroxybutyrate to acetoacetate ratio during 28 hr after x-irradiation. The solid line represents fed rats, whereas the broken line represents rats starved 24 hr before sacrifice. A single, total body dose of 690 R of x-rays was given. [From C. Streffer, *Intern. J. Radiation Biol.* **11**, 179 (1966).]

The statement is often made that "irradiation eliminates the effects of starvation," that is, that an irradiated animal is unable to adapt* to the conditions of starvation (*79*) and, therefore, continues to divert substrate to the synthesis of glycogen and lipids rather than to oxidation via the citric acid cycle. Undoubtedly there is some truth in this statement, as the following observations show: (i) The activity of glucokinase and other enzymes does not decrease in

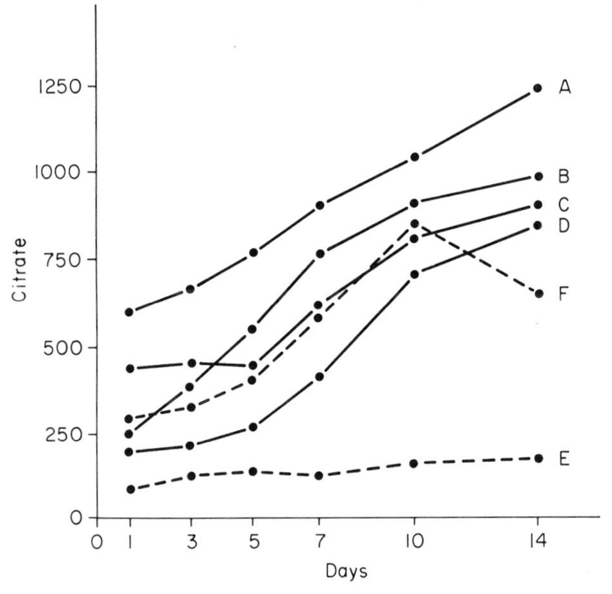

FIG. V-5. Citrate accumulation in the x-irradiated, fluoroacetate-treated rat liver. Citrate content is expressed as μgrams per gram tissue. Rats were given 200 or 400 R total body irradiation, then injected intraperitoneally with sodium fluoroacetate (3.5 mg/kg), and sacrificed 3 hr later. A, spleen, 200 R; B, thymus, 200 R; C, spleen, 400 R; D, thymus, 400 R; E, liver, 200 R; and F, liver, 400 R. Each group was composed of four animals. [From K. P. DuBois *et al.*, *Proc. Soc. Exptl. Biol. Med.* **76**, 422 (1951).]

starved animals after irradiation (*31*); and (ii) incorporation of acetate into lipids does not increase when the irradiated animals are force-fed glucose (*206*). Nevertheless, such a statement is deceptively general and should be supplemented by a detailed analysis of the metabolic changes involved.

It is self-evident that lipid synthesis and gluconeogenesis can proceed only when sufficient amounts of substrate are available. After irradiation, tissue in radiosensitive organs breaks down, with consequent liberation of glucogenic and ketogenic amino acids (*224a*, *225*). Indeed, some polypeptide frag-

* "Adapt" refers here to a change in enzymatic activity in response to an external stimulus (e.g., hormonal, dietary), whereas "induction" is used as defined elsewhere (*106*).

ments accumulate in the liver (*191*), the amino acid level in the blood may increase transiently (*7*, *86*, *89*, *103*, *148*, *236a*), increased conversion of free amino acid to glycogen and lipid may occur (*225*), and ^{14}C-labeled amino acid residues of labeled tissue protein may contribute ^{14}C activity to hepatic glycogen (*224a*). All of these observations suggest that amino acids are offered to the liver in excess-of-normal levels during the early postirradiation period.

In summary, the following intepretation of radiation-induced changes with respect to lipid and carbohydrate metabolism appears reasonable. Enhanced lipogenesis and glycogen accumulation occur after irradiation because more substrate for these metabolic processes is made available as a result of tissue breakdown. Substrate utilization for synthesis rather than for oxidation predominates during the postirradiation period because adaptation to the state of starvation in irradiated animals is restricted to nonoxidative enzymes.

Finally, it should be mentioned that very high doses of radiation (more than 5 kR) *in vivo* or *in vitro* do not cause accumulation of glycogen, but rather its degradation (*154*, *156*, *159*, *168*, *172*, *198*). This glycogenolysis may result from an increased conversion of phosphorylase a to phosphorylase b. In addition, membrane damage may result in a shift from Na^+ to K^+ in the cell, with K^+ activating phosphorylase b. An *in vitro* exposure of rat liver slices to 35 kR depresses the incorporation of glucose-^{14}C into glycogen by about 30% but has no effect on its oxidation to CO_2 (*37*). It is conceivable, therefore, that incorporation is depressed largely because newly formed ^{14}C-labeled glycogen is rapidly degraded.

C. Oxidative Metabolism

Although the oxidative activities and the P to O ratio of hepatic mitochondria are not altered immediately after irradiation, structural changes, e.g., aggregation and vacuolization, can be detected in rats and mice by means of electron microscopy about 30 minutes after exposure to several hundred roentgens (*32*). Later, in the postirradiation period, swelling of mitochondria and damage to mitochondrial membranes are also noted (*33*, *233*, *234*). The radiosensitivity of hepatic mitochondria is determined to some degree by the location of the cells harboring these organelles in the liver lobule. Thus, mitochondria in the central and median portion of the lobule swell after an exposure to 500 R, whereas those at the periphery of the lobule show no such effect after exposure to 500 R (*168*). Moreover, the number of mitochondria per cell may also decrease after radiation doses of 1000 to 9600 R (*32*). The crucial question which remains to be answered is the following: Can the early structural changes be correlated with biochemical alterations, and what, if any, influence do these changes have on the subsequent fate of the cell?

The effect of irradiation *in vivo* or *in vitro* on the oxygen consumption of liver (*215*) has captured the interest of many investigators ever since manometric techniques became available. The results of many of these studies have been reviewed on several occasions, but it is still difficult to reconcile the numerous contradictory results (*11, 203, 205, 241*). These discrepancies are probably only apparent in many instances and may be caused by differences in experimental conditions, nutritional state, etc.

Many of the reported changes of oxygen consumption and associated alterations in the activity of various respiratory enzymes may represent secondary effects of radiation or may be abscopal in nature because they respond differently to local and total body irradiation (*128, 215, 231*). Recently, the intracellular redox state in irradiated liver cells has been reviewed critically, and experiments on the formation of NADH *in vivo* during aerobic–anaerobic transitions have shown that the NAD^+ to NADH ratio is not altered, not even by supraletahal doses of radiation (*128, 305*).

The NAD^+, and, even more, the NADH content of liver decreases several days after exposure (*131, 141, 186, 235, 254, 256, 257, 305*), probably as a result of impaired biosynthesis of pyridine nucleotides (*257*). Since the NAD^+ to NADH ratio remains largely unaltered after irradiation, this part of the electron transport chain is apparently radioresistant.

D. Oxidative Phosphorylation

If one employs the P to O ratio* as a criterion of the efficiency of oxidative phosphorylation, one finds a marked depression of this ratio after *in vivo* irradiation only in radiosensitive organs, particularly those of the lymphoid group (*14, 232, 262*). Little change in P to O ratio is noted in relatively radioresistant organs like liver (*5, 231, 232, 262, 268, 272, 283*). However, some investigators found diminished P to O ratios in normal liver several hours after total body irradiation (*18, 22, 49a, 96, 114, 135c, 193, 233, 280, 290, 296, 297, 299, 300, 302, 303*) or local (*226*) irradiation and in regenerating liver 24 or 72 hours after exposure at a time when DNA is being synthesized (*290*), i.e., 24 hours after surgery. It appears, therefore, that in normal liver a depression of the P to O ratio is not detectable until the late postirradiation period and is generally attributable to the technique of isolating mitochondria and/or artifacts introduced in the course of the preparative procedures, rather than to a direct effect of irradiation. Secondary effects, such as nutrition, sex, species, length of time after exposure, or temperature of storage, may further contri-

* The P to O ratio is defined as μmoles of inorganic phosphate esterified per μatoms of oxygen consumed.

bute to the reduction in efficiency of oxidative phosphorylation. A compilation of many investigations pertaining to the effect of ionizing radiation on liver mitochondria as well as details of isolation and purification of mitochondria can be found in a recent publication (*262*).

There are only three reports which describe an early transient decrease of the P to O ratio concurrent with structural changes in mitochondria of rat (*114*, *198*) and mouse (*280*) liver, and, contrary to many other studies on mitochondria of low P to O ratio, these three investigations utilize mitochondria with a high efficiency of oxidative phosphorylation. These observations have not been confirmed (*267*).

Trauma inflicted during isolation and/or storage of mitochondria can reduce the efficiency of oxidative phosphorylation and amplify minor latent radiation damage so that it becomes detectable. The presence of trauma as a result of these factors is frequently revealed by a low P to O ratio in corresponding nonirradiated controls. Mitochondria which have sustained radiation injury might be sufficiently unstable so as to be destroyed in the course of isolation; consequently, severely damaged mitochondria might escape biochemical scrutiny. One may assume that metabolically intact mitochondria have been utilized in an investigation when (i) the yields of mitochondria from irradiated and nonirradiated liver are identical, (ii) irradiation does not alter the distribution of subcellular particulates with respect to size of enzymatic activity (*262*), and (iii) the P to O ratio in irradiated and control mitochondria approaches the theoretical value. The purity of mitochondria is another important factor in the radiation response. Diverse methods of isolation can yield products of various degrees of purity; they sometimes contain contaminants of extramitochondrial origin (nuclei, lysosomes, and microsomes). Contamination with lysosomal acid phosphatase may considerably lower the P to O ratio by reducing the concentration of phosphate acceptors (*262*); under these conditions a high P to O ratio cannot be maintained for a prolonged period of time.

Mitochondria of certain organs, especially of lymphoid tissue, show a radiation-induced decrease in P to O ratio only when they have been aged 15–30 minutes at 0°C and, thus, have been exposed to a metabolic stress (*229*). Aged mitochondria are obviously no longer normal with respect to their metabolism, particularly since they are apt to leak cytochrome *c*.

The strength of cytochrome *c* binding to mitochondria varies from one tissue to another; it is greater in mitochondria of liver and Ehrlich ascites tumor cells than in those of thymocytes (*229*). Whether this variation bears any relation to the differential response of mitochondria of thymus compared to those in liver is a matter of conjecture. Nevertheless, even in radiosensitive organs (spleen and thymus) *in vivo* irradiation is not followed by a fall of the P to O ratio unless the cytochrome *c* content of mitochondria is low; as the cytochrome *c* content decreases, the efficiency of oxidative phosphorylation declines (*262*) In general, *in vivo* exposure to radiation does not alter the permeability of liver mitochondria to NAD and other cofactors, even after exposure to high doses (*230*).

Except for two reports to the contrary (*49a*, *82*), *in vitro* irradiation of mitochondria or homogenates of liver is without effect on the P to O ratio at dose levels in the range of an LD_{50-100} (*5*, *267*, *301*), and doses as high as 10^5–10^6 R are required to uncouple oxidative phosphorylation (*215*). Not all radiation-induced changes in liver mitochondria require such a high dose. After exposure to only 5 kR, K^+ is lost when mitochondria are incubated in a medium of low K^+ content and a slightly higher *in vitro* dose (10–20 kR) blocks the uptake of Ca^{2+} by liver mitochondria incubated in a medium of low Ca^{2+} concentration (*293*). The results of these *in vitro* experiments do not provide an adequate basis for estimating the extent to which similar alterations in ion transport play a significant role under *in vivo* conditions. The transport of Ca^{2+} across the mitochondrial membranes appear to be more sensitive to *in vitro* irradiation than the oxidative phosphorylation to which Ca^{2+} is coupled.

Oxidative phosphorylation in isolated liver mitochondria after *in vivo* irradiation may be summarized as follows:

(i) The efficiency of oxidative phosphorylation remains unaltered during the first few hours after exposure to lethal or sublethal doses but decreases after 12 to 24 hr. The latter phase does not represent a direct effect of radiation but more likely is ascribable to damage suffered during isolation (*262*).

(ii) Irradiation has no effect on oxygen consumption or esterification of inorganic phosphate during the first 4 hr after exposure.

(iii) *In vivo* irradiation does not produce changes in swelling, permeability, or particle size during the very early postirradiation period, although vacuolization and aggregation of mitochondria may occur as early as 30 minutes after exposure.

(iv) *In vitro* irradiation at dose levels below 10^5 R does not affect oxidative phosphorylation (*215*).

E. Lipid Peroxides and Lysosomal Damage

Peroxide-like substances in tissues of irradiated animals have often been detected (*107*, *210a*, *304*, *306*), but they have not yet been proved to be lipoperoxides. Malonaldehyde is formed by specific fragmentation of long-chain fatty acid peroxides and can be utilized for the quantitative determination of lipoperoxides (*210a*, *294*). Radiation-induced lipoperoxidation occurs in triglycerides (*107*) but not in phospholipids (*169*) and is promoted *in vitro* by nonheme ferrous ion (*294*). The latter effect, which depends on the radiation dose, is not yet fully understood.

Lipoperoxides generally are present in tissues at a low level because water-soluble antioxidants react with them (*293*). This factor makes determination of peroxide-like substances difficult and may also be responsible for the fact that tissue concentration of lipoperoxides cannot be correlated with the degree of radiation damage.

Formation of lipoperoxides is related to the effects of irradiation on membranes of subcellular particles, especially lysosomes (*294*). Liberation of lysosomal hydrolases, as well as disruption of lysosomes, takes place after relatively low doses of *in vivo* irradiation (100–700 R) (*60*, *93*, *94*, *102*, *201*) but requires much higher doses *in vitro* (5–20 R) (*294*). Various factors influence the release of enzymes from *in vitro* irradiated lysosomes of rat liver (*294*). (i) Anaerobiosis drastically reduces, whereas high O_2 pressure promotes, enzyme liberation; (ii) ascorbic acid and ferrous ions increase enzyme release, whereas vitamine E suppresses it; (iii) the liberation of enzymes is maximal 20 hours after irradiation.

The peroxidative process antecedes the release of enzymes from *in vitro* irradiated lysosomes (*294*) and may be an obligatory step leading to the disintegration of the membrane. The lysosomal membrane is single-layered, and, perhaps for this reason, it is more radiosensitive than the double-layered membrane of mitochondria. The latter contains considerable amounts of linoleic and arachidonic acid in the phosphatidyl choline fraction, and these polyethenoid acids may afford protection to the membrane (*126*).

F. Amino Acid Metabolism

The liver has a pivotal function in the metabolism of amino acids and proteins. It takes up free amino acids from the portal circulation and converts them into other vital metabolites, disposes of them, or incorporates them into liver and serum proteins. Excessive protein catabolism, initiated primarily by the destruction of radiosensitive cells after exposure to lethal or sublethal doses of irradiation, is one of the salient features of the post-irradiation period. The disruption of the cellular structure causes activation and liberation of proteolytic enzymes (*111*), particularly of cathepsins; this leads to an *in vivo* autolysis in lymphoid tissue, for example, where interphase death after irradiation is common. There is no increased catheptic activity in liver (*111*) since irradiation does not destroy hepatic cells, although it has been claimed that extracts of liver of irradiated rats have a greater "digestive power" than those of non-irradiated animals (*111*).

The transient expansion of metabolic pools of amino acids can be asssessed *in vivo* by administering a compound which "traps" the metabolite in question and, after conjugation, is excreted in the urine. In this way, the metabolic pool of "free" glycine can be sampled as urinary hippuric acid formed by conjugation with benzoic acid in liver (*236a*). This technique has also been applied to other problems in radiation biochemistry, e.g., to the "trapping" of ribose released from RNA by means of imidazolyl acetate (*87*).

When the body proteins are labeled with glycine-^{14}C prior to irradiation, it can be demonstrated, by determination of urinary hippuric acid, that the pool of free glycine increases as a result of tissue protein degradation (*86*, *148*). These results have been confirmed in studies on hippuric acid synthesis in isolated perfused rat liver (*89*) (Table V-7).

TABLE V-7

HIPPURIC ACID SYNTHESIS IN ISOLATED PERFUSED LIVER OF NORMAL AND X-IRRADIATED RATS[a]

Experimental conditions	Hippuric acid synthesis[b]		Synthesis of urea[b]	Initial utilization of glucose[b]	Liver weight (gm)
	No glycine added	Glycine added			
Nonirradiated control	2.5 ± 0.5[c]	9.7 ± 1.2[c]	15	47	8.5
X-irradiated[d] (1000 R)	4.2 ± 0.5	6.1 ± 0.9	14	50	8.3

[a] Adapted from G. B. Gerber and J. Remy-Defraigne, *Intern. J. Radiation Biol.* **7**, 53 (1963).

[b] All values are expressed as μmole/hr/gm of liver. Only the initial decrease in glucose concentration has been used for this computation, since concentration of glucose attains a stable value in 2 to 3 hr.

[c] The differences between irradiated and nonirradiated livers with respect to their capacity for hippuric acid synthesis are significant at $P \leq 0.015$.

[d] X-irradiated 24 hr before sacrifice.

The concentration of free amino acids in liver is not as markedly altered as it is in spleen, and such increases as do occur do not exceed 150% of the control values (*259*). The hepatic concentration of serine rises and that of ethanolamine, the decarboxylation product of serine, falls after irradiation, paralleling the decline in activity of certain amino acid decarboxylases (*8*, *147*, *259*) (compare also the decrease urinary excretion of ethanolamine and phosphatidyl ethanolamine in Chapter VIII). Other aspects of changes in amino acid metabolism after irradiation are discussed in the chapter on body fluids (cf., tryptophan and taurine).

Several interesting observations pertain to the metabolism of urea in irradiated animals. On the first day after irradiation, urinary excretion of urea increases (*81*, *132*, *261*, *292*) (cf. Chapter VIII). Subsequently, the content of urea (*284*) and ornithine (*259*) decreases for several days after total body irradiation of mice with 690 R; concurrently, the ammonia content rises. If, however, ornithine is injected 15 minutes before irradiation, the ammonia content of the liver decreases, whereas urea is not affected (*284*). This effect of ornithine is, however, not demonstrable in the isolated perfused liver of an irradiated rat (*177*).

The effect of ornithine on ammonia concentration (*284*) is difficult to explain since this amino acid has only a catabolytic function in urea synthesis and is regenerated upon completion of the Krebs–Henseleit cycle. The increase in hepatic ammonia is an unexpected finding in view of the great capacity of the liver to deal with an excessive load of this metabolite, even under conditions of impaired liver function.

Ammonia could be derived from increased breakdown of urea by intestinal bacteria or from impaired urea synthesis. Degradation of urea is not increased after irradiation, even after doses that cause lethal damage to the gastrointestinal tract (cf. Chapter IV), and there is no significant accumulation of urea in tissues until the animal is moribund. Changes in urea synthesis might involve the interference by irradiation in the synthesis of carbamoyl phosphate or glutamine, but nothing is known so far concerning the effect of irradiation on these metabolic reactions. The demonstration of a rise in liver ammonia levels in species other than the mouse, to supplement the investigations on the mouse (*284*) as well as an extension of the studies on urea synthesis after irradiation, would be desirable.

The liver is the main anatomical site for the conjugation of metabolites and the detoxification of drugs and other substances foreign to the body. The reactions of conjugation involve, in many instances, the same reaction partners, e.g., sulfate, glucuronate, and acetyl. A defect in detoxification has been postulated by some investigators as a possible factor in radiation sickness. The data assembled in Table V-8 show, however, that irradiation influences only moderately the ability to conjugate. It should also be mentioned that assaying activities of these enzymes yields only an estimate of the potential capacity of the liver for conjugation. It is unlikely that demands for conjugation of such large amounts of substrate would ever arise after irradiation.

G. Aminotransferases

Aminotransferases play a central role in determining the metabolic fate of amino acids, since they permit, in conjunction with L-glutamate dehydrogenase, conversion of the amino N of α-amino acids to NH_3 and, thus, provide a means for the biosynthesis of nonessential amino acids. Aminotransferases, especially those in liver and muscle, respond readily to tissue damage and may be released into the bloodstream (see Chapter VIII). The activity of these enzymes is influenced by the nutritional state and increases markedly during

TABLE V-8

THE EFFECT OF IONIZING RADIATION ON THE CAPACITY OF MAMMALIAN LIVER TO CONJUGATE METABOLITES AND FOREIGN SUBSTANCES[a]

Type of reaction	Time, dose, and other experimental conditions	Nature of the assay system	Observations	References
Bilirubin glucuronide formation	24 hr after 1000 R total body x-irradiation	Isolated, perfused liver (rat)	Slight decrease in glucuronide formation (see also Chapter VIII)	*85b*
Total glucuronide formation involving endogenous metabolites or toxic substances	Total body γ-irradiation (^{60}Co), 1000 R	Intact rat; glucuronides excreted in urine	Total glucuronide excretion decreased immediately after irradiation; lowest level 3–4 days postirradiation (50% of controls). Recovery at 9 days; irradiation also increased sensitivity of rats to intragastric naphthalene	*48, 49*
o-Aminophenol glucuronide formation	Local x-irradiation of one liver lobe *in vivo* 400 or 1200 R; liver slices incubated with *o*-aminophenol	Rat, irradiated *in vivo*; slices tested *in vitro*	Increased capacity for conjugation on first day, followed by a slight decrease and finally "overshoot" phenomenon	*118b*
N-Acetyl-*p*-aminophenol sulfation	(1) 1000 R over right side of liver (single dose); administration (NAPAP[b]) 24 hr or 17 days postirradiation; (2) fractionated dose of 20 × 300 R on successive days	Intact rabbit—excretion of phenolic sulfate in urine after sulfate loading, 24 hr postirradiation or after the last of 20 irradiations	Decreased conjugated sulfate excretion 24 hr after irradiation; levels normal at 14 days; no retention in blood; slight increase in conjugated sulfate excretion on 17th day Multiple exposures result in slight increase 14 days postirradiation	*26*
BSP conjugation[c]	1200 R, total body x-irradiation; biliary excretion of BSP, etc., studied 6–48 hr postirradiation	Rat, intact, but with biliary fistula	Total BSP elimination unchanged, BSP-glutathione conjugate elimination uniformly decreased; free BSP increased; *in vitro* assessment of mercaptide-forming capacity in liver homogenates decreased 20% of controls	*291*

	Local irradiation of liver—single doses of 500 or 1000 R or 15–20 × 200 R; studied up to several months postirradiation	Rabbit, BSP elimination by sampling blood 3–15 min after IV[d] administration of BSP	Elimination rate decreases with doses up to 2000 R (single exposure); minimum dose 500 R for detectable effect, 4000 R for a fractionated dose	*26a*
Aromatic amine acetylation				
(1) Sulfanilamide	450 or 750 R total body x-irradiation; acetylating capacity tested	Acetylating capacity tested in rat liver homogenates	Progressive decrease of acetylating capacity, 6–12 hr after irradiation; constant to 30 days postirradiation; gradual return to normal	*113*
	450 R total body x-irradiation	Intact rat irradiated, *in vitro* test of capacity using acetate-2-^{14}C as acetylating agent precursor	Acetylation capacity significantly decreased 3 days postirradiation; cysteamine further decreased acetylating capacity	*116*
	650 or 3320 R total body x-irradiation	Intact rat, urinary excretion studied	No change with 650 R, but less total drug excreted after exposure to 3320 R (50% of controls); the amount of acetylated drug excreted did not change	*221*
(2) *p*-Aminobenzoic acid	Sublethal dose	Intact rat, urinary excretion studied	Decreased excretion of *p*-amino benzoic acid on the 3rd and 6th postirradiation day; no change in acetylated drug excretion	*75*
Conjugation of bile acids	24 hr after 1000 R total body x-irradiation	Isolated, perfused rat liver	No change	*85a*
Formation of hippuric acid	24 hr after 1000 R total body x-irradiation	Isolated, perfused rat liver	Slight decrease with excess glycine (see also Table V-7)	*89*

[a] For radiation effects on certain drug-detoxifying or -metabolizing enzymes see Table V-10. Unless otherwise stated, all exposures were single, total body x-irradiation.

[b] NAPAP, *N*-acetyl-*p*-aminophenol.

[c] BSP, Bromsulphalein.

[d] IV, intravenous.

TABLE V-9

EFFECT OF IRRADIATION ON HEPATIC AMINOTRANSFERASES AND RELATED ENZYMATIC ACTIVITIES[a]

Aminotransferases or related enzymes	Species	Nutritional status	Radiation dose (R)	Sacrifice, days after irradiation	Radiation-induced changes in enzymatic activity	References
1. Aspartate aminotransferase (E.C. 2.6.1.1)						
(a) Total activity	Rat	?	100	1, 2, 24[b]	Decreased	*211*
	Mouse	Fed	450	2, 4, 14	Increased 28%	*34*
					Increased 65%	*34*
	Mouse	Fed	530	Up to 12	No change	*271*
	Rat	Fed	600	10	Slight decrease	*35*
	Rat	Fed 60% glucose or fructose; then starved 24 hr	900	1	Decreased 16%, fructose feeding; no change, glucose feeding	*79*
	Rat	Pair-fed	1000	2, 4	Increased 40–50%	*38*
(b) Cytoplasmic activity	Mouse	Starved	690	1–12	No change	*260*
	Mouse	Fed	690	1, 6	Increased 25%	*260*
	Mouse	Starved	810	1, 3, 5	No change	*260*
	Mouse	Starved	9000	1, 3	Increased 30%	*260*
(c) Particulate activity	Mouse	Fed	690	1, 3, 6	No change	*260*
	Mouse	Starved	690	8–10	Decreased 30%, 9th day	*260*
	Mouse	Starved	810, 9000	1, 3, 5	No change	*260*
2. Alanine aminotransferase (E.C. 2.6.1.2)						
(a) Total activity	Rat	?	100	1, 2, 24[b]	Decreased	*211*
	Mouse	Fed	450	2, 4	Increased 27%	*34*
	Rat	Fed	600	10	Decreased 50%	*35*

	Rat	Fed 60% glucose or fructose, then starved 24 hr	900	1	Decreased 20–30% on 60% glucose or fructose diet	*79*
	Rat	Pair-fed	1000	2, 4	Increased 40–50%	*38*
(b) Cytoplasmic activity	Mouse	Fed	690	1, 3, 6	No change	*260*
	Mouse	Starved	690	1, 3	Increased 10%	*260*
	Mouse	Starved	810, 9000	1, 3	Increased about 15%	*250*
(c) Particulate activity	Mouse	Fed	690	1, 3, 6	No change	*260*
	Mouse	Starved	690	8–10	Decreased 16–29%	*260*
	Mouse	Starved	810, 9000	1, 3, 5	No change	*260*
3. Tyrosine aminotransferase (E.C. 2.6.1.5)	Rat	Starved	200[c]	2	No change	*9*
	Rat	Starved	600[c]	2	Increased 18%	*9*
	Rat	Starved	600[c,d]	2	Increased 23%	*9*
	Rat	Starved	1200[c–e]	2	Increased 15%	*9*
	Rat	Starved	2000[c,d,f]	2	Decreased 21%	*9*
4. Aspartate carbamoyl-transferase (E.C. 2.1.3.2)	Rat	?	600[d]	?	Increased	*173*
5. Kynurenine aminotransferase (E.C. 2.6.1.7) (see also Chapter VIII)						
(a) Cytoplasmic activity	Mouse	Starved	690	12[b]	Increased 90%	*211, 258*
	Mouse	Starved	690	1	Decreased 29%	*211, 258*
	Mouse	Starved	690	3	Increased 165%	*211, 258*
	Mouse	Starved	690	12	Increased 170%	*211, 258*
(b) Particulate activity	Mouse	Starved	690	3	Increased 65%	*211, 258*
	Mouse	Starved	690	6, 8, 9	Increased about 40%	*211, 258*
6. L-Glutamate dehydrogenase (E.C. 1.4.1.2)						
(a) Cytoplasmic activity	Mouse	Fed or starved	690	1, 3, 6	No change	*260*
	Mouse	Starved	810, 9000	1, 3, 5	No change	*260*

TABLE V-9 (contd.)

Aminotransferases or related enzymes	Species	Nutritional status	Radiation dose (R)	Sacrifice, days after irradiation	Radiation-induced changes in enzymatic activity	References
(b) Particulate activity	Mouse	Fed	690	1, 3, 6	No change	*260*
	Mouse	Starved	690	7–10	Decreased 23–37%	*260*
	Mouse	Starved	810, 9000	1, 3, 6	No change	*260*

[a] Unless otherwise stated, animals are exposed to one total body x-irradiation. The statement "No change," in most cases, is based on statistical inference.
[b] Time in hours.
[c] Partial body irradiation of abdomen.
[d] γ-radiation from ^{60}C source.
[e] Administered in two equal doses on two successive days.
[f] Administered over a period of 14 days, 10 daily doses of 200 R.

starvation (*260*). Nevertheless, starvation is not the sole factor responsible for the changes in hepatic aminotransferase activities after irradiation (see Table V-9). This is shown by the marked differences in the behavior of various aminotransferases; also time and species influence radiation-induced enzymatic changes. Alanine aminotransferase seems more radiosensitive than aspartate aminotransferase and the activity of the former, but not of the latter, enzyme increases in livers of rats fed a high glucose (60%) diet after exposure to 900–1000 R (*38*, *70*). Exposure of fed mice to 450–690 R increases aspartate and alanine aminotransferase activity (*34*). Rats given a comparable dose exhibit a decrease in these enzymatic activities 4–6 days later (*35*), whereas after higher doses (1000 R) (*38*) the activity of aminotransferases is enhanced. It is not clear whether these differences are attributable to variations in radiosensitivity of the two species, to conditions of enzymatic assay, or to the nutritional status of the animals.

Leakage of enzymes from subcellular particulates into cytoplasm appears not to be responsible for changes in activity of hepatic aminotransferases. This is deduced from the data pertaining to L-glutamate dehydrogenase, an enzyme confined to mitochondria. This enzyme is not released into the cytoplasm after irradiation. Since aminotransferase activity in mouse liver appears to diminish when expressed "per gram of tissue" and since there is a concomitant decrease in protein content (*260*), one may conclude that these changes reflect a decrease in the number of organelles rather than diminished enzymatic activity per organelle.

Are the changes in hepatic aminotransferases after irradiation significant with regard to amino acid metabolism? It appears questionable that these changes are important under *in vivo* conditions since they are small, species dependent, and do not affect ammonia formation (*284*).

H. Protein Metabolism

Total body irradiation has been reported to enhance incorporation of labeled amino acids into proteins of normal liver (*6*, *8a*, *28*, *42*, *119*, *120a*, *121*, *122*, *134*, *166*, *219*, *220*, *227*), but agreement on this point is by no means universal (*125*, *197*). In regenerating liver, a decrease in amino acid incorporation into proteins has been found. For example, the *in vitro* incorporation of D,L-leucine-1-^{14}C into nuclear proteins is depressed 24 hours after partial hepatectomy if irradiation is given 2–6, but not 16 hours, after surgery (*242*). Similar observations have been made for *in vivo* incorporation of L-leucine-1-^{14}C into cytoplasmic proteins of regenerating liver (*227*). However, *in vitro* x-irradiation of liver homogenates with doses of 5 kR does not affect the incorporation of D, L-alanine-^{14}C (*237*). It should be pointed out that data on specific activities of amino acids are often not available for *in vivo* studies. It is, therefore, difficult to determine whether changes in the level of incorporation of labeled precursors involve alterations in protein synthesis, amino acid pools, or both. Moreover, there exists a considerable variety of experimental conditions which have been employed in the study of radiation effects on protein synthesis; under these conditions it is uncertain whether protein synthesis proceeds at optimal levels.

Some efforts have been made to shed further light on radiation-induced changes in protein synthesis by exploring radiation effects at the level of the ribosomes (*8a*, *122*, *297a*). In normal, nonregenerating liver of rats or guinea pigs x-irradiation with doses of 600–6000 R, the ratio of polysomes to monomeric ribosomes increases at 12–24 hours after irradiation (*8a*, *122*). At 24 hours after exposure the heavy polysome fraction increases by about 10% (*8a*), presumably at the expense of the lighter particles. The capacity of these polysome-enriched fractions from x-irradiated liver to incorporate labeled amino acids into protein is enhanced (*8a*, *122*) to the extent of 50% above nonirradiated control levels after exposure to 1000 and 3000 R but only to the extent of 22% after exposure to 4000 R (*122*).

According to one group of investigators (*8a*), protein synthesis in both regenerating liver and nonregenerating liver is similarly affected by total body x-irradiation, but this conclusion is not shared entirely by another group (*297a*) who report that, under slightly different experimental conditions with respect to radiation dose and temporal relationships, x-irradiation suppresses the appearance of the excess heavy polysomes associated with regenerating liver when rats are exposed to 1500 or 6000 R total body x-irradiation at 6, 9, or 12 hours after partial hepatectomy and sacrificed 24 hours after surgery. If a longer period, e.g., 36 hours, intervenes between the time of irradiation and sacrifice, recovery in the form of heavy polysome reappearance is observed (*297a*). It is unlikely, at least in nonregenerating liver, that starvation is responsible for the changes observed after irradiation since starvation per se brings about an increase of monomeric ribosomes (*122*). It appears that there is evidence for increased incorporation of amino acids into liver proteins as a result of irradiation, at least for the nonregenerating liver. As for the regenerating liver, the divergent results (*8a*, *297a*), may, in part, be attributable to differences in experimental conditions.

Synthesis of plasma proteins, except the γ-globulins, represents one of the major metabolic functions of mammalian liver. The concentration and metabolism of plasma proteins have frequently been studied after irradiation since these macromolecules are readily accessible (Fig. V-6).

Radiation-induced changes in plasma proteins are also considered in Chapter VIII of this book, and the present discussion is limited to those aspects germane to liver, namely, hepatic synthesis and catabolism of these proteins. Recent experiments have established that net biosynthesis of fibrinogen and

α_1-acid glycoproteins increases in isolated rat liver perfused with rabbit blood 4–6 days after irradiation (*178*). This was demonstrated by measuring simultaneously the uptake of L-lysine-1-^{14}C into these proteins and the quantity of the rat proteins by specific immunological techniques. The increase in net synthesis of α_1-acid-glycoprotein is of considerable interest since it might serve as a useful indicator of radiation damage. In the isolated perfused rat liver, incorporation of tryptophan (*216*) or tyrosine (*228*) is enhanced after irra-

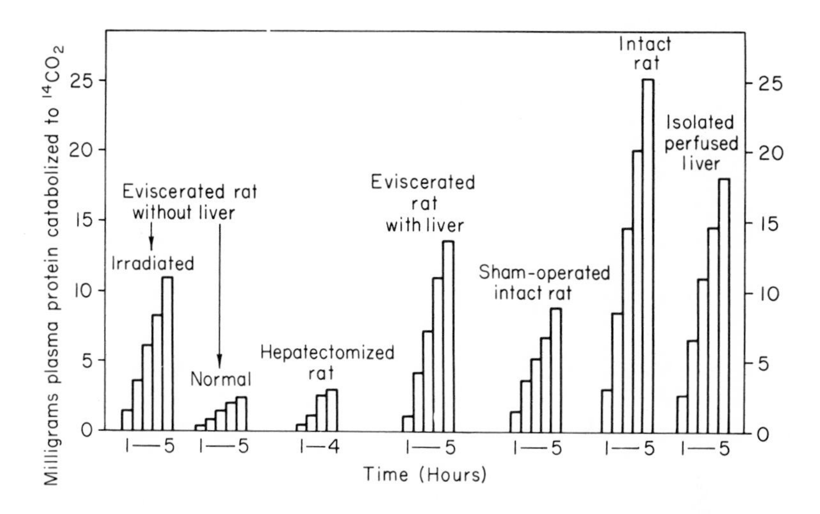

FIG. V-6. Rats were eviscerated 1 to 2 days after x-irradiation (750 R, total body) and injected with leucine-1-^{14}C-labeled plasma protein immediately after evisceration. These animals were sacrificed within 1 to 5 hours after injection of plasma protein. From L. L. Miller, "Entstehungsorte, Speicherorte und Umsatz der Serumproteine." *Verh. Dtsch. Ges. inn. Med.* **66**, pp. 250–260. München: J. F. Bermann 1960 (*176*) and (*177*).

diation with respect to proteins which remain at normal or elevated levels, whereas no change in labeling occurs with respect to proteins which fall to a lower level.

Catabolism of serum proteins in liver appears not greatly altered after irradiation, as shown by experiments in which the isolated perfused liver was perfused for 12 hours with a mixture of ^{3}H-phenylalanine and ^{131}I-labeled serum proteins (*91*, *217*). Since neither synthesis nor catabolism in liver can explain the decrease in serum proteins after irradiation, extrahepatic catabolism, or loss into the extracellular space, must be responsible. Indeed, labeled serum proteins disappear more rapidly from the circulation after irradiation (*217*, *243*) and ^{14}C proteins are degraded to CO_2 more rapidly in irradiated eviscerated preparations (*178*) than in nonirradiated ones.

TABLE V-10

EFFECTS OF IONIZING RADIATION AND OTHER INHIBITORS ON THE SUBSTRATE- AND HORMONE-INDUCED INCREASE IN THE ACTIVITY OF SEVERAL HEPATIC ENZYMES[a]

Enzyme	Stimulant	Effect of ionizing radiation	Effect of actinomycin D	Effect of puromycin	Experimental conditions and related aspects	References
L-Serine hydro-lyase (deaminating) (E.C. 4.2.1.13) L-serine dehydratases	Casein hydrolysate (stomach tube) to protein-depleted dogs	Inhibition[b] if irradiated within 1 hr after casein feeding; no inhibition at 7 hr	Inhibition	—	Rats exposed total body to γ-radiation (^{60}Co); dose range: 400–3200R; enzyme radiosensitive only in the early postinduction period; radioresistent 6 or 16 hr after casein	*212–214*
L-Ornithine: 2- oxo-acid aminotransferase (E.C. 2.6.1.13) ornithine-keto acid aminotransferase	Casein hydrolysate	Inhibition[b] (except at low dose levels)	Inhibition	Inhibition	Rats exposed to γ-radiation as above	*212–214*
Glucokinase (see also Table V-4)	High glucose diet	Inhibition (except after hypophysectomy)	—	—	Rats (1000 R)	*31*
L-Tryptophan: oxygen oxidoreductase (E.C. 1.13.1.12) tryptophan oxygenase ("pyrrolase")	None	Increased activity: 4 times control level 4 hr and 50–100% 72 hr postirradiation	—	—	Young rats (900 R)	*45*
	Tryptophan	Increased activity if irradiated immediately after tryptophan administration	—	—	Induction in liver homogenate of irradiated rats *in vitro* decreases enzymatic activity	*45*

None	Increased activity: 4 times control level 4 hr after total body x-irradiation (1000 R); normal at 48–72 hr;	—	—	Rats; effect dose-dependent to 5000 R; minimal inhibitory dose: 200 R; response abolished after adrenalectomy, although cortisone and hydrocortisone can increase enzymatic activity; enzyme response to tryptophan not affected by irradiation (600 R)	*264–266*
Tryptophan	15-fold increase in irradiated rats 4 days postirradiation	—	—		
Tryptophan	Inhibition in intact or regenerating liver	Inhibition	—	Rats, total body x-irradiated with 500 R; 6 hr postirradiation increases activity to 6 hr; 4 days postirradiation decreases inhibition in the intact liver, minimum at 15 days, normal at 25 days	*140*
Cortisone	No effect	Inhibition only in adrenalectomized rats	—		*140*
Tryptophan, α-Methyl tryptophan	Inhibition abolished by 1600 R	No effect	Inhibition	Adrenalectomized rats (800 or 1600 R) 48 or 24 hr before sacrifice and 5 or 11 hr after hormone- or substrate-type inducer injection; neither radiation dose affects hydrocortisone action	*105*
Hydrocortisone	No effect in fed, adrenalectomized rats	Inhibition	Inhibition		*105*

TABLE V-10 (contd.)

Enzyme	Stimulant	Effect of ionizing radiation	Effect of actinomycin D	Effect of puromycin	Experimental conditions and related aspects	References
L-Tryptophan: oxygen oxidoreductase (E.C. 1.13.1.12) tryptophan oxygenase ("pyrrolase")	Tryptophan	No effect	—	—	Five-day post-adrenalectomy rats injected with hormone- or substrate-type inducer 0–1 hr before γ-irradiation (1600 or 3200 R)	*213*
	Hydrocortisone	No effect				
	Tryptophan	No effect	—	—	Rats, 24 hr after sham- or x-irradiation (800R), and starvation, hydrocortisone or tryptophan injected IP; decreased enzymatic activity followed for 13 hr	*179*
	Hydrocortisone	Inhibition	—	—		
	Tryptophan	No effect 2–3 days postirradiation; 4–16 days thereafter activity increased 2–3-fold in irradiated group	—	—	Rat and rabbit, total body γ-irradiation (400 R, over 4–5 hr)	*108*
L-Tyrosine: 2-oxoglutarate aminotransferase (E.C. 2.6.1.5) tyrosine aminotransferase	α-Methyl tryptophan or pyridoxine	Inhibition, compared with starved rats (controls), dose: 800 R	No inhibition	Inhibition	Rats, fed or starved	*105*
	Hydrocortisone	No effect in fed or starved, irradiated rats (corrected for starvation effect)	—	—		

	Hydrocortisone	Stimulation (3200 R)	Radiation-induced stimulation of hormone induction inhibited	—	Rats, γ-irradiation followed by cortisone and/or actinomycin; enzymatic activity assayed 4 hr later	*213*
	Cortisone	No effect on cortisone induction	—	—	Rats (650 R)	*202a*
	Cortisone	Increased enzymatic activity after 600 R γ-irradiation, not after 600 R x-irradiation	—	—	Rats, x- or γ-irradiation of abdomen only; divided doses up to 2000 R (cumulative); enzymatic assays 48 hr after (last) irradiation and 18 hr after starting fast	*9*
L-Alanine: 2-oxo-glutarate aminotransferase (E.C. 2.6.1.2) alanine aminotransferase	None	Enhancement of activity 2 and 4 days after irradiation	—	—	Rats (1000–1500 R), pair-fed	*38*
ATP: thymidine-monophosphate phosphotransferase (E.C. 2.7.4.9) TMP kinase	Thymidine	No effect on activity after postirradiation (dTR administration)	Same as for ionizing radiation	—	Rats, (1500 R) 45 min before dTR administration; sacrificed 1 hr after dTR injection	*1*
Nucleosidetriphosphate: RNA nucleotidyltransferase (E.C. 2.7.7.6) Mg^{2+} activated, RNA nucleotidyltransferase	None	Activity 40% above controls, 24 hr postirradiation; thereafter decline below control level at 10 days	—	—	Liver nuclei prepared from total body, x-irradiated, fed rats (650 or 1000 R); similar results with fasted, but not with adrenalectomized rats; after correction for starvation effect, fed rats (600 R) are comparable to those cited (*202a*)	*202a* *10*

TABLE V-10 (contd.)

Enzyme	Stimulant	Effect of ionizing radiation	Effect of actinomycin D	Effect of puromycin	Comments: experimental conditions and related aspects	Reference
Phosphorothioate oxidase	None	Inhibition of enzyme development in 23-day-old male rats	—	—	Rats, given partial or total body x-irradiation (200 R): Shielding liver and testes has little effect but shielding head and testes reduces the inhibitory effect of irradiation, as does hypophysectomy; irradiation of head only inhibits enzymes	*122a*
	Phenobarbital	Inhibition in weanling rats, not in older animals (100–400 R)	—	—	Steroids (e.g., cortisone) may act like phenobarbital, this type of induction not being inhibited by irradiation. Similar data for hexabarbital-metabolizing enzymes (*189*) available	*62*

[a] Unless otherwise stated, all exposures are single, total body x-irradiations.

[b] The enzymes tryptophan pyrrolase, tyrosine transaminase, ornithine transaminase, and serine dehydratase can be induced by administration of glucagon. γ-Irradiation as well as actinomycin D inhibit glucagon induction (*213*).

I. Effect of Irradiation on the Formation and Activity of Adaptive ("Induced") Enzymes

Liver provides a superb arena for investigating the effects of radiation on synthesis of specific proteins induced by substrates or hormones. This approach makes it possible to focus on the synthesis of an enzyme protein and, thus, on a specific "messenger" RNA (mRNA). Moreover, one may compare the effect of irradiation with that of inhibitors at certain specific steps in protein synthesis and thus define more closely the site of action of radiation (Table V-10).

Mammalian liver contains several inducible enzymes and of these tryptophan pyrrolase has been studied most extensively. An increase in tryptophan pyrrolase activity can be induced by (i) certain hormones and (ii) tryptophan or its analogs. It is generally assumed that induction of enzymes by hormones entails increased synthesis of mRNA whereas induction by substrate acts on already existing mRNA (*104*). Actinomycin suppresses induction by hormones (*44*, *104*, *125*, *140*) but has little or no effect on induction by substrate. By way of contrast, puromycin, a general inhibitor of protein synthesis, inhibits induction of the enzyme by hormone as well as by substrate.

Total body irradiation blocks induction of tryptophan pyrrolase by substrate in normal and adrenalectomized rats, but apparently it has no effect on the induction by hormone (*125*, *140*, *264*, *265*). It must be pointed out, however, that radiation-induced inhibition of induction by hormone, but not by substrate, has also been reported (*179*). Induction by substrate is somewhat enhanced for about 6 hours after irradiation with 900 R (*44*, *266*) then diminishes markedly (*105*, *140*), and reaches its lowest value about 15 days after exposure to 500 R (*105*). This observation makes it unlikely that the inhibition of induction is a direct effect of radiation on mRNA.

Observations on other inducible enzymes complement those on tryptophan pyrrolase. Induction of tyrosine-α-ketoglutarate aminotransferase in rat liver by hydrocortisone is inhibited by actinomycin and puromycin but slightly activated in normal and more extensively (several-fold) in adrenalectomized rats after total body exposure to 50–3200 R (*214*); L-serinedehydratase induced by feeding casein hydrolysate is inhibited by actinomycin, as well as by total body exposure (400–1600 R). At certain periods at which actinomycin does not block enzyme induction, irradiation also is ineffective (*214*). These periods are thought to be related to changes in stability of mRNA during which the inductive process is resistant to the influence of actinomycin and irradiation. Synthesis of L-ornithine-δ-aminotransferase, like that of serine dehydratase, is controlled by intake and responds to actinomycin and irradiation in a way similar to L-serine dehydratase. However, the observations on induction of tryptophan pyrrolase and tyrosine aminotransferase by substrate or hormone indicate that the mechanisms of action of actinomycin and irradiation are not identical. Two major questions remain unanswered:

(1) By what mechanisms does irradiation inhibit substrate induction and how does irradiation affect the synthesis sites on existing templates?

(2) Why does irradiation not inhibit induction by hormones when this process is inhibited by actinomycin?*

J. Nucleic Acids

Most cells in normal adult liver are in interphase and do not divide. Consequently, incorporation of metabolic precursors into nuclear DNA proceeds at a very low level (*72*, *199*). However, as was recognized only recently, a detectable degree of DNA turnover exists in normal liver and is associated with a metabolically active species present in mitochondria (*46*, *84a*, *86a*, *90*, *98a*). The scarcity of parenchymal cells synthesizing DNA parallels the very low activity of cytoplasmic enzymes related to pyrimidine metabolism (*249*) and DNA synthesis. Nevertheless, irradiation of normal liver depresses incorporation of precursors into DNA in those few cells which are in S phase and into DNA of mitochondria in hepatocytes. The biosynthesis of DNA in mitochondria is, however, less radiosensitive than that process in nuclei (*46*, *84a*, *98a*). Nuclear DNA differs from mitochondrial DNA (mDNA) in the degree of utilization of labeled precursors and this may explain, in part, why x-irradiation reduces the incorporation of thymidine into total DNA to a greater extent than that of deoxycytidine. Another factor contributing to this phenomenon may be related to the higher adenine–thymine content of mDNA as compared with nuclear DNA (cited in Chang and Looney, *46*).

Cells irradiated in interphase, nevertheless, sustain radiation injury which becomes manifest only when the cells are stimulated to divide. On the basis of different rates of repair, at least two types of such latent injury may be distinguished: (i) delay in the onset of regeneration after partial hepatectomy is abolished within a few weeks (*4*, *44*, *124*, *125*, *175a*) and (ii) chromosome breaks, visible during induced cell division (regeneration), require several months to repair (cf. also Volume I, Fig. III, 44d) (*4*, *157*, *286*).

Extensive regeneration of the liver can occur when part of the organ is destroyed by chemical or surgical means. After excision of two-thirds of the liver, the remaining parenchymal cells begin to divide 24–36 hours later and the organ is restored to nearly its original mass and cell number within 2–3 weeks (*39*). Since regenerating liver represents a fairly uniform, semisynchronous cell population of rapid and controlled growth—about 90% of the parenchymal cells divide 24–28 hours after surgery—it is a very useful tool for studying the effects of radiation on the processes leading the cell from its resting stage, G_0,

* Recently, M. B. Yatvin and H. C. Pitot (in press) suggested that inducer availability is affected by radiation and/or actinomycin D action on gastrointestinal function and absorption from the application site. This factor may explain the divergent reports on hormonal and dietary induction.

TABLE V-11

BIOCHEMICAL PROCESSES IN REGENERATING LIVER AND THEIR ALTERATION BY IRRADIATION AT DIFFERENT TIMES AFTER PARTIAL HEPATECTOMY[a]

Experimental conditions	0–6 hr	6–12 hr	12–18 hr	18–24 hr	24 hr and later
Normal pattern of regeneration	Latency period: no significant biochemical change	Increased activity of RNA polymerase and synthesis of nuclear RNA; synthesis of cytoplasmic RNA enhanced later	Synthesis of enzymes needed for DNA synthesis (dTR and dTMP phosphotransferases, dCMP aminohydrolase, DNA polymerase)	DNA synthesis (S phase)	Mitosis, eventually followed by another cell cycle
Pattern of regeneration when irradiated at some time during the period indicated	Low doses (200 R) delay all regeneration-related reactions	Low to intermediate doses delay enzyme formation and DNA synthesis and depress synthesis of nuclear RNA and later of cytoplasmic RNA	Intermediate doses delay enzyme and DNA synthesis; no effect on enzymes already formed	High doses (1000 R) immediately depress DNA synthesis and delay mitosis; no immediate effect on enzymes or primer activity	Intermediate doses delay mitosis and affect subsequent cell cycle

[a] For reviews see reference *39*, *39a*, *99*, *99a*, *145ab*.

TABLE V-12

EFFECTS OF IONIZING RADIATION ON DNA METABOLISM IN MAMMALIAN LIVER[a]

Labeled precursor[b]	Injection of precursor (hr before sacrifice)	Species	Radiation dose (R)	Experimental conditions	Irradiation (hr before or after partial hepatectomy)[c]	Sacrifice (hr after irradiation)	Radiation-induced changes[b]	References
				Regenerating Liver				
$^{32}P_i$	3	Rat	150	*In vivo*; high protein diet 48 hr before surgery; amytal 30 min before irradiation of remaining lobe of liver, rest of the body shielded	+12 to +24	8–24	Irradiation at 12 hr delays DNA synthesis; had no effect at 18 hr; delayed second peak of DNA synthesis at 24 hr	*44*
	0.5, 1	Rat	200, 400	*In vivo*	+6, +12	12 or 18	Delay of DNA synthesis; accumulation of dCR	*127*
	3	Rat	450	*In vivo*, conditions as above for 150 R	+12 to +24		Irradiation at 12 hr abolishes first peak of DNA synthesis, at 18 hr, causes a marked decrease and at 24 hr a delayed second peak of DNA synthesis	*44*
	3	Rat	450	*In vivo*	+12	12	Normally expected peak delayed 10 hr; exposure before surgery has the same effect as exposure 12 hr after surgery	*124*, *125*
	3				+24 to +27	12	DNA synthesis barely started	
	3				+48	60	Delayed, decreased DNA synthesis (low dose effect)	

3	Rat	750 (head only, or total body)	*In vivo*	+2	17	Decreased DNA synthesis after total body or head irradiation; pituitary hormones after irradiation causes a further decrease in DNA synthesis and head shielding has a protective effect on DNA formation	*90a*
2	Mouse	800	*In vivo*; hepatic regeneration following CCl_4 administration	0 to +96 after CCl_4 administration		No changes when irradiated 24 and 48 hr after CCl_4; decreased S.A. DNA at 0, 12, 72, or 96 hr after exposure	*135a*
	Rat	800	*In vivo*	+24	15	Decreased S.A. and content of DNA	*263*
2	Rat	1000	*In vivo*; starved 24 hr before surgery; glucose feeding 24 hr after surgery	+28, +42, +47	2	Decreased DNA metabolism; at 42, 47 hr, irradiated RL is 61% and at 28 hr 26% of control	*204*
	Rat	1000	*In vivo*; one-half the radiation dose given at +2; remainder at +20 hr			No change in ^{32}P incorporation	*118a*
1	Mouse	2000	*In vivo*	+34, +70	1	Inhibition from 66 to 50% (34–72 hr)	*129*
2	Mouse	2000	*In vivo*; CCl_4-induced regeneration			No change in S.A. DNA	*135a*
3	Rat	2000, 2200	*In vivo*	+19 +24 +46	4 3 3	50% inhibition of DNA synthetic rate in the first and third, but 37% in the second case; no inhibition delays	*125*
	Rat	3000	*In vivo*; one-half the radiation dose given at +2; remainder +20 hr after surgery			Decreased incorporation of ^{32}P into DNA	*118a*

TABLE V-12 (contd.)

Labeled precursor[b]	Injection of precursor (hr before sacrifice)	Species	Radiation dose (R)	Experimental conditions	Irradiation (hr before or after partial hepatectomy)[c]	Sacrifice (hr after irradiation)	Radiation-induced changes[b]	References
Orotate-6-^{14}C	4	Rat	365	*In vivo*; fed	0, −28, −48 days	All sacrifices 27 hr after surgery	Inhibition of incorporation is 54, 71, 68% (of controls); degree of incorporation dose dependent	*4*
	24 or 30	Rat	375, 750	*In vivo*; starved 6 hr before surgery until sacrifice	+6 to +12 +12 to +18 +24 to +30	24	Marked inhibition of DNA labeling	*17*
	4	Rat	600	*In vivo*, fed	0, −6 days	All sacrifices 27 hr after surgery	Incorporation of orotate-^{14}C 19 and 34% of control, resp., at 0 or 6 days between irradiation and surgery	*4*
	24 or 30	Rat	1500, 3000	*In vivo*; starved 6 hr before surgery and thereafter until sacrifice	+6 to +12, +12 to +18 +24 to +30	24	Complete inhibition of DNA synthesis	*17*
	5–6	Rat	3000		−6 to −10.5	20–24	DNA synthesis decreases, similar to decrease when RL irradiated after surgery	
d-Thymidine (dTR-^{3}H)		Rat	800	Irradiated *in vivo*; nuclei incubated *in vitro* with dTR-^{3}H	+24	Up to 24	Progressive decrease of incorporation into DNA with time after exposure; 24 hr after irradiation response to 200–3000 R is the same	*152a*
	1	Rat	800	*In vivo*; starved after irradiation	+24	0–24	Decreased incorporation of dTR-^{3}H to below the level of nonirradiated controls	*152*

					+24	4.5–23.5	Rate of incorporation of dTR-^{3}H into DNA is more rapid in nonirradiated RL than in its irradiated counterpart	*153*
		Rat	1000	Isolated, perfused RL; Blood and liver donors irradiated immediately after surgery; perfused for 2 hr	Immediately before surgery	24	No effect on catabolism of dTR; decreased incorporation into DNA **(dTR-^{3}H, -CH_3-^{14}C)**	*90*
		Rat	1500	*In vivo* irradiation and *in vitro* incubation of subcellular fragments (nuclei, cytoplasm)	−24	6 or 24	Decreased incorporation of dTR-^{3}H into DNA; inhibition is greater in the 48-hr than in the 6-hr RL and proportional to log(dose); 700 and 3000 R were also tried	*142–145*
	1–2	Rat	1500	*In vivo* irradiation; mitochondrial and nuclear functions isolated	+21	0 or 8	Decreased dTR-^{3}H incorporation into mDNA and nDNA, the latter is more radiosensitive; at 6000 R, decrease in incorporation is immediate and greater than with the lower dose: S.A. mDNA 56%, S.A. nDNA 91% of controls	*45a*
	2	Rat	6000		+21			
Formate-^{14}C	3	Rat	2200	Local irradiation of remaining liver lobe, the rest of the rat shielded	+17 to +51	3	Decreased DNA synthesis at all stages of mitosis; purine and pyrimidine labeling equally affected	*169*
Formaldehyde-^{14}C		Rat	1500	Irradiated *in vivo*; response tested *in vitro*	+6 to +7 +15 to +16	17.5–18.5 9.5–8.5	Decreased utilization of formaldehyde (25% of controls)	*15*

TABLE V-12 (contd.)

Labeled precursor[b]	Injection of precursor (hr before sacrifice)	Species	Radiation dose (R)	Experimental conditions	Irradiation (hr before or after partial hepatectomy)[c]	Sacrifice (hr after irradiation)	Radiation-induced changes[b]	References
dCMP-^{3}H		Rat	1500	Irradiated *in vivo*; nuclei incubated *in vitro*	+24	+24	Decreased incorporation into DNA of nuclei	*144, 145*
dTMP-^{14}C, -^{3}H, -^{32}P dTTP-^{14}C -^{3}H, -^{32}P		Rat	600–1500	Irradiated *in vivo*; incubated *in vitro*. See also Table V–14	6–24	—	Substrates for enzymes in supernatant fraction of homogenates from x-irradiated animals	*29, 73, 144, 145* *29, 73, 145, 278*
Nonregenerating Liver								
$^{32}P_{i}$	2–4	Mouse	300, 800	*In vivo*; starved 24 hr before irradiation		0–6 days	Decreased S.A. DNA-independent of dose; return to control level 4–5 days after exposure; S.A. values very low, even in control DNA	*135b*
	2	Mouse	600	*In vivo*		2.3	Decreased S.A. of DNA	*209*
	2	Mouse	800	*In vivo*		1–48	Decreased S.A. of DNA to 20% of control values at 48 hr after exposure; at 1–3 hr decreased to 60%	*19*
		Rat	950	*In vivo*		6	Decreased incorporation into DNA soon after exposure	*120a*
	2	Rat	1000	*In vivo*; starved 2.5 hr before exposure		2	Decreased incorporation into DNA	*219*
				In vitro; uptake into DNA of isolated nuclei			Decreased incorporation into DNA in contrast with glycine	

	4	Rat	2440	*In vivo*; starved at time of irradiation	22	Decreased S.A. of DNA	*209*
	4.5		2500		28	Decreased S.A. of DNA	
	2–4	Mouse	2500	*In vivo*	0–6 days	Decreased S.A. of DNA	*135b*
d-Thymidine		Rabbit	300	Nuclei irradiated *in vitro*; incubated for 2 hr with dTR-^{3}H		Decreased uptake of dTR-^{3}H, 50% of control at the end of incubation	*160*
	1–30 days	Rat	500, 1000	*In vivo*; autoradiographic study supplementing radiochemical data		No significant change in labeling patterns	*88*
	1	Rat	800	Irradiation *in vivo*; incubation of nuclei *in vitro*	Up to 24	No effect on dTR-^{3}H incorporation	*152a*
		Rat	1000	Isolated, perfused liver; blood and liver donors irradiated immediately before transfusion	24	No significant changes	*90*
	2	Rat	1500	*In vivo*; mitochondrial and nuclear DNA isolated	0 or 8	A slight decrease in incorporation 8 hr after exposure	*45a*
5-Amino-4-imidazole carboxamide-5-^{14}C	24	Mouse	400	*In vivo*	108	S.A. of DNA purines slightly lower than their control values	*55*
D, L-Ureidosuccinic acid (ureido-^{14}C)	24	Mouse	400	*In vivo*	44, 68, 92	Decreased incorporation into DNA pyrimidines	*298*
Glycine-2-^{14}C	2	Rat	1000	*In vivo*; starved 2.5 hr before glycine given	2	Decreased uptake into DNA of isolated liver nuclei	*219*
Acetate-1-^{14}C	6	Rat	950	*In vivo*	6	Incorporation into purines depressed: one-half nonirradiated control value	*120a*

TABLE V-12 (contd.)

Labeled precursor[b]	Injection of precursor (hr before sacrifice)	Species	Radiation dose (R)	Experimental conditions	Irradiation (hr before or after partial hepatectomy)[c]	Sacrifice (hr after irradiation)	Radiation-induced changes[b]	References
Formate-^{14}C		Rat	5000	Irradiated *in vivo*; tissue minces incubated; rats starved 24 hr before exposure		17	Increased formate metabolism: increased labeling of nucleoprotein and acid soluble fractions	*215a*
Adenine-8-^{14}C		Rat	300	Isolated nuclei or homogenates irradiated *in vitro*; incubated for 90 min		2	Incorporation in DNA decreased 50%	*160, 161*
Adenine-4, 6-^{14}C	2	Mouse	800	*In vivo*; precursor injected 1–48 hr before exposure			S.A. of DNA decreased to <20% of control 48 hr after exposure, decreased to 60% at 1–3 hr	*19*

[a] Unless otherwise stated, all exposures are single total body x-irradiation. Unless mentioned otherwise, all partial hepatectomies are surgical.

[b] Abbreviations: P_i, inorganic phosphate; cRNA, cytoplasmic RNA; nRNA, nuclear RNA; mDNA, mitochondrial DNA; nDNA, nuclear DNA; S.A., specific activity; and RL, regenerating liver.

[c] + indicates that x-irradiation follows partial hepatectomy; − that partial hepatectomy follows x-irradiation.

via DNA synthesis, to mitosis. Of these processes preparation of the cell for DNA synthesis and DNA synthesis itself have been investigated most frequently.

The biochemical reactions associated with different phases of regeneration and effects of irradiation administered at various times after partial hepatectomy are summarized in Table V-11. In the following, we shall discuss certain of these aspects in more detail: The effect of irradiation (a) on the delay of regeneration when the liver is irradiated before or after surgery; (b) on the synthesis of RNA; (c) on the synthesis of enzyme proteins; and (d) on the synthesis of DNA on cells in S phase.

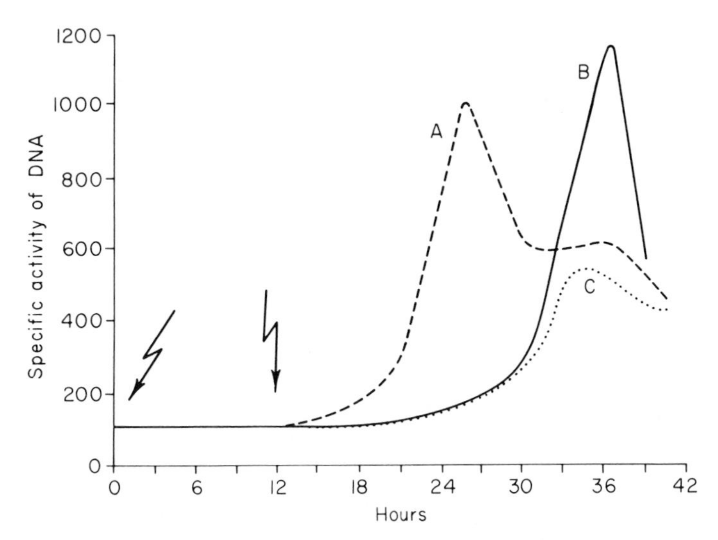

FIG. V-7. The rate of DNA synthesis in regenerating rat liver after total body x-irradiation. A, nonirradiated control; B, irradiated 12 hr after partial hepatectomy; C, irradiated shortly before partial hepatectomy; arrows, time of x-irradiation (450 R). Specific activity of DNA: cpm/0.1 mg DNA. Abscissa: Hours after hepatectomy. [From B. E. Holmes, *Ciba Found. Symp. Ionizing Radiation Cell Metab.*, p. 225 (1957).] (Fig. 1.)

(a) Exposure of the total body or of the region of the liver at the time of partial hepatectomy depresses the incorporation of a variety of precursors* into DNA 18–24 hours later, reduces the mitotic index at the time of maximal mitotic activity 24–36 hours after surgery (*157*), and causes chromosome breaks and mitotic irregularities (*4*, *157*, *286*). Other changes observed at appropriate times after partial hepatectomy and irradiation are an increase in cell size and a decrease in the number of cells, as well as a greater variation in the DNA content of the hepatocytes (*286*). When DNA synthesis and mitotic index are

* ^{32}P, thymidine-^{3}H, thymidine-^{14}C, adenine, etc. (see Table V-12).

TABLE V-13

EFFECTS OF IONIZING RADIATIONS ON RNA METABOLISM IN MAMMALIAN LIVERS[a]

Labeled precursor	Injection of precursor (hr before sacrifice)	Species	Radiation dose (R)	Experimental conditions	Irradiation (hr before or after partial hepatectomy)	Sacrifice (hr after irradiation)	Radiation-induced changes	References
				Regenerating Liver				
$^{32}P_i$	3	Rat	150, 450 (local irradiation of lobes)	*In vivo*; high protein diet 48 hr before partial hepatectomy; Amytal given ½ hr before exposure			RNA synthesis is unaffected except for an increase in 21-hr RL; incorporation into nRNA inhibited during early stages of regeneration	*44*
	2	Rat	700	*In vivo*	+2, +4, +6	6–10	Incorporation into nRNA inhibited during early stages of regeneration	*289*
	1–18	Rat	800 ^{60}Co γ-radiation	*In vivo*	+24	48	Increased incorporation into cRNA as well as into mitochondrial and microsomal RNA	*263*
	2	Rat	1000	*In vivo*	+28, +42, +47	2	Irradiation has little effect on cRNA and nRNA metabolism	*204*
	0.3	Rat	1500	*In vivo*	0 to +18	6	Increased ^{32}P uptake by rapidly labeled nRNA 0–6 and 12–24 hr, decreased uptake 6–12 hr after surgery	*279*
Orotate-6-^{14}C	0.3	Rat	600	*In vivo*	0 to +18	6	No inhibition of synthesis	*279*
	1	Rat	1000	*In vivo*	−24	30	No effect on cRNA and nRNA	*73*
	0.3	Rat	1500	*In vivo*	0 to +18	6	Greatest inhibition of orotate incorporation 24 hr	*279*

	0.17–0.5	Rat	600, 1000, 1500	*In vivo*; surgery 18–24 hr before sacrifice	+18	6	Decrease in S.A. of rapidly labeled RNA not as marked at lower dose levels as at 1500 R	*17, 278*
		Rat	1000	*In vivo*; 10 min pulse for nRNA, 45 min pulse for cRNA; 12, 15, 18 hr regenerating liver	−1		Maximal inhibition at 6–12 hr; at 15 hr marked decrease of cRNA synthesis; nRNA change less marked	*20, 21, 99a*
^{3}H-CTP		Rat	1000	*In vitro* application as part of nuclear RNA polymerase system; rats starved before surgery and x-irradiation	−1 or +4.5		Delayed increase in substrate utilization compared to control	*281*
^{14}C-ATP		Rat	1000, 1500	*In vitro* incubation of 6 hr RL after total body exposure	−24	6	No significant difference between S.A. RNA of irradiated and nonirradiated nuclear fractions of liver	*73*
				Nonregenerating Liver				
$^{32}P_i$		Mouse	510	*In vivo*; CMP, AMP, GMP, and UMP isolated and ^{32}P activity measured		0.5–6	Increased ^{32}P incorporation into RNA mononucleotides beginning 0.5 hr after irradiation and lasting until 6 hr	*244*
	2	Mouse	600	*In vivo*; starved 24 hr before irradiation		2.3	Increased ^{32}P incorporation into cRNA; decreased incorporation into nRNA	*209*
	3.5	Rat	700	*In vivo*; isolation of and 2 nRNA fractions		7.5	Decreased ^{32}P incorporation into nRNA I(tRNA?); increased incorporation into nRNA II (rRNA?)	*137*

TABLE V-13 (contd.)

Labeled precursor	Injection of precursor (hr before sacrifice)	Species	Radiation dose (R)	Experimental conditions	Irradiation (hr before or after partial hepatectomy)	Sacrifice (hr after irradiation)	Radiation-induced changes	References
$^{32}P_i$	2	Rat	1000	*In vivo*	−24	2	S.A. of c- and nRNA not altered	*204*
	4 4.5	Rat Rat	2440 2500	*In vivo*, starved from exposure until sacrifice		22 28	Increased ^{32}P incorporation into cRNA; decreased incorporation into n-RNA	*209*
Orotate-6-^{14}C		Rabbit	300	*In vitro* incubation of nuclei for 2 hr		2	Uptake by nuclei inhibited 40%	*160*
	3.5	Rat	850	*In vivo*; fractionation into nuclear, mitochondrial, and cytoplasmic fractions		4, 20	Increased S.A. UMP, decreased S.A. CMP	*210*
	0.3	Rat	1500	*In vivo*		6	No effect on rapidly labeled nRNA	*279*
Adenine-8-^{14}C		Rabbit	300	*In vitro* irradiation of nuclei and cytoplasmic components before incubation for 90 min			Uptake of adenine-^{14}C by nuclei or microsomes inhibited 45%	*160*
		Rat	300	Same as above, except incubation time was 120 min			Uptake of adenine mainly during the first 30 min of incubation; after 2 hr inhibition about 50% of control value for nuclei or microsomes	*161*
Adenine-4, 6-^{14}C	2	Mouse	800	*In vivo*; adenine injected 1 to 48 hr after irradiation		2	Adenine incorporation into DNA inhibited as much as ^{32}P incorporation	*19*

Uracil-2-^{14}C		Rabbit	300	*In vitro* irradiation of cell fractions as above (adenine) incubation time 120 min		U-2-^{14}C uptake by nuclei or microsomes inhibited 45% of controls	*160*
Ureido-succinic acid (ureido-^{14}C)	24 (after last injection)	Mouse	400	*In vivo*; 5 equal doses of precursor injected at 12-hr intervals beginning 20 hr after x-irradiation	44, 68, 92	Increased uptake by nuclei and incorporation into pyrimidines of nRNA, particularly during the late postirradiation period	*298*
5-Amino-4-imidazole carboxamide-5-^{14}C	24 (after last injection)	Mouse	400	*In vivo*; fed multiple doses of precursor as for the preceding compound	84	Increased incorporation into RNA purines	*55*

[a] Unless otherwise stated, all exposures were total body x-irradiation. Unless mentioned, all partial hepatectomies are surgical.
[b] See Footnote *b* of Table V-12 for abbreviations.

studied at various times after surgery and irradiation, it is evident that after doses lower than 2 kR the regenerative response is delayed but not abolished (Fig. V-7), whereas after doses which cause extensive reproductive death of hepatocytes, the liver no longer regenerates to its original size. Moreover, the time interval between maximal DNA synthesis and mitotic activity increases

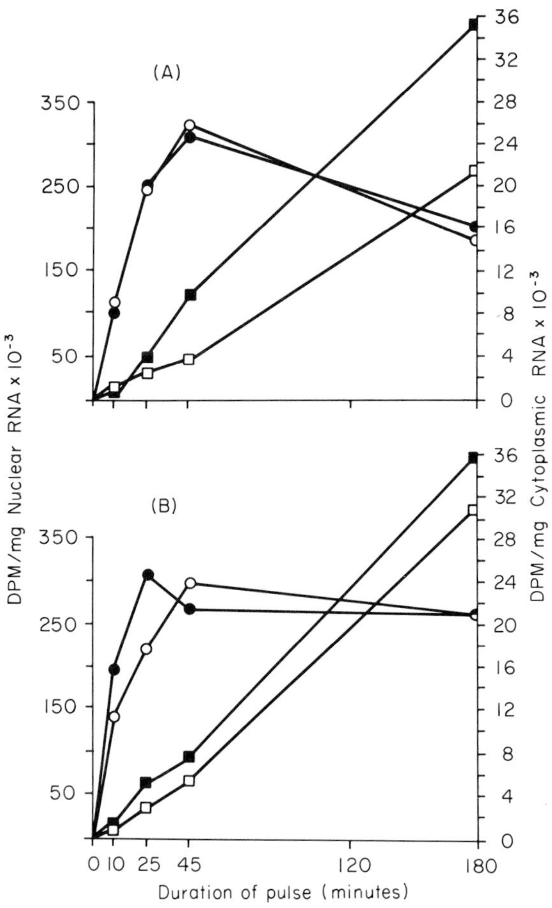

FIG. V-8. Incorporation of orotate-6-^{14}C into nuclear and cytoplasmic RNA of regenerating rat liver after total body x-irradiation. (A) Fifteen hours after partial hepatectomy, and (B) 12 hr after partial hepatectomy. (●) Nuclear RNA, nonirradiated control; and (○) nuclear RNA; irradiated with 1000 R 1 hr after partial hepatectomy. (■) Cytoplasmic RNA; nonirradiated control. (□) Cytoplasmic RNA, irradiated with 1000 R 1 hr after partial hepatectomy. Orotate-^{14}C (5 μCi) was injected 12 hr after hepatectomy (for a 3-hr pulse the precursor was injected 2 hr earlier). The rats were killed after injection at times as indicated. Each point on the curve represents the mean value obtained with three rats. [From T. L. Berg and R. Goutier, *Arch. Intern. Physiol. Biochim.* **75**, 49 (1967)] (see also *99a*).

after irradiation, indicating that delay of mitosis occurs during the G_2 period (*125*, *135*) (see also Volume 1). When irradiation is carried out 2–4 weeks before partial hepatectomy, the liver recovers its regenerative ability (Fig. V-7) and DNA synthesis proceeds as in nonirradiated liver.

(b) Formation of mRNA precedes *de novo* synthesis of proteins. One might expect, therefore, that synthesis of nuclear RNA is stimulated in regenerating liver before the enzymes required for DNA synthesis are elaborated. Indeed, one of the earliest signs of regeneration is an accumulation of acid-soluble ribonucleotides and an enhanced activity of RNA polymerase (*39*, *39a*). Subsequently, incorporation of precursors into nuclear RNA increases, and, later, synthesis of the enzymes required for DNA synthesis follows. Administration of actinomycin, which interferes with the synthesis of mRNA, also suppresses the formation of many enzymes required for regeneration (RNA polymerase, dTMP phosphotransferase, dTR phosphotransferase, and dCMP aminohydrolase), but it does not affect certain other enzymes (dTMP synthetase, dTDP phosphokinase) (*72*, *73*, *92*, *100*, *171*, *277*).

During the past 20 years, the action of radiation on the synthesis of RNA in normal and regenerating liver has been studied repeatedly although not as

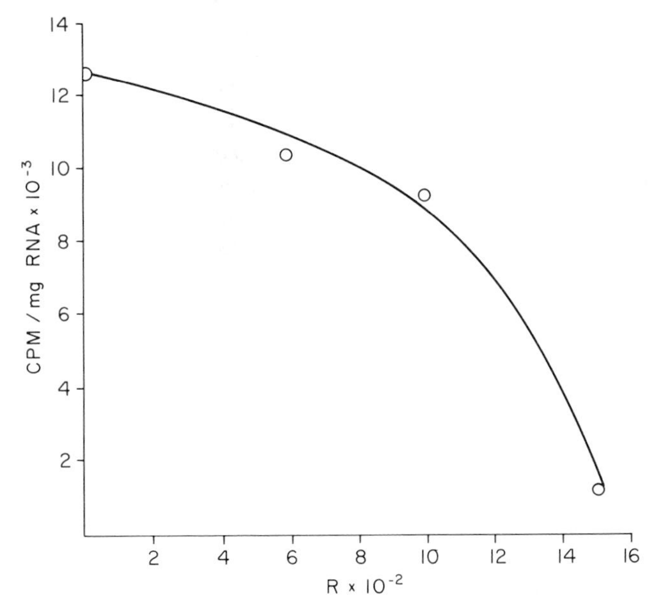

FIG. V-9. Incorporation of orotate-^{14}C into rapidly labeled, nuclear RNA as a function of radiation dose. The radiation dose in roentgens (R) is plotted against specific activity of RNA. Rats were hepatectomized 24 hr and x-irradiated (total body) 6 hr before sacrifice. Each point on the curve represents an average value obtained with three pooled livers. [From T. Uchiyama *et al.*, *Arch. Biochem. Biophys.* **110**, 191 (1965).]

often as that of DNA. In earlier studies using ^{32}P, no influence of radiation on total RNA incorporation by regenerating or normal liver was observed. Most of this work is now obsolete, since new techniques for separating different types of RNA in the cell have become available. Recent studies, using mostly orotate-^{14}C as precursor, show that the stimulation of nuclear RNA synthesis 6–12 hours after partial hepatectomy (Figs. V-8 and V-9) is retarded when the liver is irradiated shortly before or after surgery (*20*, *279*) (see Table V-13). Some investigators failed to note any effect of irradiation on the synthesis of nRNA, which, in many instances, was probably the result of choosing an irradiation time after surgery that was unsuitable for demonstrating the radiosensitivity of the biosynthetic process (*44*, *73*). Moreover, disagreement still exists concerning the right time for maximal stimulation of nRNA synthesis, but differences in experimental conditions, e.g., the length of the labeling period and/or the route of precursor administration, may be responsible for the divergent results (*20*, *21*).

Synthesis of cytoplasmic RNA is but little affected early in the period of regeneration, later (5 hours after surgery), the specific activity of cytoplasmic RNA diminishes markedly, whereas that of nuclear RNA increases (Fig. V-8) (*20*).

This effect could result from a delay in transfer of RNA from the nucleus to the cytoplasm, but it may also be related to the turnover of the pCp CA terminal site of cytoplasmic RNA.

The high molecular weight fractions exhibit the greatest stimulation during regeneration and the most pronounced inhibition of synthesis after irradiation as shown by centrifugal fractionation of nuclear RNA in a sucrose gradient. The relative amounts of the different nuclear RNA fractions are, however, not altered, as determined on the basis of their ultraviolet absorption (*20*, *279*). However, labeling by orotate after irradiation of cytoplasmic RNA is diminished to about the same extent in all fractions of cytoplasmic RNA.

Decreased incorporation of precursors into RNA could result from (i) changes in the precursor pool due, for example, to a dilution from breakdown products of RNA in radiosensitive organs; (ii) decreased formation of nucleotide triphosphates; (iii) diminished activity of RNA polymerase; and (iv) structural changes of chromosomal sites at which the RNA is assembled. A meaningful statement can be made only with regard to (iii), namely, that the formation of RNA polymerase is delayed markedly after irradiation (*20*, *21*, *137*, *278*, *279*, *281*, *289*).

(c) The activity of certain cytoplasmic enzymes involved in the synthesis of DNA or of pyrimidine precursors increases several-fold shortly before DNA synthesis commences, whereas the activity of most of the other enzymes changes but little during this time. Of the enzymes exclusively located in the nucleus only NAD pyrophosphorylase increases markedly during regeneration.

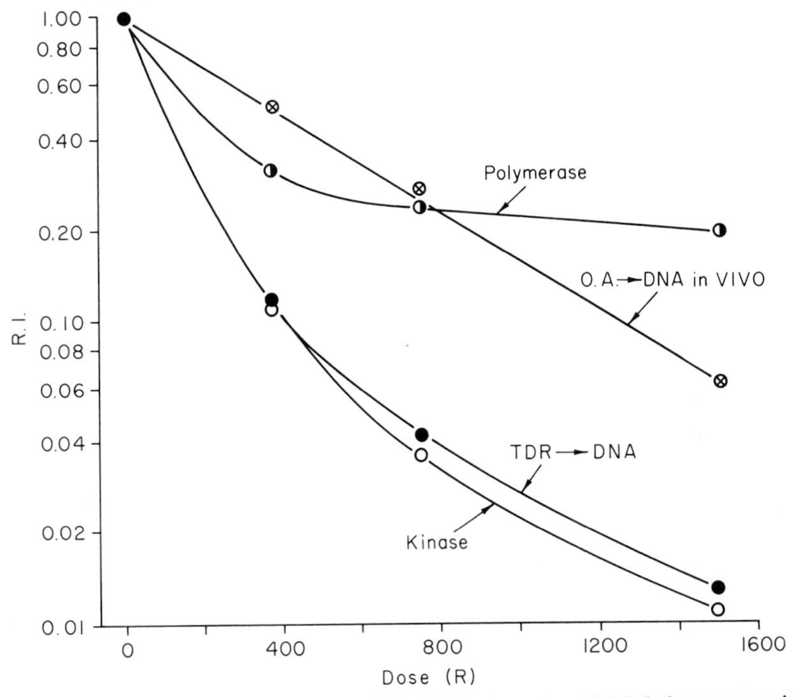

FIG. V-10. Comparison of *in vivo* inhibition of incorporation of labeled precursors into DNA with *in vitro* assay of enzymatic activity. The dose–effect relationship of *in vivo* incorporation of orotate-^{14}C into DNA as compared with the *in vitro* assays of enzymatic activities 24 hr after hepatectomy and 18 hr after x-irradiation associated with DNA synthesis, in terms of dTR-^{3}H incorporation. R.I. stands for relative incorporation. [From F. J. Bollum *et al.*, *Cancer Res.* **20**, 138 (1960) and R. E. Beltz *et al.*, *Cancer Res.* **17**, 688 (1957).]

Irradiation prior to enzyme formation (0–15 hours after partial hepatectomy) delays the *de novo* synthesis (Table V-14) but has no effect on enzymes already present in liver. This inhibition of enzyme synthesis occurs only in the exposed lobe of a partially irradiated liver (*187*). Differences exist with respect to the radiosensitivity of the various enzymes; formation of thymidine kinase is very radiosensitive (*278*, *290*). The radiation-induced inhibition of the biosynthesis of the dTR-kinase parallels the extent of the depression of DNA synthesis after irradiation; DNA polymerase synthesis, however, is less radiosensitivie (*17*, *29*) (see Fig. V-10).

(d) If liver is irradiated at a time when the hepatocytes have already started DNA synthesis, i.e., 18 hours after surgery or later, incorporation of precursors into DNA is diminished immediately after exposure to 500 R or more (see Fig. III-44d, Volume 1). A possible mechanism for this immediate effect on S phase cells is discussed extensively elsewhere in this book (see Chapter III),

TABLE V-14

EFFECT OF *in Vivo* X-IRRADIATION ON ENZYMES FORMED IN REGENERATING LIVER[a]

Enzyme	Time after irradiation (hours before or after partial hepatectomy)[b]	Time of sacrifice (hours after partial hepatectomy)	Dose of x-rays (R)	Effect of irradiation	References
Thymidine kinase (E.C. 2.7.1.21) ATP: thymidine 5′-phosphotransferase	+6–7	24–25	1500	97; 99% inhibition	*15, 29*
	+6–7	24–25	750	96% inhibition	*29*
	+6–7	24–25	375	89% inhibition	*29*
	+12–13, 5	24–25	1500	92% inhibition	*16*
	+15–16	24–25	1500	85% inhibition	*15*
	+16	24–25	375, 700, 1500	No inhibition	*29*
	−2	24	1000	90% inhibition	*100*
	−2	36	1000	42% inhibition	*100*
	−2	48	1000	No inhibition	*100*
Thymidine kinases (dTR→dTTP) and DNA polymerase (combined systems)	+12	24	800[c]	Almost complete inhibition	*200*
	+24	40	800	52% inhibition	*153*
	+24	48	800	72% inhibition	*153*
	+24	30	3000	74% inhibition	*143, 145*
	+24	30	1500	69% inhibition	*142, 143*
	+24	30	700	61% inhibition	*143*
	+24	48	3000	93% inhibition	*143*
	+24	48	1500	86% inhibition	*143*
	+24	48	800[c]	72% inhibition	*153*
	+24	48	700	72% inhibition	*143*
	−24, −48	24	1000	Marked inhibition	*13*
	−96	24	1000	Significant inhibition	*13*

Thymidine monophosphate kinase (E.C. 2.7.4.9) ATP: thymidinemonophosphate phosphotransferase	+18	24	1500	Marked inhibition	*278*
	+18	24	600	No inhibition	*278*
	−24	24	1000	Marked inhibition	*73*
	−24	36	1000	Marked inhibition	*73*
	−24	24 (Up to 14 days)	600	No inhibition	*73*
Thymidine phosphate kinases and DNA polymerase	+14	20	1500	Marked inhibition	*144*
	+24	48	1500	Marked inhibition	*144*
Thymidine monophosphate synthetase	+6–7	24.5	1500	25% inhibition	*15*
	+15–16	24.5	1500	25% inhibition	*15*
Thymidine phosphate kinases	−12	24	500	Marked inhibition	*13*, *100*
	−24, −48	24	500	Slight to moderate inhibition	*13*
dTR→dTDP+dTTP	−24 to −96	24	1000	Almost complete inhibition	*13*
dTR→dTMP+dTDP+dTTP					
	+6, +12	24	1000	Significant inhibition	*13*
	+17	24	1000	No effect	*13*
	+7, +12	24	1500	Marked inhibition	*13*
	+18	24	1500	Significant inhibition	*13*
dTR→dTDP+dTTP	−1.5	12 or 23	1000	Complete inhibition	*13*
dTR→dTMP+dTDP+dTTP		39–46	1000	45–55% inhibition	*13*
Deoxycytidylate aminohydrolase, (E.C. 3.5.4.12)	+5	Not stated	500	Delayed enzyme synthesis	*253*
	+24	35–40	600	50–60% inhibition	*188*
	+48	59–64	600	50–60% inhibition	*188*
	Immediately before	11–16	600	50–60% inhibition	*188*
	Immediately before	24	1500	58% inhibition	*187*
	Immediately before	48	1500	77% inhibition	*187*
	Immediately before	72	1500	74% inhibition	*187*
		72	1500[c]	58% inhibition	*187*

TABLE V-14 (contd.)

Enzyme	Time after irradiation (hours before or after partial hepatectomy)[b]	Time of sacrifice (hours after partial hepatectomy)	Dose of x-rays (R)	Effect of irradiation	References
DNA nucleotidyltransferase	+6	24	1500	81% inhibition	*29*
(E.C. 2.7.7.7) deoxynucleo-	+6	24	750	77% inhibition	*29*
sidetriphosphate: DNA	+6	24	375	69% inhibition	*29*
deoxynucleotidyltransferase	+12–13	25	1500	50% inhibition	*16*
	+16	24	375, 750 1500	No inhibition	*29*
	+18	24	600	No inhibition	*278*
	+18	24	1500	Slight inhibition	*278*
	−24	24	600	No inhibition	*73*
	−24	36	600	No inhibition	*73*
Thymidine phosphorylase	Immediately before	24	1500	None	*187*
(E.C. 2.4.2.4.) thymidine:		48	1500	21%	*187*
orthophosphate deoxyribosyl-		72	1500	38%	*187*
transferase		72	1500[c]	27%	*187*
NAD pyrophosphorylase	Immediately before	24	1500	28%	*187*
(E.C. 2.7.7.1) ATP: NMN		48	1500	28%	*187*
adenylyltransferase		72	1500	28%	*187*
		72	1500[c]	27%	*187*
RNA nucleotidyltransferase	−1	6–18	1000	Marked prolonged	*281*
(E.C. 2.7.7.6) nucleosidetri-				inhibition	*21*
phosphate: RNA nucleotidyl-	+4	8–18	1000	Marked inhibition, less protracted than under preceding conditions	*99a*
transferase					
	+2–+12	12–18	750 or 1500	No inhibition in nuclei and extracts	*45a*

[a] Unless otherwise stated, the rat was the species employed and all exposures are a single dose of total body x-irradiation.
[b] + indicates x-irradiation after partial hepatectomy and − indicates x-irradiation before partial hepatectomy.
[c] Local irradiation of the liver area only.

and only the following aspects pertaining to DNA synthesis will be discussed in this chapter, since they pertain uniquely to regenerating liver: (1) the activity of the enzymes synthesizing DNA, (2) the structure of the nucleoproteins and the formation of their protein moieties, (3) the amounts of deoxyribonucleotides present, and (4) the activity of nucleolytic enzymes.

(1) As already mentioned, exposure to doses which reduce incorporation of precursors into DNA immediately after irradiation does not affect the activity of enzymes already present at the time of irradiation. A few hours after

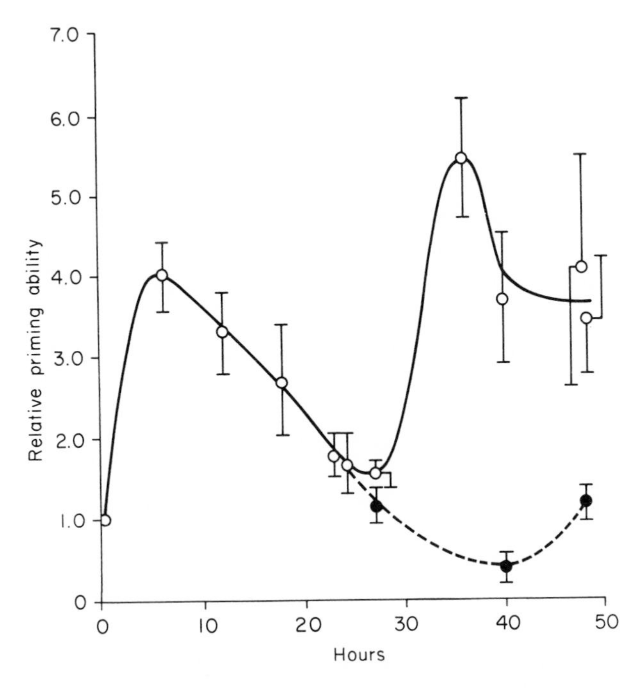

FIG. V-11. Altered priming ability of isolated rat liver nuclei after partial hepatectomy and later total body x-irradiation. (○) Nuclei of nonirradiated, regenerating liver; and (●) nuclei of regenerating livers after 800 R total body x-irradiation administered 24 hr after partial hepatectomy. [From S. M. Lehnert and S. Okada, *Intern. J. Radiation Biol.* **8**, 75 (1964).]

irradiation of parenchymal cells in S phase, enzymatic activity decreases when fewer cells have undergone mitosis and the flow of cells from G_1 into S phase has been restricted. It is unlikely, moreover, that enzymatic activities are the rate-limiting factors in DNA synthesis of S phase cells, since, for example, the activity of dTR phosphotransferase or DNA polymerase does not parallel DNA synthesis during the course of liver regeneration (*29*).

(2) The ability of DNA in liver nuclei to act as primer for DNA synthesis *in vitro* if precursors and enzymes are provided in excess has been utilized to assess structural changes of nucleoproteins during regeneration and after irradiation (*151*, *152*). Priming activity during regeneration is highest before DNA synthesis begins and diminishes during S-phase before rising again (Fig. V-11). Irradiation 24 hours after partial hepatectomy has no immediate effect on priming activity but abolishes the subsequent increase, presumably because progress through the cell cycle is blocked after irradiation. These data suggest that primer activity determined under *in vitro* conditions bears no relation to the rate of DNA synthesis *in vivo* (*252*), and, therefore, neither primer activity nor activity of enzymes synthesizing DNA can provide an explanation for the immediate effect of radiation on DNA synthesis.

Nuclear proteins represent another factor influencing DNA synthesis, but little is known yet of the mechanisms involved here. Most investigators report that irradiation *in vitro* or *in vivo* has no effect on synthesis of proteins in nuclei (*42*, *84*, *219*), but one group of experimenters (*160*, *161*) claims to have found striking changes in incorporation of amino acids in irradiated nuclei. Histone synthesis is not altered when liver cells are irradiated during S phase (*151*, *152*), but acid-insoluble nuclear proteins incorporate less lysine-^{14}C (*153*) after irradiation. Since synthesis of some proteins continues after DNA synthesis has declined, irradiated nuclei contain more protein in relation to DNA than do normal nuclei.

A transient loss of histones and nuclear globulins from nuclei of normal rat liver is observed 10–24 hours after an exposure to 1000 R (*69–71*, *112*).* This loss of nuclear proteins is irreversible in lymphoid organs (see Chapter III) and antecedes interphase death, but, in liver, normal concentrations are reestablished after 2–3 days. Solubility in salt solution of nucleoproteins increases in lymphoid organs but not in regenerating liver (*99a*). The question whether the bond between histones and DNA is rendered more labile by irradiation of the liver requires further investigation.

(3) The deoxyribonucleotide content of regenerating liver increases shortly before DNA synthesis begins (*236*), and this increase is abolished when liver is irradiated at the time of surgery (*127*), presumably because certain enzymes needed for deoxyribonucleotides are not formed after irradiation. Earlier, e.g., 6–12 hours after exposure, deoxyribonucleosides, especially dCR, accumulate in normal or even higher than normal amounts in regenerating liver (see Table V-15) (*251*), and this increase coincides with the enhanced excretion of dCR in urine (*61*) (see Chapter VIII). These observations indicate that liver (especially the smaller regenerating liver) is incapable of handling all the dCR liberated as a result of DNA breakdown in radiosensitive organs.

(4) Activity of lysosomal enzymes is enhanced in many organs where cell death and necrosis prevail after irradiation. Both DNase II in lysosomes and

* This could not be confirmed recently (*261a*).

TABLE V-15

EFFECTS OF X-IRRADIATION ON NUCLEOSIDE CONTENT IN REGENERATING LIVER[a,b]

Group	Deoxy-cytidine	Methylde-oxycytidine	Deoxy-uridine	Thymidine
Control[c]	18	1.1	2.0	2.8
Irradiated[c]	26	0.5	0.17	0.4
Partially hepatectomized	24	1.9	0.76	4.6
Partially hepatectomized and irradiated	31	0.4	1.30	2.3

[a] Adapted from J. Soška *et al.*, *Folia Biol.* (*Prague*), **8**, 239 (1962).

[b] Nucleoside content expressed as millimicromoles per gram of tissue. Rats were exposed to 600 R x-rays (total body) 24 hr before partial hepatectomy and killed 24 hr after surgery.

[c] Nonregenerating, normal liver.

DNase I in mitochondria increase in regenerating liver a few hours before DNA synthesis begins (*101, 102*). This time pattern is not altered by irradiation, but a portion of the DNase II complement is released into the supernatant fraction, perhaps as a result of greater fragility of lysosomes after irradiation (*102*). A similar increase of DNase II in the supernatant fraction is found also in irradiated normal liver and may be related to concomitant changes in pH as shown in the pH activity curve (*101*). DNase I activity in normal liver is little affected by total body x-irradiation of mice with 500 R (*74*) or 780 R (*140a*). The total activity of ribonuclease in rat liver increases 2–3 days after total body exposure to 600 R and then diminishes later during the postirradiation period (*224*). Changes in the concentration of an inhibitor (*223*) were thought responsible for this effect, but more adequate techniques for assaying RNase and its inhibitor should be applied in order to confirm these findings.

REFERENCES

1. Adelstein, S. J., and Kohn, H. I., *Biochim. Biophys. Acta* **138**, 163 (1967).

1a. Agarwal, S. K., and Mehrotra, R. M. L., *Indian J. Med. Res.* **52**, 1073 (1064).

2. Agostini, C., and Sessa, A., *Intern. J. Radiation Biol.* **5**, 155 (1962).

3. Agostini, C., Sessa, A., Fenaroli, A., and Ciccarone, P., *Radiation Res.* **23**, 350 (1964).

4. Albert, M. D., and Bucher, N. L. R., *Cancer Res.* **20**, 1514 (1960).

5. Altenbrunn, H. J., and Kobert, E., *Acta Biol. Med. Ger.* **7**, 251 (1961).

6. Altman, K. I., Casarett, G. W., Noonan, T. R., and Salomon, K., *Federation Proc.* **8**, 349 (1949).

7. Altman, K. I., Gerber, G. B., and Hempelmann, L. H., *Intern. J. Radiation Biol.* Spec. Suppl. 1, p. 277 (1960).

8. Anderson, C. F., Blaschko, H., Burn, J. H., and Mole, R. H., *Brit. J. Pharmacol.* **6**, 342 (1951).

8a. Baeyens, W., and Goutier, R., *Arch. Intern. Physiol. Biochim.* **75**, 875 (1967).
9. Barnabei, O., Mazzone, A., and Rava, G., *Ric. Sci.* **34**, (II-B) 179 (1964).
10. Barnabei, O., Romano, B., Bitonto, G. D., Tomasi, V., and Sereni, F., *Arch. Biochem. Biophys.* **113**, 478 (1966).
11. Barron, E. S. G., *in* "Biological Effects of External and Internal X and Gamma Radiation" (R. E. Zirkle, ed.), pp. 412–428; McGraw-Hill, New York, 1954.
12. Bartsch, G. G., and Gerber, G., B. *J. Lipid Res.* **7**, 204 (1966).
13. Baugnet-Mahieu, L., Goutier, R., and Semal, M., *Radiation Res.* **31**, 808 (1967).
14. Bekkum, D. W. Van, *Ciba Found. Symp. Ionizing Radiations Cell Metab.* pp. 77–91 (1956).
15. Beltz, R. E., *Biochem. Biophys. Res. Commun.* **9**, 78 (1962).
16. Beltz, R. E., and Applegate, R. L., *Biochem. Biophys. Res. Commun.* **1**, 298 (1959).
17. Beltz, R. E., Lancker, J. L. Van, and Potter, V. R., *Cancer Res.* **17**, 688 (1957).
18. Benjamin, T. L., and Yost, H. T., Jr. *Radiation Res.* **12**, 613 (1960).
19. Bennett, E. L., Kelly, L., and Kreuckel, B., *Federation Proc.* **13**, 181 (1954).
20. Berg, T. L., and Goutier, R., *Arch. Intern. Physiol. Biochim.* **74**, 909 (1966).
21. Berg, T. L., and Goutier, R., *Arch. Intern. Physiol. Biochim.* **75**, 49 (1967).
22. Berndt, T. J., *Intern. J. Radiation Biol.* **13**, 187 (1967).
23. Berndt, J., and Gaumert, R., *Intern. J. Radiation Biol.* **11**, 593 (1966).
24. Berndt, J., and Gaumert, R., *Naturwissenschaften* **53**, 307 (1966).
24a. Berndt, J., and Gaumert, R., *Intern. J. Radiation Biol.* **12**, 453 (1967).
25. Berndt, J., Gaumert, R., and Ulbrich, O., *Biochim. Biophys. Acta* **137**, 43 (1967).
25a. Berndt, J., and Ulbrich, O., *Strahlentherapie* **133**, 288 (1967).
26. Birzle, H., Beck, K., and Dieterich, E., *Strahlentherapie* **133**, 602 (1967).
26a. Birzle, H., and Franzius, E., *Strahlentherapie* **126**, 119 (1967).
27. Blokhina, V. D., and Demin, N. N., *Biokhimiya* **24**, 663 (1959).
28. Blokhina, V. D., and Romantsev, E. F., *Dokl. Akad. Nauk. SSSR* **168**, 454 (1966).
29. Bollum, F. J., Anderegg, J. W., McElya, A. B., and Potter, V. R., *Cancer Res.* **20**, 138 (1960).
30. Bonsignore, D., Canonica, C., and Lenzerini, L., *Ital. J. Biochem.* **14**, 1 (1965).
31. Borrebaek, B., Abraham, S., and Chaikoff, I. L., *Biochim. Biophys. Acta*, **90**, 451 (1964).
32. Braun, H., *Strahlentherapie* **117**, 134 (1962).
33. Braun, H., *in* "Strahlenpathologie der Zelle" (E. Scherer and H. S. Stender, eds.) p. 233. Thieme, Stuttgart, 1963.
34. Braun, H., Hornung, G., and Heise, E. R., *Strahlentherapie* **122**, 211 (1963).
35. Braun, H., Hornung, G., and Neumeyer, M., *Stahlentherapie* **126**, 454 (1965).
36. Bresciani, F., Blasi, F., and Fragomele, F., *Commun. 7th Congr. Soc. Ital. Patol.*, (1961).
37. Bresciani, F., Blasi, F., and Puca, G. A., *Radiation Res.* **26**, 451 (1965).
38. Brin, M., and McKee, R. W., *Arch. Biochem. Biophys.* **61**, 384 (1956).
39. Bucher, N. L. R., *Intern. Rev. Cytol.* **15**, 245 (1963).
39a. Bucher, N. L. R., *New Engl. J. Med.* **277**, 686 and 738 (1967).
40. Bucher, N. L. R., McGarrahan, K., Gould, E., and Loud, A. V., *J. Biol. Chem.* **234**, 262 (1959).
41. Burke, W. T., and Miller, L. L., *Cancer Res.* **19**, 148 (1959).
42. Butler, J. A. V., Cohn, P., and Crathorn, A. R., *in* "Advances in Radiobiology" (G. C. de Heresy, A. G. Forssberg, and J. D. Abbatt, eds.) p. 33. Oliver & Boyd, Edinburgh and London, 1957.
43. Carrol, H. W., *Radiation Res.* **16**, 557 (1962).

44. Cater, D. B., Holmes, B. E., and Mee, L. K., *Acta Radiol.* **46**, 655 (1956).
45. Cha, H. S., *Bull. Yamaguchi Med. School* **8**, 73 (1961); *Chem. Abstr.* **56**, 6327i (1962).
45a. Chambon, P., Mandel, P., Weill, J. D., and Busch, S., *Life. Sci.* **1**, 167 (1962).
46. Chang, L. O., and Looney, W. B., *Intern. J. Radiation Biol.* **12**, 187 (1967).
47. Chevallier, A., and Burg, C., *Ann. Nutr. Aliment.* **7**, 81 (1953).
48. Chiriboga, J., *Nature* **198**, 803 (1963).
49. Chiriboga, J., *Puerto Rico Nucl. Center* **PRNC 80** (1965).
49a. Clark, J. B., *European J. Biochem.* **2**, 19 (1967).
50. Coniglio, J. G., Cate, D. L., and Hudson, G. W., *Federation Proc.* **18**, 206 (1959).
51. Coniglio, J. G., Cate, D. L., Benson, B., and Hudson, G. W., *Am. J. Physiol.* **197**, 674 (1959).
52. Coniglio, J. G., Darby, W. J., Wilkerson, M. C., Stewart, R., Stockell, A., and Hudson, G. W., *Am. J. Physiol.* **172**, 86 (1953).
53. Coniglio, J. G., Kirschman, J. C., and Hudson, G. W., *Am. J. Physiol.* **191**, 350 (1957).
54. Coniglio, J. G., McCormick, D. B., and Hudson, G. W., *Am. J. Physiol.* **185**, 577 (1956).
54a. Coniglio, J. G., Hudson, G. W., and McCormick, D. B., *Fed. Proc.* **14**, 30 (1955).
55. Conzelman, G. M., Jr., Mandel, H. G., and Smith, P. K., *Cancer Res.* **14**, 100 (1954).
56. Cornatzer, W. E., Davison, J. P., Engelstad, O. D., and Simonson, C., *Radiation Res.* **1**, 546 (1954).
57. Dancewicz, A. M., Mazanowska, A., and Panfil, B., *Nukleonika* **10**, 783 (1965).
57a. Dancewicz, A. M., and Panfil, B., *Nukleonika* **12**, 255 (1967).
58. Del Favero, A., Pasotti, C., and Nicrosini, F., *Minerva Radiol. Fisoil, Radiobiol.* **10**, 380 (1965).
59. Denson, J. R., Gray, E. J., Gray, J. L., Herbert, F. J., Tew, J. T., and Jensen, H., *Proc. Soc. Exptl. Biol. Med.* **82**, 707 (1953).
60. Desai, I. D., Sawant, P. L., and Tappel, A. L., *Biochim. Biophys. Acta* **86**, 277 (1964).
61. Dienstbier, Z., *Acta Univ. Carolinae, Med.* **11**, 3 (1965).
62. DuBois, K. P., *Radiation Res.* **30**, 342 (1967).
63. DuBois, K. P., Cochran, K. W., and Doull, J., *Proc. Soc. Exptl. Biol. Med.* **76**, 422 (1951).
64. Eberhagen, D., Hartung, R., and Zöllner, N., *Strahlentherapie* **126**, 601 (1965); **132**, 441 (1967).
64a. Eichel, H. J., and Spirtes, M. A., *Proc. Soc. Exptl. Biol. Med.* **88**, 412 (1955).
65. Elko, E. E., and Di Luzio, N. R., *Radiation Res.* **11**, 1 (1959).
66. Elko, E. E., Wooles, W. R., and Di Luzio, N. R., *Radiation Res.* **21**, 493 (1964).
67. Ellinger, F., *Radiology* **44**, 241 (1945).
68. Ellinger, F., "Medical Radiation Biology." Thomas, Springfield, Illinois, 1957.
69. Ernst, H., *Naturwissenschaften* **48**, 575 (1961).
70. Ernst, H., *Z. Naturforsch.* **17b**, 300 (1962).
71. Ernst, H., *Naturwissenschaften* **50**, 227 and 333 (1963).
72. Fausto, N., and Van Lancker, J. L., *J. Biol. Chem.* **240**, 1247 (1965).
73. Fausto, N., Uchiyama, T., and Van Lancker, J. L., *Arch. Biochem. Biophys.* **106**, 447 (1964).
74. Fellas, V. M., Meschan, I., Day, P. L., and Douglass, C. D., *Proc. Soc. Exptl. Biol. Med.* **87**, 231 (1954).
75. Fendrich, Z., and Grossmann, V., *Česk. Fysiol.* **15**, 414 (1966).
76. Filippova, V. N., Soldatenkova, L. I., and Seits, I. F., *Radiobiologiya* **5**, 806 (1965)

77. Fischel, E., *Schweiz. Med. Wochschr.* **71**, 764 (1941).
78. Fischer, P., *Arch. Intern. Physiol. Biochim.* **62**, 134 (1953).
79. Fitch, M. F., Chaikoff, I. L., and Hill, R., *Arch. Biochem. Biophys.* **94**, 387 (1961).
80. Florsheim, W. H., and Morton, M. E., *Am. J. Physiol.* **176**, 15 (1954).
81. Friedman, N. B., *A.M.A. Arch. Pathol.* **34**, 749 (1942).
82. Fritz-Niggli, H., and Bührer, G., *Fortschr. Gebiete Röentgenstrahlen Nuklearmed.* **92**, 343 (1960).
83. Garattini, S., Paoletti, P., and Paoletti, R., *in* "Biochemistry of Lipids" (G. Popjak, ed.) p. 184. Pergamon, Oxford, 1960.
84. Gerbaulet, K., Maurer, W., and Brückner, J., *Biochim. Biophys. Acta* **68**, 462 (1963).
84a. Gerber, G. B., personal communication (1968).
85. Gerber, G. B., Altman, K. I., Gerber, G., and Deknudt, G., *Strahlentherapie* **137**, 620 (1969).
85a. Gerber, G. B., and Remy-Defraigne, J., *Intern. J. Radiation Biol.* **10**, 141 (1966).
85b. Gerber, G. B., and Remy-Defraigne, J., *Intern. J. Radiation Biol.* **7**, 59 (1963).
86. Gerber, G. B., Gerber, G., Altman, K. I., and Hempelmann, L. H., *Radiation Res.* **9**, 119 (1958).
86a. Gerber, G. B., Gerber, G., and Altman, K. I., *J. Biol. Chem.* **235**, 1433 (1960).
87. Gerber, G. B., Gerber, G., Altman, K. I., and Hempelmann, L. H., *Intern. J. Radiation Biol.* **1**, 277 (1959).
88. Gerber, G. B., Gerber, G., Altman, K. I., and Hempelmann, L. H., *Intern. J. Radiation Biol.* **6**, 17 (1963).
89. Gerber, G. B., and Remy-Defraigne, J., *Intern. J. Radiation Biol.* **7**, 53 (1963).
90. Gerber, G. B., and Remy-Defraigne, J., *Arch. Intern. Physiol. Biochim.* **74**, 785 (1966).
91. Gerber, G. B., Reuter, A., Kennes, F., and Remy-Defraigne, J., *Conf. on Problems Connected with Preparation and Use of Labelled Proteins in Tracer Studies, Pisa 1966* p. 165. 1966.
92. Giudice, G., and Novelli, G. D., *Biochem. Biophys. Res. Commun.* **12**, 383 (1963).
93. Glavind, J., and Faber, M., *Intern. J. Radiation Biol.* **11**, 445 (1966).
94. Glavind, J., Hansen, L., and Faber, M., *Intern. J. Radiation Biol.* **9**, 409 (1965).
95. Goetze, T., and Goetze, E., *Acta Biol. Med. Ger.* **15**, 353 (1965).
96. Goldfeder, A., *in* "Progress in Radiobiology" (B. Mitchell and B. E. Holmes, eds.) p. 77. Thomas, Springfield, Illinois, 1956.
96a. Gould, D. M., Floyd, K. W., Whitehead, R. W., and Sanders, J. L., *Nature* **192**, 1309 (1961).
97. Gould, R. G., Bell, V L., and Lilly, E. H., *Am. J. Physiol.* **196**, 1231 (1959).
98. Gould, R. G., and Popjak, G., *Biochem. J.* **66**, 51P (1957).
98a. Goutier, R., personal communication (1968).
99. Goutier, R., *Progr. Biophys.* **11**, 53 (1961).
99a. Goutier, R., "Irradiation, radioprotection et synthèse de l'acide désoxyribonucléique." Editions Arscia, Brussels, 1967.
100. Goutier, R., Baugnet-Mahieu, L., and Semal, M., *Arch. Intern. Physiol. Biochim.* **71**, 130 (1963).
101. Goutier, R., Goutier-Pirotte, M., and Ciccarone, P., *Intern. J. Radiation Biol.* Spec. Suppl. 1 p. 93 (1960).
102. Goutier-Pirotte, M., and Goutier, R., *Radiation Res.* **16**, 728 (1962).
103. Graevskaya, B. M. and Shchedrina, R. N., *Radiobiologiya*, **3**, 168 (1963).
104. Greengard, O. *Advan. Enzyme Regulation* **1**, 61 (1963).
105. Greengard, O., *Proc. Natl. Acad. Sci. U.S.* **56**, 1303 (1966).

106. Greengard, O. *Enzymol. Biol. Clin.* **8**, 91 (1967).
107. Gromov, V. A., Zhulanova, Z. I., Romantsev, E. F., Smolin, D. D., and Sokolova, G. N., *Radiobiologiya* **4**, 378 (1954).
108. Gros, P., Talwar, G. P., and Coursaget, J., *Bull. Soc. Chim. Biol.* **36**, 1569 (1954).
109. Grossi, E., Paoletti, P., and Paoletti, R., *Minerva Nucl.* **2**, 343 (1958).
110. Grossi, E., Poggi, M., and Paoletti, R., *in* "Response of the Nervous System to Ionizing Radiation" (T. J. Haley and R. S. Snider, eds.), p. 388. Little & Brown, Boston, Massachusetts, 1964.
111. Hagen, U., *Strahlentherapie* **106**, 277 (1958).
112. Hagen, U., *Nature* **187**, 1123 (1960).
113. Hagen, U., Koch, R., and Langendorff, H., *Fortschr. Gebiete Röentgenstrahlen Nuklearmed.* **84**, 604 (1956).
114. Hall, J. C., Goldstein, A. L., and Sonnenblick, B. P., *J. Biol. Chem.* **238**, 1137 (1963).
115. Hanel, H. K., Hjort, G., Johansen, G., and Purser, P. R., *Acta Radiol.* **48**, 227 (1957).
116. Hanel, H. K. and Wilian-Ulrich, I., *Intern. J. Radiation Biol.* **1**, 366 (1959).
117. Hansen, H. J. M., Hansen, L., and Faber, M., *Intern. J. Radiation Biol.* **9**, 25 (1965).
118. Hansen, L., *Intern. J. Radiation Biol.* **12**, 367 (1967).
118a. Harrington, H., and Lavik, P. S., *J. Cellular Comp. Physiol.* **46**, 503 (1955).
118b. Hartiala, K., Näntö, V., and Rinne, U. K., *Acta Physiol. Scand.* **45**, 231 (1959).
119. Hempelmann, L. H., Carr, S., Frantz, I. D., Masters, R., and Lamdin, E. *Federation, Proc.* **9**, 183 (1950).
120. Hevesy, G., *Nature* **158**, 268 (1946).
120a. Hevesy, G., *Nature* **163**, 869 (1949).
121. Hevesy, G., *Nature* **164**, 269 (1949).
121a. Hevesy, G., and Forssberg, A., *Nature* **168**, 692 (1951).
122. Hidvégi, E. J., Holland, J., Bölöni, E., Lónai, P., and Várterész, V., *Acta Biochim. Biophys. Biochem. J.* **109**, 495 (1968).
122a. Hietbrink, B. E., and DuBois, K. P., *Radiation Res.* **27**, 669 (1966).
123. Hill, R., Kiyasu, J., and Chaikoff, I. L., *Am. J. Physiol.* **187**, 417 (1956).
124. Holmes, B. E., *Ciba Found. Symp. Ionizing Radiation Cell Metab.* p. 225 (1957).
125. Holmes, B. E., and Mee, L. K., *in* "Radiobiology Symposium" (Z. N. Bacq and P. Alexander, eds.), p. 220. Academic Press, New York, 1955.
126. Jackson, K. L., and Entenman, C., *USNRDL* **TR-504**, 45 (1961).
127. Jaffe, J. J., Lajtha, L. G., Lascelles, J., Ord, M. G., and Stocken, L. A., *Intern. J. Radiation Biol.* **3**, 241 (1959).
128. Jamieson, D., *Nature* **209**, 361 (1966).
129. Jardetzky, C. D., Barnum, C. P., and Vermund, H., *J. Biol. Chem.* **222**, 421 (1956).
130. Kainova, A. S., *Biokhimyia* **25**, 540 (1960).
131. Kalashnikova, V. P., *Med. Radiol.* **8**, 58 (1963).
132. Kay, R. E., Early, J. C., and Entenman, C., *Radiation Res.* **6**, 98 (1957).
133. Kay, R. E., and Entenman, C., *Proc. Soc. Exptl. Biol. Med.* **91**, 143 (1956).
134. Kay, R. E., and Entenman, C., *Arch. Biochem. Biophys.* **62**, 419 (1956).
135. Kelly, L. S., *Progr. Biophys.* **8**, 143 (1957).
135a. Kelly, L. S., Hirsch, J. D., Beach, G., and Palmer, W., *Cancer Res.* **17**, 117 (1957).
135b. Kelly, L. S., Hirsch, J. D., Beach, G., and Payne, A. H., *Radiation Res.* **2**, 490 (1955).
135c. Khanson, K. P., *Radiobiologiya* **5**, 44 (1965).

135d. Kivy-Rosenberg, E., Cascarano, J., and Zweifach, B. W., *Radiation Res.* **20**, 668 (1963).
136. Kivy-Rosenberg, E., Cascarano, J., and Zweifach, B. W., *Radiation Res.* **23**, 310 (1964).
137. Klouwen, H. M., *Biochim. Biophys. Acta* **42**, 366 (1960).
138. Koch, R., and Bohländer, A., *Strahlentherapie* **85**, 331 (1951).
139. Koch, R., and Bohländer, A., *Strahlentherapie* **85**, 599 (1951).
140. Kröger, H., and Greuer, B., *Biochem. Z.* **341**, 190 (1965).
140a. Kurnick, N. B., Massey, B. W., and Sandeen, G., *Radiation Res.* **11**, 101 (1959).
141. Kusnets, E., *Dokl. Akad. Nauk SSSR* **130**, 657 (1960).
142. Lancker, J. L. Van, *Biochim. Biophys. Acta* **33**, 587 (1959).
143. Lancker, J. L. Van, *Biochim. Biophys. Acta* **45**, 57 (1960).
144. Lancker, J. L. Van, *Biochim. Biophys. Acta* **45**, 63 (1960).
145. Lancker, J. L. Van, *Federation Proc.* **21**, 1118 (1962).
145a. Lancker, J. L. Van, *Lab. Invest.* **15**, 192 (1966).
145b. Lancker, J. L. Van, *in* "Biochemistry of Cell Division" (B. Baserga, ed.) Chap. X. Thomas, Springfield, Illinois, 1969.
146. Lane, J. L., Wilding, J. L., Rust, J. H., Trum, B. F., and Schoolar, J. C., *Radiation Res.* **2**, 64 (1955).
147. Langendorff, H., Scheuerbrandt, G., Streffer, C., and Melching, H.-J., *Naturwissenschaften* **51**, 290 (1964).
148. Lauenstein, K., Haberland, G. L., Hempelmann, L. H., and Altman, K. I., *Biochim. Biophys. Acta* **26**, 421 (1957).
149. Leblond, C. P., Thoai, N.-V., and Segal, G., *Compt. Rend. Soc. Biol.* **130**, 1557 (1939).
150. Lee, T. C., Salmon, R. J., Loken, M. K., and Mosser, D. G., *Radiation Res.* **17**, 903 (1962).
151. Lehnert, S. M., and Okada, S., *Nature* **199**, 1108 (1963).
152. Lehnert, S. M., and Okada, S., *Intern. J. Radiation Biol.* **8**, 75 (1964).
152a. Lehnert, S. M., and Okada, S., *Intern. J. Radiation Biol.* **5**, 323 (1962).
153. Lehnert, S. M., and Okada, S. *Intern. J. Radiation Biol.* **10**, 601 (1966).
154. Lelièvre, P., *Compt. Rend. Soc. Biol.* **149**, 1296 (1955).
155. Lelièvre, P., *Compt. Rend. Soc. Biol.* **151**, 412 (1957).
156. Lelièvre, P., and Betz, E. H., *Compt. Rend. Soc. Biol.* **149**, 2034 (1955).
157. Leong, G. F., Pesotti, R. L., and Krebs, J. S., *J. Natl. Cancer Inst.* **27**, 131 (1961).
158. Lerner, S. R., Warner, W. L., and Entenman, C., *Federation Proc.* **12**, 5 (1953).
159. Levy, B., and Rugh, R., *Proc. Soc. Exptl. Biol. Med.* **82**, 223 (1953).
159a. Lim, J. K., Piantadosi, C., and Snyder, F., *Proc. Soc. Exptl. Biol. Med.* **124**, 1158 (1967).
160. Logan, R., *Biochim. Biophys. Acta* **35**, 251 (1959).
161. Logan, R., Errera, M., and Ficq, A., *Biochim. Biophys. Acta* **32**, 147 (1959).
162. Lourau, M., *Compt. Rend.* **236**, 422 (1953).
163. Lourau, M., and Lartigue, O., *J. Physiol.* (*Paris*) **43**, 593 (1951).
164. Lourau-Pitres, M., *Arkiv. Kemi* **7**, 211 (1954).
165. Maass, H., *Z. Ges. Exptl. Med.* **133**, 97 (1960).
166. Maass, H., Jenner, E., and Hölzel, F., *Z. Naturforsch.* **22b**, 634 (1967).
167. Maass, H., and Timm, M., *Strahlentherapie* **123**, 64 (1964).
168. MacCardle, R. C., and Congdon, C. C., *Am. J. Pathol.* **31**, 725 (1955).
169. McCay, P. B., and Macfarlane, M. G., *Nature* **198**, 98 (1963).
170. McKee, R. W., and Brin, M., *Arch. Biochem. Biophys.* **61**, 390 (1956).

171. Maley, G. F., Lorenson, M. G., and Maley, F., *Biochem. Biophys. Res. Commun.* **18**, 364 (1965).
172. Mandel, P., *Progr. Nucleic Acid Res.* **3**, 299 (1964).
173. Marchetti, M., Barbiroli, B., and Caldareva, C. M., *Boll. Soc. Ital. Biol. Sper.* **39**, 1819 (1953).
174. Maros, T., Seres-Sturm, L., and Rettegi, C., *Fiziol. Normala Patol.* (*Bucharest*) **9**, 19 (1963).
175. Masoro, E. J., *J. Lipid Res.* **3**, 149 (1962).
175a. Mee, L. K., *in* "Progress in Radiobiology" (J. S. Mitchell, B. E. Holmes, and C. L. Smith, eds.), p. 12. Oliver & Boyd, London and New York, 1956.
176. Miller, L. L., personal communication of data to appear in the proceedings of the Protein conference held at Rutgers University, 1966 (Rutgers Univ. Press, New Brunswick, New Jersey). All but the radiation data were published in *Verhand. Deut. Ges. Inn. Med.* **66**, 250–265 (1960).
177. Miller, L. L., personal communication (1968).
178. Miller, L. L., John, D. W., and Cloutier, P. F., personal communication (1900).
179. Mishkin, P. E., and Shore, M. L., *Radiation Res.* **33**, 437 (1968).
180. Mole, R. H., *Brit. J. Exptl. Pathol.* **37**, 528 (1956).
181. Morehouse, M. G., and Searcy, R. L., *Federation Proc.* **13**, 267 (1954).
182. Morehouse, M. G., and Searcy, R. L., *Science* **122**, 158 (1955).
183. Morehouse, M. G., and Searcy, R. L., *Federation Proc.* **15**, 316 (1956).
184. Morehouse, M. G., and Searcy, R. L., *Radiation Res.* **4**, 175 (1956).
185. Morehouse, M. G., and Searcy, R., *Federation Proc.* **16**, 223 (1957).
186. Myers, D. K., *Nature* **187**, 1124 (1960).
187. Myers, D. K., *Can. J. Biochem. Physiol.* **40**, 619 (1962).
188. Myers, D. K., Hemphill, C. A., and Townsend, C. M., *Can. J. Biochem. Physiol.* **39**, 1043 (1961).
189. Nair, V., and Bau, D., *Proc. Soc. Exptl. Biol. Med.* **126**, 853 (1967).
190. Nims, L. F., and Geisselsoder, J. L., *Radiation Res.* **5**, 58 (1956).
191. Nims, L. F., and Sutton, E., *Am. J. Physiol.* **177**, 51 (1954).
192. Nims, L. F., and Thurber, R. E., *Endocrinology* **70**, 589 (1962).
193. Nitz-Litzow, D., and Bührer, G., *Strahlentherapie* **113**, 201 (1960).
194. North, N., and Nims, L. F., *Federation Proc.* **8**, 119 (1949).
195. Notario, A., Di Marco, N., Doneda, G., Zanetti, A., and Slaviero, G., *Minerva Med.* **9**, 237 (1964).
196. Notario, A., Slaviero, G., and Santagati, G., *Minerva Nucl.* **9**, 79 (1965).
196a. Notario, A., Slaviero, G., and Santagati, G., *Minerva Nucl.* **9**, 84 (1965).
197. Notario, A., Slaviero, G., Doneda, G., and Santagati, G., *Experientia* **21**, 264 (1965).
198. Noyes, P. P., and Smith, R. E., *Exptl. Cell. Res.* **16**, 15 (1959).
199. Nygaard, O. F., and Potter, R. L., *Radiation Res.* **10**, 462 (1959).
200. Okada, S., and Hempelmann, L. H., *Intern. J. Radiation Biol.* **1**, 305 (1959).
201. Okada, S., and Peachey, L. D., *J. Biochem. Biophys. Cytol.* **3**, 239 (1957).
202. Oliva, L., Misurale, F., Pomini, P., and Valli, P., *Minerva Med.* **49**, 2861 (1958).
202a. Omata, S., Ichii, S., and Yago, N., *J. Biochem.* (*Tokyo*) **63**, 695 (1968).
203. Ord, M. G., and Stocken, L. A., *Physiol. Rev.* **33**, 356 (1953).
204. Ord, M. G., and Stocken, L. A., *Biochem. J.* **63**, 3 (1956).
205. Ord, M. G., and Stocken, L. A., *in* "Mechanisms in Radiobiology" (M. Errera and A. Forssberg, eds.), Vol. 1, p. 285. Academic Press, New York, 1960.

206. Pande, S. V., Ramaiah, A., and Venkitasubramanian, T. A., *Biochim. Biophys. Acta* **65**, 516 (1962).
207. Paoletti, P., and Paoletti, R., *Minerva Nucl.* **2**, 341 (1958).
208. Paoletti, R., Poggi, M., Vertua, R., and Farkas, T., *Progr. Biochem. Pharmacol* **1**, 186 (1965).
209. Payne, A. H., Kelly, L. S., and Entenman, C., *Proc. Soc. Exptl. Biol. Med.* **81**, 698 (1952).
210. Petrović, S., *Biochim. Biophys. Acta* **61**, 842 (1962).
210a. Philpot, J. S. L., *Radiation Res.* **3**, Suppl., 55 (1963).
211. Pikulev, A. T., and Yakubovick, L. S., *Nauch.* No. 1, 90 (1966).
212. Pitot, H. C., and Peraino, C., *J. Biol. Chem.* **239**, 1783 (1964).
213. Pitot, H. C., Peraino, C., and Lamar, C., Jr., *Radiation Res.* **33**, 599 (1968).
214. Pitot, H. C., Peraino, C., Lamar, C., Jr., and Lesher, S., *Science* **150**, 901 (1965).
215. Rajewsky, B., Gerber, G., and Pauly, H., *Strahlentherapie* **102**, 517 (1957).
215a. Rappoport, D. A., Seibert, R. A., and Collins, V. P., *Radiation Res.* **6**, 148 (1957).
216. Reuter, A. M., Gerber, G. B., Altman, K. I., Deknudt, G., and Kennes, F., *Klin. Wochschr.* **44**, 1310 (1966).
217. Reuter, A. M., Gerber, G. B., Kennes, F., and Remy-Defraigne, J., *Radiation Res.* **30**, 725 (1967).
218. Rhoades, R. P., *in* "Histopathology of Irradiation from External and Internal Sources" (W. Bloom, ed.), Chapter 11, p. 541. McGraw-Hill, New York, 1948.
219. Richmond, J. E., Ord, M. G., and Stocken, L. A., *Biochem. J.* **66**, 123 (1957).
220. Ricotti, V. and Notario, A., *Minerva Nucl.* **7**, 18 (1962).
221. Romantsev, E. F. and Zhulanova, Z. I., *Biokhimiya* **21**, 685 (1956).
222. Ross, M. H., and Ely, J. O., *J. Cellular Comp. Physiol.* **37**, 163 (1951).
223. Roth, J. S., *Arch. Biochem. Biophys.* **60**, 7 (1956).
224. Roth, J. S., Eichel, H. J., Wase, A., Alper, C., and Boyd, M. J., *Arch. Biochem. Biophys.* **44**, 95 (1953).
224a. Rushmer, H. N., and Altman, K. I., unpublished experiments (1967).
225. Rust, J. H., LeRoy, G. V., Spratt, J. L., Ho, G. B., and Roth, L. J., *Radiation Res.* **20**, 703 (1963).
226. Ryser, H., Schmidli, B., Zuppinger, A., and Aebi, H., *Helv. Physiol. Acta* **13**, 270 (1955).
227. Sarkar, N. K., Devi, A., and Hempelmann, L. H., *Nature* **192**, 179 (1961).
228. Sassen, A., Gerber, G. B., Kennes, F., and Remy-Defraigne, J., *Radiation Res.* **25**, 158 (1965).
229. Scaife, J. F., *Can. J. Biochem.* **44**, 433 (1966).
230. Scaife, J. F., and Alexander, P., *Radiation Res.* **15**, 658 (1961).
231. Scaife, J. F., and Hill, B., *Can. J. Biochem. Physiol.* **40**, 1025 (1962).
232. Scaife, J. F., and Hill, B., *Can. J. Biochem.* **41**, 1223 (1963).
233. Scherer, E., *Strahlentherapie* **99**, 230 (1956).
234. Scherer, E., *Strahlentherapie* **99**, 242 (1956).
235. Schneider, M., O'Rourke, F., and Redetzki, H. M., *Intern. J. Radiation Biol.* **5**, 183 (1962).
236. Schneider, W. C., and Brownell, L. W., *J. Natl. Cancer Inst.* **18**, 579 (1957).
236a. Schreier, K., DiFerrante, N., Gaffney, G. W., Hempelmann, L. H., and Altman, K. I., *Arch. Biochem. Biophys.* **50**, 417 (1954).
237. Schubert, G., Bettendorf, G., Künkel, H., Maass, H., and Rathgen, G. H., *Strahlentherapie* **106**, 483 (1958).

238. Schwarz, H. P., Dreisbach, L., Barrionuevo, M., Kleschick, A., and Kostyk, I., *Arch. Biochem. Biophys.* **92**, 133 (1961).
239. Schwarz, H. P., Dreisbach, L., and Kleschick, A., *Arch. Biochem. Biophys.* **101**, 103 (1963).
240. Schwarz, H. P., Dreisbach, L., Polis, E., Polis, B. D., and Soffer, E., *Arch. Biochem. Biophys.* **111**, 422 (1965).
241. Scott, C. M., *Med. Res. Council, Spec. Rep. Ser.* **223**, 74 (1937).
242. Šestan, N., *Nature* **205**, 615 (1965).
243. Shaber, G. S. and Miller, L. L., *Proc. Soc. Exptl. Biol. Med.* **113**, 346 (1963).
244. Sherman, F. G., and Almeida, A. B., *in* "Advances in Radiobiology" (G. C. de Hevesy, A. G. Forssberg, and J. D. Abbatt, eds.), p. 49. Oliver & Boyd, Edinburgh and London, 1957.
245. Sherman, F. G., and Dwyer, F. M., *Federation Proc.* **15**, 169 (1956).
246. Simonov, V. V., *Radiobiologiya* **5**, 907 (1965).
247. Skalka, M., *Folia Biol.* (*Prague*) **5**, 446 (1959).
248. Skalka, M., *Folia Biol.* (*Prague*) **6**, 152 (1960).
249. Smith, K. C., and Low-Beer, B. V. A., *Radiation Res.* **6**, 521 (1957).
250. Sols, A., Sillero, A., and Salas, J., *J. Cellular Comp. Physiol.* **66**, 23 (1965).
251. Soška, J., Skalka, M., and Bezdek, M., *Folia Biol.* (*Prague*), **8**, 239 (1962).
252. Stacey, K. A., *Biochem. Biophys. Res. Commun.*, **5**, 486, (1961).
253. Stevens, L., and Stocken, L. A., *Biochem. Biophys. Res. Commun.*, **7**, 315 (1962).
254. Streffer, C., *Radiol. Clin. Biol.* **35**, 358 (1966).
255. Streffer, C., *Intern. J. Radiation Biol.* **11**, 179 (1966).
256. Streffer, C., *Ger. Med. Monthly* **12**, 35 (1967).
257. Streffer, C., *J. Vitaminol.* (*Kyoto*) **14**, 130 (1968).
258. Streffer, C., and Langendorff, H.-U., *Intern. J. Radiation Biol.* **11**, 455 (1966).
259. Streffer, C., and Melching, H.-J., *Strahlentherapie* **125**, 341 (1964).
260. Streffer, C., and Melching, H.-J., *Strahlentherapie* **128**, 406 (1965).
261. Syromyatnikova, N. V., *Dokl. Akad. Nauk SSSR* **111**, 730 (1956).
261a. Telia, H., Reuter, A., Gerber, G. B., and Kennes, F., *Intern. J. Radiation Biol.* **16**, 389 (1969).
262. Thomson, J. F., *Radiation Res.* **21**, 46 (1964).
263. Thomson, J. F., Carttar, M. S., and Tourtelotte, W. W., *Radiation Res.* **1**, 165 (1954).
264. Thomson, J. F., and Klipfel, F. J., *Proc. Soc. Exptl. Biol. Med.* **95**, 107 (1957).
265. Thomson, J. F., and Mikuta, E. T., *Endocrinology* **55**, 232 (1954).
266. Thomson, J. F., and Mikuta, E. T., *Proc. Soc. Exptl. Biol. Med.* **85**, 29 (1954).
267. Thomson, J. F., Nance, S. L., and Bordner, L. F., *Radiation Res.* **29**, 121 (1966).
268. Thomson, J. F., and Rahman, Y. E., *Radiation Res.* **17**, 573 (1962).
269. Thurber, R. E., and Nims, L. F., *Endocrinology* **70**, 595 (1962).
270. Tom, S. S., Ph.D. Thesis, University of Chicago (1956).
271. Tonazy, N. E., White, N. G., and Umbreit, W. W., *Arch. Biochem. Biophys.* **28**, 36 (1950).
272. Toropova, G. P., *Dokl. Akad. Nauk SSSR* **117**, 266 (1957).
273. Toropova, G. P., *Med. Radiol.* **4**, 89 (1959).
274. Tretyakova, K. A., *Byul. Eksperim. Biol. i Med.* **57**, 47 (1964).
275. Tretyakova, K. A., *Vopr. Med. Khim.* **12**, 502 (1966).
276. Tretyakova, K. A. and Grodzenskii, D. E., *Biokhimiya* **25**, 399 (1960).
277. Tsukada, K., and Lieberman, I., *J. Biol. Chem.* **239**, 2952 (1964).

278. Uchiyama, T., Fausto, N., and Van Lancker, J. L., *Arch. Biochem. Biophys.* **110**, 191 (1965).
279. Uchiyama, T., Fausto, N., and Van Lancker, J. L., *J. Biol. Chem.* **241**, 991 (1966).
280. Urakami, H., *Okayama Igakkai Zasshi* **76**, 13 (1965); cited and abstracted by *Chem. Abstr.* **63**, 18607 (1965).
281. Vandergoten, R. and Goutier, R., *Intern. J. Radiation Biol.* **11**, 449 (1966).
282. Vinogradova, M. F., *Vestn. Leningr. Univ., Ser. Biol.* **3**, 102 (1962).
282a. Vittorio, P. V., Spence, W. P., and Johnston, J., *Can. J. Biochem. Physiol.* **37**, 787 (1959).
283. Waldschmidt, M., *Strahlentherapie* **128**, 586 (1965).
284. Waldschmidt, M., *Intern. J. Radiation Biol.* **11**, 429 (1966).
285. Waldschmidt, M., *Strahlentherapie* **132**, 463 (1967).
285a. Waldschmidt, M., *Strahlentherapie* **135**, 695 (1968).
286. Webber, M. M., and Stich, H. F., *Can. J. Biochem.* **43**, 817 (1965).
287. Weber, G., and Cantero, A., *Am. J. Physiol.* **197**, 1284 (1959).
288. Weinman, E. O., Lerner, S. R., and Entenman, C., *Arch. Biochem. Biophys.* **64**, 164 (1956).
289. Welling, W., and Cohen, J. A., *Biochim. Biophys. Acta* **42**, 181 (1960).
290. Wen-Mei, S., *Radiobiologiya* **3**, 159 (1963).
291. Wernze, H., Braun, H., and Gabor, M., *Klin. Wochschr.* **40**, 1220 (1962).
292. White, J., Burr, B. E., Cool, H. T., David, P. W., and Ally, M. S., *J. Natl. Cancer Inst.* **15**, 1145 (1955).
293. Wills, E. D., *Intern. J. Radiation Biol.* **11**, 517 (1966).
294. Wills, E. D., and Wilkinson, A. E., *Biochem. J.* **99**, 657 (1966).
295. Wooles, W. R., *Radiation Res.* **30**, 788 (1967).
296. Yang, F.-Y., Lin, C.-H., and Li, S.-K., *Sci Sinica (Peking)* **12**, 1595 (1963).
297. Yang, F.-Y., Lin, C.-H., and Li, S.-K., *Shih Yen Sheng Wu Hsueh Pao* **9**, 261 (1964); *Nucl. Sci. Abstr.* **19**, 26033 (1965).
297a. Yatvin, M. B., and Lathrop, T. P., *Biochem. Biophys. Res. Commun.* **25**, 535 (1966).
298. Yen, C. Y., Anderson, E. P., and Smith, P. K., *Radiation Res.* **11**, 7 (1959).
299. Yost, H. T., Jr., Glickman, R. M., and Beck, L. H., *Biol. Bull.* **127**, 173 (1964).
300. Yost, H. T., Jr., Richmond, S. S., and Beck, L. H., *Biol. Bull.* **127**, 526 (1964).
301. Yost, H. T., Jr. and Robson, H. H., *Biol. Bull.* **116**, 498 (1959).
302. Yost, H. T., Yost, M. T., and Robson, H. H., *Biol. Bull.* **133**, 697 (1967).
303. Yost, M. T., Robson, H. H., and Yost, H. T., *Radiation Res.* **32**, 187 (1967).
304. Zícha, B., Beneš, J., Lejsek, K., and Simek, J., *Naturwissenschaften* **53**, 613 (1966).
305. Zícha, B., Beneš, J., and Dienstbier, Z., *Strahlentherapie* **135**, 467 (1968).
306. Zícha, B., Lejsek, K., Beneš, J., and Dienstbier, Z., *Experientia* **22**, 712 (1966).

Chapter VI

RADIATION BIOCHEMISTRY OF MISCELLANEOUS ORGANS

A. Introduction

This chapter deals with several organs for which information on metabolic changes after irradiation is limited. Radiation biochemistry has followed the lead of general biochemistry in neglecting organs other than liver, bone marrow, intestine, or lymphoid organs. Nevertheless, it must be emphasized that these other organs represent the largest part of the body (more than three-quarters) and exhibit pronounced changes in weight after irradiation (Table VI-1). Moreover, certain compounds are found mainly in one or another of these organs. To mention only a few examples, proteins of connective tissue and muscle (mostly collagen, elastin, actin, or myosin) represent about 80% of all body proteins; muscle and brain contain more than 90% of all creatine, 85% of all taurine, and 60% of all K^+ in the body; certain lipids occur only in brain, etc. It is evident that even small changes in uptake, synthesis, or loss from damaged cells of such compounds could greatly influence body pools, i.e., levels in blood and urine. Radiobiologists, who often tend to think of muscle, bone, skin, connective tissue, or brain as being radioresistant, should not exclude these organs from consideration if the overall metabolic changes in irradiated animals are to be interpreted.

TABLE VI-1

TISSUE WEIGHT CHANGES IN NORMAL AND X-IRRADIATED RATS (AFTER 700 R WHOLE-BODY EXPOSURE)[a]

Tissue	Nonirradiated controls	Time after exposure								
		3 hr	12 hr	1 day	2 days	3 days	4 days	6 days	8 days	10 days
Total body	100	98.9	98.2	98.6	92.2	88.2	87.0	87.9	90.5	94.5
Muscle	52.3 ± 1.5	96.7[b]	94.4[c]	96.2[c]	91.3[c]	83.3[c]	180.2[c]	85.3[c]	85.4[c]	87.6[c]
Supportive	8.82 ± 0.52	99.2	103.1	103.4	113.1[c]	100.7	113.9[c]	105.8[b]	102.4	106.7[b]
Skin	20.7 ± 1.6	97.0	97.4	101.4	93.8	86.4	91.1	87.7	94.3	98.2
Liver	4.77 ± 0.53	101.3	95.8	96.4	90.8	97.2	94.7	94.6	96.1	106.5
Lungs	0.64 ± 0.10	110.8	90.2	101.9	99.4	99.4	112.6	97.0	94.7	97.3
Heart	0.332 ± 0.030	101.5	105.4	104.8	103.0	97.3	98.8	98.8	97.3	106.6
Kidneys	0.87 ± 0.15	106.2	104.6	99.0	100.9	86.4	92.1	98.5	94.2	103.3
Testes	1.08 ± 0.13	99.7	115.0	105.6	108.3	103.1	107.3	88.4	90.2	101.5
Spleen	0.245 ± 0.034	79.6[c]	64.2[c]	57.3[c]	47.5[c]	37.4[c]	36.7[c]	40.9[c]	43.1[c]	60.7[c]
Brain	0.632 ± 0.046	103.8	101.6	105.7	107.4[b]	99.1	106.8[b]	104.0	97.2	102.1
Stomach	0.577 ± 0.038	97.1	110.7[b]	107.5	103.0	106.4	101.1	104.4	86.4[c]	105.3
Small intestine	2.22 ± 0.27	98.0	88.7[b]	83.9[c]	72.6	90.1	121.0[c]	111.2[b]	86.4[b]	105.5
Large intestine	1.02 ± 0.11	87.4	82.7[b]	90.0	86.5[b]	105.1	120.6[b]	116.0[b]	97.5	97.8
Thymus	0.143 ± 0.025	101.8	86.4	42.2[c]	11.5[c]	22.2[c]	22.9[c]	23.5[b]	26.7[c]	43.4[c]

[a] The rats of Holzman strain weighing 300–350 gm were exposed to 700 R whole-body irradiation and allowed free access to food and water. The data for normal organ weights are expressed as percent of total body weight (after substraction of depot fat and gastrointestinal content). The changes in total body and organ weights after exposure are given in percent of the respective nonirradiated controls. Note the decrease in weight of lymphoid tissues and muscle. [From W. O. Caster, *et al.*, *Proc. Soc. Exptl. Biol. Med.* **91**, 122 and 126 (1956).]

[b] Significant *L* 0.05.

[c] Significant *L* 0.01.

B. Connective Tissue

Connective tissues, which represent a substantial part of the total body mass, are present not only in gross supporting structures, but are also distributed throughout the interstitial spaces of all parenchymal organs. Characteristic of connective tissue is that cells—mainly fibrocytic elements—occupy only a small volume compared with that of the extracellular structures. The latter consist of collagen or elastin fibers embedded in a ground substance composed mainly of mucopolysaccharides. Peptide chains of these proteins are elaborated inside cells, and the fibers become more and more metabolically inert and insoluble as soon as they are secreted into extracellular space and linked to other constituents of connective tissue (collagen, elastin, mucopolysaccharides). Certain other effects of irradiation, such as depolymerization of the ground substance (p. 313) and the spreading of dyes and proteins (p. 314) are considered elsewhere in this book, and those dealing with bone and cartilage form a separate section.

1. Collagen Metabolism

Collagen metabolism after irradiation has been studied in different organs as well as in granulomas induced by implanted polyethylene sponges or by trauma. It should be remembered that the metabolism of collagen differs from that of other nonfibrous proteins in that collagen is inhomogeneous from a metabolic and chemical point of view. This fact must be considered when interpreting data obtained in tracer experiments on collagen metabolism. Newly formed collagen is soluble in salt solution and exhibits a high specific activity shortly after injection of a labeled precursor. Subsequently, this soluble collagen is either degraded or converted into more insoluble and metabolically inactive forms of collagen.

The collagen content of skin decreases 4–9 days after a whole-body exposure to 700 R (*164*), but increases 15 weeks after 3 kR of localized irradiation (*136*), possibly as a result of chronic inflammation and scarring. In irradiated muscle, soluble collagen increases at the expense of the insoluble fraction, suggesting a defect in the conversion of soluble to insoluble collagen (*29*). Similar observations have been made on collagen in the irradiated bladder (*232*).

Less collagen, especially the soluble form, is laid down in the implanted sponges after whole-body exposure to 450 R or after local exposure (*200*, *201*), whereas the hexosamine content in the granulomas is not altered after irradiation. In contrast, in the nonirradiated sponges, only collagen (not hexosamine) increases during the experimental period. Less collagen is also produced in fresh wounds during the first week after an 800 rad whole-body exposure of rats, whereas the content of methionine and cystine and the incorporation of sulfate-^{35}S in such wounds are not modified (*133*). Experiments measuring variations in collagen content over a period of several days are, however, not suitable to determine whether the synthesis of collagen in a given cell is altered or whether these changes reflect migration, division, or differentiation of collagen-forming fibrocytes after irradiation.

Studies of free amino acid pools suggest that tissue breakdown constitutes an important source of "free" urinary amino acids after irradiation (*155, 240*). While searching for the origin of these breakdown products, it was observed that excess radioactive glycine or phenylalanine injected prior to exposure is lost from muscle collagen 1–2 days after irradiation with 750–1000 R (*87, 102*). Studies with precursors more specific for collagen (proline, which gives rise to hydroxyproline, an amino acid, found exclusively in collagen) demonstrated that more soluble collagen, especially in skin, bone, and muscle (*90*), but not in tendon (*142*), is degraded after whole-body irradiation, and that the conversion of soluble to insoluble collagen is impaired. Starvation influences collagen metabolism in a different manner, namely, by decreasing collagen synthesis (*90*). Granulomas formed in implanted sponges labeled with proline also exhibit a similar defect in conversion to insoluble collagen after whole-body, but not local, exposure to 750 R (*142*). Urinary pyrrole carboxylic acid, a catabolite of hydroxyproline and, therefore, of collagen, increases in specific activity 1 day after whole-body, but not local, irradiation (*142*). These two observations indicate that collagen metabolism is also influenced by abscopal effects as well as by direct action of radiation.

2. Collagen and Aging

Animals exposed to irradiation age more rapidly and die prematurely from nonspecific diseases (*38*). In this connection, it should be mentioned that collagen is also modified by age: it becomes metabolically inert and increases in quantity relative to the ground substance; the ratio of soluble to insoluble collagen diminishes; thermal contraction becomes stronger, etc. Attempts to correlate the "accelerated aging" of animals after irradiation with such changes in connective tissues, however, have not yielded entirely conclusive results. The degree of thermal contraction (up to 17 weeks after 900 R) (*276*; see also *53a*) and the ratio of soluble to insoluble collagen (3 months after 425 R) (*283*) are unaltered after irradiation. Observations which suggest that irradiation accelerates aging processes are: collagen increases relative to hexosamine 15–18 months after whole-body exposure of rats to 550 R (*250*); insoluble collagen increases in quantity and its thermal shrinkage is modified after fractional irradiation (*26*); the isotope distribution in collagen and mucopolysaccharides after injection of labeled glycine and glucose and subsequent irradiation resembles that seen in older animals (*141*).

3. Elastin and Mucopolysaccharides

Replacement of elastin in skin is delayed after irradiation with 1000 R (*89*). Short-term uptake, i.e., permeability to sulfate-^{35}S in the soft tissues of the knee joint, is enhanced 4 days after local exposure to 2 kR. After the 10th week after exposure, the rapidly turning over (5.5 hr) mucopolysaccharides in the knee joint incorporate more sulfate-^{35}S whereas those with a low turnover rate display only slight changes (*289*). Incorporation of sulfate-^{35}S into the ground substance surrounding locally irradiated tumors diminishes (*66a*), and replacement of sulfate and hexosamine in mucopolysaccharides of rat skin and bone is delayed after a whole-body irradiation with 100 R (*91*). Changes in mucopolysaccharides in blood and urine during radiation therapy of cancer patients may reflect the radiation reaction of the irradiated tumor (*125*).

C. Blood Vessels and Adipose Tissue

1. Blood Vessels

Incorporation of sulfate-^{35}S into the aorta increases shortly after 500 R, but diminishes later (*146, 147*). Several months after an irradiation of the kidney, glomerulosclerosis may develop (p. 214), perhaps as a result of metabolic changes in blood vessels. Indeed, excess cholesterol is deposited in atheromatous lesions of the rat's aorta after whole-body irradiation (*154*), but this effect depends on species, strain (*9*), and nutrition (*195*). Irradiation of the aorta by high doses of superficial β-rays (30 kR) leads to necrosis, a decrease in mucopolysaccharides, and, eventually, to scarification with an increase in collagen (*147*).

2. Adipose Tissue

Few data are available on the metabolism of irradiated fatty tissue. Lipase activity in fatty tissue increases 2 hr after whole-body irradiation with 700 to 800 R (*168*). Cholesterol granulomas formed in muscle of animals on a high cholesterol diet are mobilized more rapidly after whole-body α-irradiation from incorporated radon (*157*). The increase in lipolytic activity induced by growth hormone is eliminated by irradiation of the fat cells (*71c*). Depot fats do not seem to participate significantly in the generalized disturbance of lipid metabolism in whole-body irradiated animals (*71*).

D. Cartilage and Bone

In young animals, bone is formed mainly from epiphyseal cartilage. Chondroblasts divide, arrange themselves in columns, and form a matrix of collagen and mucopolysaccharides in which hydroxyapatite is deposited. This primary bone structure is subsequently remolded by osteoblasts which deposit and by osteoclasts which remove bone. The latter reactions proceed at a reduced rate throughout adult life.

Cell renewal in chondroblasts is easily affected by irradiation. Recently,

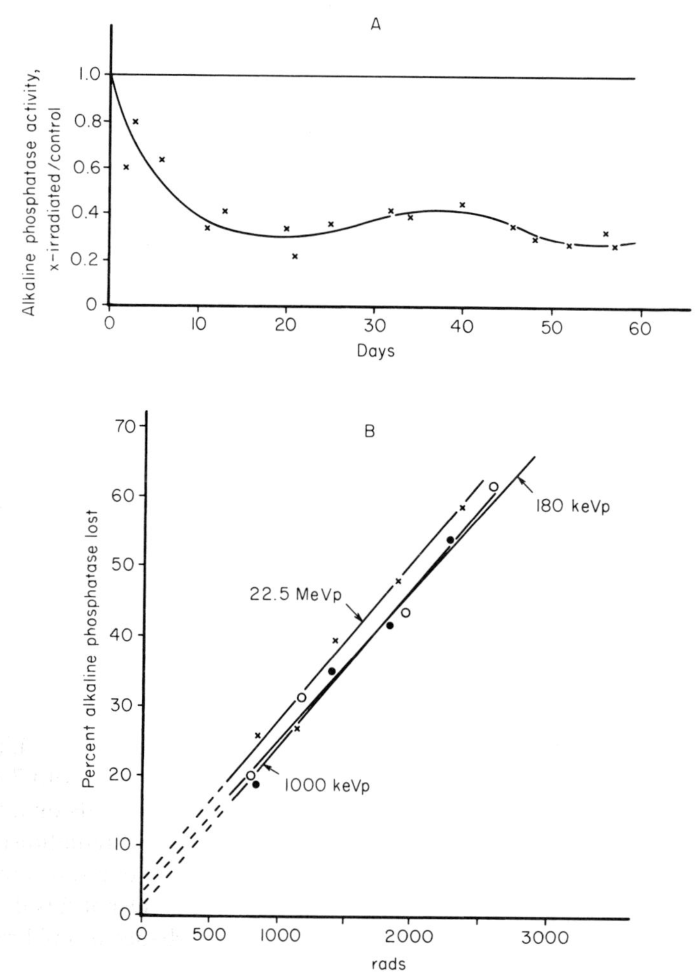

FIG. VI-1. Alkaline phosphatase in mouse tibia (A) [ratio to that of nonirradiated leg (straight line)] at different times after local irradiation with 2610 rads of x-rays (180 keVp), and alkaline phosphatase in mouse tibia (B) (per cent of activity lost) 50 days after irradiation with different doses of x-rays (180 keVp) or γ-rays (1 MeVp, 22.5 MeVp). The female mice weighed 20 gm at the time of irradiation. Note that the rapid and irreversible decrease in alkaline phosphatase which is a linear function of the dose and is only slightly dependent on radiation quality. [From H. Q. Woodard, and J. S. Laughlin, *Radiation Res.* 7, 236 (1957).]

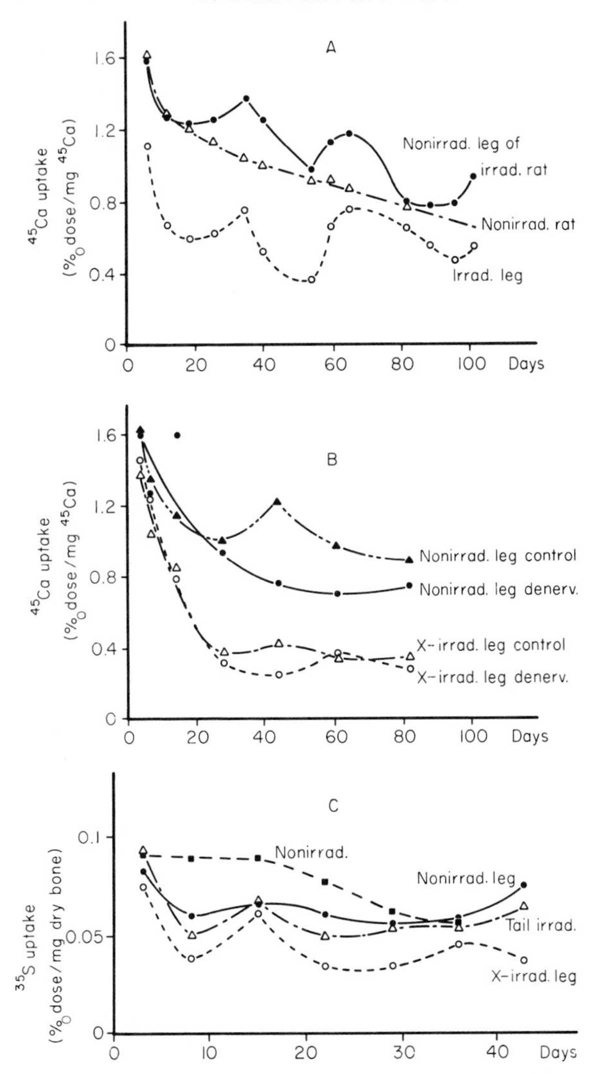

FIG. VI-2. Uptake of ^{45}Ca or ^{35}S 48 hr after administration in the proximal tibial epiphysis of young rats (120–130 gm) at different times after local irradiation with 2 kR. Note that in graph A, the ^{45}Ca uptake in the x-irradiated leg diminishes by 1 week after exposure, then follows a cyclical pattern. Similar cycles, even resulting in stimulation (compared to nonirradiated animals) are seen in the nonirradiated leg of an irradiated rat. Graph B demonstrates that the waves in ^{45}Ca uptake in the normal and x-irradiated legs are suppressed if the irradiated leg is denervated prior to exposure. Graph C demonstrates that uptake of ^{35}S decreases in the x-irradiated and the nonirradiated tibia of rats. Even irradiation of the proximal ends of the tails affects ^{35}S metabolism in the tibia. Controls were not irradiated. [From A. Babicky, J. Kolár, and L. Vyhnánek, *Strahlentherapie* **126**, 449 (1965); **130**, 349 (1966).]

the radiobiological characteristics of chondroblasts were assayed by clonal techniques and a D_0 of 165 rads and an extrapolation number n of 6 were found. In anoxia, these values become $D_0 = 340$ rads, and $n = 20$ (*128*). Since one column in the epiphyseal plate consists of about fifteen chondroblasts, at least one cell in each column must remain intact after exposure to less than 800 R for growth to be merely delayed without being permanently impaired. After exposure to higher doses (up to about 2 kR) the architecture of the epiphyseal plate is disturbed, and growth is arrested for several weeks.

Alkaline phosphatase activity in bone diminishes progressively beginning 24 hr after exposure and returns to normal levels when regeneration of bone starts (*44*, *259b*, *292*, *293*) (Fig. VI-1). This decrease may be related to the elevation of phosphatase activity in the blood (see p. 269) (*259b*). Histochemical investigations reveal that alkaline phosphatase activity (*185*, *237*) and glycogen (*185*) are lost mainly from the area of hyperblastic chondroblasts in the epiphyseal plate. Alkaline and acid phosphatases in osteoblasts show little change in activity and distribution, but the activity of the latter enzyme is augmented in areas where damaged epiphyseal tissue is resorbed (*237*).

Incorporation of sulfate-^{35}S into mucopolysaccharides of the epiphysis is depressed 12 days after exposure to 900 R (*67*) and earlier after higher doses (Fig. IV-2) (*12*, *13*). Incorporation of sulfate-^{35}S into chondroblastomas is elevated temporarily after exposure and then declines (*50*). On the other hand, the metabolic replacement of mucopolysaccharides in rat bone labeled with sulfate-^{35}S or glucose-^{14}C prior to exposure appears to be delayed as early as 1–2 days after whole-body irradiation with 500 or 1000 R (*91*). Bone collagen of young rats labeled with proline before irradiation undergoes the same change as that in connective tissue of other organs (p. 197), namely, impaired conversion of soluble to insoluble collagen and excessive degradation (*90*).

Mineralization of the bone is affected by the second week after irradiation, i.e., a few days later than incorporation of sulfate-^{35}S is impaired (*12*), but the time and extent of inhibition vary with the dose and the age of the animals (*13*, *288*). Accretion of Ca in the bone diminishes, whereas the easily exchangeable Ca fraction increases within 12 days after 500 R of whole-body or 2000 R of local irradiation (*45*); this effect is not seen at lower dose levels (*206*). A plot of Ca uptake by bone vs. radiation dose reveals two components, one radiosensitive (about 30%), and the other radioresistant (*298*). These components might represent bone formation in the epiphysis and osteoblasts, respectively. Incorporation of phosphate-^{32}P into the knee joint also diminishes by the second week postirradiation (*288*), following the pattern of Ca metabolism. On the other hand, uptake of ^{32}P by the skeleton (*77*, *151*) and the distribution and excretion of Ca and Sr have been shown to be altered (e.g., *171*) as early as one day after whole-body irradiation. Most likely, such early effects are secondary to metabolic changes occurring outside the bone.

Modification of growth and incorporation of sulfate-^{35}S and ^{45}Ca depend not only on the direct action of radiation on bone, but also on systemic factors. Thus, local exposure to a given dose is much less effective in inhibiting growth than whole-body irradiation. Systemic factors have been estimated to cause about 50% of the growth inhibition after 450 R whole-body exposure (*211*) and about 70% (*47*) after 600 R. Uptake of ^{45}Ca (*14, 27, 140*) and the incorporation of sulfate-^{35}S (*13*) are not only depressed in the irradiated leg, but undergo cyclical variations with time in irradiated as well as in nonirradiated bone (Fig. VI-2B) (*13, 14, 140*). If the nerves to the irradiated leg are cut within a certain time after exposure, cyclical variations in all other bones disappear (Fig. VI-2B), whereas denervation of a nonirradiated leg eliminates the cycles in this leg only (*14*). Pregnancy also suppressed temporarily these systemic reactions in bone (*14a*). Undoubtedly, regulation by the central nervous system plays a part in these phenomena, but other possible factors must also be considered. Anorexia, an important factor early in retardation of bone growth, is probably unimportant at later times. Adrenal activity may affect osteoblastic activity, and the parathyroids may influence osteoblastic as well as osteoclastic activity. These factors need to be studied in more detail.

Although epiphyseal growth is most susceptible to the effects of irradiation, the osteoblast–osteoclast system in mature bone cannot be neglected. After low doses, stimulation of osteoblastic activity (*185*) and more rapid healing of fractures are observed (*95*). High doses of radiation delivered by plutonium deposited in bone depress the alkaline phosphatase activity of osteoblasts (*70*). Whole-body irradiation delays the removal of ^{89}Sr deposited in bone (*74*). Differentiation of osteoblasts is prevented by irradiating internally with plutonium deposited in mature rat bone (*60*).

E. Skin

Changes in skin were the first radiation-induced injuries reported in man, and largely determined the limits and possibilities of radiation therapy until the introduction of modern techniques of therapy. Nevertheless, information concerning the relationship between pathological aspects, cell renewal, and biochemistry of the irradiated skin is still fragmentary. Skin is a composite organ and consists of an outer epidermal layer, a cutaneous and subcutaneous connective tissue layer, and various accessory organs, such as hairs, nails, exocrine glands, and sensory receptors. All of these structures participate to varying degrees in the reaction of skin to radiation exposure.

1. Epidermis

The epidermal layer represents a cell renewal system akin to the intestinal epithelium, but cell replacement and transit times are slower. Only about 2%

of all cells are renewed daily in the epidermis as contrasted with 50% in the intestine (*208*, *209*). The epidermal cells divide in the germinal stratum, differentiate by keratinization, and are finally shed from the surface of the skin. The length of the cell cycle and the transit time depend largely on the anatomical site, age, strain, and functional state of the skin. Thus, transit time in the epidermis varies from about 8 to 30 days (e.g., guinea pigs, 8.5 days; man, 14–17 days; mice and rats, 10–30 days) (*252*).

One to two days after exposure to 800 R or more, the skin may redden temporarily. During this early erythema the blood vessels are congested and edema occurs in the subcutaneous layer, whereas the epidermis appears normal except for mitotic arrest (*139*) and a depression in thymidine incorporation (*56*). Histamine, heparin, serotonin, or other vasoactive substances (see p. 307) may be liberated during this phase and affect blood vessels and connective tissue directly or by way of activation of proteases (*119*) or hyaluronidase (*135*). Direct depolymerization of connective tissue ground substance has also been observed (see p. 313). These changes in permeability permit proteins (see p. 314) or dyes to diffuse more rapidly from the blood vessels into subcutaneous tissue.

A latent period of about a week follows the early erythema and then the main erythematous reaction commences and usually lasts from 10 until 20–30 days after exposure. The epidermal layer flattens or is even completely destroyed. Mitoses and DNA synthesis are often abnormal, and, after administration of thymidine, autoradiography reveals intensively labeled giant cells. The cell nucleus may be retained during keratinization. These reactions may correspond to the abortive recovery of the type seen in bone marrow, intestine, etc. The main erythema reaction involves not only the epidermis, but also the underlying strata of skin, especially the blood vessels. When the basal layers of the epidermis regenerate, the erythema disappears, but may reappear in a wave-like manner and lead to ulcerative or dry dermatitis. Hyperplastic regeneration of the irradiated skin could involve an altered control mechanism of cell proliferation (*252*).

Recently, the radiosensitivity of epidermal stem cells was estimated directly on the basis of the number of skin nodules formed in denuded skin after a high dose of irradiation; the D_0 was 135 rads in air and 350 rads under anoxic conditions (*290*). The extrapolation number n may be estimated from split-dose experiments (n = about 6), but the type of dose-effect curve (type 1 or 2 in Volume I, Appendix 2, A) remains uncertain (*113*). Ulceration of the skin occurs when less than 10–20 cells/cm^2 of skin are able to recover their ability to divide (*290*).

Since biochemical changes in irradiated skin must be evaluated in relation to the morphological lesions, attention must be given to the type of exposure. Thus, irradiation with 3 krep of β-rays from ^{90}Sr, which reach only the superficial layers to a depth of a few millimeters, causes hyperkeratinization and an increase in epidermal protein content (Fig. VI-3). On the other hand, x-irradiation affects the deeper layers of the skin as well, and diminishes the amount of skin proteins (*164*). Various biochemical observations in irradiated skin are summarized in Table VI-2.

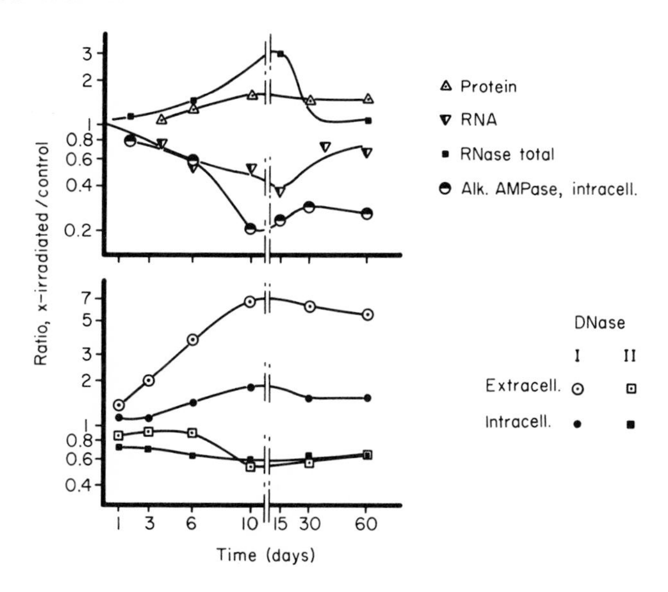

FIG. VI-3. Biochemical changes in guinea pig epidermis irradiated with a skin-surface dose of 3 krep β-rays from a ^{90}Sr source. Note that DNase I increases, particularly outside the cell, whereas DNase II diminishes after irradiation. The increase in extracellular DNase represents one of the earliest changes after irradiation, and could reflect cell death or repair of radiation damage. It is, however, more likely related to the altered permeability of the skin, since it can also be elicited by trauma. [J. Tabachnick *et al.*, *J. Albert Einstein Med. Center* **15**, 71 (1967).] RNA, intra- and extracellular AMPase, and alkaline pyrophosphatase also decrease (of these enzymes, only intracellular AMPase is shown in the figure). On the other hand, RNase activity is enhanced as a result of the removal of some unknown inhibitor. The cell protein content rises as the epidermis becomes hyperplastic. An increase in activity of acid phosphatase has been described, especially around the thirtieth day after exposure (*262*, *265*, *266*).

2. HAIR

Hair is formed by rapidly dividing cell populations in the follicles. Hair growth is not continuous, but occurs in cycles and may be initiated by plucking the hair. Mitotic activity in follicle cells stops even after exposure to low doses of radiation. DNA synthesis diminishes immediately after exposure and declines further with time as fewer cells enter the S phase (*41a*). Although DNA synthesis can resume several hours or days later, the structure of the hair

TABLE VI-2

BIOCHEMICAL ALTERATIONS IN THE IRRADIATED SKIN

Function	Conditions	Early period (1–2 days)	Latent period (2–10 days)	Erythema (10–30 days)	Late changes (30 days and later)
DNA content	x-rays	Decrease (*164*, *278*)[a]	Decrease (*85*)[b]	—	—
	3 krep β-rays, guinea pig	—	—	Decrease max. 15 days [also RNA (*261*)]	—
Protein content	3 krep β-rays, guinea pig	—	Progressive increase to day 10; decrease after 700 R x-rays (*164*) or 14 krep β-rays (*174*, *175*)	Continues to remain high to day 60 (*266*)	
Free amino acids	3 krep β-rays, guinea pig	—	Biphasic increase from day 10 to day 25 (also urocanic acid) (*263*, *264*)		—
Glycogen	x-rays	Increased synthesis 800 R (*3*)	Increase, then decrease in concentration (>1 kR) (*18*, *36*)		Decrease 800 R x-rays (*80*), but increase in linoleic acid 2 kR (*282*)
Fat	β- and x-rays	—	Decrease 14 krep β-rays (*175*)	—	
DNaseI	3 krep β-rays, guinea pig	Extracellular increase from day 1 (less intracellular) (*265*) (Fig. VI-3)	—	—	—
DNase II	3 krep β-rays, guinea pig	Intracellular decrease (*265*); extracellular no change	—	—	—

RNase (alkaline)	3 krep β-rays, guinea pig	—	—	Increase maximal day 15 (*262*)	—
Acid phosphatase	3 krep β-rays, guinea pig	—	—	Intracellular increase, remains high to day 60 (*266*); extracellular, slight decrease, increase day 30	
	3 kR x-rays, guinea pig	Increase (*37*)	—	—	—
Alkaline phosphatase	3 kR x-rays, guinea pig	Increase (*266*); increase 14 krep β-rays (*174*); increase at 2 days, decrease at 4 days, later normal (*180*)			—
AMPase	3 krep β-rays, guinea pig	Intra- and extracellular decrease (*266*) (similar PPase)		—	Low levels to day 60 (*266*)
5-Nucleotidase	2.4 kR x-rays, rat	No change (*17*)	—	—	—
	400 R x-rays, guinea pig	Increase (1 to 3 days) (*111*)	—	—	—
Esterase and lipase	14 krep β-rays, rat	—	Decrease (*174*)	—	—
Hexokinase	700 R x-rays, rat	—	Decrease (*179a*)	—	—
O_2 consumption	—	*In vitro:* decrease, 24 hr (*82*) (> 10 kR)	—	—	—
		In vivo: increase, 2 days (*175*) 14 krep β-rays; decrease 7 days, increase 2nd week after 700 R whole-body exposure (*180b*); A + low doses, oscill. changes parallel erythema (*130*)			—
Glycolysis	—	*In vitro:*	—	—	—
		decrease, 9 hr (*82*) (> 10 kR)	—	—	—
		In vivo: increase, 2 days (*175*) 14 krep β-rays			
Dehydrogenases	1 kR x-rays, mice	—	Increase, then decrease (*118*)	—	—
	200–800 R, x-rays, rabbit	Decrease, [also lipoamide dehydrogenase (*122b*)]		—	—

[a] 700 R, mice.
[b] 3 kR, rat.

follicle is often damaged in such a way that the hair cannot grow in an orderly fashion. Later the hair becomes dysplastic (*98*, *271*) and may eventually be lost either temporarily or permanently. Moreover, hair of rodents turns gray after irradiation as a result of damage to the clear follicle cells (*83*). Loss of hair has been frequently utilized for the study of radioprotective mechanisms. Similarly, protection against epilation is obtained by the local injection of compounds such as hyaluronidase or chondroitin sulfate (*19*, *272*) which cause liberation of vasoactive protective substances. Slight local trauma, which liberates serotonin from blood platelets, is sufficient to produce this result (*273*).

F. Muscle

1. Immediate and Early Effects of High Doses

Very high doses of radiation (about 100 kR) are needed to damage the function of the isolated muscle immediately after exposure and to cause structural changes within a few hours. The resting and action potentials as well as the chronaxie of isolated frog or mammalian muscle are reduced after a dose of the order of 100 kR (*22*, *25*, *53*, *92a*, *108*, *219*, *221*) dependent on dose rate. Isolated muscle can contract spontaneously during such high-intensity exposure (Fig. VI-4), but later may become refractory to excitation (*25*). Yet, *in vivo*, muscle functions normally for 22 hr after local exposure to 73 kR (*287*). Muscle with spontaneous activity may be more responsive to irradiation, although it appears difficult to differentiate between the action of radiation on the muscle proper and on its autonomic innervation. Rhythmically contracting Purkinje fibers show a diminished action potential after exposure to 25 kR (*108*), and muscular preparations from leeches contract after about 20 kR (*117*).

Oxygen consumption (*21*, *31*, *220*, *221*) and glycolysis (*59*) of isolated muscle increase immediately after exposure to about 100 kR and decrease after still higher doses (Fig. VI-5). Potassium is lost (*22*, *53*, *220*) (Fig. VI-6), but excess sodium enters the cell (*220*, *221a*). Incorporation of ^{32}P into AMP and ATP (Fig. VI-5) may increase initially (*59*), but to a smaller extent, than O_2 consumption, thereby uncoupling oxidative phosphorylation (*21*, *220*). More ATP is found in frog heart after exposure to 100 kR (*59*), but under other experimental conditions, ATP levels are reduced (*21*, *220*) and ATPase activated (*221*). No direct effect of radiation on enzymes in muscle, such as pyruvate kinase (*23*), is expected with dose levels below 300 kR; rather, radiation may damage the muscle membrane, impair the sodium pump, and subsequently uncouple oxidative phosphorylation.

A second phase of radiation damage to isolated muscle commences 1–3 hr after exposure (*59*). Respiration deteriorates rapidly, as do glycolysis (*31*) and other metabolic functions, such as utilization of acetate (*59*) (Fig. VI-5). However, muscle irradiated *in vivo* may behave in a manner different from that irradiated and maintained *in vitro*, since metabolites can be transported *in vivo* to and from other organs. Thus, potassium is not lost after irradiation of

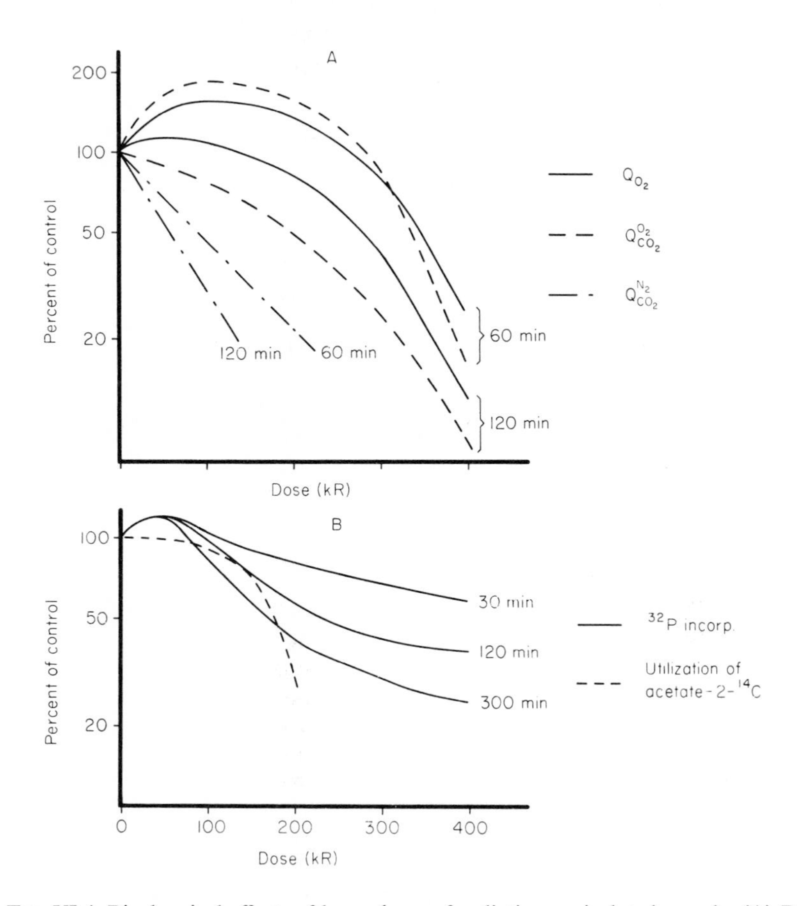

FIG. VI-4. Biochemical effects of large doses of radiation on isolated muscle. (A) Rat diaphragm (assayed at 38°C); (B) frog heart (assayed at 20°C). Note that 1 hr after 100–200 kR, respiration (Q_{CO_2}), and even more, aerobic glycolysis (Q^{O2}_{CO2}) are enhanced in rat diaphragm. Later, both reactions, especially aerobic glycolysis, diminish. Anaerobic glycolysis (Q^{N2}_{CO2}) is depressed at all times after irradiation. In the frog heart, incorporation of phosphate into ATP and AMP is enhanced early after exposure to 100–200 kR, but the heart contains more ATP and less creatine phosphate at this time. Later, or after higher doses, incorporation of phosphate diminishes. Utilization of acetate for the various reactions of the citric acid cycle is also depressed 2 hr after exposure. [From F. Bresciani *et al.*, *Naturforsch.* **14b**, 158 (1959) and K. Dose and F. Bresciani, *Biochem. Z.* **331**, 172 and 187 (1959).]

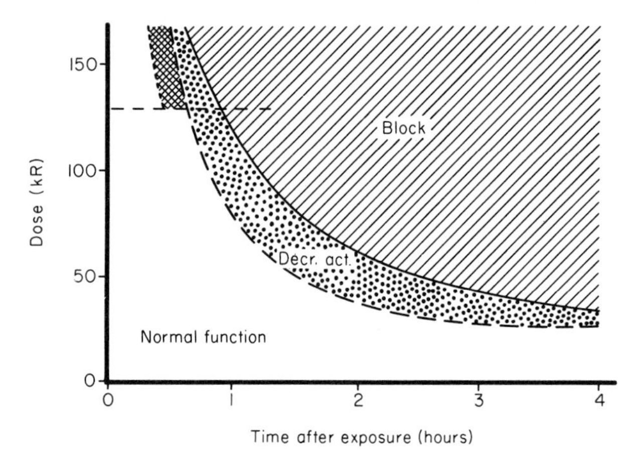

FIG. VI-5. Effect of x-irradiation (2.8 kR/min) on electrical activity of the frog gastrocnemius muscle stimulated by the sciatic nerve. Note that doses below 30 kR have no effect. Higher doses decrease activity and lead to a total block after a time interval dependent on the dose. Very high doses induce spontaneous contraction (crosshatched area). [From R. M. Bergström, R. Blafield, and A. Salmi, *Intern. J. Radiation Biol.* **4**, 351 (1962).]

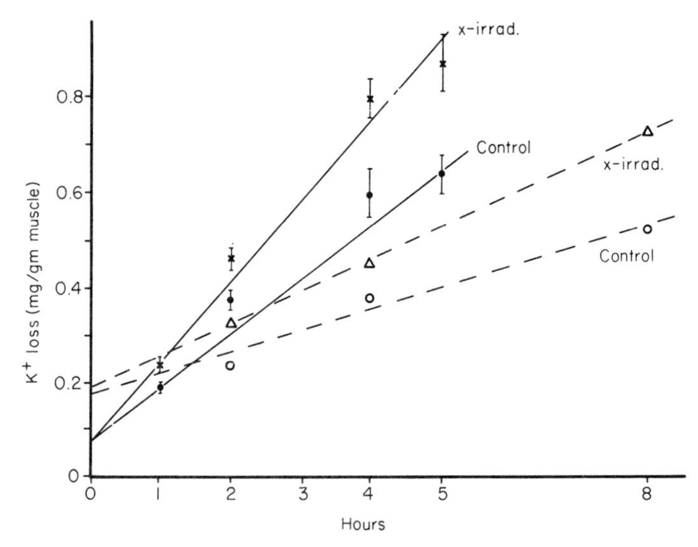

FIG. VI-6. Potassium loss from frog muscle as a function of time after irradiation. Isolated hind leg muscles were soaked for 2 hr in K^+-free Ringer's solution and then irradiated with 10 kR at 6 kR/min. Note the increased K^+ loss from irradiated muscle. More K^+ is lost at 25°C (solid lines) than at 0°C (dashed lines), but the relative increase in K^+ loss on irradiation is about the same at both temperatures. K^+ loss appears to be a consequence of damage to the sodium pump. [From A. Portela, J. C. Perez, P. Stewart, M. Hines, and V. Reddy, *Exptl. Cell Res.* **29**, 527 (1963).]

the hind limb with 73 kR (*287*) and, although glycolysis is unaltered after whole-body exposure to 60 kR, lactate accumulates (*158*).

Cardiac function is rather resistant to the immediate effects of irradiation. Atrioventricular transmission in mouse heart is blocked only after exposure to 2 MR, and ventricular fibrillation begins only after still higher doses (*183*). Function and survival of the isolated frog heart are affected only slightly by irradiation with 800 kR, although an accelerated uptake of Rb indicates changes in the cell membrane (*190*). On the other hand, the isolated perfused rabbit heart loses K^+ after exposure to only 500 R (*71a*). Bradycardia observed after exposure to 2–5 kR most likely reflects vagal stimulation (*205*), whereas changes in the length of ST waves in the ECG after 50 kR may indicate increased potassium content of the blood (*224*). In dogs, the ECG is not altered immediately after local exposure to 100 kR, but doses of 5–10 kR eventually cause myocardial necroses with corresponding changes in the ECG (*212*).

2. Changes in Acute Radiation Sickness

During the acute radiation sickness after whole-body exposure, the metabolism and function of skeletal and heart muscle are also modified (see data on composition and enzymatic activity of muscle, Table VI-3). Muscle weight decreases significantly within 3–6 hr after 700 R (see Table VI-1). The uptake of glucose by the heart diminishes within a few hours after 1000 R (*148*) and the extracellular space increases (*111a*). In skeletal muscle, the K^+ content decreases (Fig. VI-7) (*61*, *111a*) and the Na^+ increases during the first day after exposure, whereas in heart muscle the reverse is observed (*111a*). Later, the K^+ content of skeletal muscle in normal, but not in adrenalectomized rats may rise above normal levels (*32*).

Incorporation of precursors into muscle proteins is altered markedly after irradiation. Less glycine (*6*), alanine (*109*), and methionine (*259*) are incorporated into muscle proteins within the first hours after exposure. Incorporation of phosphate-^{32}P into the phosphoproteins of muscle is enhanced, whereas the phosphoprotein concentration diminishes (*251*). Moreover, incorporation of ^{59}Fe (*28a*) and glycine (*4a*) into the heme of myoglobin in rat muscle is depressed 2–3 days after a whole-body irradiation, whereas incorporation of glycine into the globin is unaltered. Changes in incorporation could, however, result not only from an altered synthesis of proteins but also from modifications of the pool size of the precursor or of its uptake by the cell. Indeed, uptake of glycine (*78*) and ^{32}P (*77*) by the muscle is altered in radiation sickness. Incorporation of acetate-^{14}C into proteins of rabbit muscle after whole-body exposure to 400–800 R diminishes at 3 hr, increases to above normal levels at 18 hr, and declines thereafter (*297*).

TABLE VI-3

BIOCHEMICAL CHANGES IN SKELETAL AND HEART MUSCLE AFTER IRRADIATION

Enzyme or substance	Change	
	Skeletal muscle	Heart muscle
Glycolysis	—	No change 2 hr, 4 kR (*148*)
Succinoxidase	—	Increase, rats 0.5–1.5 days, 1 kR (*68*)
Dehydrogenases	Succinate dehydrogenase and diaphorase increase 1 day, 400 R (*120*), later decrease (*239a*); lactate dehydrogenase decreases 3 days, 1000 R (*7*)	Lactate dehydrogenase shift in isozymes after 600 R (*114*)
Hexokinase	Increase 240 R (*245*), decrease 600 R (*236*)	No change
Aldolase	No change 240 R (*245*)	No change
Aminotransferases	—	Glutamic-pyruvic amino transferase decreases, glutamic-oxalacetic amino transferase increases 4 days, 600 R (*30*); see also (*214*, *215*)
Alkaline phosphatase	Increase 3 days, 1000 R (*196*)	—
ATPase	Increase 1 day, 400 R (*120*), then decrease; decrease 3 weeks, 20 kR (*284*)	—
Acid phosphatase	Increase 3 days, 1000 R (*196*) or 400 R (*143*) (in endothelial cells	—
Esterases	Thiolesterase decreases, E600 resistant esterase increases 400 R, 1 day (*143*)	—
Glycogen	Slight decrease 2–4 days, 1000 R (*42*, *150*, *202*)	Decrease 2 hr, 4 kR (*148*), increase later (*96*)
ATP	Increase, then decrease 700 R (*42*, *152*); no immediate change 30 kR (*189*)	Increase 700 R (*152*)
Lactate	Decrease 600 R (*236*)	—

When proteins of muscle are labeled prior to exposure by injecting a ^{14}C-labeled amino acid, metabolic replacement of soluble proteins and elastin occurs to the same extent in normal and irradiated animals (*89*). Since this type of experiment averages the metabolic replacement over a period of 1–2 days, changes of short duration could escape detection. However, muscle collagen in excess of normal amounts is degraded after whole-body irradiation (*90*) and may contribute significantly to the pools of free amino acids (*87*, *155*). The protein composition of skeletal (*177*, *297*) and heart (*203*) muscle is altered after whole-body irradiation, but the meaning of these changes is not clear.

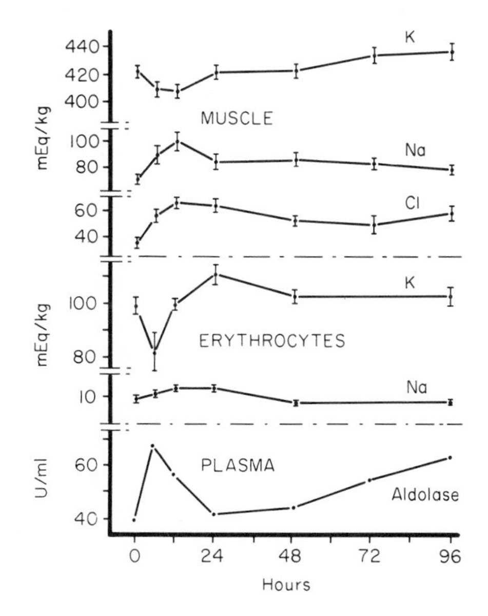

FIG. VI-7. Electrolytes (K^+, Na^+, and Cl^-) in muscle and erythrocytes, and aldolase in plasma after irradiation of the hindquarters of rats with 3 kR. The values for muscle are related to dry, fat-free tissue. Note that K^+ in muscle and in erythrocytes diminishes slightly and Na^+ in muscle increases during the first day after exposure. Simultaneously, aldolase activity in plasma is enhanced. [From R. M. Dowben, and L. Zuckerman, *Nature* **197**, 400 (1963).]

Nucleic acids in irradiated muscle have rarely been investigated. RNA present in muscle at the time of irradiation is degraded more rapidly than that in normal muscle, and the metabolites formed can be detected in the urine (*88*).

Muscular fatigue (*103a*) is often experienced in patients undergoing radiation therapy, and has been related to changes in creatine metabolism, as is discussed in more detail elsewhere (p. 260) (*149*). The ability of the muscle (dog) to respond to electrical stimulation of the sciatic nerve is reduced 3–4 days after exposure to 700 R and the glycogen content of these muscles also diminishes during work, simultaneous with an increased flow of blood through the muscle (*165*).

G. Kidney

1. Function of the Kidney

The reaction of the kidney to irradiation resembles that of the liver in many respects. In both cases, we are dealing with an organ of highly specialized functions in which a normally slow cell renewal system can be stimulated by the removal of part of the organ. The physiological and biochemical functions of mammalian kidney, like those of the liver, are rather radioresistant, but the conditioned cell renewal system is radiosensitive. In contrast to the liver, the kidney irradiated locally with large doses of radiation develops characteristic late changes, i.e., glomerulosclerosis accompanying or following hypertension.

Doses below 1000 R do not seem to affect the histological appearance of adult kidney. Exposure to several kR causes cloudy swelling in the tubules, and sometimes exudate is present in the glomeruli accompanied by albuminuria; doses above 20 kR produce renal necrosis (*73*). Renal function is impaired seriously only if the radiation dose exceeds 1 kR. For example, the rate of glomerular filtration, when determined by means of creatinine or inulin clearance, as well as renal blood flow are not significantly altered in the dog after whole-body or local irradiation at dose levels of 800 R or less (*107*, *116*, *166*). The ability to concentrate urine during starvation (*20*) decreases somewhat after the second postirradiation day, whereas the reabsorption of Na^+ and Cl^- increases (*166*). K^+ retention and perhaps *p*-aminohippurate (PAH) accumulation (*116*) by kidney slices incubated *in vitro* are impaired during radiation sickness.

The failure of glomerular and, to a greater extent, of tubular function in kidneys irradiated locally at a dose level of several kR is distinctly evident soon after exposure, increases in extent several days after exposure (*43*, *75*, *187* but compare *296*), and is sometimes preceded by a transient rise in renal output. Furthermore, the Na^+ content of the kidney diminishes whereas the K^+ increases 3 days after 2000 R (*32*). Much higher doses, however, are needed to damage kidney function immediately after exposure. Accummulation of PAH (*188*) and glycolysis (*94*) are reduced after 50 kR. K^+ is lost (*37a*) and oxygen consumption in the presence of α-ketoglutarate increases in kidney slices after doses exceeding 100 kR (*33*), whereas endogenous respiration decreases (*210*).

2. Biochemistry of the Kidney

In general, enzymatic activites, as assayed by histochemical or chemical methods in the irradiated kidney, are unaltered if the dose is below 1000 R.

Thus, total dehydrogenase activity is not modified by whole-body exposure to 800 R (*66*, *159*), but the ratio between different isozymes changes (*114*, *123*). After 1000 R or more, the activity of various dehydrogenases (especially malate dehydrogenase) is reduced (*86a*), and the enzyme may be excreted in the urine (*124*). Formazan deposits become coarsely granular 2–7 days after 2–10 kR; irradiation causes enzymes to be bound to cellular structure in a different manner compared to that normally encountered (*228*).

An increase, or no change, in alkaline phosphatase is reported after irradiation of the kidney with low doses (*103*, *173*, *184*, *194*, *196*). Exposure to several kR produces loss of alkaline phosphatase from the tubules 1–30 days later, depending on the dose (Fig. VI-8) (*63*, *72*), but the distribution of alkaline

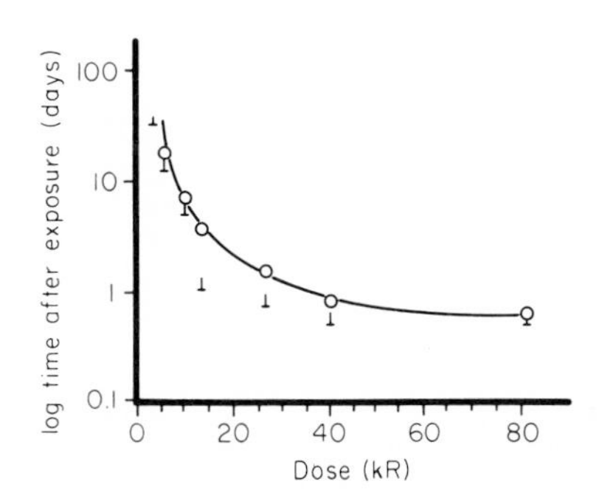

FIG. VI-8. Decrease in alkaline phosphatase activity in rat kidney after local irradiation. Time of earliest decrease (to 80% of control): inverted T's; time of decrease to 50% of the controls: circles. Note that alkaline phosphatase diminishes when the dose exceeds 2.5–5 kR. The time lag before the decrease in phosphatase activity appears to depend on the dose. [From U. Feine, and G. B. Gerber, *Strahlentherapie* **102**, 563 (1957).]

phosphatase in the cells changes little until necrosis occurs (*72*). Acid phosphatase (*184*), amino transferase (*30*), and peptidase (*127*) increase early after low doses of radiation (*127*); later, catalase (*277d*), esterase (*131*), and peptidase (*127*) diminish.

Less than the normal amount of thymidine is incorporated into kidney DNA after whole-body irradiation (*167*). Since few cells in the kidney are normally in S phase, this effect may reflect the action of radiation on the turnover of "metabolic" or mitochondrial DNA. Incorporation of precursors into proteins is reduced to some extent during the first post-irradiation days (*69*, *259*, *259a*), but is increased later on (*227*). The concentration of serine (*62*), basic amino acids (*71b*), ethanolamine, and aromatic amino acids in irradiated kidney decreases at first and then increases at 5–10 days (*255*). These changes may be related to the activity of pyridoxal-5-phosphate coenzymes. More formate (*226*) and less ^{32}P (*48*) is incorporated into lipids of irradiated as compared to nonirradiated kidney; the deposition of lipids in the tubules is visible on microscopic inspection.

Erythropoietin formation by the kidney could influence the recovery of bone marrow in radiation sickness. Whole-body irradiated animals produce more erythropoietin in the absence of a stimulus (*101*) and about as much as normal animals stimulated by hypoxia (500 R) (*115*). Kidneys irradiated locally with 3.6 kR synthesize less than normal quantities of erythropoietin (*75*).

3. Regeneration of the Kidney

Unilateral nephrectomy gives rise to a compensatory hypertrophy of the contralateral kidney, accompanied by weight increase and a proportionately smaller increase in DNA content. One day after surgery, DNA synthesis (*176*) and, somewhat later, mitotic activity are enhanced. This early response is delayed if the area of the kidney is irradiated shortly before or not more than 24 hr after unilateral nephrectomy (*176*, *268*) (Fig. VI-9). If the time interval between irradiation and operation is longer than 2–3 weeks, the cells appear to recover their ability to form the DNA-synthesizing enzymes and to divide (*46*). A rise in thymidine kinase activity produced by injecting a large amount of thymidine is not prevented by irradiation (*4*). The augmentation of kinase activity is thought to result from stabilization of existing enzyme apoprotein by its substrate rather than from *de novo* synthesis.

The extent of compensatory hypertrophy attained during longer intervals between operation and irradiation is not exclusively determined by the early response of DNA synthesis. Thus, local exposure with doses up to 20 kR did not influence the compensatory weight gain of mouse kidney (*254*). The weight gain of the irradiated kidney in these experiments may have been due, in part, to swelling of the organ because of the influx of water and electrolytes, while the DNA content remained unchanged in irradiated and nonirradiated kidney. Local exposure to 1 kR does not influence the final increase in weight after unilateral nephrectomy (*281*), and 2–4 kR reduces it only slightly. Whole-body irradiation with 500–800 R severely limits the compensatory hypertrophy (*46*, *81*, *268*, *279*), even if the kidney is shielded from direct radiation (*280*). The shielded kidney of an irradiated nephrectomized rat incorporates as much thymidine shortly after exposure as a nonirradiated organ (*281*), but, at a later time, anorexia may reduce the mitotic activity (*280*). Food deprivation, restriction of growth (*213*), and perhaps systemic factors (*281*) could, therefore, be responsible for the abscopal effects of radiation on compensatory kidney hypertrophy.

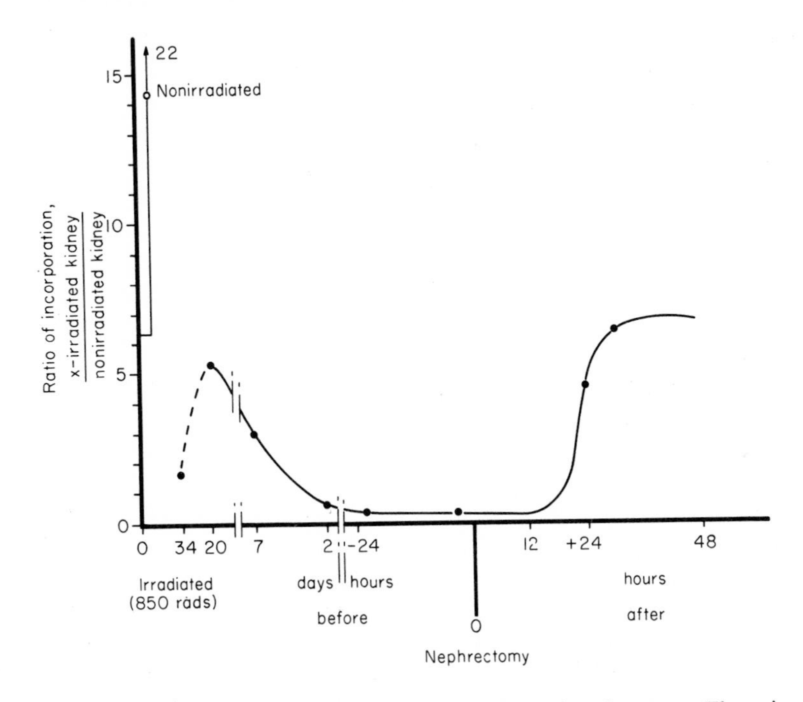

FIG. VI-9. Inhibition of kidney regeneration after unilateral nephrectomy. The animals were irradiated at various times before or after operation, and thymidine incorporation into the acid-soluble fraction (i.e., thymidine kinase and DNA polymerase) was measured 48 hr after the operation. The values represent the ratio of regenerating irradiated to non-regenerating nonirradiated kidney. The range of values for non-irradiated regenerating kidney is shown in the left side of the figure. Note that irradiation after exposure interferes with DNA synthesis only if it is carried out before the enzymes are synthesized, i.e., less than 18 hr after operation. Irradiation before the operation becomes less effective when the interval between the treatments is longer than 7 days; the irradiated cells must, therefore, have recovered in this interval. [From R. K. Main, L. J. Cole, and E. R. Walwick, *Intern. J. Radiation Biol.* 7, 45 (1963).]

H. Eye

Investigations of the irradiated eye have been concerned mainly with changes in the retina, the cornea, or the lens.

1. Retina

Metabolism in the retina has been likened to that in tumor cells. Similarly, glycolysis and respiration of the retina appear more radiosensitive than in most other tissues, and are depressed soon after exposure to 2 and 5 kR, respectively (*172*). Glycogen in the retina increases after irradiation with a few kR (*2*, *129*), and the distribution of acid phosphatase is altered (*186*). One hour after exposure to 200 R, incorporation of leucine into proteins is enhanced in the conducting elements but is diminished in the sensory cells of the retina (*181*, *182*). Of course, x-irradiation in very minute doses, such as a few mR, at a high dose rate must cause biochemical changes in the sensory elements of the retina (*163*), since the radiation is perceived as light.

2. Lens

The mammalian lens consists of clear fibers which are continuously formed from epithelial cells at the anterior surface of the lens. The state of dehydration and thus the transparency of the lens is maintained by energy derived mainly from glucose-degrading reactions. The lens has no proper blood supply, but depends on diffusion for the transport of metabolites. Irradiation causes an immediate cessation of mitoses (*209*). Depression of DNA synthesis (*277c*) becomes particularly striking when cell division has been previously stimulated by trauma to the eye (*105*). Although DNA synthesis resumes 3–24 hr after exposure, the fibers produced by the newly formed cells are damaged and opacities (cataracts) in the lens often develop.

Incorporation of ^{32}P or adenine-^{14}C into insoluble (but not into soluble or microsomal) RNA is enhanced (*160*, *161*) immediately after exposure of the lens to 1.5 kR; the UV absorption of RNA increases less on heating with formaldehyde than does that from nonirradiated lens (*162*), indicating a change in the physicochemical structure of RNA. A few weeks after exposure, the metabolism of albuminoid RNA returns to normal, but the DNA (*277b*) and soluble RNA (*55*) content of the lens decrease. Furthermore, soluble proteins diminish, whereas the insoluble proteins increase as they do during aging (*58*, *207*, *233*).

Glutathione and other SH compounds (*51*, *217*, *274*), as well as glutathione reductase (*216*, *217*, *260*) decrease by the first day after exposure to about 1 kR. The permeability of the lens to Rb^+ (*153*) and ^{32}P phosphate (*112a*) is also modified at this time (*153*), and lactate and amino acids may leak into the aqueous humor (*144*). Hexokinase activity diminishes, whereas that of ATPase increases 2 days after exposure to 1.4 kR (*180a*). Other enzymatic alterations develop later, but before cataracts become manifest. The levels of acid phosphatase (*228a*), glucose-6-phosphate dehydrogenase (*260*), and other dehydrogenases (*139a*) are reduced 2–3 weeks after exposure, but ATP remains at the normal level until cataracts develop (*217*). Several weeks after exposure, glycolysis decreases (*218*), and permeability to Na^+ and I^- (*277b*) increases, as do certain enzymatic activities, e.g., those of nucleoside phosphorylase, NADP dehydrogenase, and cytochrome *c* reductase (*260*). Glycolysis or respiration in the lens diminish immediately after exposure only when the dose is of the order of 100 kR (*112*, *112a*); these doses produce cataracts rather promptly.

I. Brain

1. Physiological Functions

If rats or mice are exposed to more than 10–15 kR, symptoms of damage to the central nervous system (CNS syndrome) supersede those of the gastrointestinal tract (p. 3). After a certain delay, the animals become atactic, develop convulsions, and finally die in coma. The time interval after which these symptoms arise is a function of the dose (see p. 2), but not all species are equally susceptible to CNS death. Burros develop this syndrome after exposure to a few hundred roentgens, whereas mice do so only after 20 kR. Irradiation of the head can usually provoke the same symptoms as a whole-body exposure. Histological examination of the brain from animals suffering the CNS syndrome reveals numerous pycnotic nerve cells, perivascular infiltrations, meningitis, and edema. Death of nerve cells occurs only above a certain threshold dose, and probably is of the interphase type; cell death can also be a consequence, in part, of the increased intracranial pressure.

Much lower doses than those needed to evoke the CNS syndrome can also affect the function of the nervous system. The following summary is proposed only as a guideline for understanding the biochemical effects of radiation on the brain; the reader may, therefore, be referred to several excellent recent reviews (*84*, *104*, *104a*, *132*, *156*, *163*) for more information and for the literature on this subject.

During exposure of the brain to low doses of radiation (10–100 R), certain functions of the CNS are altered. Sleeping animals can be aroused, conditioned reflexes modified, and aversion to saccharin, salt solutions, or the site of exposure may be induced by the irradiation. In these responses, radiation could act via quasiphysiological mechanisms either on such peripheral receptors as the eye and sensory organs of lower animals, or directly on certain brain areas such as the olfactory bulb.

After irradiation ceases, various functional alterations occur in the CNS. The electrical activity of the cortex, i.e., the EEG, is modified after exposure of the brain to a few hundred roentgen's or more. Conditioned reflexes may be facilitated after low doses, but can be abolished after higher doses, as is seen later in the course of radiation disease. Performance of complicated tasks is impaired by irradiation, although ability to learn is not affected. For example, the threshold for electroshock or audiogenic seizures is diminished and the seizures are shorter. Exposure to several kiloroentgens, however, may reduce convulsions provoked by drugs or electroshock, and can reinforce the action of narcotics.

Areas of the CNS other than the cortex are also affected by irradiation, especially the brain stem. Neurosecretory granules are lost in the hypothalamus (see p. 291) and electrical potentials in the brain stem are abnormal after exposure to 100 R and more. Cerebellar tonic and spinal reflexes, as well as spinal synaptic potentials, are modified after exposures to about 1000 R. Reflexes controlling blood pressure (see p. 312), respiration, temperature, and fever in rabbits (*188a*), etc., may also be altered following irradiation.

2. Water and Electrolytes

The blood–brain barrier has been studied frequently in animals after irradiation, since changes in uptake or loss of substances might severally influence the function of the brain. Nevertheless, the results of studies of irradiation effects in the brain barrier are contradictory. Uptake of ^{32}P, ^{131}I-albumin, dyes, etc., is reported to be increased after a whole-body irradiation with doses of 1000 R or less. On the other hand, local irradiation of the head with 8 kR only slightly modified the uptake of sulfate-^{35}S, but that of albumin-^{131}I was not affected (*197*). Still higher doses (20 kR) lower the barrier between blood and brain beginning 24 hr after exposure, as shown by an increased uptake of labeled phosphate (*24*).

Edema of the brain is apparent on histological examination, but is difficult to verify by chemical analysis. After whole-body irradiation with 700 R, the water content of the brain decreases for 1 day (*41*), and increases within 3–4 days with a higher dose (1000 R) (*97*). Yet, no change in water content was observed after exposure of the heads of rabbits or dogs to 2–5 kR (*97*, *231*). Exposure to doses as high as 14 kR resulted in a transient decrease in brain water, and this was followed by an increase (*93*). Brains contain more sodium 1–4 days after local exposure to doses above 1 kR (*97*, *231*). On the other hand, less sodium is found after whole-body irradiation with 700 R (*41*). No significant changes in potassium content in the brain have been reported after irradiation (*41*, *97*, *231*), but isolated nerve cells display an enhanced permeability for K^+ (*104b*).

3. Nucleic Acids

In evaluating the biochemistry of the irradiated brain, it must be kept in mind that the brain is not a homogeneous structure. Ideally, the interpretation of gross biochemical changes should, therefore, always be supplemented by histochemical data, but such combined investigations have rarely been carried out. Moreover, analysis of the brain presents special problems, in view of its large content of interfering substances such as lipids and carbohydrates. For example, determination of nucleic acids by phosphate analysis is inadequate, and the following results showing large changes in DNA and RNA contents, and in incorporation of ^{32}P after irradiation should be regarded with caution.

The DNA content was reported to decrease (*41*) 4–10 days after exposure to 700 R, and 1 day after 2 kR (*145*, *253*). Others failed, however, to confirm this observation (*64*). Incorporation of ^{32}P into DNA was depressed one day after whole-body irradiation with 2 kR (*253*), but not after 4.5 kR (*57*). Autoradiography reveals that only certain endothelial (*106a*), subependymal, and glia cells incorporate thymidine (*106*), that DNA synthesis is reduced significantly after irradiation (*106*), and that labeled DNA is lost from the brain (*188b*).

The RNA content of the brain was diminished as early as 30 min after irradiation (*145*, *243*, *253*), and remains low for several days after whole-body doses of 2 kR (*253*, spinal cord *244*). RNA bound to subcellular structures also shows changes in affinity for dyes soon after irradiation (*242*). Incorporation of ^{32}P into RNA is reduced after 1 day and again 3–5 days after exposure to 2 kR (*253*). On the other hand, cytidine-^{3}H injected immediately after exposure to 500 R or 15 kR is incorporated into RNA at a normal rate, as is shown by autoradiographic studies. The newly formed labeled RNA, however, is transferred more slowly from the nucleus to the cytoplasm, and the cytoplasmic RNA is more rapidly degraded after irradiation (*294*).

4. Proteins, Carbohydrates, and Lipids

An inconstant increase in the protein content of the brain stem (*231*) has been reported after high-dose irradiation, perhaps as a result of inflammatory reactions. The protein composition of the brain changes after 1 kR as shown by electrophoretic separation (*42a*, *293a*). Phosphoproteins diminish in quantity, although in guinea pigs they incorporate more ^{32}P after an exposure to 2 kR (*253a*). Incorporation of leucine into proteins of the brains of rats is diminished (*198*), and that of methionine (*5*) and acetate (*110*) enhanced after irradiation. but these differences may be due to variations in the uptake of amino acids by the irradiated brain tissue. The oxygen pressure in the brain increases after head irradiation of rabbits (*249*), but not of rats (*270b*), an observation which may reflect the lowered oxygen consumption of the irradiated brain. Oxygen uptake is depressed in mitochondria isolated from the irradiated cortex or pons 2 hr after exposure to a dose of 20 kR, but not in tissues of other parts of the brain (*169*). Moreover, oxidative phosphorylation in brain mitochondria is diminished 2 days after an exposure of the brain to 3–9 kR (*41b*, *246*) or immediately after 90 kR (*247*).

Glycogen (*42*, *79*, *192*, *291*) and mucopolysaccharides (*79*) in the brain increase after exposure to moderate and high doses of radiation (Fig. VI-10). A decrease in brain glycogen has been reported for guinea pigs 6 days after exposure to 200 and 400 R, and less glucose was incorporated into "bound" but more into "free" glycogen (*34*). Deposition of glycogen could result from a lowered utilization of glucose in pathways other than glycogen synthesis or from diminished degradation of glycogen. The first alternative appears most likely, since less glucose is oxidized by the irradiated brain than by the non-irradiated one (*126*). On the other hand, glycolysis is normal 1–3 days after irradiation with 1.5 kR (*126*) or even enhanced later (*76*). Blood flowing to and from the brain after exposure to 1 kR contains the same concentration of lactate and glucose as the controls (*295*). Glycolysis is, however, impaired 2 days after exposure to higher doses (9 kR) (*246*) or at once after 90 kR (*247*), and lactic acid accumulates in the brain immediately after 60 kR (*158*).

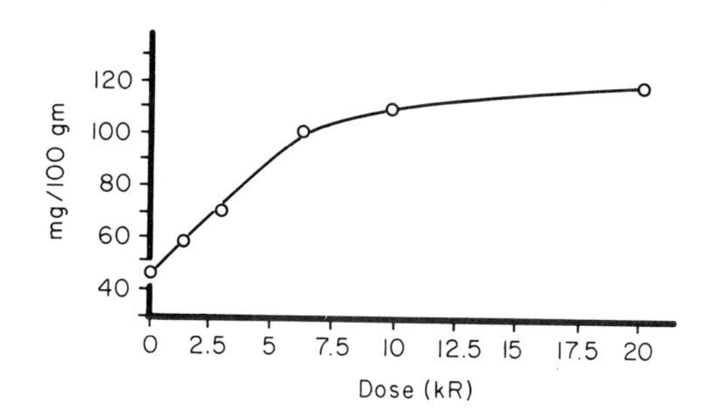

FIG. VI-10. Glycogen in rat brain 24 hr after irradiation of the dorsal parts of the cerebrum and the cerebellum. Glycogen content was elevated 5 hr after exposure and attained maximum values from 10–30 hr, before returning to normal. [From J. Miquel, P. R. Lundgren, and J. O. Jenkins, *Acta Radiol., Therapy, Phys. Biol.* [*N.S.*] **5**, 123 (1966).]

Total lipids in rabbit brain increase 7 days after 2 kR (*231*) and less acetate is incorporated 1 day after exposure to about 800 R (*277*). Gangliosides are reduced in rat brain 2 days after irradiation (*193*), and synthesis of cholesterol (*99*, *100*) from mevalonate-^{14}C and of sulfatides from ^{35}S is decreased (*49*). Incorporation of phosphate into phospholipids decreases in whole-body irradiated rats after a transient rise (*86*).

Data on other enzymatic activities and on content of various low-molecular compounds are fragmentary. Citric acid utilization (*65*), dehydrogenase activity (*66*), and distribution of LDH isozymes (*28*) are unaltered after moderate levels of irradiation, whereas oxidation of succinate is enhanced (*104b*). Formazan, however, is deposited in coarse granules in the irradiated brain, as shown by histochemical assay of dehydrogenases (*35*). The localization of acid phosphatase in neurons is altered (*35*, *122a*, *204*), and the activity of alkaline and acid phosphatase increases (*293a*). The activity of aldolase, hexokinase (*245*, *270a*), nucleoside deamination (*54*), and NADH diaphorase (*120a*) diminish slightly. Nucleoside phosphorylase (*248*, *225*) is unchanged and ATPase (*223*) decreases in the irradiated brain. Levels of ATP are normal (*42*) or slightly reduced by the second day (*223*, *270a*, *275*). Ammonia in the brain increases only after exposure to very high doses (*256*, *257*).

5. Neurohormones

Substances involved in excitation and synaptic transfer of nervous impulses are of particular importance in explaining the action of radiation on brain function. γ-Amino butyric acid (GABA), a compound which interferes with synaptic transfer, increases in rabbit brain, especially in the cortex and the cerebellum, 2–8 hr after exposure to 1000 R (*10*); glutamate decarboxylase activity is enhanced at this time (*11*). Later, GABA returns to normal levels in the rabbit brain, whereas in mice an increase in GABA and glutamate is found only later, i.e., 2–4 days after 900 R (*121*). The ratio of the specific activity of GABA to glutamate after addition of labeled glutamate increases at this time, suggesting *de novo* synthesis as the source of excess GABA (*120b*, *122*). Furthermore, other pathways degrading glutamic acid may be affected after irradiation by 500 R, as shown by a decrease in glutaminase activity (*8*). In monkeys, GABA content in the brain falls shortly after exposure to 800 R and rises after 3 days (*199*). Glutamate content decreases after exposure to 1–2 kR (*191*), and the content of reduced glutathione increases after 600 R (*270*). Acetylcholine, the neurohormone liberated and degraded during propagation of nerve impulses, diminishes in the sympathetic ganglion shortly after irradiation (*137*), whereas acetylcholinesterase activity in the brain is elevated a few days after exposure (*1*, *104b*, *234*). Noradrenaline increases and adrenaline decreases in brain after exposure to as low a dose as 40 R (*235*), and in the sympathetic ganglion after 400 R (*138*). Monamine oxidase, the enzyme degrading serotonin, noradrenaline, etc., decreases in the rat brain 4–7 days after whole-body exposure to 700 R (*52*, *258*), and serotonin is liberated and degraded in the brain on the first day after exposure, as is discussed elsewhere (see p. 309).

These changes in the content of neurohormones and enzymatic activities are difficult to interpret at the present time, since they seem to depend not only on the time after exposure, but also on the species. Careful studies relating the content of different neurohormones to the functional state of the brain are needed. One might conjecture from the present data that early after irradiation, substances favoring excitation are liberated, whereas later, sympathetic transfer and excitation become depressed.

6. Developing Brain

Brain irradiated before or shortly after birth appears to be more sensitive than mature brain. Growth is retarded, RNA synthesis is depressed (*43a*), the Na/K ratio is disturbed (*286*), and the formation of dehydrogenases (*54a*) and acetylcholinesterase is depressed (*178*, *179*). Rats irradiated at an age of 2 days have a high O_2 consumption in the cortex (but not in the hypothalamus) and a lowered CO_2 production when they reach adulthood, as compared with non-irradiated rats (*269*). Differentiation to the mature pattern of proteins is also delayed, as shown by amino acid analysis of the proteins (*285*). On the other hand, some time after neonatal irradiation, LDH isoenzymes in brain irradiated shortly after birth attain their mature pattern more rapidly (*28*). Exposure to 750 R delays formation of myelin in the brain stem of rats irradiated at an age of 2 days, so that more cholesterol (and less phospholipids) can be detected at an age of 2 weeks (*238*). Brain weight at a later time (60 days) is reduced markedly, due to a lack of synthesis of cerebrosides, sphingomyelins, and unsaturated fatty acids (*238*). More cerebrosides are, however, reported after a prenatal irradiation due to glial proliferation (*275a*). Apparently, myelin formed after irradiation differs from normal in that it contains less unsaturated fatty acids (*239*). Lower doses (400 R) seem to have no significant influence on myelin formation (*286*), and myelin synthesis in the spinal column is impaired only after exposure to more than 1 kR (*229*).

7. Nervous Function

The function of the peripheral nerves appears rather resistant to radiation damage. The velocity of conduction, the maximal amplitude, the propagation

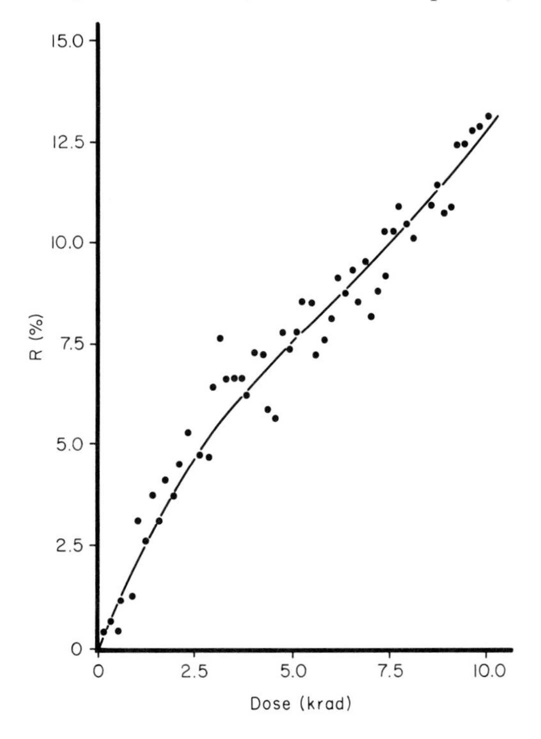

of the impulse, chronaxie, and rheobase are usually diminished only after exposure to doses of 10 kR or even more (*15*, *92*, *92b*, *230*). On the other hand, a change in chronaxie and action potential with rapid stimulation was reported after 5 rep of α-rays or 60 R of x-rays (*277a*). Sodium efflux (*170*) from the nerve increases after exposure to 6 kR. Potassium (*53b*) is lost after 10 kR or more. The excitability of the nerve is heightened after irradiation with a few kiloroentgens (*15*, *16*, *134*, *222*, *230*) and the action potential increases upon irradiation when stimulation is not already maximal (*241*) (Fig. VI-11). It seems likely that irradiation causes changes in permeability of the nerve membrane which are dependent on the dose, the dose rate, the ionic milieu, the speed of stimulation, etc., and which may facilitate or, after high-level exposure, impede propagation of the nervous impulse.

REFERENCES

1. Abbadessa, S., Avellone, S., and La Grutta, G., *Boll. Soc. Ital. Biol. Sper.* **41**, 1408 (1965).
2. Abdullaev, V. M., *Arkhiv. Patol.* **27**, 45 (1965).
3. Adachi, K., Chow, D. C., and Rothman, S., *Science* **135**, 216 (1962).
4. Adelstein, S. J., and Kohn, H. I., *Biochim. Biophys. Acta* **138**, 163 (1967).

4a. Åkeson, Å., Ehrenstein, V. G., Hevesy, G., and Theorell, H., *Arch. Biochem. Biophys.* **91**, 310 (1960).

5. Alexandrov, S. N., *Dokl. Akad. Nauk SSSR* **106**, 153, 363, and 569 (1956).
6. Altman, K. I., Casarett, G. W., Noonan, T. R., and Salomon, K., *Federation Proc.* **8**, 349 (1949).
7. Amoroso, C., Lorenz, W., Lossen, H., and Oswald, H., *Strahlentherapie* **121**, 269 (1963).
8. Arabei, L., Schian, S., Capilin, S., Ghizari, E., Stoenescu, R., and Stefanescu, P., *Nature* **196**, 1223 (1962).
9. Artom, C., Lofland, H. B., and Clarkson, T. B., *Radiation Res.* **26**, 165 (1965).
10. Avellone, S., Abbadessa, S., La Grutta, G., and Ortolani, E. A., *Boll. Soc. Ital. Biol. Sper.* **41**, 1415 (1965).
11. Avellone, S., Abbadessa, S., and La Grutta, G., *Boll. Soc. Ital. Biol. Sper.* **41**, 1419 (1965).
12. Babicky, A., and Kolár, J., *Strahlentherapie* **120**, 300 (1963).
13. Babicky, A., Kolár, J., and Vyhnanek, L., *Strahlentherapie* **126**, 449 (1965); **130**, 349 (1966).
14. Babicky, A., and Kolár, J., *Radiation Res.* **27**, 108 (1966).

FIG. VI-11. Effect of x-irradiation (2 krads/min) on the submaximal action potential of the isolated frog sciatic nerve. The data are presented as the ratio $R = [(SM/M)_2 - (SM/M)_1]/(SM/M)_1$, where SM refers to the submaximal, M to the maximal action potential, and the subscripts 1 and 2 denote conditions before and during irradiation, respectively. Note that a significant increase in the submaximal action potential is observed after exposure to 500 rads. Recovery occurs within 15 min after exposure to 10 krads, and the minimal dose rate to produce an effect is of the order of 500 rads/min. [From R. Seymour, and K. B. Dawson, *Intern. J. Radiation Biol.* **12**, 1 (1967).]

14a. Babicky, A., Kolár, J., Ostadalova, I., and Parizek, J., *Strahlentherapie* **137**, 170 (1969).

15. Bachofer, C. S., and Gautereaux, M. E., *J. Gen. Physiol.* **42**, 723 (1959); *Radiation Res.* **12**, 575 (1960).

16. Bacq, Z. M., and Beaumariage, M. L., *Arch. Intern. Physiol. Biochim.* **70**, 111 (1962).

17. Bankowski, Z., and Vorbrodt, A., *Folia Morphol.* **9**, 49 and 57 (1958).

18. Barboglio, F., and Bugliani, D., *Radiol. Fis. Med.* **5**, 30 (1938).

19. Beaumariage, M. L., and Van Caneghem, P., *Radiation Res.* **33**, 74 (1968).

20. Berech, J., Jr., and Curtis, H. J., *Radiation Res.* **22**, 95 (1964).

21. Bergeder, H. D., and Hockwin, O., *Naturwissenschaften* **47**, 161 (1960); *Strahlentherapie* **113**, 208 (1960).

22. Bergeder, H. D., *Strahlentherapie* **114**, 406 (1961); **115**, 90 (1961).

23. Bergeder, H. D., and Geissler, G., *Strahlentherapie* **119**, 513 (1962).

24. Bergen, J. R., Seay, H. D., Levy, C. K., and Koella, W. P., *Proc. Soc. Exptl. Biol. Med.* **117**, 459 (1964).

25. Bergström, R. M., Blafield, R., and Salmi, A., *Intern. J. Radiation Biol.* **4**, 351 (1962).

26. Bjorksten, J., Andrews, F., Bailey, J., and Trenk, B., *J. Am. Geriat. Soc.* **8**, 37 (1960).

27. Blackburn, J., and Wells, A. B., *Brit. J. Radiol.* **36**, 505 (1963).

28. Bonavita, V., Amore, G., Avellone, S., and Guarneri, R., *J. Neurochem.* **12**, 37 (1965).

28a. Bonnichsen, R., Hevesy, G., and Åkeson, Å., *Nature* **175**, 723 (1955).

29. Bottero, M., Vecchioni, R., Galucci, V., and Ruffato, C., *Boll. Soc. Ital. Biol. Sper.* **38**, 484 (1962).

30. Braun, H., Hornung, G., and Neumeyer, M., *Strahlentherapie* **126**, 454 (1965).

31. Bresciani, F., Dose, K., and Rajewsky, B., *Naturforsch.* **14b**, 158 (1959).

32. Breuer, H., Parchwitz, H. K., and Winkler, C., *Strahlentherapie* **105**, 579 (1958).

33. Breuer, H., and Parchwitz, H. K., *Strahlentherapie* **110**, 451 (1959).

34. Brodskaya, N. I., *Radiobiologiya* **5**, 465 (1965).

35. Brownson, R. H., Shanklin, W. M., and Suter, D. B., *J. Neuropathol. Exptl. Neurol.* **23**, 660 (1964).

36. Bruni, L., *Minerva Dermatol.* **38**, 53 (1963).

37. Bruni, L., and Mazza, A., *Minerva Dermatol.* **37**, 104 (1964).

37a. Burdin, K. S., and Parkhomenko, I. M., *Nauch. Dokl. Vyssh. Shkoly, Biol. Nauk.* **1**. 94 (1966).

38. Casarett, G. W., *Advan. Gerontol. Res.* **1**, 109 (1964).

39. Caster, W. O., Poncelet, J., Simon, A. B., and Armstrong, W. D., *Proc. Soc. Exptl. Biol. Med.* **91**, 122 (1956).

40. Caster, W. O., and Armstrong, W. D., *Proc. Soc. Exptl. Biol. Med.* **91**, 126 (1956).

41. Caster, W. O., Redgate, E. S., and Armstrong, W. D., *Radiation Res.* **8**, 92 (1958).

41a. Cattaneo, S. M., Quastler, H., and Sherman, F. G., *Radiation Res.* **12**, 587 (1960).

41b. Cenciotti, L., and Sanna, G., *Radiobiol., Radioterap. Fis. Med.* **21**, 390 (1966).

42. Cherkasova, L. S., Koldobskaya, F. D., Kukushkina, V. A., Mironova, T. M., Remberger, V. G., Taits, M. Yu., and Fomichenko, K. V., *Radiobiologiya* **6**, 179 (1966).

42a. Chernevskaya, Kh. I., *Vopr. Biokhim. Golovn. Mozga* **14** (1966).

43. Coburn, J. W., Rubini, M. E., and Kleeman, C. R., *J. Lab. Clin. Med.* **67**, 209 (1966).

43a. Cohan, S., and Ford, D., *Acta Neurol. Scand.* **45**, 53 (1969).

44. Cohn, S. H., and Gong, J. K., *Am. J. Physiol.* **173**, 115 (1953).
45. Cohn, S. H., *Radiation Res.* **15**, 355 (1961).
46. Cole, L. J., and Rosen, V. J., Jr., *Exptl. Cell Res.* **23**, 416 (1961).
47. Conard, R. A., *Ann. N. Y. Acad. Sci.* **114**, 335 (1960).
48. Cornatzer, W. E., Davison, J. P., Engelstad, O. D., and Simsonson, C., *Radiation Res.* **1**, 546 (1954).
49. Cornatzer, W. E., and Chally, C. H., *Nature* **202**, 1224 (1964).
50. Correa, J. N., Swarm, R. L., Andrews, J. R., and Walker, C. D., *Radiation Res.* **25**, 182 (1965).
51. Daiseley, K. W., *Biochem. J.* **60**, X (1955).
52. Danysz, A., Kocmierska-Grodzka, D., Wisniewski, K., and Killar, M., *Strahlentherapie* **135**, 90 (1968).
53. Darden, E. B., *Am. J. Physiol.* **198**, 709 (1960).
53a. Darden, E. B., and Upton, A. C., *J. Gerontol.* **19**, 62 (1964).
53b. Dawson, K. B., Seymour, R., and Mutton, D. E., *Radiation Res.* **37**, 83 (1969).
54. Davydova, I. N., *Vopr. Radiobiol.* **2**, 110 (1957).
54a. De Vellis, J., and Schjeide, O. A., *Biochem. J.* **107**, 259 (1968).
55. Devi, A., and Banerjee, R., *Radiation Res.* **22**, 713 (1964).
56. Devik, F., *Intern. J. Radiation Biol.* **5**, 59 (1962).
57. Dikovenko, E. A., *Med. Radiol.* **3**, 2, 42 (1958).
58. Dische, Z., Elliot, J., Pearson, E., and Meriam, G. R., *Am. J. Ophthalmol.* **47**, 368 (1959).
59. Dose, K., and Bresciani, F., *Biochem. Z.* **331**, 172 and 187 (1959).
60. Doty, S. B., Yates, C. W., Lotz, W. E., Kisecleski, W., and Talmage, R. V., *Proc. Soc. Exptl. Biol. Med.* **119**, 77 (1965).
61. Dowben, R. M., and Zuckerman, L., *Nature* **197**, 400 (1963).
62. Dragoni, G., *Aminoacidosi* (*Torino*) **7**, 85 (1959).
63. Dragoni, G., and Paolini, F., *Radioterap., Radiobiol. Fis. Med.* **14**, 226 (1959).
64. Dubinskii, A. M., Likhotkin, I. P., and Vodop'Yanova, L. G., *Mekh. Biol. Deistviya Ioniz. Izluch* (*Lvov*) p. 60 (1965).
65. Dubois, K. P., Cochran, D. W., and Doull, J., *Proc. Soc. Exptl. Biol. Med.* **76**, 422 (1951).
66. Dubois, K. P., and Raymund, A. B., NP-11966, pp. 65–78 (1961).
66a. Duhig, J. T., *Cancer Res.* **21**, 485 (1961).
67. Dziewiatkowski, D. D., and Woodard, H. Q., *Lab. Invest.* **8**, 202 (1959).
68. Easly, J. F., Shirley, R. L., and Davis, G. K., *Quart. J., Florida Acad. Sci.* **28**, 285 (1965).
69. Edwards, C. H., Gadsden, E. L., and Edwards, G. A., *Radiation Res.* **22**, 116 (1964).
70. Elkina, N. L., *Radiobiologiya* **3**, 351 (1963).
71. Elko, E. E., and Di Luzio, N. R., *Radiation Res.* **14**, 760 (1961).
71a. Ellinwood, L. E., Wilson, J. E., and Coon, J. M., *Proc. Soc. Exptl. Biol. Med.* **94**, 129 (1957).
71b. Erede, P., Durst, F., and Podesta, A. M., *Pathologica*, **56**, 841 (1964).
71c. Fain, J. N., *Science* **157**, 1062 (1967).
72. Feine, U., and Gerber, G. B., *Strahlentherapie* **102**, 563 (1957).
73. Feine, U., *Strahlentherapie* **108**, 408 (1959).
74. Finston, R. A., and Woodard, H. Q., *Radiation Res.* **25**, 188 (1965).
75. Fisher, J. W., and Roh, B. L., *Acta Haematol.* **31**, 90 (1964).
76. Fomichenko, K. V., *Vestsi Akad. Navuk Belarusk. SSR, Ser. Biyal. Navuk* **1**, 73 (1962).

77. Forssberg, A., and Hevesy, G., *Arkiv Kemi* **5**, 93 (1953).
78. Forssberg, A., *Intern. J. Radiation Biol.* **6**, 358 (1963).
79. Franke, H. D., and Lierse, W., *Strahlentherapie, Sonderbaende* **64**, 179 (1967).
80. Franklin, T. J., *Intern. J. Radiation Biol.* **3**, 467 (1961).
81. Franklin, T. J., *Nature* **193**, 289 (1962).
82. Gahlen, W., and Stüttgen, G., *Strahlentherapie* **97**, 259 (1955).
83. Galbraith, D. B., and Chase, H. B., *Science* **135**, 96 (1962).
84. Gangloff, H., and Hug, O., *Adv. Biol. Med. Phys.* **10**, 1 (1965).
85. Garazsi, M., and Deme, I., *J. Invest. Dermatol.* **27**, 49 (1956).
86. Gasteva, S. V., and Chetverikov, D. A., *Dokl. Akad. Nauk SSSR* **142**, 1180 (1962).
86a. Genazzani, E., and Montemagno, U., *Arch. Ostet. Ginec.* **64**, 441 (1959).
87. Gerber, G. B., Gerber, G., Altman, K. I., and Hempelmann, L. H., *Intern. J. Radiation Biol.* **1**, 277 (1959).
88. Gerber, G. B., Gerber, G., and Altman, K. I., *Intern. J. Radiation Biol.* **4**, 67 (1961).
89. Gerber, G. B., Gerber, G., Altman, K. I., and Hempelmann, L. H., *Intern. J. Radiation Biol.* **4**, 615 (1962).
90. Gerber, G. B., Gerber, G., Altman, K. I., and Hempelmann, L. H., *Intern. J. Radiation Biol.* **4**, 609 (1962).
91. Gerber, G. B., Gerber, G., Altman, K. I., and Hempelmann, L. H., *Intern. J. Radiation Biol.* **5**, 427 (1962).
92. Gerstner, H. B., *Am. J. Physiol.* **184**, 333 (1956).
92a. Gerstner, H. B., Lewis, R. B., and Richey, E. O., *J. Gen. Physiol.* **37**, 445 (1954).
92b. Gerstner, H. B., Orth, J. S., and Richey, E. O., *Am. J. Physiol.* **180**, 232 (1955).
93. Gerstner, H. B., Brooks, P. M., Vogel, T. S., and Smith, S. A., *Radiation Res.* **5**, 318 (1956).
94. Goldfeder, A., and Fershing, J. L., *Radiology* **31**, 81 (1938).
95. Golub, F. M., and Britun, A. I., *Eksperim. Khirurg. i Anesteziol.* **8**, 44 (1963).
96. Golubentsev, D. A., *Med. Radiol.* **4**, 36 (1959).
97. Gregersen, M. I., Pallavicini, C., Chien, S., Bates, N., and Temnikov, V., *Radiation Res.* **17**, 226 (1962).
98. Griem, M. L., and Malkinson, F. D., *Radiation Res.* **30**, 431 (1967).
99. Grossi, E., Paoletti, P., and Paoletti, R., *Arch. Intern. Physiol. Biochim.* **67**, 651 (1959).
100. Grossi, E., Poggi, M., and Paoletti, R., *in* "Response of the Nervous System to Ionizing Radiation" (T. J. Haley and R. S. Snider, eds.) pp. 388. Little, Brown, Boston, 1964.
101. Gutnisky, A., Nohr, M. L., and Van Dyke, D., *Calif. J. Nucl. Med.* **5**, 595 (1964).
102. Haberland, G. L., Schreier, K., Altman, K. I., and Hempelmann, L. H., *Biochim. Biophys. Acta* **25**, 237 (1957).
103. Hajduković, S., and Rasković, D., *Bull. Inst. Nucl. Sci. "Boris Kidrich" (Belgrade)* **9**, 173 (1959).
103a. Haley, T. J., Flesher, A. M., Komesu, U. M., McCullah, E. F., and McCormick, W. G., *Am. J. Physiol.* **193**, 355 (1958).
104. Haley, T. J., and Snider, R. S. *in* "Response of the Nervous System to Ionizing Radiation" (T. J. Haley, and R. S. Snider, eds.), Little, Brown, Boston, 1964.
104a. Haley, T. J., and Snider, R. S., *in* "Response of the Nervous System to Ionizing Radiation" (T. J. Haley and R. S. Snider, eds.), Academic Press, New York, 1962.
104b. Hallén, O., Hamberger, A., Rosengren, B., and Hockert, H., *J. Neuropathol. Exptl. Neurol.* **26**, 327 (1967).

105. Harding, C. V., Thayer, M. N., Eliashof, P. A., and Rugh, R., *Radiation Res.* **24**, 305 (1965).
106. Hassler, O., *J. Neuropathol. Exptl. Neurol.* **25**, 97 (1966).
106a. Haymaker, W., *in* "Effects of Ionizing Radiation on the Nervous System" (A. V. Lebedinskii and Z. N. Nakhil'nitskaya, eds.) p. 309. IAEA, Vienna, 1962.
107. Heidland, A., Braun, H., Klütsch, K., and Ammon, K., *Z. Ges. Exptl. Med.* **134**, 242 (1961).
108. Heistracher, P., Pillat, B., and Kraupp, O., *Intern. J. Radiation Biol.* **4**, 555 (1962).
109. Hempelmann, L. H., Carr, S., Frautz, J. D., Masters, R., and Lamdin, E., *Federation Proc.* **9**, 183 (1950).
110. Hevesy, G., and Dreyfus, G., *Arkiv. Kemi* **4**, 337 (1951).
111. Hilf, R., Adachi, R., and Eckfeldt, G., *Radiation Res.* **15**, 86 (1961).
111a. Hochrein, H., Braun, H., Reinert, M., and Kriegsmann, B., *Strahlentherapie* **120**, 587 (1963).
112. Hockwin, O., *Exptl. Eye Res.* **1**, 422 (1962).
112a. Hockwin, O., *Fortschr. Med.* **80**, 659 (1962).
113. Hornsey, S., *in* "Radiation Research" (G. Silini, ed.), p. 587–603. North-Holland Publ., Amsterdam, 1967.
114. Hornung, G., Lehmann, F.-G., and Braun, H., *Strahlentherapie* **128**, 595 (1965).
115. Høst, H., and Skjaelaaen, P., *Scand. J. Haematol.* **3**, 154 (1966).
116. Huang, K. C., Almand, J. R., and Hargan, L. A., *Radiation Res.* **1**, 426 (1954).
117. Hug, O., and Schliep, G. I., *in* "The Initial Effects of Ionizing Radiation on Cells" (R. T. C. Harris, ed.), pp. 272–300, Academic Press, New York, 1961.
118. Iversen, O. H., and Devik, F., *Intern. J. Radiation Biol.* **4**, 277 (1961).
119. Jolles, B., and Harrison, R. G., *Brit. J. Radiol.* **39**, 12 (1965).
120. Jonek, J., Kosmider, S., and Gajos, S., *Ann. Radiol.* **7**, 57 (1964).
120a. Jonek, J., Kosmider, S., and Jonek, T., *Polski Przegl. Radiol.* **30**, 307 (1966).
120b. Jovanovic, M., and Cordic, A., *Strahlentherapie* **134**, 533 (1967).
121. Jovanovic, M., and Svecenski, N., *Strahlentherapie* **125**, 588 (1964).
122. Jovanovic, M., and Svecenski, N., *Strahlentherapie* **129**, 446 (1966).
122a. Kagan, E. H., Brownson, R. H., and Suter, D. B., *Arch. Pathol.* **74**, 195 (1962).
122b. Kärcher, K. H., *in* "Strahlenpathologie der Zelle" (E. Scherer and H. S. Stender, eds.), pp. 317–333. Thieme, Stuttgart, 1963.
123. Kärcher, K. H., and Kato, H. T., *Strahlentherapie* **125**, 548 (1964).
124. Kärcher, K. H., and Schulz, G., *Strahlentherapie* **131**, 395 (1966).
125. Kärcher, K. H., *Strahlentherapie, Sonderbaende* **64**, 169 (1967); *Strahlentherapie* **129**, 80 (1966); **132**, 559 (1967).
126. Kay, R. E., and Chan, H., *J. Neurochem.* **14**, 401 (1967).
127. Keller, M., Tanka, D., and Fiam, B., *Magy. Radiol.* **16**, 191 (1964).
128. Kember, N. F., *Brit. J. Radiol.* **40**, 496 (1967); *Nature* **207**, 501 (1965).
129. Kent, S. P., *Radiation Res.* **10**, 380 (1959).
130. Kihn, L., *Arch. Physik Therapie* **14**, 343 (1964).
131. Kikuchi, Y., Hornung, G., and Braun, H., *Strahlentherapie* **130**, 567 (1966).
132. Kimeldorf, D. J., and Hunt, E. L., "Ionizing Radiation". Academic Press, New York, 1965.
133. Kinnamon, K. E., and Sutter, J. L., *Radiation Res.* **25**, 566 (1965).
134. Kirzon, M. V., and Pshennikova, M. G., *Biofizika* **2**, 686 (1957).
135. Kiselev, P. N., and Karpova, E. V., *Med. Radiol.* **10**, 154 (1965).
136. Kitagawa, T., Glicksman, A. S., Tyree, E. B., and Nickson, J. J., *Radiation Res.* **15**, 761 (1961).

137. Klimovskaya, L. D., and Maslova, A. F., *Byul. Eksperim. Biol. i Med.* **61**, 38 (1966).
138. Klimovskaya, L. D., and Maslova, A. F., *Byul. Eksperim. Biol. i Med.* **61**, 50 (1966).
139. Knowlton, N. P., and Hempelmann, L. H., *J. Cellular Comp. Physiol.* **33**, 73 (1949).
139a. Koch, H. R., and Hockwin, O., *Indian. J. Ocul. Path.* 4 (1967).
140. Kolár, J., and Babicky, A., *Strahlentherapie* **120**, 98 (1963); **124**, 434 (1964).
141. Konno, K., Altman, K. I., and Hempelmann, L. H., *Proc. Soc. Exptl. Biol. Med.* **101**, 320 (1959).
142. Konno, K., Traelnes, K. R., and Altman, K. I., *Intern. J. Radiation Biol.* **8**, 367 (1964).
143. Kosmider, S., Jonek, J., and Grzybek, H., *Radiobiol. Radiotherap.* **5**, 685 (1964).
144. Kubena, K., *Cesk. Oftalmol.* **20**, 324 (1964).
145. Kucherenko, N. E., and Sopin, E. F., *Ukr. Biokhim. Zh.* **38**, 50 (1966).
146. Kunz, J., Braselmann, H., and Keim, O., *Acta Biol. Med. Ger.* **13**, 228 (1964).
147. Kunz, J., and Braselmann, H., *Acta Biol. Med. Ger.* **14**, 570 (1965).
148. Kurata, N., *Nippon Junkanki Gakushi* **25**, 360 (1961).
149. Kurohara, S. S., Rubin, P., and Hempelmann, L. H., *Radiology* **77**, 804 (1961).
150. Kuzin, A. M., and Silaev, M., *Radiobiologiya* **3**, 1 (1963).
151. Kyo, S. K., Nam, S. B., and Ung, Y. B., *J. Korean Med. Assoc.* **7**, 955 (1964).
152. Labanova, N. M., *Vestsi Akad. Navuk Belarusk. SSR, Ser. Biyal. Navuk* **3**, 114 (1964).
153. Lambert, B. W., and Kinoshita, J. H., *Invest. Ophthalmol.* **6**, 624 (1967).
154. Lamberts, H. B., and de Boer, W. G. R. M., *Intern. J. Radiation Biol.* **6**, 343 (1963).
155. Lauenstein, K., Haberland, G. L., Hempelmann, L. H., and Altman, K. I., *Biochim. Biophys. Acta* **26**, 421 (1957).
156. Lebedinskii, A. V., and Nakhil'nitskaya, Z. N., *in* "Effects of Ionizing Radiation on the Nervous System", Elsevier, Amsterdam, 1963.
157. Leites, F. L., *Med. Radiol.* **8**, 54 (1963).
158. Lelievre, P., *Compt. Soc. Rend. Biol.* **151**, 412 and 631 (1957).
159. Le May, M., *Proc. Soc. Exptl. Biol. Med.* **77**, 337 (1951).
160. Lerman, S., Devi, A., and Banerjee, R., *Invest. Ophthalmol.* **1**, 95 (1962).
161. Lerman, S., and Zigman, S., *Invest. Ophthalmol.* **2**, 626 (1963).
162. Lerman, S., *Physiol. Rev.* **45**, 98 (1965); *N.Y. State J. Med.* **62**, 3075 (1962).
163. Levy, C. K., *Current Topics Radiation Res.* **3**, 95 (1967).
164. Lipkan, N. F., *Ukr. Biokhim. Zh.* **29**, 90 (1957).
165. Liu, C. T., and Overman, R. R., *Radiation Res.* **24**, 452 (1965).
166. Liu, C. T., and Overman, R. R., *Radiation Res.* **25**, 55 (1965).
167. Lohmann, W., Denny, W. F., Perkins, W. H., Moss, A. J., and Fowler, C. F., *Acta Radiol.* **4**, 3 (1966).
168. Lomova, M. A., *Radiobiologiya* **5**, 908 (1965).
169. Lott, J. R., and Hines, J. F., *Texas T. Sci.* **19**, 391 (1967).
170. Lott, J. R., and Yang, C. H., *in* "Response of the Nervous System to Ionizing Radiation" (T. J. Haley and R. S. Snider, eds.) p. 271. Little, Brown, Boston, 1964.
171. Lowrey, R. S., and Bell, M. C., *Radiation Res.* **23**, 580 (1964).
172. Lucas, D. R., *Intern. J. Radiation Biol.* **10**, 551 (1966).
173. Ludewig, S., and Chanutin, A., *Am. J. Physiol.* **163**, 648 (1950).
174. Lushbaugh, C. C., Storer, J. B., and Hale, D. B., *Cancer* **6**, 671 (1953).
175. Lushbaugh, C. C., and Hale, D. B., *Cancer* **6**, 683 and 886 (1953).

176. Main, R. K., Cole, L. J., and Walwick, E. R., *Intern. J. Radiation Biol.* **7**, 45 (1963).
177. Makhlina, A. M., *Radiobiologiya* **6**, 174 (1966).
178. Maletta, G. J., and Timiras, P. S., *J. Neurochem.* **13**, 75 (1966).
179. Maletta, G. J., Vernadakis, A., and Timiras, P. S., *J. Neurochem.* **14**, 647 (1967).
179a. Mandel, P., and Rodesch, J., *J. Physiol.* (*Paris*) **47**, 234 (1955).
180. Mandel, P., and Rodesch, J., *J. Physiol.* (*Paris*) **48**, 637 (1956).
180a. Mandel, P., and Schmitt, M. L., *Experientia* **12**, 223 (1956).
180b. Mandel, P., Gros, C., and Rodesch, J., *in* "Radiology Symposium," (Z. M. Bacq and P. D. Alexander, eds.), p. 210. Academic Press, New York, 1955.
181. Maraini, G., Franguelli, R., Marcato, M., and Santori, M., *Ann. Ottalmol. Clin. Oculist* **91**, 583 (1965).
182. Maraini, G., Santori, M., Marcato, M., and Franguelli, R., *Ann. Ottalmol. Clin. Oculist* **91**, 590 (1965).
183. Mason, H. C., and Moos, W. S., *Atompraxis* **12**, 347 (1966).
184. McKelvey, R. W., and Geoffrey, H. B., *Ann. Histochim.* **8**, 339 (1963).
185. Melanotte, P. L., and Follis, R. H., *Am. J. Pathol.* **39**, 1 (1961).
186. Melik-Musyan, A. B., *Radiobiologiya* **6**, 88 (1966).
187. Mendelsohn, M. L., and Caceres, E., *Am. J. Physiol.* **173**, 351 (1953).
188. Mendelsohn, M. L., and Weimer, E. Q., *Am. J. Physiol.* **180**, 599 (1955).
188a. Meredith, O. M., and Finnegan, C., *Nature* **194**, 592 (1962).
188b. Merits, I., and Cain, J., *Biochim. Biophys. Acta* **174**, 315 (1969).
189. Milch, L. J., Stinson, J. V., and Albaum, H. G., *Radiation Res.* **4**, 321 (1956).
190. Millo, A., and Debenedetti, A., *Riv. Biol.* (*Perugia*) **57**, 151 (1964).
191. Minayev, P. F., and Skvortsova, R. I., *Vopr. Biokhim. Nervnoi. Sistemy, Akad. Nauk Ukr. SSR, Inst. Biokhim., Tr. 2-i Nauchn. Konf., Kiev* p. 289 (1957).
192. Miquel, J., Lundgren, P. R., and Jenkins, J. O., *Acta Radiol., Therapy, Phys., Biol.* [N.S.] **5**, 123 (1966).
193. Mkhitaryan, V. G., and Badalyan, G. E., *Biol. Zh. Arm.* **19**, 54 (1966).
194. Mollura, J. L., and Goldfeder, A., *Am. J. Physiol.* **186**, 224 (1956).
195. Morrison, L. M., Bernick, S., Alfin-Slater, R. B., Patek, P. R., and Ershoff, B. H., *Proc. Soc. Exptl. Biol. Med.* **123**, 904 (1966).
196. Muto, V., Lombardi, S., Crocco, E. M., and Egineta, M., *Rass. Med. Sper.* **10**, 415 (1964).
197. Nair, V., and Roth, L. J., *Radiation Res.* **23**, 249 (1964).
198. Nakazawa, S. Y., Takao, H., and Komei, U., *Can. J. Biochem.* **43**, 1091 (1965).
199. Nguyen, H. C., and Sytinskii, I. A., *Probl. Neirokhim., Akad. Nauk SSSR* p. 162 (1966).
200. Nimni, M. E., Gerth, N., and Bavetta, L. A., *Radiation Res.* **26**, 221 (1965).
201. Nimni, M. E., Lyons, C., and Bavetta, L. A., *Proc. Soc. Exptl. Biol. Med.* **122**, 134 (1966).
202. Nims, L. F., and Geisselsoder, J. L., *Radiation Res.* **12**, 251 (1960).
203. Oganesjan, S. S., and Zaminjan, T. S., *Dokl. Akad. Nauk. Arm. SSR* **34**, 207 (1962).
204. Ojak, M., *Patol. Polska* **15**, 561 (1964).
205. Onkelinx, C., *Nature* **195**, 302 (1962).
206. Onkelinx, C., *Compt. Rend. Soc. Biol.* **158**, 912 (1964).
207. Orekhovich, V. N., Firfarova, K. F., Kedrova, E. M., and Levdikova, G. A., *Exptl. Eye Res.* **1**, 324 (1962).
208. Patt, H. M., and Quastler, H., *Physiol. Rev.* **43**, 357 (1963).
209. Patt, H. M., and Quastler, H., *Physiol. Rev.* **43**, 357 (1963).
210. Pauly, H., and Rajewsky, B., *Strahlentherapie* **99**, 383 (1956).

211. Phillips, R. D., and Kimeldorf, D. J., *Am. J. Physiol.* **210**, 1096 (1966).
212. Phillips, S. J., Reid, J. A., and Rugh, R., *Am. Heart J.* **68**, 524 (1964).
213. Phillips, T. L., and Vaeth, J. M., *Proc. Soc. Exptl. Biol. Med.* **121**, 161 (1965).
214. Pikulyev, A. T., and Ronyaeva, M. P., *Ukr. Biokhim. Zh.* **38**, 258 (1966).
215. Pikulyev, A. T., and Smelyanskaya, G. N., *Ukr. Biokhim. Zh.* **38**, 383 (1966).
216. Pirie, A., van Heyningen, R., and Boag, J. W., *Biochem. J.* **54**, 682 (1953).
217. Pirie, A., van Heyningen, R., and Howard Flanders, P., *Biochem. J.* **61**, 341 (1955).
218. Pirie, A., and Howard Flanders, P., *Am. J. Ophthalmol.* **57**, 849 (1952).
219. Portela, A., Hines, M., Perez, J. C., Brandes, D., Bourne, G. H., Stewart, P. and Groth, D. P., *Exptl. Cell Res.* **21**, 468 (1960).
220. Portela, A., Perez, J. C., Stewart, P., Nelson, L., and Hines, M., *Radiation Res.* **20**, 303 (1963).
221. Portela, A., Perez, J. C., Stewart, P., Hines, M., and Reddy, V., *Exptl. Cell Res.* **29**, 527 (1963).
221a. Portela, A., Perez, J. C., Stewart, P., and Hines, M., *Exptl. Cell Res.* **31**, 281 (1963).
222. Pshennikova, M. C., *Biofizika* **3**, 680 (1958).
223. Rabassini, A., Rossi, C. S., and Gregolin, C., *Giorn. Biochim.* **9**, 217 (1960); **10**, 84 (1961).
224. Rajewsky, B., Wilhelm, G., and Frankenberg, D., *Strahlentherapie, Sonderbaende* **62**, 187 (1966).
225. Rappoport, D. A., *Radiation Res.* **12**, 465 (1960).
226. Rappoport, D. A., Seibert, R. A., and Collins, V. P., *Radiation Res.* **6**, 148 (1957).
227. Revis, V. A., and Titov, V. N., *Med. Radiol.* **9**, 1154 (1964).
228. Rich, J. G., Glagov, S., Larsen, K., and Spargo, B., *Arch. Pathol.* **72**, 388 (1961).
228a. Richards, R. D., Schocket, S. S., and Michaelis, M., *Radiation Res.* **38**, 140 (1969).
229. Rodgers, C. H., *Exptl. Neurol.* **11**, 502 (1965).
230. Rosen, D., and Dawson, K. B., *Radiation Res.* **12**, 357 (1960); *Intern. J. Radiation Biol.* **5**, 535 (1962).
231. Rowley, G. R., Dellenback, R. J., and Temnikov, V., *Radiation Res.* **17**, 244 (1962).
232. Ruffato, C., Gallucci, V., Vecchioni, R., and Bottero, M., *Boll. Soc. Ital. Biol. Sper.* **38**, 486 (1962).
233. Ruppe, C. O., Monsul, R. W., and Koenig, V. L., *Radiation Res.* **8**, 265 (1958).
234. Sabine, J. C., Miller, H. M., and Nickolai, D. J., *Am. J. Physiol.* **187**, 275 (1956).
235. Sacukevič, V. B., *Dokl. Akad. Nauk Belorussk. SSR* **10**, 506 (1966).
236. Sahasrabudhe, M. B., Nerurkar, M. K., and Baxi A. J., *Intern. J. Radiation Biol.* **1**, 52 (1959).
237. Sams, A., *Intern. J. Radiation Biol.* **10**, 123 (1966).
238. Schjeide, O. A., and Haack, K., *Acta Radiol.* **5**, 185 (1966).
239. Schjeide, O. A., Lin, I-San R., and de Vellis, J., *Radiation Res.* **33**, 107 (1968).
239a. Schmidt, K. J., Walther, G., and Ladner, H. A., *Strahlentherapie* **136**, 365 (1968).
240. Schreier, K., Di Ferrante, N., Gaffney, G. W., Hempelmann, L. H., and Altman, K. I., *Arch. Biochem. Biophys.* **50**, 417 (1954).
241. Seymour, R., and Dawson, K. B., *Intern. J. Radiation Biol.* **12**, 1 (1967).
242. Shabadash, A. L., Agracheva, N. D., and Zelikina, T. I., *Radiobiologiya* **2**, 105 (1962).
243. Sharobaiko, V. I., *Tsitologiya* **6**, 101 (1964).
244. Shungskaya, V. E., Zarkh, E. N., and Nikolaeva, N. D., *Radiobiologiya* **6**, 353 (1966).
245. Silyayeva, M. F., *Vestsi Akad. Navuk Belarusk. SSR, Ser. Biyal. Navuk* **1**, 101 (1964).

246. Skvortsova, R. I., *Radiobiologiya* **2**, 189 (1962).
247. Smith, D. E., and Thomson, J. F., *Radiation Res.* **11**, 198 (1959).
248. Smith, K. C., and Low-Beer, B. V. A., *Radiation Res.* **6**, 521 (1957).
249. Snezhko, A. D., *Biofizika* **1**, 585 (1956); **2**, 70 (1957).
250. Sobel, H., Gabay, S., and Bonoppis, G., *J. Gerontol.* **15**, 253 (1960).
251. Sokolov, I. I., *Ukr. Biokhim. Zh.* **38**, 264 (1966).
252. Song, C. W., and Tabachnick, J., *Nature* **217**, 650 (1968); *Intern. J. Radiation Biol.* **15**, 171 (1969).
253. Sopin, E. F., and Kucherenko, M. Ye., *Ukr. Biokhim. Zh.* **37**, 251 (1965).
253a. Sopin, E. F., and Kucherenko, M. Ye., *Ukr. Biokhim. Zh.* **39**, 269 (1967).
255. Streffer, C., and Melching, H. J., *Strahlentherapie* **125**, 341 (1964).
256. Strelkov, R. B., and Semenov, L. F., *Radiobiologiya* **6**, 377 (1966).
257. Strelkov, R. B., *Vopr. Med. Chem.* **13**, 73 (1967).
258. Strubelt, O., *Strahlentherapie* **124**, 570 (1964).
259. Sukhomlinov, B. F., Datskiv, M. Z., Chaika, Ya. P., Velikii, N. N., Kapis, M. V., and Shevchuk, Ya. G., *Mekh. Biol. Deistviya Ioniz. Izluch.* (*Lvov*) p. 157 (1965).
259a. Sukhomlinov, B. F., and Datskiv, M. Z., *Ukr. Biokhim. Zh.* **39**, 654 (1967).
259b. Sviderskaya, T. A., and Liberman, A. N. *Radiobiologiya* **7**, 334 (1967).
260. Swanson, A. A., *Exptl. Eye Res.* **4**, 231 (1965).
261. Tabachnick, J., and Weiss, C., *Radiation Res.* **11**, 684 (1959).
262. Tabachnick, J., *Radiation Res.* **15**, 785 (1961).
263. Tabachnick, J., *J. Albert Einstein Med. Center* **11**, 17 (1963).
264. Tabachnick, J., *Radiation Res.* **22**, 242 (1964).
265. Tabachnick, J., Perlish, J. S., and Freed, R. M., *Radiation Res.* **23**, 594 (1964).
266. Tabachnick, J., Perlish, J. S., Chang, L. F., and Freed, R. M., *Radiation Res.* **32**, 293 (1967).
267. Tabachnick, J., Song, C. W., Freed, R. M., and Perlish, J. S., *J. Albert Einstein Med. Center* **15**, 71 (1967).
268. Threlfall, G., Cairnie, A. B., Taylor, D. M., and Buck, A. T., *Radiation Res.* **27**, 559 (1966).
269. Timiras, P. S., Moguilevsky, J. A., and Geel, S., *in* "Response of the Nervous System to Ionizing Radiation" (T. J. Haley and R. S. Snider, eds.) pp. 365–376. Little, Brown, Boston, 1964.
270. Tomana, M., *Radiobiol. Radiotherap.* **7**, 249 (1966).
270a. Totskii, V. N., *Radiobiologiya* **5**, 65 (1965).
270b. Vacek, A., Davidová, E., and Hošek, B., *Intern. J. Radiation Biol.* **8**, 499 (1964).
271. Van Caneghem, P., Beaumariage, M. L., and Lachapelle, J. M., *Radiation Res.* **26**, 44 (1965).
272. Van Caneghem, P., *Intern. J. Radiation Biol.* **8**, 541 (1965).
273. Van Caneghem, P., Beaumariage, M. L., and Lachapelle, J. M., *Intern. J. Radiation Biol.* **10**, 43 (1966).
274. Van Heyningen, R., and Pirie, A., *Biochem. J.* **56**, 372 (1954).
275. Veksler, Ya. I., *Vopr. Med. Khim.* **12**, 26 (1966).
275a. Vernadakis, A., Casper, R., and Timiras, P. S., *Experientia* **24**, 237 (1968).
276. Verzár, F., *Gerontologia* **3**, 163 (1959).
277. Vinogradova, M. F., *Radiobiologiya* **2**, 695 (1962).
277a. von Kroebel, W., and Krohm, G., *Atomkernenergie* **4**, 280 (1959).
277b. von Sallmann, L., and Locke, B., *A.M.A. Arch. Ophthalmol.* **45**, 149 and 431 (1951).
277c. von Sallmann, L., *Invest. Ophthalmol.* **4**, 471 (1965).

277d. Vodicka, I., *Strahlentherapie* **137**, 730 (1969).
278. Vorbrodt, K., Godlewski, H., and Dux, A., *Folia Morphol.* **9**, 39 (1958).
279. Wachtel, L. W., and Cole, L. J., *Radiation Res.* **25**, 78 (1965).
280. Wachtel, L. W., Cole, L. J., and Rosen, V. J., Jr., *Intern. J. Radiation Biol.* **10**, 75 (1966).
281. Wachtel, L. W., Phillips, T. L., and Cole, L. J., *Radiation Res.* **28**, 647 (1966).
282. Wagner, H., Wagner, O., and Bernhard, K., *Helv. Physiol. Pharmacol. Acta* **19**, C113 (1961).
283. Walford, R. L., Sjaarda, J. R., and Anderson, R. E., *Exptl. Gerontol.* **1**, 117 (1965).
284. Weiner, N., Albaum, H. G., and Milch, L. J., *A.M.A. Arch. Pathol.* **60**, 621 (1955).
285. Wender, M., and Waligora, Z., *J. Neurochem.* **11**, 583 (1964).
286. Wender, M., *Bull. Soc. Amis Sci. Lettres Poznan* **C13**, 13 (1964); *Folia Biol.*, (*Warsaw*) **13**, 323 (1965).
287. Wilde, W. S., and Sheppard, C. W., *Proc. Soc. Exptl. Biol. Med.* **88**, 249 (1955).
288. Wilson, C. W., *Brit. J. Radiol.* **29**, 86 and 571 (1956); **31**, 384 (1958).
289. Wilson, C. W., *Brit. J. Radiol.* **38**, 271 (1965).
290. Withers, H. R., *Brit. J. Radiol.* **40**, 187 and 335 (1967).
291. Wolfe, L. S., Klatzo, I., Miquel, J., Tobias, C., and Haymaker, W., *J. Neurochem.* **9**, 213 (1962).
292. Woodard, H. Q., and Spiers, F. W., *Brit. J. Radiol.* **26**, 38 (1953).
293. Woodard, H. Q., and Laughlin, J. S., *Radiation Res.* **7**, 236 (1957).
293a. Wrage, D., Lierse, W., and Franke, H. D., *Strahlentherapie* **137**, 320 (1969).
294. Yamamoto, Y. L., Feinendegen, L. E., and Bond, V. P., *Radiation Res.* **21**, 36 (1964).
295. Zakharov, S. V., *Med. Radiol.* **4**, 77 (1959).
296. Zaruba, K., *Radiation Res.* **25**, 1 (1965) and **35**, 661 (1968).
297. Zhurbin, G. I., *Ukr. Biokhim. Zh.* **38**, 235 and 241 (1966).
297a. Zubkova, S. R., and Chernavskaya, N. M., *Dokl. Akad. Nauk SSSR* **126**, 1114 (1959).
298. Zuppinger, A., and Minder, W., *Strahlentherapie* **117**, 63 (1962).

Chapter VII

RADIATION BIOCHEMISTRY OF TUMORS

A. Introduction

Few, if any, biochemical characteristics are common to all tumor cells; biochemical specialization may be as great among different tumors as it is among normal tissues. The reason for considering the radiation biochemistry of all the different tumor cells in one chapter is mainly a practical one: much information on biochemical changes in irradiated tumor cells has been gathered with the general goal of discovering what makes tumors more radiosensitive than normal tissue, and what happens when a tumor becomes radioresistant. Moreover, most biochemical investigations have been restricted to only a few types of tumor cells, namely, to tumors growing in ascites fluid, to lymphoma and leukemia cells, and to certain solid tumors, such as hepatoma, Walker carcinoma, mammary carcinoma, and Jensen sarcoma.

B. Radiosensitivity and Composition of Tumor Cells

Even the early radiation therapists noted that not all malignant tumors are equally susceptible to irradiation, and that a tumor may become resistant during therapy. An earlier investigator (*19*) reported that radioresistant tumors often have a higher ash content than sensitive ones. Most later studies have, however, been concerned with the relation between the radiosensitivity of a tumor and its DNA content (or ploidy), its concentration of sulfhydryl (SH) compounds, its number of subcellular particles, its enzymatic composition, and the length of its cell cycle. Yet, it is difficult to pinpoint the factor(s) responsible for sensitivity to radiation, since several properties, not only one, usually change when a tumor becomes radioresistant.

As discussed in detail elsewhere (Vol. 1, p. 103), the radiosensitivity of different types of cells can be correlated with DNA content, ploidy, or chromosome volume. Comparison of DNA content and radiosensitivity in different tumors has, however, yielded inconsistent results, even when the strains of cells compared were genetically related. The reason for such inconclusive findings may be that the tumor cells do not always duplicate all chromosomes as their DNA content increases, and, apparently, polyploid strains lack the genetic information needed for greater radioresistance. Diploid and tetraploid tumor cells [leukemia (*9*) and Ehrlich ascites cells (*106*)] have about the same D_0 (Appendix, Vol. 1), but tetraploid strains have a higher extrapolation number (see also *18*, *28*). Cultures of two Ehrlich ascites tumor strains that exhibit marked differences in radiosensitivity *in vivo* do not exhibit these differences *in vitro*; cell cycles in *in vitro* cultures are shortened, and the modal chromosome number is almost doubled (*10a*). However, a consistent relationship between radiosensitivity and DNA content of tumor cells is not always observed (e.g., *1*, *5*, *69*, *90*), and radioresistant tumor strains with less than the average DNA content are known (*5*).

Almost 20 years ago, Barron suggested that destruction of SH compounds plays an important role in the action of radiation on cells (*6a*). Consequently, the presence of SH compounds in a cell was thought to be associated with radioresistance. Indeed, certain resistant tumor cells contain more nonprotein-bound SH groups (*84a*, *91*) and, in some cases, also protein-bound SH groups (*5*, *18*), when compared with the radiosensitive tumor cells. Moreover, cellular SH groups increase in quantity in tumor cells protected by sulfhydryl compounds (*45*, *92*, *93*) and diminish in those treated with radiosensitizers (*46*).

Fewer mitochondria are present in the cells of a radiosensitive spindle cell tumor than in those of a resistant epithelial tumor (*42*, *43*), an observation which one may relate to the fact that the cells most susceptible to interphase death, i.e., adult lymphocytes, also contain few mitochondria. Other changes noted in radioresistant tumors are: a higher activity of deoxycytidylate aminohydrolase (*84a*) or catalase (*83a*) and a greater concentration of histidine (*5*). The cell cycle is shortened in certain tumors as they become radioresistant (*1*). It is probable, as noted above, that more than one factor is involved in the development of radioresistance by a previously radiosensitive tumor cell. One such mechanism might involve shift in balance between interphase and reproductive death (compare Vol. 1, p. 282), reflecting minor changes in cell composition and metabolism.

C. Metabolism of Macromolecules

Incorporation of different precursors into DNA of tumor cells diminishes immediately after irradiation, as it also does in other types of cells (Fig. VII-1). This fact was first discovered by von Euler and von Hevesy (*117, 118*) in Jensen sarcoma using labeled phosphate, and has since been documented amply for cells growing in ascites fluid using ^{32}P (*8, 54, 56, 74*), thymidine (*8, 35*), glycine, or formate (*36*); for lymphoma cells (*6*) and for solid tumors using ^{32}P (*20, 21,*

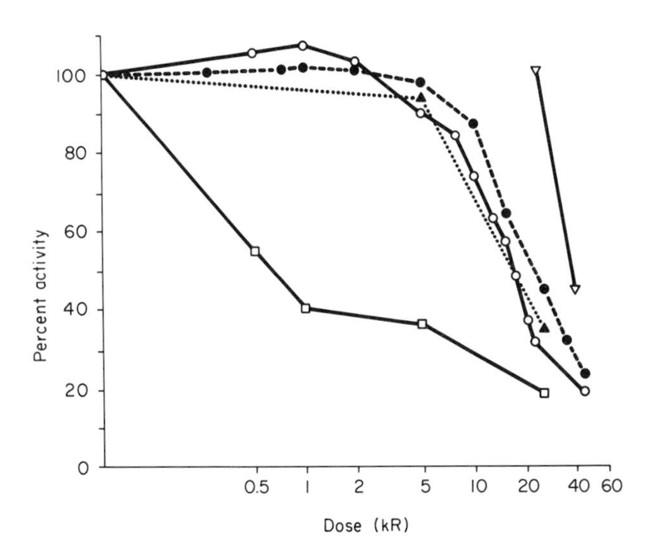

FIG. VII-1. Incorporation of ^{32}P into DNA: squares; incorporation of alanine-^{14}C into proteins: solid triangles; anaerobic glycolysis: solid circles; NAD concentration: open circles; and activity of triosephosphate dehydrogenase: open triangles in Yoshida ascites tumor cells after x-irradiation. Note that incorporation into DNA is most radiosensitive, whereas glycolysis, loss of NAD, and protein synthesis display approximately the same dose-effect curves. Triosephosphate dehydrogenase is inactivated after still higher doses. [From H. Maass and H. A. Künkel, *Intern. J. Radiation Biol.* **2**, 269 (1960).]

22, 115) or iododeoxyuridine (*23*). The dose-effect curve for this depression of incorporation resembles that seen in other irradiated cells, and reveals two or more components with different radiosensitivities (*6, 8*). When the incorporation into DNA is plotted *vs* the log of the dose, straight lines differing in slope and origin can be drawn for various tumors (*21*). Incorporation of precursors in nuclear as well as mitochondrial DNA is depressed, with the level depending on the precursor used (*19a*).

Several factors influence the incorporation of precursors into DNA of irradiated tumor cells, e.g., the temperature (*7*) and the incubation medium (*7*).

Nicotinamide, which prevents the loss of NAD from irradiated tumor cells, also protects DNA synthesis (*60*). Apparently, DNA synthesis is diminished but not blocked completely (*27*), and, eventually, premitotic levels of DNA are attained by most of the irradiated cells (*17, 38, 74, 77*). Reduced incorporation does not result from changes in the uptake of thymidine or its phosphorylation, but, rather, involves the final step of incorporation of thymidine triphosphate into DNA (*27*). Moreover, the effect on DNA synthesis can be elicited not only by exposure of the nucleus but, also, of the cytoplasm (at a tenfold higher dose), as is shown by local α-irradiation of HeLa cells (*113*). One factor thought (*103*) to distinguish the action of radiation on tumor (HeLa and L) cells from that on normal cells is that in these cells, DNA synthesis need not be initiated at the end of the G_1 period, since tumor cells synthesize nuclear proteins (and perhaps even the key enzyme needed for DNA synthesis) throughout their cell cycle. This property may be shared with other rapidly dividing cells, although the available data are insufficient to prove such a hypothesis.

The structure of nucleic acids present in tumor cells is altered by irradiation. For example, separation of DNA from Walker carcinoma into two fractions, DNA I and DNA II, shows that, after irradiation, DNA II diminishes to a greater extent than DNA I (*4, 49*). Since DNA II occurs almost exclusively in tissues with a rapid cell turnover, the DNA composition of irradiated tumors approaches that of nondividing cells, e.g., of normal liver. Nucleoproteins present in euchromatin of irradiated (20 kR) nuclei of ascites cells are more active as primers of RNA polymerase and bind less actinomycin than nuclei of nonirradiated cells. Conversely, the priming action of heterochromatin fractions is impaired by irradiation (*116*).

Incorporation of precursors into total RNA is much less affected after irradiation than is that into DNA (*25, 50, 55, 110, 115*). One exception has been reported—incorporation of glycine, which diminishes to about the same extent for both DNA and RNA (*36*), and even shows pronounced alteration for that into lipids. Changes in the uptake of glycine and its utilization in the many different pathways employing this amino acid may be responsible for the discrepant behavior. Studies using autoradiography or chromatographic separation of RNA, however, reveal that not all types of RNA are equally resistant. Thus, incorporation of cytidine into nuclear RNA and, even more, into nucleolar RNA of HeLa cells diminishes, and transfer of labeled RNA from the nucleus to the cytoplasm is delayed after exposure to only 100 R, as is shown by autoradiography (*11*). A similar effect on nuclear RNA is found in ascites tumor cells after exposure to higher dose levels (*50–53*; cf. *99*). Another investigator failed to observe a marked change in incorporation of uridine into nuclear RNA even after 10 kR but noted "oscillations" in the uptake (*80*). Chromatographic fractionation of nuclear RNA, however, shows that only certain components of nucleolar RNA are affected by irradiation (*52, 111*).

In addition to changes in nuclear RNA, a decrease in incorporation of ^{32}P was also reported in the soluble and ribosomal RNA of ascites tumor cells after exposure to 1000 R (*84*).

Incorporation of precursors into total proteins is modified only after doses which also depress glycolysis (Fig. VII-1) (*10, 14, 15, 67, 83, 84, 101*). In Yoshida tumor cells, glycolysis appears, however, to return to normal 1 hr after exposure, whereas inhibition of protein synthesis is irreversible (*63*). Incorporation of methionine (*67*) or glycine (*85*) into histones continues at a normal rate after exposure to 1–2 kR, even though DNA synthesis is impaired. Later, DNA synthesis is blocked at premitotic levels but synthesis of RNA and proteins in the nucleus continues for some time (*17, 38, 75, 76, 104*). Nevertheless histone synthesis must eventually cease, since the histone/DNA ratio (*95*) is the same for normal ascites cells and for those exposed to 1.2 kR 24 hr before.

D. Glycolysis, NAD Metabolism, and Other Biochemical Changes in Irradiated Tumor Cells

Crabtree's classic experiments (*26*), reported in 1935, demonstrated that exposure of tumor cells to about 10 kR markedly diminishes glycolysis, but has little effect on respiration (see also the earlier work of Frick, *37*). This observation has since been confirmed for many different tumor tissues (*14, 15, 29–31, 33, 40, 41, 61, 62, 66, 70, 81, 86, 88*). Aerobic glycolysis is less depressed by irradiation than anaerobic glycolysis. Respiration (*86*) and aerobic glycolysis (*16*) (compare *79a*) may even be stimulated when the tumor cells are irradiated under certain conditions (Fig. VII-2). Such an effect could result from increased permeability of the cell or cellular organelles.

Warburg (*119–121*) suggested in 1958 that the inhibition of glycolysis (and perhaps reproductive death) is caused by the hydrogen peroxide formed by radiolysis of cell water (see Vol. 1, p. 95). The following arguments have been adduced (*119–123*; see also *13*) to support this hypothesis. (a) When hydrogen peroxide is added to a tumor cell suspension, glycolysis diminishes, triosephosphate dehydrogenase is inactivated (*119–121*), and NAD is lost from the cell (*68*). In order to obtain this effect on glycolysis, hydrogen peroxide must be added in concentrations four times larger than that amount formed by a given x-ray dose in the cell. Earlier experiments reported by Warburg (*120*) suggest that cell concentration influences the course of radiation-induced inactivation of glycolysis, indicating that H_2O_2 formed by radiation in extracellular space may be important. Traces of heavy metals that have formed complexes in the medium through the action of EDTA abolish this effect (*122*). (b) Catalase added to the medium diminishes the effect of radiation (*13, 120, 122*) on glycolysis, but does not modify transplantability (*102*).

During recent years, most investigators have tended, however, to discount the role played by H_2O_2 in the inhibition of glycolysis for the following reasons. (a) Glycolysis is more extensively inhibited when the cells are irradiated in bicarbonate buffer than in phosphate buffer (*29*, *31*, *63*, *107*). Addition of phosphate (*29*, *31*) can reverse the radiation-induced depression of glycolysis to a certain extent; yet, the amount of H_2O_2 formed by irradiation must be about the same in both cases. (b) The pattern of accumulation of glycolytic

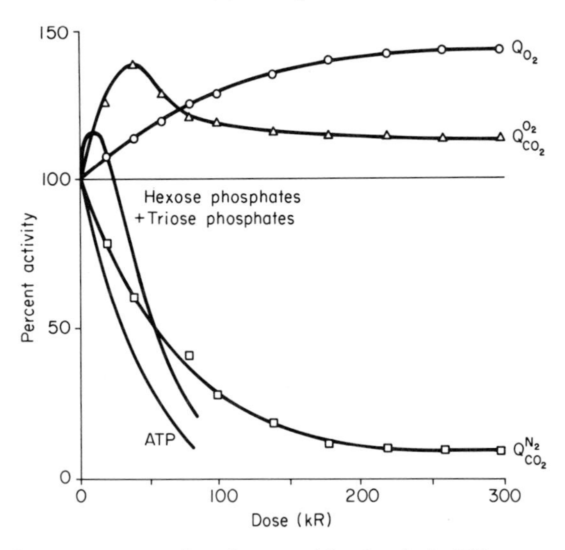

FIG. VII-2. Oxygen consumption Q_{O_2}, aerobic glycolysis $Q_{CO_2}^{O_2}$, anaerobic glycolysis $Q_{CO_2}^{N_2}$, ATP concentration, and concentration of triosephosphates plus hexosephosphates after irradiation of Ehrlich ascites cells. Irradiation and assays were carried out in Ringer's bicarbonate solution. The first three parameters are taken from A. Caputo and B. Giovanello, *Radiation Res.* **13**, 809 (1960), the others from K. Dose and U. Dose, *Intern. J. Radiation Biol.* **4**, 85 (1962). Note that respiration and aerobic glycolysis increase, whereas anaerobic glycolysis decreases markedly. Acid-soluble phosphates decrease with glycolysis at about the same rate.

intermediates in irradiated cells suggests that inhibition occurs at the level of dehydrogenation of triosephosphates (*31*, *83*, *124*). Whereas inactivation of triosephosphate dehydrogenase was formerly held responsible for the inhibition of glycolysis, it is now evident that loss of NAD is the important causal factor (*2*, *58*, *59*, *60*, *79*, *83*, *89*, *100*). This is demonstrated by experiments comparing inactivation of triosephosphate dehydrogenase, loss of NAD and inhibition of glycolysis under the same conditions (see Fig. VII-1) (*83*). Further evidence is derived from experiments in which glycolysis is poisoned by iodoacetate (*83*). (c) H_2O_2 causes a loss of NAD from tumor cells, as does irradiation, and this loss can also be prevented by addition of nicotinamide (*68*); but it seems likely that radiation and H_2O_2 do not react directly with NAD but, rather, initiate enzymatic degradation of NAD in the cell, as is discussed in more detail below. The nonspecificity of this reaction is shown by the observation that cytostatic agents also cause loss of NAD (*68*) and depression of glycolysis. Moreover, nicotinamide does not act as scavenger for H_2O_2 in the cell, since it is still effective if added immediately after exposure (*60*), and the protective effect of nicotinamide can be explained satisfactorily on the basis of inhibition of enzymatic break-down of NAD (*59*, *60*). (d) Exposure to H_2O_2 increases endogenous catalase activity of tumor cells, and these cells are less radiosensitive with respect to glycolysis than non-treated cells (*64*, *102*).

Next must be considered the mechanism of the loss of NAD from tumor cells and the significance of this loss for the fate of the irradiated cells. A decreased concentration of intracellular NAD could result from diminished synthesis, leakage of NAD from the cell, or from increased catabolism. Any direct effect of radiation on NAD may be excluded for the reasons mentioned above. Synthesis of NAD is not diminished early after irradiation (*2*, *58*, *59*), and may be stimulated after low doses (*96*), as is shown by the studies on the incorporation of labeled nicotinic acid into NAD. Moreover, the normal half-life of NAD in tumor cells is about 30–45 min, whereas most of the NAD is lost within 10 min after irradiation.

Loss of NAD from the cell, then, results from increased catabolism. Indeed, radioactive nicotinamide is released into the surrounding medium by irradiated ascites cells whose NAD has been labeled previously with nicotinamide-^{14}C (*58*). One could hardly expect enzymes catabolizing NAD to have been synthesized *de novo* within such a short time after exposure; rather, already existing enzymes may have been activated. NAD nucleosidase in the intact cell is normally inhibited by small concentrations of endogenous nicotinamide (10^{-5} to 10^{-4} *M*) (*58*, *59*). If nicotinamide leaked out of the cell after irradiation, NAD nucleosidase would have been activated and could have attacked cellular NAD (*2*, *59*, *60*). This explanation seems (*2*, *59*, *60*) reasonable in view of the fact that nicotinamide, even when added after exposure (*58–60*, *79*), prevents loss of NAD and, therefore, the inhibition of glycolysis.

Loss of NAD affects glycolysis primarily at the level of triosphosphate dehydrogenase. Irradiation may also interfere with other glycolytic reactions: fructose-1,6-diphosphate accumulates as a result of phosphofructokinase activation, pyruvate kinase activity is depressed, and the "Crabtree effect" is abolished after irradiation of Ehrlich tumor cells (50 kR) (*124*). The radiation-induced inhibition of glycolysis seems closely related to that of incorporation of precursors into proteins and DNA, because these reactions behave in a similar manner in tumor strains of different radiosensitivity (*58*, *60*) and can be protected by addition of nicotinamide. On the other hand, transplantability has characteristics different from those of the other reactions

TABLE VII-1

ACTION OF RADIATION ON RESPIRATION, NAD CONTENT, AND DNA SYNTHESIS IN DIFFERENT STRAINS OF TUMORS CELLS[a]

Tumor cell strain[b]	DNA content/ cell (9×10^{-12})	NAD content after 5 kR (% of control)	Respiration after 5 kR (% of control)	Incorporation of thymidine into DNA (% of control)
Ehrlich cells I	1.02	22	45	26
Ehrlich cells II	0.93	75	90	78
Yoshida cells	0.85	80	100	78

[a] Data from (*58a*).
[b] Transplantability was most radiosensitive in Yoshida tumor cells.

(Table VII-1) (*58a*), and is more radiosensitive than NAD loss (*64*), a pattern which is the reverse of the sensitivity to H_2O_2. Moreover, some tumor strains, like Yoshida cells, are highly radiosensitive with respect to transplantability, but much less so than to NAD loss (*58a*). In certain ascites tumors, transplantability is impaired only after exposure to doses which depress glycolysis.

Most likely, as more strains of tumor cells are investigated, the relations between inactivation of cells via reproductive death and other mechanisms, such as damage to glycolysis, DNA, protein synthesis, and cell membranes,

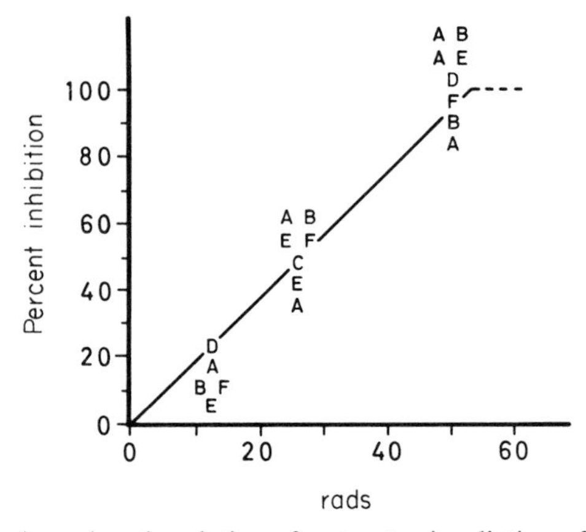

FIG. VII-3. Nuclear phosphorylation after *in vitro* irradiation of nuclei isolated from various human malignant tumors. (A) Adenocarcinoma of the cervix; (B) squamous tumor of the cervix; (C) papillary adenocarcinoma of the ovary; (D) adenocarcinoma of the breast; (E) adenocarcinoma of the stomach; (F) adenocarcinoma of the colon. Note that inhibition decreases with dose to about the same extent for all tumors. [From A. R. Fahmy and W. J. Williams, *Brit. J. Cancer* **19**, 501 (1965).]

will become apparent; all of these mechanisms may be indicative of early death in interphase.

Studies of irradiated tumor cells dealing with aspects of metabolism other than synthesis of macromolecules, glycolysis, and respiration are fragmentary. Phosphorylation of AMP in nuclei or in mitochondria is more radiosensitive in tumor cells which undergo interphase death, for example, than it is in those which do not (*78*) (see also Fig. III-9). "Nuclear phosphorylation" is also readily depressed by irradiation *in vitro* of nuclei isolated from different human cancerous tissues (see Fig. VII-3) (*34*). The significance of "nuclear

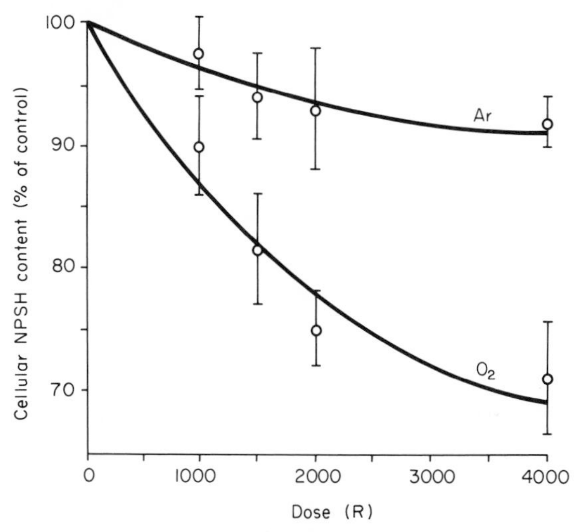

FIG. VII-4. Acid-soluble, nonprotein-bound sulfhydryl groups (NPSH) in Ehrlich ascites tumor cells measured immediately after irradiation in argon or oxygen. Note that nonprotein-bound sulfhydryl groups diminish as a function of the dose. Simultaneously, an increase in nonprotein-bound disulfide groups is observed. [From L. Révész and H. Modig, *Ann. Med. Exptl. Biol. Fenniae* (*Helsinki*) **44**, 333 (1966).]

phosphorylation" *in vivo* is, however, still a matter of controversy, as is discussed elsewhere (p. 63). The specific activity of ATP after injection of ^{32}P (*97*) or glycine-^{14}C (*77*) diminishes slightly after irradiation; yet, it is unlikely that the supply of ATP could become critical in tumor cells irradiated with doses of 5 kR or less, for ATP content decreases only when glycolysis is impaired (*12a*, *24*, *82*, *83*).

Glycogen in leukemia cells (*3*) or uterus carcinoma (*114*) diminishes a few hours after irradiation, as do also the activities of succinate dehydrogenase (*109*), lactate dehydrogenase (*39*, *72*, *73*), and glutamate dehydrogenase (EC 1.4.1.2) (*112*). These changes probably reflect the incipient death of the irradiated cells, since an immediate effect of radiation on the succinate oxidase system in hepatoma mitochondria is (*87*) not observed after less than 500 kR.

Nonprotein-bound (free) and protein-bound SH groups decrease in Ehrlich ascites cells immediately after exposure to 1000 R or more, whereas disulfide groups increase simultaneously (*94*) (Fig. VII-4). Lower dose levels have no influence on the SH content (*98*).

As in other damaged tissues, lysosomal enzymes in tumor cells may become activated after irradiation, and the number of lysosomes may increase (*86a*). Thus, acid RNase (*39*) in uterus carcinoma, β-glucuronidase and arylsulfatase in mammary carcinoma (*86a*), and acid phosphatase (*12*, *32*), and DNase II in ascites cells (*105*) increase after irradiation. An altered intracellular distribution, but no change in total activity, was reported for RNase in lymphoma cells (*48*) and for catalase in spindle tumor cells (*44*).

Damage to the cell membrane may be an important factor in the loss of nicotinamide (see above). Indeed, leakage of K^+ (*35a*) and pyruvate (*31*) parallels other biochemical alterations when irradiated cells are maintained in potassium-free medium. Uptake of glycine by ascites cells also diminishes shortly after exposure to more than 1 kR and is dependent on the cell concentration (*47*). Changes in the electrophoretic mobility of irradiated tumor cells (*65*, *108*) may also be regarded as circumstantial evidence for radiation damage to the cell membrane. Nevertheless, direct action of radiation on exchange of K^+ requires doses of several hundred kiloroentgens (*57*), i.e., well beyond those needed to affect, for example, loss of NAD.

REFERENCES

1. Alexander, P., *Nature* **192**, 572 (1961).
2. Altenbrunn, H. J., Steebeck, L., Ehrhardt, E., and Schmaglowski, S., *Intern. J. Radiation Biol.* **9**, 43 (1965).
3. Astaldi, G., and Meardi, G., *Boll. Soc. Ital. Biol. Sper.* **33**, 250 (1957).
4. Backmann, R., and Harbers, E., *Biochim. Biophys. Acta* **16**, 604 (1955).
5. Balmukhanov, S. B., Efimov, M. L., and Kleinbock, T. S., *Nature* **216**, 709 (1967).
6. Barnum, C. P., Scheller, M. S., and Herman, N. P., *Cancer Res.* **24**, 1155 (1964).

6a. Barron, E. S. G., *in* "Radiation Biology" (A. Hollaender, ed.), Vol. 1, p. 283. McGraw-Hill, New York, 1954.

7. Beer, J. Z., Lett, J. T., and Alexander, P., *Nature* **199**, 193 (1963).
8. Berry, R. J., Hell, E., Lajtha, L. G., and Ebert, M., *Nature* **186**, 563 (1960); *Intern. J. Radiation Biol.* **4**, 61 (1962).
9. Berry, R. J., *Radiation Res.* **18**, 236 (1963).
10. Bettendorf, G., and Maass, H., *Strahlentherapie* **106**, 263 (1958).

10a. Bhaskaran, S., and Dittrich, W., *Atomkernenergie* **14**, 51 (1969).

11. Boudnitskaya, E. V., Brunfaut, M., and Errera, M., *Biochim. Biophys. Acta* **80**, 567 (1964).
12. Brandes, D., Sloan, K., Anton, E., and Bloedorn, F., *Cancer Res.* **27**, 731 (1967).

12a. Bresciani, F., and Dose, K., *Radiobiol., Radioterap. Fis. Med* **15**, 313 (1960); *Atti Soc. Ital. Patol.* **6**, 695 (1959).

13. Burk, D., and Woods, M., *Radiation Res. Suppl.* **3**, 212 (1963).
14. Cammarano, P., *Nature* **194**, 591 (1962).
15. Cammarano, P., *Radiation Res.* **18**, 1 (1963).
16. Caputo, A., and Giovanella, B., *Radiation Res.* **13**, 809 (1960).
17. Caspersson, O., Klein, E., and Ringertz, N. R., *Cancer Res.* **18**, 857 (1958).
18. Caspersson, O., and Révész, L., *Nature* **199**, 153 (1963).
19. Cathie, J. A. B., *J. Pathol. Bacteriol.* **48**, 1 (1939).
19a. Chang, L. O., Looney, W. B., and Morris, H. P., *Intern. J. Radiation Biol.* **14**, 75 (1968).
20. Clifton, K. H., and Vermund, H., *Radiation Res.* **18**, 516 (1963).
21. Clifton, K. H., and Vermund, H., *Radiation Res.* **18**, 528 (1963).
22. Clifton, K. H., and Drapers, N. R., *Intern. J. Radiation Biol.* **7**, 515 (1963).
23. Clifton, K. H., Szybalski, W., Heidelberger, C., Gollin, F. F., Ansfield, F. J., and Vermund, H., *Cancer Res.* **23**, 1715 (1963).
24. Coe, E. L., Garcia, E. N., Ibsen, K. H., and McKee, R. W., *Radiation Res.* **20**, 586 (1963).
25. Conzelman, G. M., Jr., Mandel, H. G., and Smith, P. K., *Cancer Res.* **14**, 100 (1954).
26. Crabtree, H. C., *Biochem. J.* **29**, 2334 (1935).
27. Crathorn, A. R., and Shooter, K. V., *Intern. J. Radiation Biol.* **7**, 575 (1963).
28. De, N., *Cancer* **15**, 54 (1961).
29. Dose, K., and Dose, U., *Biochem. Z.* **332**, 214 (1959).
30. Dose, K., *Strahlentherapie* **119**, 419 (1962).
31. Dose, K., and Dose, U., *Intern. J. Radiation Biol.* **4**, 85 (1962).
32. Dragoni, G., and Raimondi, L., *Med. Nucl. Radiol. Lat.* **9**, 351 (1966).
33. Ebina, T., and Kurosu, M., *J. Natl. Cancer Inst.* **20**, 1023 (1958).
34. Fahmy, A. R., and Williams, W. J., *Brit. J. Cancer* **19**, 501 (1965).
35. Feine, U., *Nucl. Med. Suppl.* **3**, 119 (1965).
35a. Flemming, K., Mehrisch, J. N., and Napier, J. A. F., *Intern. J. Radiation Biol.* **14**, 175 (1968).
36. Forssberg, A., and Klein, G., *Exptl. Cell Res.* **7**, 480 (1954).
37. Frick, K., and Posener, K., *Fortschr. Geblete Roentgenstrahlen Ergdb.* **34**, 46 (1926).
38. Gardella, J. W., and Lichtler, E. J., *Cancer Res.* **15**, 529 (1955).
39. Goldberg, D. M., Ayre, H. A., and Pitts, J. F., *Cancer* **20**, 1388 (1967).
40. Goldfeder, A., *Am. J. Cancer* **36**, 603 (1939).
41. Goldfeder, A., *Cancer Res.* **10**, 89 (1950).
42. Goldfeder, A., *Intern. J. Radiation Biol.* **3**, 155 (1961).
43. Goldfeder, A., and Miller, L. A., *Intern. J. Radiation Biol.* **6**, 575 (1963).
44. Goldfeder, A., Selig, J. N., and Miller, L. A., *Intern. J. Radiation Biol.* **12**, 13 (1967).
45. Graevskii, E. Y., Konstantinova, M. M., Nekrasova, I. V., Sokolova, O. M., and Tarasenko, A. G., *Nature* **212**, 475 (1966).
46. Gronow, M., *Intern. J. Radiation Biol.* **9**, 123 (1965).
47. Hagemann, R. F., and Evans, T. C., *Radiation Res.* **33**, 371 (1968).
48. Hamilton, L. D. G., Furlan, M., and Alexander, P., *Proc. 2nd Intern. Congr. Radiation Res., Harrogate, Engl., 1962* p. 205.
49. Harbers, E., and Backmann, R., *Exptl. Cell Res.* **10**, 125 (1956).
50. Harbers, E., and Heidelberger, C., *J. Biol. Chem.* **234**, 1249 (1959).
51. Harbers, E., *Strahlentherapie* **112**, 333 (1960).
52. Harbers, E., *Radiation Res.* **14**, 789 (1961).
53. Harbers, E., *Strahlentherapie, Sonderbaende* **57**, 182 (1964).
54. Harrington, H., Rauschkolb, D., and Lavik, P. S., *Radiation Res.* **3**, 231 (1955).

55. Harrington, H., Rauschkolb, D., and Lavik, P., *Cancer Res.* **17**, 34 (1957).
56. Harrington, H., and Lavik, P., *Cancer Res.* **17**, 38 (1957).
57. Hasl, G., Pauly, H. and Rajewsky, B., *Strahlentherapie, Sonderbaende* **62**, 194 (1966).
58. Hilz, H., Hubmann, B., Oldekop, M., Scholz, M., and von Gossler, M., *Biochem. Z.* **336**, 62 (1962).
58a. Hilz, H., Maass, H., Hölzel, F., Hirsch-Hoffmann, A., and Oldekop, M., *Biochem. Z.* **339**, 198 (1963).
59. Hilz, H., Hlavica, P., and Bertram, B., *Biochem. Z.* **338**, 283 (1963).
60. Hilz, H., and Berndt, H., *Z. Krebsforsch.* **66**, 155 (1964).
61. Höhne, G., Künkel, H. A., Rathgen, G. H., and Uhlmann, G., *Naturwissenschaften* **42**, 630 (1955).
62. Höhne, G., Künkel, H. A., Maass, H., Rathgen, G. H., and Uhlmann, G., *Z. Krebsforsch.* **61**, 666 (1957).
63. Hölzel, F., Maass, H., and Schmack, W., *Strahlentherapie* **119**, 194 (1962).
64. Hölzel, F., *Intern. J. Radiation Biol.* **5**, 405 (1962).
65. Hollingshead, S., and Thomason, D., *Nature* **195**, 1217 (1962).
66. Holmes, B. E., *Proc. Roy. Soc.* **B127**, 223 (1939).
67. Holmes, B. E., and Mee, L. K., *Brit. J. Radiol.* **25**, 273 (1952).
68. Holzer, H., and Frank, S., *Angew. Chem.* **70**, 570 (1958).
69. Horikawa, M., Doida, Y., and Sugakawa, T., *Radiation Res.* **22**, 478 (1964).
70. Inouye, K., *Klin. Wochsch.* **15**, 613 (1936).
71. Ito, T., *Nippon Sanf. Zasshi* **17**, 982 (1965).
72. Kärcher, K. H., *Strahlentherapie* **119**, 23 (1962).
73. Kärcher, K. H., Toshi, H., and Schleich, A., *Strahlentherapie* **127**, 229 (1965).
74. Kelly, L. S., Hirsch, J. D., Beach, G., and Petrakis, N. L., *Proc. Soc. Exptl. Biol. Med.* **94**, 83 (1957).
75. Killander, D., Ribbing, C., Ringertz, N. R., and Richards, B. M., *Exptl. Cell Res.* **27**, 63 (1962).
76. Killander, D., Richards, B. M., and Ringertz, N. R., *Exptl. Cell Res.* **27**, 321 (1962).
77. Klein, G., and Forssberg, A., *Exptl. Cell Res.* **6**, 211 (1954).
78. Klouwen, H. M., Appelman, A. W. M., and de Vries, M. J., *Nature* **209**, 1149 (1966).
79. Koch, R., and Klemm, S., *Arzneimittel-Forsch.* **12**, 465 (1962).
79a. Kohen, E., Kohen, C., and Thorell, B., *Exptl. Cell Res.* **49**, 169 (1968).
80. Kuzin, A. M., and Kalendo, G. S., *Radiobiologiya* **6**, 30 (1966).
81. Loew-Beer, A., and Reiss, M., *Strahlentherapie* **42**, 157 (1931).
82. Maass, H., and Höhne, G., Künkel, H. A., and Rathgen, G. H., *Z. Naturforsch.* **12b**, 553 (1957).
83. Maass, H., and Künkel, H. A., *Intern. J. Radiation Biol.* **2**, 269 (1960).
83a. Magdon, E., *Radiobiol. Radiotherap.* **5**, 249 (1964).
84. Mantyeva, V. L., Rapoport, E. A., Tulykes, S. G., and Zbarsky, I. B., *Vopr. Med. Khim.* **12**, 407 (1966).
84a. Matsudaira, H., Sekiguchi, T., Orii, H., Sekiguchi, F., and Nagasawa, H., *Gann* **58**, 415 (1967).
84b. Mukerjee, H., and Goldfeder, A., *Intern. J. Radiation Biol.* **13**, 305 (1967).
85. Ontko, J. A., and Moorehead, W. R., *Biochim. Biophys. Acta* **91**, 658 (1964).
86. Ontko, J. A., and Moorehead, W. R., *Radiation Res.* **23**, 135 (1964).
86a. Paris, J. E., Brandes, P., and Anton, E. J., *J. Natl. Cancer Inst.* **42**, 383 (1969).
87. Rajewsky, B., Gerber, G., Parchwitz, K. H., and Pauly, H., *Z. Naturforsch.* **11b**, 415 (1956).

88. Rathgen, G. H., and Maass, H., *Strahlentherapie* **106**, 266 (1958).
89. Rathgen, G. H., Höhne, G., Künkel, H. A., and Maass, H., *Klin. Wochschr.* **34**, 1094 (1956).
90. Révész, L., Glas, U., and Hilding, G., *Nature* **198**, 260 (1963).
91. Révész, L., Bergstrand, H., and Modig, H., *Nature* **198**, 1275 (1963).
92. Révész, L., and Bergstrand, H., *Nature* **200**, 594 (1963).
93. Révész, L., and Modig, H., *Nature* **207**, 430 (1965).
94. Révész, L., and Modig, H., *Ann. Med. Exptl. Biol. Fenniae (Helsinki)* **44**, 333 (1966).
95. Ringertz, N. R., *Exptl. Cell Res.* **32**, 401 (1963).
96. Rüter, J., Vachek, H., Oldekop, M., Wueppen, I., and Hilz, H., *Biochem. Z.* **344**, 153 (1966).
97. Salmon, R. J., Loken, M. K., Mosser, D. G., and Marvin, J. F., *Cancer Res.* **20**, 292 (1960).
98. Scaife, J. F., *Can. J. Biochem.* **42**, 1717 (1964).
99. Scaife, J. F., *Can. J. Biochem.* **43**, 119 (1965).
100. Schneider, M., O'Bourke, F., and Redetzki, H. M., *Intern. J. Radiation Biol.* **5**, 183 (1962).
101. Schubert, G., Bettendorf, G., Künkel, H. A., Maass, H., and Rathgen, G. H., *Strahlentherapie* **106**, 483 (1958).
102. Schubert, G., *Arch. Geschwulstforsch.* **15**, 119 (1959).
103. Seed, J., *Nature* **192**, 944 (1961).
104. Seed, J., *J. Cell Biol.* **28**, 257 (1966).
105. Sekiguchi, T., and Yoshikawa, H., *Rinsho Seirik. Shinpoz.* **2**, 63 (1959).
106. Silini, G., and Hornsey, S., *Intern. J. Radiation Biol.* **5**, 147 (1962).
107. Silk, M. H., Hawtrey, A. O., and MacIntosh, I. J. C., *Cancer Res.* **18**, 1257 (1958).
108. Stein, G., Seaman, G. V. F., Mehrishi, J. N., and Simon-Reuss, T., *Intern. J. Radiation Biol.* **10**, 251 (1966).
109. Sullivan, W. D., O'Brien, E. M., and Frawley, H. E., *Broteria* **35**, 129 (1966).
110. Sung, S.-C., *Can. J. Biochem.* **43**, 859 (1965).
111. Takamori, Y., Akaboshi, M., and Honjo, I., *Ann. Rept. Sci. Works, Fac. Sci. Osaka Univ.* **12**, 55 (1964).
112. Trandafirescu, M., and Abbei, L., *Acta Biol. Med. Ger.* **17**, 250 (1966).
113. Vainson, A. A., and Kuzin, A. M., *Dokl. Akad. Nauk SSSR* **165**, 933 (1965).
114. Verhagen, A., *Strahlentherapie* **119**, 38 (1962).
115. Vermund, H., Barnum, C. P., Huseby, R. A., and Stenstrom, K. W., *Cancer Res.* **13**, 633 (1953).
116. Vogt, M., and Harbers, E., *Strahlentherapie* **133**, 426 (1967).
117. von Euler, H., and von Hevesy, G., *Kgl. Danske Videnskab. Selskab, Biol. Medd.* **17**, 3 (1942).
118. von Hevesy, G., *Nature Rev. Mod. Phys.* **17**, 102 (1945).
119. Warburg, O., Gawehn, K., Geissler, A. W., Schroeder, W., Gewitz, H. S., and Voelker, W., *Arch. Biochem. Biophys.* **78**, 573 (1958).
120. Warburg, O., Schroeder, W., Gewitz, H. S., and Voelker, W., *Z. Naturforsch.* **13b**, 59 (1958).
121. Warburg, O., Gawehn, K. W., Geissler, A. W., Schroeder, W., Gewitz, H. S., and Voelker, W., *Naturwissenschaften* **48**, 225 (1959).
122. Warburg, O., Gawehn, K., Geissler, A. W., and Lorenz, G., *Z. Naturforsch.* **18b**, 654 (1963).
123. Warburg, O., *Naturwissenschaften* **51**, 373 (1964).
124. Yamada, T. J., *Radiation Res. (Tokyo)* **9**, 41 (1968).

Chapter VIII

CHANGES IN THE BIOCHEMISTRY OF BODY FLUIDS AFTER IRRADIATION

A. Introduction

Exposure to radiation alters the metabolism and physiological function of specific organs in the body, and thus leads to changes in the biochemistry of the body fluids. The possible mechanisms involved in such changes are summarized in Table VIII-1. Naturally, these mechanisms operating *in vivo* can and do overlap, and are not always sharply defined or clearly distinguishable from one another. For example, irradiation may so damage the cell membrane that the cell not only loses a certain compound but also becomes incapable of taking it up from the surrounding medium.

An alteration in the amount of a metabolite excreted can result from direct damage to an excreting organ, but such an alteration is more frequently a reflection of a corresponding change in the amount of this substance in the blood. Moreover, urinary excretion may sometimes exaggerate the changes in the blood, as when a substance is completely removed by the kidney or when a threshold level exists below which excretion does not take place. Later in this chapter, we shall deal with the metabolism of serum proteins, urinary taurine, creatine, and metabolites of tryptophan and DNA, and then discuss in more detail several mechanisms by which changes in body fluids can occur.

TABLE VIII-1

POSSIBLE MECHANISMS OF ACTION OF IRRADIATION CAUSING CHANGES IN CONCENTRATION OF SUBSTANCES IN BODY FLUIDS

Factors contributing to changes in concentration	Cause of change	Example
Release from a tissue	Temporary damage to cell membrane	Increased level of serum enzymes due to leakage from cells
	Abnormal cellular metabolism and subsequent release of metabolites into the blood	Excess excretion of kynurenic and xanthurenic acids due to inhibition of kynureninase
	Destruction of cells	Loss of taurine or nucleic acid metabolites from lymphoid tissue into the blood and urine
Uptake by a tissue	Temporary damage to cell membrane	Creatinuria due to decreased uptake of creatine by the muscle
	Destruction of cells normally using the metabolite	Increase in the blood iron due to decreased uptake by the bone marrow depleted of erythropoietic cells
Excretion from organism	Polyuria	Increased excretion of Cl^- on the first day after irradiation
	Damage to excreting organ	Uremia after irradiation of the kidney; loss of Na^+ by the intestine after whole-body irradiation
Uptake by organism	Starvation	Changes in concentration of many metabolites in blood and urine due to decreased absorption of food and impairment in the adaptation to starvation
	Changes in gastrointestinal motility and abnormal bacterial growth	Excess excretion of indoxylsulfate from indole formed by intestinal bacteria
Stress and systemic responses	Release of hormones and biogenic amines	Excess excretion of 17-ketosteroids due to enhanced activity of the adrenals
		Excess excretion of 5-OH-indoleacetic due to release from subcellular particles and degradation of serotonin

B. Fluid and Electrolyte Metabolism

1. Blood and Plasma Volume

The principal factors which determine the changes in water metabolism after irradiation are: loss of water into the extravascular space and the gastrointestinal tract, as well as changes in urinary excretion of water (polyuria). A slight, transient increase in plasma volume, perhaps the result of an expansion of the extravascular space, is observed in many species shortly after an exposure to a LD_{50}, e.g., rabbits (*250*); monkeys (*89*); dogs (*181*, *189*, *244*); rats (*121*, *129*, *257*); and man (*178*). Later, as the acute stage of radiation illness develops, and particularly just before death, dehydration occurs and the plasma volume decreases (Fig. VIII-1).

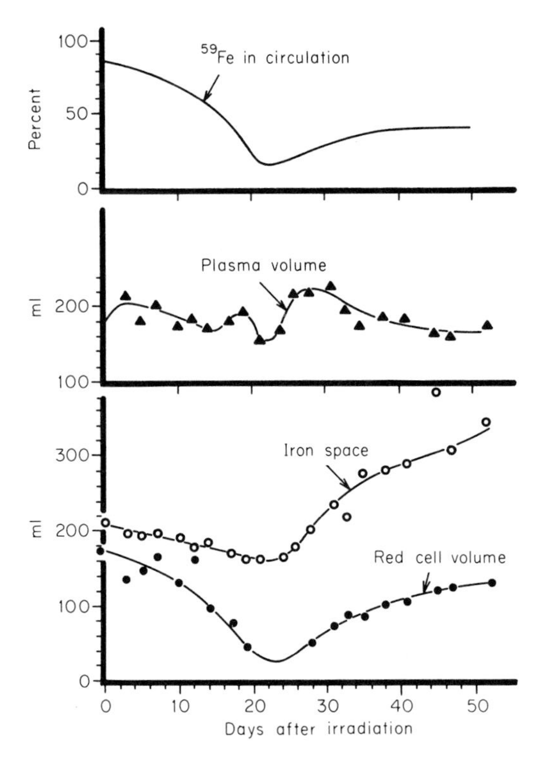

Fig. VIII-1. Plasma volume, red cell mass, iron space (after injection of inorganic ^{59}Fe), and Fe taken up by erythrocytes after whole-body irradiation of rhesus monkeys with 520 rads. Note that plasma volume increases slightly during the first day after exposure and again during recovery. The red cell mass decreases until the twentieth day after exposure, then returns to normal. [From C. W. Gilbert, E. Paterson, M. V. Haigh, and R. Schofield, *Intern. J. Radiation Biol.* **5**, 9 (1962).]

The change in the total blood volume results primarily from changes in the number of red cells, and secondarily from those in plasma volume. After irradiation, fewer erythrocytes are produced by the bone marrow and more are lost from hemorrhage, etc. Thus, the mass of the red cells and the blood volume usually decrease steadily until death (*89*, *250*, *264*). The decrease in plasma and blood volume is, however, more severe after doses which cause gastrointestinal death, because diarrhea and vomiting cause acute dehydration (*38*, *182*, *258*).

2. Electrolytes in Blood and Urine

The above-mentioned factors which bring about changes in plasma volume

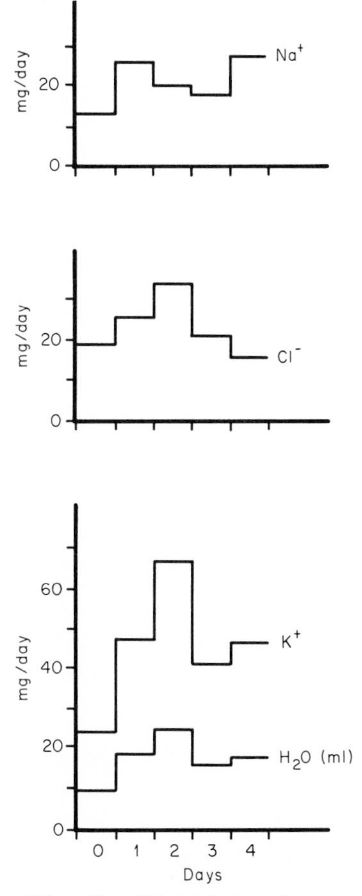

Fig. VIII-2. Excretion of Na^+, Cl^-, K^+, and water (urine volume) after a whole-body irradiation of rats (600 R). Note that polyuria after irradiation is accompanied by increased excretion of K^+ and Cl^-. [From V. Holoček, J. Petrášek, V. Schrieber, Z. Dienstbier, and V. Kmentova, *Physiol. Bohemoslov.* **11**, 119 (1962).]

must also be considered when interpreting changes in electrolyte metabolism. In addition, radiosensitive organs, e.g., lymphoid tissues, and perhaps even radioresistant ones, e.g, muscle (*121*) and brain, may lose potassium after large doses of radiation. (See also p. 211.)

The levels of electrolytes in serum have been widely studied and, in general, remain within normal limits after doses in the range of a LD_{50} (*24, 25, 97, 157,*

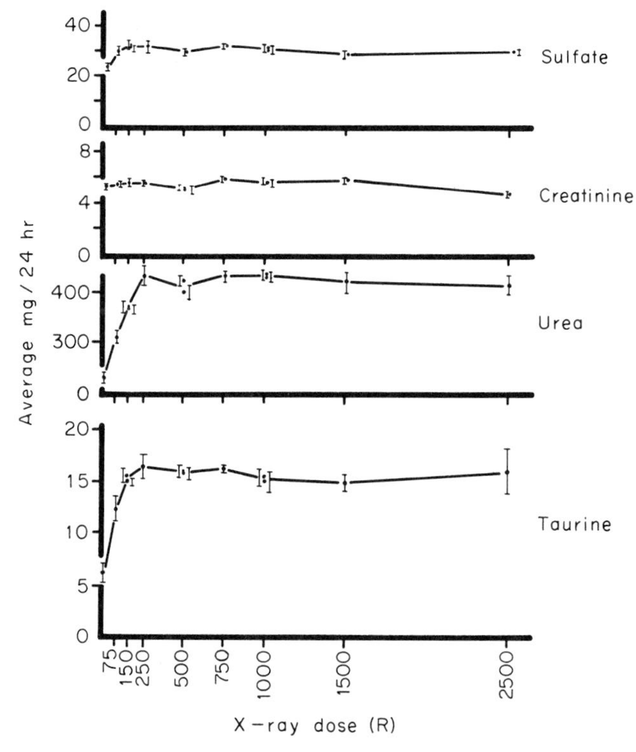

FIG. VIII-3. Excretion of taurine, urea, sulfate, and creatinine by rats as a function of the whole-body dose. Note that excretion of taurine and urea increase up to a dose of about 250 R. [From R. E. Kay, J. C. Early, and C. Entenman, *Radiation Res.* **11**, 375 (1959).]

158, 189, 224, 244). Although the concentration of Na^+ in the serum and the sodium space (*121*) is not altered significantly, some Na^+ and Cl^- may be retained in certain organs after irradiation (*244*). Blood levels of Cl^- decrease slightly and the excretion of Na^+, Cl^-, and particularly of K^+ is elevated during the period of polyuria, i.e., 1–2 days after irradiation (Fig. VIII-2) (see *33, 34, 125, 189*). Changes in electrolyte metabolism have been reported in patients during radiation therapy (*58*), but, most likely, these reflect the underlying malignant disease processes rather than the influence of radiation. Man is, however, more prone to vomit after irradiation than most mammalian species, and this factor, as well as diarrhea, gives rise to a depletion of electrolytes, especially of Na^+ and Cl^- after an accidental exposure. Irradiation with doses causing gastrointestinal death produces severe diarrhea in mammals and could, in synergism with decreased reabsorption of Na^+, result in an excessive loss of Na^+ (*133*; see p. 96).

Although the potassium level in serum is only slightly or not at all altered after exposure to low doses of radiation (*25*, *97*), K^+ may be lost from radiosensitive organs like the spleen (*29*, *189*). It has been claimed, although not confirmed by all investigators, that exposure to very high doses, which causes central-nervous death, liberates K^+ from many organs and thereby results in hyperpotassiemia (*202*). Levels of phosphate, calcium, and magnesium (*105*) in the serum or urine, show little or no changes after whole-body irradiation, but the excretion of sulfate (*150*) (Fig. VIII-3) increases slightly. Serum magnesium is decreased, and serum copper (*105*, *106*), zinc (*219b*), and iron (*20*, *37*, *88*, *106*) are elevated after irradiation. The increase in iron may result from decreased synthesis of hemoglobin in the bone marrow and from changes in the transferrin level. Alkali reserve and pH of the blood apparently decrease shortly after exposure, but this acidosis is later superseded by an alkalosis (*59*, *186*). Conductivity, viscosity, and surface tension of the blood, also show significant alterations during the first hours after exposure (*48*, *120a*).

C. Compounds of Low Molecular Weight in Blood and Urine after Irradiation

1. General Observations

Irradiation causes changes in blood level and urinary excretion of many compounds of low molecular weight, the most important of which are tabulated in Tables VIII-2 and VIII-3. Most of the early work on glucose and cholesterol in the blood of patients during radiation therapy is not mentioned in Table VIII-2 since it is now obsolete and is reviewed elsewhere (*58*). An increase in blood glucose, perhaps related to changes in glycogen metabolism, is frequently observed during the first hours after irradiation of man or animal. Blood cholesterol, phospholipids, fatty acids, and lipoproteins are elevated after whole-body irradiation (Fig. VIII-4). The increase in plasma lipids is probably a result of an increased *de novo* synthesis, since labeled palmitate is removed from the blood at an unchanged rate after irradiation, indicating a normal utilization (*57*), whereas incorporation of precursors in lipids is known to be enhanced (p. 104). Moreover, mobilization of lipids from adipose tissue is also unaltered after irradiation (*56*).

Breakdown of tissue after irradiation and starvation may result in a negative nitrogen balance (*61*, *67*, *94*, *101*, *103*, *107*, *137*, *224*, *274*). Urea degradation in irradiated animals is lower than that in starved controls (*86e*, *227a*). Concentration of urea in tissues was, however, found to be diminished, and that of

TABLE VIII-2

MODIFICATIONS IN CONCENTRATION OF VARIOUS COMPONENTS IN THE BLOOD AFTER IRRADIATION

Compound	Species	Change	Remarks
Proteins (total nitrogen)	Man (therapy or accidents)	Decrease from day 2–4	Albumin/globulin ratio decreases; albumin, prealbumin, and γ-globulins decrease
	Various species	Decrease from day 2–4	α- and β-globulins increase; several lipoproteins also increase, dependent on species
Nonprotein nitrogen	Man and various species	Increase 1–3 days	Nitrogen balance negative; possible increase in urea and purines (uric acid) (*41*, *107*, *112*, *150*, *152*, *157*)
Amino acids	Man, rat, rabbit, and, dog	Inconsistent increase for several days	Increase in rat (500–1000 R) (*16*, *149*); tryptophan decreases (*21*), tyrosine decreases in rabbit (*218*), no change in dog (*163*); decrease in man of several amino acids (*173*)
Creatine	Rat and dog	Increase 1st day	Increases in rat (*171*) and dog (*1*), especially in erythrocytes (*171*)
Organic acids	Man and various species	Increase for several days	Lactate increases in monkeys (*59*) and man (*232*), but not in rats (*13*)
Glucose	Man and various species	Increase 1st day	A temporary increase in blood glucose has often been observed during radiation therapy in man and in whole-body irradiated animals (*58*, *148*, *157*, *158*)
			Plasma polysaccharides increase in rats 6–8 days after 700 R (*236*); mucopolysaccharides return to normal during radiation therapy (*145*); protein-bound hexosamines increase 8–12 days after 600 R (*265b*)
Lipids	Man and various species	Increase for several days	Total lipids increase in rabbit 1–2 days after 1000 R (*50*, *56*) due to increased hepatic synthesis (*57*); increase also in phospholipids (*56*, *60*) and in plasmalogen level (*128*)
Cholesterol	Man and various species	Increase 1st day	An increase in serum cholesterol in rats preceded by a decrease at 3 hr (*54a*) has been observed often during radiation therapy and in whole-body irradiated animals (*50*, *55*, *56*, *58*, *158*, *237a*)
Bilirubin	Rat	Increase 1st day	After 500–950 R (*180*)

ammonia enhanced after irradiation (*267a*). Increased levels of amino acids in blood and urine have been reported by some authors but denied by others. Evidently, the problem of nonprotein nitrogen metabolism after irradiation should be studied with special attention to the influence of dietary conditions.

Some of the early investigators (*112*, *179*) detected an increase in uric acid in blood and urine. Since that time, several other products of nucleic acid

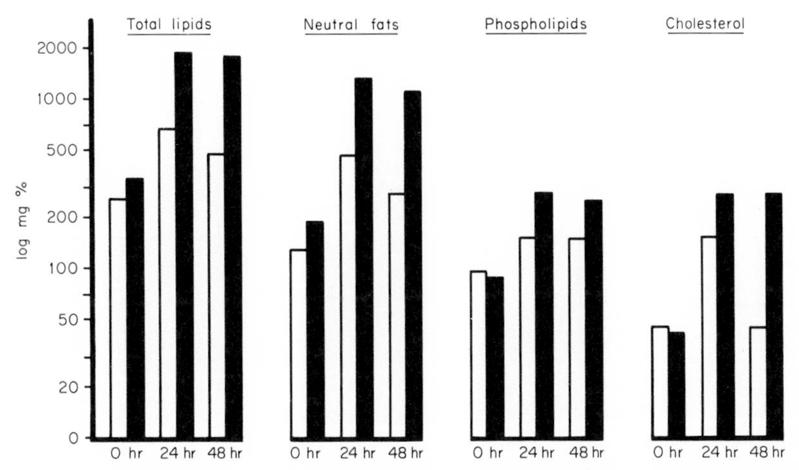

FIG. VIII-4. Changes in the levels of several lipids in serum of irradiated rabbits (1000 R). Note that total lipids, neutral fats, cholesterol, and phospholipids, increase in the serum after irradiation. The changes are more severe in the animals that died (black columns) than in those that survived (white columns). [From N. R. Di Luzio, and K. A. Simon, *Radiation Res.*, **7**, 79 (1957).]

metabolism have been found in the urine of irradiated mammals. Significant changes also occur in the excretion of hormones, biogenic amines, and their metabolites but these are treated in detail in Chapter IX. The mechanism responsible for the many changes in metabolites in the urine has been studied in detail for only a few compounds which will be discussed below.

2. EXCRETION OF TAURINE AFTER IRRADIATION

During the first days after a whole-body exposure, excess taurine is excreted in the urine by many species, i.e., man (*18*, *35*, *114*, *147*) rats (*2–4*, *150*, *152*), and mice (*175*, *192–195*, *251*), but not by dogs or guinea pigs (*10*) (Fig. VIII-5). The amount of taurine excreted increases with the dose (Fig. VIII-3), but above a dose level of 250 R to 500 R in rats (*4*, *150*) and about 200 R in mice (*175*, *192*, *251*) taurinuria does not increase further. The urinary excretion reaches a maximum on the first and second days after exposure. Later, normal and subnormal values prevail. A second peak of taurinuria may ensue 7–10 days after irradiation (*175*, *192*).

TABLE VIII-3

EXCRETION OF VARIOUS SUBSTANCES INTO URINE AFTER EXPOSURE TO IONIZING RADIATION

Substance	Species	Change	References and remarks
Deoxyribose derivatives (Dische positive substances)	Rat and mouse	Increase 1st day	(*64*, *98*, *160*, *162*), also in dogs but not in rabbits (*65*); no change in cancer therapy (*93a*)
Deoxycytidine	Rat	Dose-dependent increase 1st day	(*101ab*, *211*, *212*, *265*, *279*) in blood (*101c*), increase in thymidine and deoxyuridine (*70a*, *188*, *278a*)
	Man	Increase 1st day	(*17*), also increase in thymidine (*238*)
β-Aminoisobutyric acid	Man	Increase during several days	(*75*, *227*), but no consistent increase in patients (*241*); increase in rat urine (*140*) and liver (*188a*)
Pseudouridine	Rat	Dose-dependent increase 1st day	(*53*, *62*, *70a*), also increase in man (*54*, *63*)
Uric acid	Rat, dog, and man	Increase 1–2 days	Man (*179*), rat (*134*), dog (*132*); other purines also increase, especially allantoin and xanthine (*132*)
Urea	Rat and man	Increase 1–2 days	(*127*, *150*, *252*, *274*)
Amino acids and related compounds	Man	Increase for several days	(*90*, *147*), especially glycine, tryptophan, cystine, proline (*72*, *73*), hydroxyproline (*147*), and serine (*167*)
	Other species	Inconsistent increase	Increase in rats, especially glycine, alanine, glutamate (*149*, 191) tyrosine (*184*), creatine (p. 260, taurine (p. 255), and tryptophan metabolites (p. 264), no increase in total amino acids found in monkey (*131*) or dog (*163*)

Pyrrolcarboxylic acid	Rat	Increase for several days	(*161*), also in man after accidents (*75*)
Indoxylsulfate	Man	Increase for several days	(*242*, *242a*), perhaps as a result of intestinal bacteria
Glucuronides	Rat	Increase 1st day (1000 R)	(*39*), decrease due to starvation can mask the increase due to irradiation
Oxalate	Mouse	Decrease 1st day (700 R)	(*210*)
2-Aminoethanol	Man	Decrease 4th day (300–900 R, local)	(*36*)
Mucoproteins	Rat	Increase second week (800 R)	(*42*)
Vitamins of the B group	Rat	Increase (500 R)	(*213*), also increased excretion of NAD metabolites, *N*-methylnicotinamide (*169*, *205a*) [intestine (*46*)], and pyridone (*86c*, *280*)
Ascorbic acid	Rat	Increase until day 3, then decrease (600 R)	(*156a*)
Porphyrins (coproporphyrin)	Dog	Increase	(*234*)

Taurine in the body is derived from sulfur-containing amino acids. A small fraction of the total body taurine, primarily that in liver, blood, and lymphoid tissues, turns over rapidly, whereas the larger fraction, predominantly in muscle, turns over slowly. Some taurine conjugated with bile acids undergoes enterohepatic circulation, during which part of it is degraded by the intestinal bacteria (Fig. VIII-6). Excess taurine excreted after irradiation could be derived from either increased synthesis, decreased catabolism, or a shift between different pools of taurine.

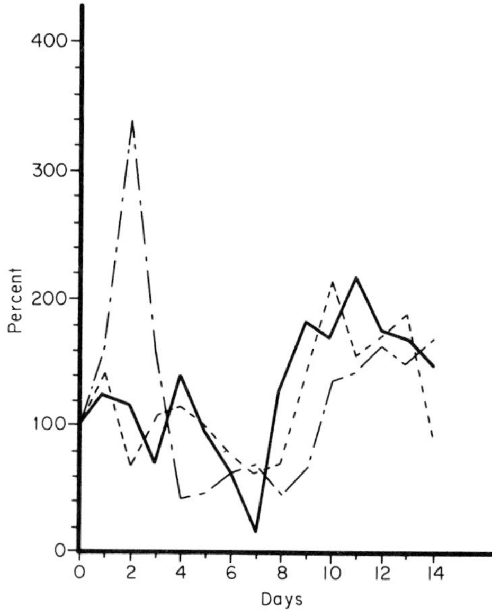

FIG. VIII-5. Excretion of taurine by mice after whole-body irradiation with different doses. Note the excess excretion on days 1–2 and the subnormal values on days 4–10 after exposure. [From H. J. Melching, *Strahlentherapie* **120**, 34 (1963).] Solid line: 200 R; dashed line: 400 R; dot-dashed line: 690 R.

Conversion of radioactive cysteine (*118*, *245*) or methionine (*151*, *206*) to taurine appears to remain unaltered during the maximal urinary excretion 1 day after irradiation, but increases later when the taurine stores in the body are depleted. Experiments on pyridoxal-deficient rats, which have an impaired synthesis of taurine, also suggest that synthesis does not increase immediately after exposure. These deficient animals are still able to develop taurinuria after irradiation, as is shown by injecting taurine prior to irradiation in order to replenish the depleted stores (*71*, *207*). If radioactive taurine is injected into rats, urinary taurine has the same specific activity in normal as in x-irradiated rats. This observation indicates that taurine excreted after irradiation originates from pools having the same metabolic characteristics as those in normal rats (*23*, *71*).

Therefore, the excess taurine excreted cannot have been synthesized *de novo* or been released by the muscle (see also *98a*), but must represent material (*22*, *23*, *71*) either (a) released from radiosensitive tissues, (b) not taken up by the muscle, (c) not conjugated with bile acids, or (d) not catabolized by the intestinal bacteria.

Uptake of taurine by the muscle is not significantly altered after exposure, as shown by its specific activity after injection of taurine-^{35}S (*23a*). The ability

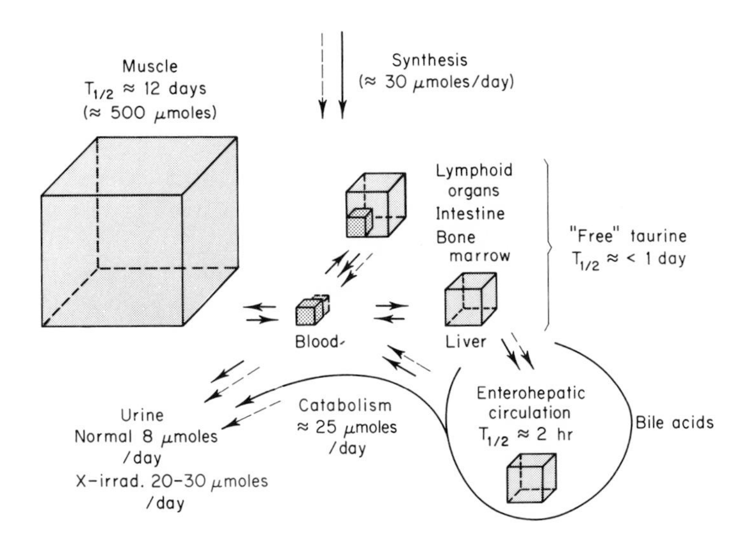

FIG. VIII-6. Simplified semiquantitative scheme of taurine metabolism in the rat. The size of the cubes corresponds approximately to the size of the pools, and the length of the arrows to the rate of reaction. Possible changes in irradiated rats are drawn as dashed lines for the rates and as darker cubes for the pools. The scheme illustrates the most likely mechanism for taurinuria after irradiation, i.e., loss of taurine from lymphoid organs, decreased uptake by muscle, changes in enterohepatic circulation, and decrease in catabolism.

to synthesize and to excrete taurine-conjugated bile acids is not affected after irradiation (*82*, *216*). More taurine is, however, available for excretion (*23a*) due to the elevated plasma level, and is removed more efficiently by the liver after irradiation (*82*). Moreover, since intestinal reabsorption of conjugated bile acids is not modified (*86a*, *256*), more taurine is degraded by intestinal bacteria and excreted as sulfate (*23a*).

Destroyed lymphoid tissue could provide an important source of excess taurine immediately after irradiation (*18*, *23a*, *233*). Although the concentration of taurine in the spleen remains constant (*247*) or even increases slightly

(*192*, *196*), lymphoid tissues lose sufficient numbers of cells to account for at least part of the taurine excreted [about 4 μmoles/100 gm from the spleen and 12 μmoles/100 gm from the lymphatic tissue (*23a*)]. Taurinuria has been reported to decrease in splenectomized (*193*, *197*) or pancreatomized (*152*) animals, but not all investigators have confirmed these observations (*214*). If one injects cortisone (*270*) or mitomycine (*23a*) prior to irradiation in order to destroy lymphoid tissues, taurinuria fails to develop. On the other hand, animals with radiosensitive myeloma show an increased excretion of taurine (*23a*). Intestine also can liberate about 14 μmoles/100 gm taurine after irradiation, and the release from radiosensitive organs thus suffices to account for all taurine excreted (Fig. VIII-6).

3. Excretion of Creatine after Irradiation

Increased excretion of creatine after irradiation has been observed in rats (*7*, *64*, *98a*, *104*, *169*, *276*), monkeys (*168*), and dogs (*7*), as well as in man during radiation therapy (*170*, *172*) or after accidents (*75*). In rats, excretion of excess creatine is a function of the total dose (*74*) (Fig. VIII-7). Elevated creatine concentrations also occur in the blood (*1*, *276*), especially in erythrocytes (*171*).

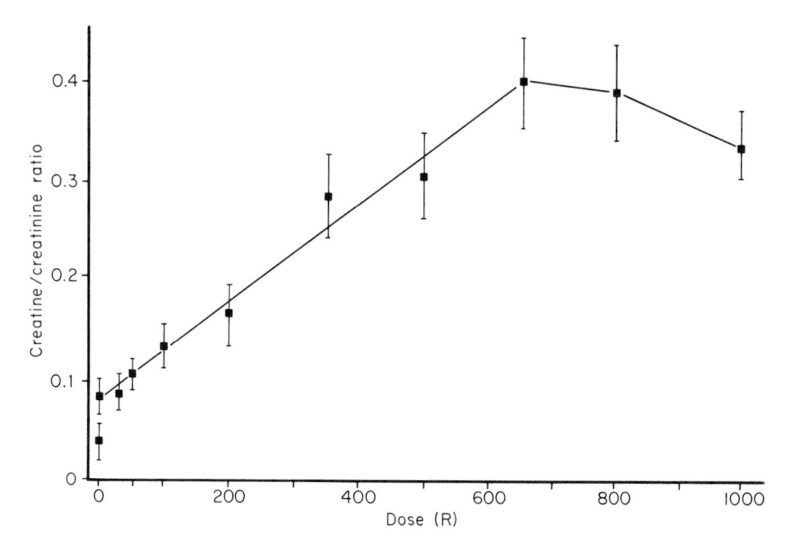

Fig. VIII-7. Excretion of creatine (creatine/creatinine ratio) as a function of the radiation dose during the first 4 days after whole-body irradiation of rats. The vertical bars represent standard errors. The intercept of the curve at dose 0 represents the value of sham-irradiated animals. The range and means below this intercept represent the average of all rats before irradiation. Note that creatinuria is dose-dependent and increases up to a dose level of 650 R. Sham irradiation also causes slight creatinuria. [From G. B. Gerber, P. Gertler, K. I. Altman, and L. H. Hempelmann, *Radiation Res.* **15**, 307 (1961).]

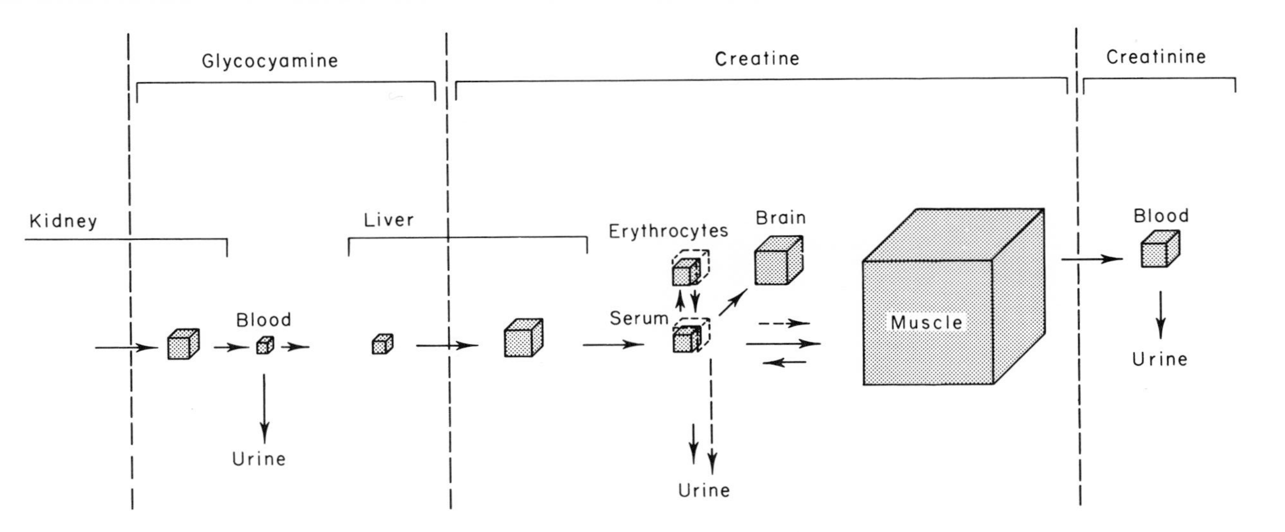

FIG. VIII-8. Simplified semiquantitative scheme of creatine metabolism in the rat. The size of the cube shows the relative pool size, and the length of the arrow the relative rate of reaction. Changes in pool sizes and rates of reaction in irradiated rats are drawn as dashed lines. The picture illustrates how a diminished uptake of creatine by the muscle will increase blood creatine and cause creatinuria. [From G. B. Gerber, *Strahlenschutz Forsch. Praxis* **4**, 239 (1964).]

Creatine is synthesized from methionine and glycocyamine, mainly in the liver, whereas the latter compound is elaborated from glycine and arginine by the kidney. A large part of the creatine in the body is present in muscle and brain as creatine phosphate, and is there converted slowly to creatinine, the end product of creatine metabolism. Smaller creatine pools with more rapid turnover (free creatine) are located in other organs and in the blood (Fig. VIII-8).

The synthesis of creatine is not altered after irradiation, as indicated by direct determination in the perfused liver (*77*) and by measurement of the sum of the excreted creatine and creatinine plus the total creatine in the body of irradiated and control rats (*7*, *78*; but see also *205*). The muscle of an irradiated rat takes up less radioactive creatine than that of a normal one, and the ratio of specific radioactivity of "free" creatine to creatine in muscle decreases after irradiation (*76*). On the other hand, the muscle of an irradiated rat does not seem to release large amounts of radioactive creatine already present in the muscle. If one injects radioactive creatine into a rat several days before exposure, so that at the time of irradiation the radioactivity of urinary creatine is derived from muscle creatine, no increase in specific activity of urinary creatine is observed (*78*). The impaired uptake of creatine by the irradiated muscle is an abscopal effect and depends on the total volume of tissue irradiated (*79*). The various pools of creatine, the rates of pertinent metabolic reactions, as well as the changes after irradiation are shown schematically in a scale drawing (Fig. VIII-8). This scheme shows that excess creatine after irradiation represents creatine synthesized at a normal rate but not taken up by the muscle. The excess creatine in the blood is, therefore, excreted.

4. Excretion of Nucleic Acid Metabolites after Irradiation

Several substances related to nucleic acid metabolism are excreted in increased amounts after irradiation, i.e., deoxycytidine, pseudouridine, purines (uric acid and xanthine), β-amino isobutyric acid, deoxyuridine, and thymidine (see Fig. VIII-3).

Excretion of deoxycytidine is dose dependent in rats after irradiation (Fig. VIII-9) (*11*, *48*, *101a*, *211*, *212*). Excess deoxycytidine could originate either from cells destroyed in radiosensitive tissues or from precursor material not utilized for DNA synthesis (*86b*, *261b*). When tissue DNA is labeled with deoxycytidine 1 day prior to irradiation, deoxycytidine excreted has a higher specific activity than that of nonirradiated rats (*86b*), indicating that labeled DNA has been the source of the extra deoxycytidine. Indeed, loss of radioactive DNA from lymphoid tissue (*101a*) and other radiosensitive organs could easily account for all deoxycytidine excreted. Not only is more radioactive deoxycytidine excreted after irradiation, but also deoxyuridine and thymidine (*86b*).

Catabolism of deoxycytidine varies, however, from one species to another (*261, 261a*). Man excretes much less deoxycytidine in urine than the rat because he possesses high levels of deoxycytidine aminohydrolase (*282*), and apparently excretion in man increases only slightly after irradiation (*131b*). Therefore, more extensive studies on deoxycytidine metabolism in different species after irradiation are needed.

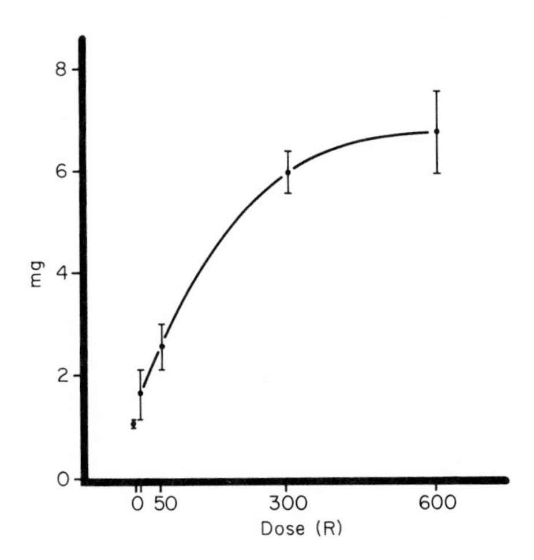

FIG. VIII-9. Excretion of deoxycytidine in rats during the first day after exposure as a function of the dose. Note that the excretion increases with the dose up to 600–1000 R. [From J. Pařízek *et al.*, *Nature* **182**, 721 (1958).] (See *11*.)

β-amino isobutyric acid (BAIBA), a metabolite derived directly from thymidine, is formed and degraded in the liver, but the ability to degrade BAIBA differs from one species to another. In rat liver, formation (*83*) and degradation (*85*) of BAIBA occur very rapidly. Furthermore, formation of BAIBA is not affected (*83*) and degradation of BAIBA only slightly (*14*, *85*) affected in the irradiated rat. BAIBA injected into irradiated rats is, however, removed more slowly from the blood (*242b*). BAIBA levels in the liver are enhanced after irradiation (*188a*). Excess excretion of BAIBA after irradiation of man, observed by some investigators (*75*, *227*) but not by others (*240*, *241*), may thus result from saturation of BAIBA catabolism or from release by the liver. BAIBA excreted shortly after irradiation is derived from DNA of destroyed cells (as are the other products of nucleic acid metabolism); later, material not used for DNA synthesis also contributes to BAIBA in urine. This conclusion is based on experiments in which DNA was labeled with radioactive thymidine prior to exposure (*78a*).

5. Excretion of Tryptophan Metabolites after Irradiation

Irradiated mice, rats, and rabbits excrete excess xanthurenic acid (*108*, *119*, *174*, *208*), kynurenic acid, and kynurenine on the first day and again from the fifth to the ninth days after irradiation (Fig. VIII-10) (*175*, *192*, *193*). Excretion

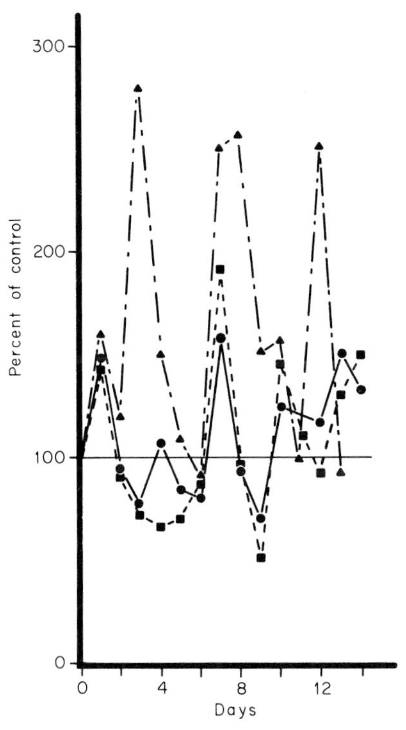

Fig. VIII-10. Excretion of kynurenic acid (triangles), anthranilic acid (circles) and kynurenine (squares) by mice after whole-body irradiation (690 R). Note that excretion of kynurenic acid and, to a smaller extent, of kynurenine and anthranilic acid increases on the first day and again from days 6–8 after exposure. [From H. J. Melching, C. Streffer, and U. Allert, *Strahlentherapie* **124**, 288 (1964).]

of anthranilic acid is insignificantly altered and that of 3-hydroxyanthranilic acid unaltered under these conditions. Radiation-induced changes in urinary concentration are also reported for other indol compounds, i.e., for serotonin (see p. 306), 5-hydroxyindoleacetate (see p. 310), and indoxylsulfate (*242*, *242a*).

Liver is the major site for catabolizing tryptophan by way of kynurenine (Fig. VIII-11). Changes corresponding to the excretion pattern are found among the different hepatic enzymes related to tryptophan catabolism, mainly for kynureninase activity, which diminishes from days 1–4 and again from days 1–10 after irradiation of mice (Table VIII-4) (*253a*, *254*, *254a*). There also

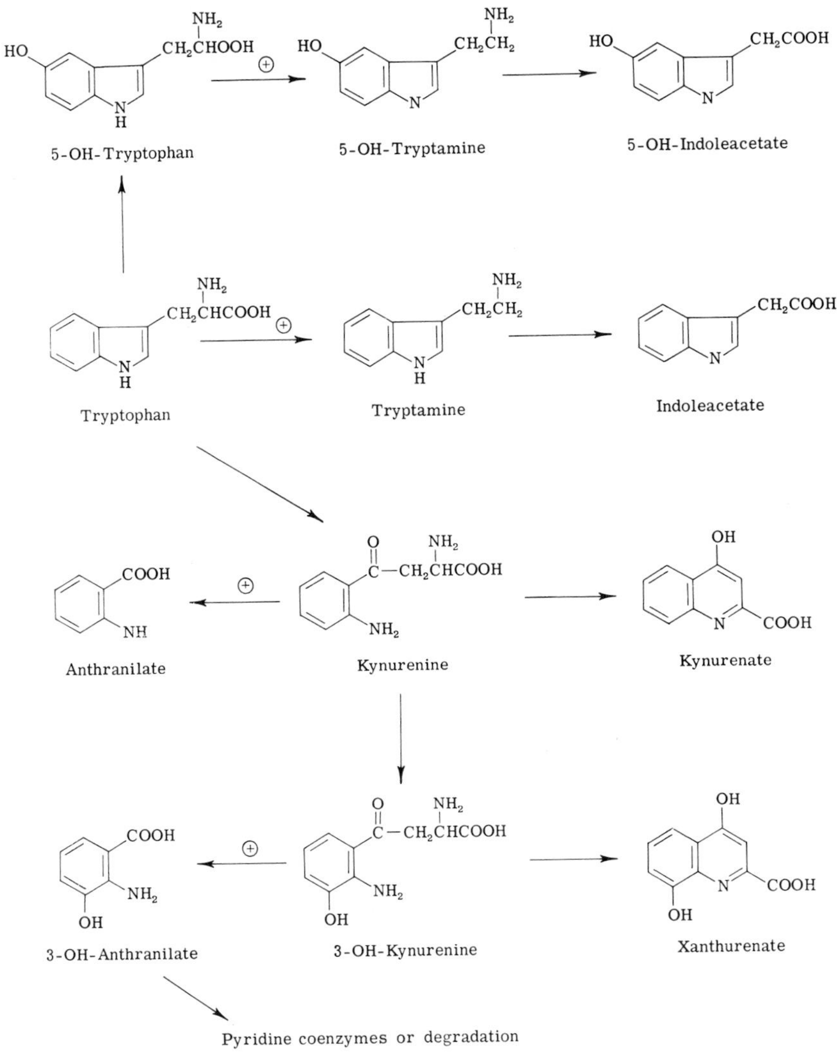

5-OH-Tryptophan → 5-OH-Tryptamine → 5-OH-Indoleacetate

Tryptophan → Tryptamine → Indoleacetate

Anthranilate ← Kynurenine → Kynurenate

3-OH-Anthranilate ← 3-OH-Kynurenine → Xanthurenate

Pyridine coenzymes or degradation

FIG. VIII-11. Important pathways of tryptophan metabolism in the mammal. Reactions in which pyridoxal-5-phosphate is involved as coenzyme are marked with a cross.

exists a block in tryptophan catabolism in vivo, as shown by the lowered incorporation of tryptophan into NAD of liver after irradiation (*254b*).

It has been suggested (*175*, *192–196*, *254a*) that the reduction of the activity of certain enzymes requiring pyridoxal-5-phosphate as coenzyme is responsible for the radiation-induced changes in the excretion of tryptophan metabolites and taurine. The following evidence has been adduced to support this contention. (1) Excretion of xanthurenic acid, kynurenic acid, and kynurenine increases, but not that of anthranilic acid. This would indicate a block of the conversion of kynurenine or 3-hydroxykynurenine to the respective anthranilic acids. Indeed, kynureninase, which performs these conversions and requires pyridoxal-5-phosphate as its coenzyme, decreases in activity after irradiation.

TABLE VIII-4

TRYPTOPHAN CATABOLISM IN THE X-IRRADIATED MOUSE LIVER

Function affected	Enzyme activity[a]		p Value
	Nonirradiated, fed mice	7 Days after whole-body x-irradiation with 810 R	
Tryptophan pyrrolase (EC 1.3.1.12)	2.94 ± 0.64	3.08 ± 0.30	0.9–0.8
Kynurenine-3-hydoxylase (EC 1.14.1.4)	1.18 ± 0.15	1.05 ± 0.16	0.7–0.8
Kynurenine aminotransferase (EC 2.6.1.7)	1.29 ± 0.20	1.67 ± 0.25	0.1–0.5
Kynureninase (EC 3.7.1.3)	1.29 ± 0.10	0.45 ± 0.04	0.001
NAD content, μmole/gm of tissue	1.02 ± 0.04	0.67 ± 0.05	0.01–0.005
Incorporation of tryptophan-^{14}C into NAD, relative values	1.00	0.63	0.01–0.005

[a] ± = Standard deviation. Enzyme activities are expressed as micromoles of substrate metabolized per hour per gram of tissue. p Values were calculated according to the Student t test. Female mice were given 810 R of total body x-irradiation. [From C. Streffer, *Ger. Med. Monthly* **12**, 35 (1967); *J. Vitaminol.* **14**, 130 (1968).]

(2) The activity of this same enzyme [and of amino acid decarboxylases (*176*)] is also lowered in vitamin B_6 deficiency (*253, 254a*), but, in both situations, aminotransferases (*251a, 254a*) are unaltered. (3) Protection of the irradiated animals by biogenic amines influences the changes in the excretion and enzymatic activity in a characteristic manner (*192–195, 253*).

Nevertheless, questions remain as to whether inhibition of kynureninase is the factor solely responsible for the changes in tryptophan catabolism and whether these changes are related to a lack of pyridoxal-5-phosphate. The normal substrate of kynureninase is 3-hydroxykynurenine, not kynurenine, and anthranilic acid is found in urine only after loading with tryptophan. Contrary to expectation, excretion of 3-hydroxykynurenine and 3-hydroxyanthranilic acid into the urine of irradiated animals is unaltered (*268*). One may argue, however, that lack of NADP could restrict hydroxylation of kynurenine even when the enzymatic activity of kynurenine hydroxylase *in vitro* is unaltered. Also, no apparent deficiency of vitamin B_6 exists in radiation disease, and addition of the vitamin does not restore tryptophan catabolism to normal (*176*, *254a*).

In order to clarify these points, it would be desirable to study in more detail (a) the excretion of conjugated metabolites, which up to now has been entirely neglected, although irradiation as well as vitamin B_6 deficiency might affect the various reactions of conjugation, and although the conjugates represent a significant portion of the total amount of tryptophan metabolite excreted; (*b*) the concentrations of 3-hydroxykynurenine and 3-hydroxyanthranilic acid in liver in parallel with the enzymatic activities; (c) the alternate pathway of kynurenine degradation, by which CO_2 is formed via glutaryl CoA.

Catabolism of tryptophan, when increased, could be the source of metabolites in the urine. Indeed, as a result of radiation-induced protein breakdown in radiosensitive tissues, tryptophan levels in the liver increase (*253a*, *268*). The irradiated liver is probably capable of handling this extra tryptophan, since the activity of tryptophan pyrrolase is normal or enhanced. For example, an increase in activity of this enzyme is found in starved mice 12 and 48 hr after exposure to 690 or 810 R (*254a*), and in rats and mice 4 hr after exposure to 900–1000 R (*262*, *263*). In the latter case, the increase was as much as 2–4-fold. In irradiated rabbit liver, formation of kynurenine is also found to be enhanced (*208*).

On the other hand, tryptophan catabolism in isolated perfused liver is altered little at normal substrate levels and delayed at high substrate levels after irradiation (*86*). A defect in the induction of tryptophan pyrrolase is also detectable after irradiation by enzymatic assay (*16a*) although there has been some discussion whether induction by hormone (*99a*, *99b*, *168a*) or by substrate (*200a*) is susceptible to radiation injury (see p. 161). In any case, the amount of tryptophan liberated after irradiation appears to be not large enough to induce enzyme production. Moreover, increased catabolism of tryptophan is not a prerequisite for excess excretion of metabolites, since tryptophan pyrrolase activity is normal at the time of maximal excretion [from 5–9 days after exposure (Table VIII-4) (*254a*, *253*)].

Other pathways of tryptophan catabolism (decarboxylation to tryptamine and hydroxylation to hydroxytryptophan) are not as prominent quantitatively as that via kynurenine, but fulfill important physiological functions. As discussed elsewhere (p. 307), the hydroxylation of tryptophan and the subsequent decarboxylation to serotonin is diminished, and degradation and liberation of serotonin are enhanced after irradiation. Formation of indoleacetate in the liver is stimulated after low doses, but depressed after higher ones (*97a*).

Synthesis of NAD from labeled tryptophan, but not from labeled nicotinic acid, is diminished after irradiation (*86d*, *141*, *254b*, *281*), and levels of NAD in liver decrease (see p. 138). In addition, metabolites of NAD are excreted in excess the first day after irradiation (*86c*, *280*). These metabolites are the result of increased catabolism of NAD after irradiation, as is demonstrated in animals in which NAD has been labeled prior to exposure (*86c*).

D. Enzymes in Body Fluids

Serum enzymes have their origin in various organs of the body from which they are released into the serum. Changes after irradiation may represent release of enzymes from radiosensitive tissues or may be due to changes in synthesis. A synopsis of radiation-induced modifications of enzymatic activities in the serum and urine of irradiated mammals is presented in Table VIII-5. Transaminases and lactic acid dehydrogenases have been most widely studied and show significant alterations during the first day after exposure. Various species behave in a different manner, and certain inconsistencies in the published observations might also result from different time schedules studied. Starvation as well as the partial starvation subsequent to radiation exposure play an important role in the level of serum enzymes, especially those involved in carbohydate and lipid metabolism; starved animals display similar alterations in certain enzymatic activities, as do irradiated ones (*5*).

Lactate dehydrogenase (LDH) in the serum can be separated into several isozymes; fractions 1–2 prevail in organs with mainly "aerobic" metabolism (heart), the other fractions are found in organs with mainly "anaerobic" metabolism (muscle). The aerobic LDH isozymes increase markedly after irradiation but not the anaerobic ones (*110*, *126*, *144*).

E. Serum Proteins after Irradiation

1. Quantitative Changes

Total proteins in the serum decrease from 2–4 days after irradiation, but sometimes a short transient rise precedes this decrease. The concentration of albumin diminishes much more than that of globulins, and, as a result, the ratio of albumin to globulins diminishes (e.g., *41*, *115*, *157*, *158*, *219*, *244*, *246*, *257*).

The techniques of electrophoresis permit a more precise definition of the changes in serum proteins. Rats (*66*, *92*, *122*, *123*, *146*, *215*, *225a*, *229*, *273*, *277*), hamsters (*52*), mice (*99*, *187*, *231*), goats (*269*), monkeys (*177*), dogs (*96*, *203*),

ENZYMES IN SERUM AND URINE AFTER IRRADIATION

Enzyme	Species	Change	References and remarks
Aminotransferases (transaminases)	Rabbit	Serum, increase 1st day (700 R)	Serum glutamate oxalacetate transaminase (SGOT) (*27*, *28*, *154*, *200*); also in sheep (*219a*)
	Rat	Serum, inconsistent values 1st day	Increase SGOT (*15*, *30*, *118ab*, *130*, *217*, *265b*); no effect (*28*); 2nd day decrease (*26*); no change 3 hr–3 days (*209*)
	Mouse	Serum, increase 6–8 hr (500–1000 R)	(*6*, *12b*, *239*), SGOT and glutamate pyruvate transaminase (SGPT)
Lactate dehydrogenase (LDH)	Rat	Serum, inconsistent increase 1–2 days	Transient increase rabbit 750 R (*5*); sheep (*219a*) increase after combined external and internal irradiation (*267*); no change in total LDH but shift in isozyme pattern (2 days 600–800 R): increase LDH fractions 4–5, decrease LDH fractions 1–2 (*93*, *126*, *127*, *143*, *144*)
	Monkey	Increase 1st day, 670 rads	Increase total activity monkey (*44*), particularly LDH fractions 4–5 (*110*); malate dehydrogenases change in the blood of radiation-therapy patients in relation to tumor destruction (*146a*)
Peptide hydrolases	Dog	Increase 4–5 days	Also in urine before death (*40*, *51*); rabbit (*45*); no effect in man (*100*)
Deoxyribonucleate-3-nucleotido hydrolases (DNAse II)	Man and Rat	Urine increase 1–2 days	Rats (*138*, *155*, *164*, *165*, *265a*); also plasma (*138*, *166*); man (*139*, *153*); no change, however, found in radiation therapy (*93a*, *271a*)
β-Glucuronidase	Man	Urine increase 1–3 weeks	(*271*)
Glycerol–ester hydrolases (lipases)	Rat	Serum increase 2nd day, 600 R	(*156*); changes in isozymes in mice (*131a*); clearing factor lipase increases in rats after 800 R (*12*, *201*)
Acylcholine acylhydrolases (cholinesterases)	Guinea pig	Serum decrease 1–2 days, 400 R	(*183*, *124*, *237*)
Ketose-1-phosphate aldolase (and related enzymes)	Rabbit	Serum increase 1st day, 750 R	(*5*, *204*) in rat under continuous irradiation (*109*) and in guinea pigs 750 R (*43*); other serum enzymes decrease: creatine kinase, a glycerophosphate dehydrogenase
Creatine phosphokinase	Rabbit	Serum increase 1–10 days, 800 R	(*227b*)

rabbits (*190*), and man (see accidents, p. 274) all display a decrease in albumin and a variable increase in α- and β-globulins (Fig. VIII-12). Although a slight fall in albumin and a rise in α_2-globulins (*92*) has been noted as early as 24 hr after exposure, the changes in serum proteins are usually not pronounced until the 4th day after exposure. γ-Globulins, in general, decrease after radiation, but an infection or autoimmune response (*87*, *187*, *229*) can occasionally bring about a rise in this globulin fraction (and in prealbumin). Prealbumin, which represents different types of proteins in various species, usually decreases in mice and rats (*229*, *231*), but little is known about its behavior in other species.

Immunoelectrophoresis allows a still more detailed analysis of serum proteins, but certain radiation-induced changes reported by some investigators using immunoelectrophoresis on a given strain could not be confirmed by other investigators. Thus, new types of proteins were allegedly observed in irradiated mice by some experimenters (*99*, *146*, *185*) but not by others (*231*).

An increase in glycoproteins (*60a*, *91*) and especially in lipoproteins [rabbits (*8*, *111*, *116*, *275*), dogs (*60*, *117*) and mice (*220*)] has been observed in irradiated animals. The opalescence of the serum in irradiated rabbits suggests an increase in lipoproteins and is an early prognostic sign of death (*226*). The increase

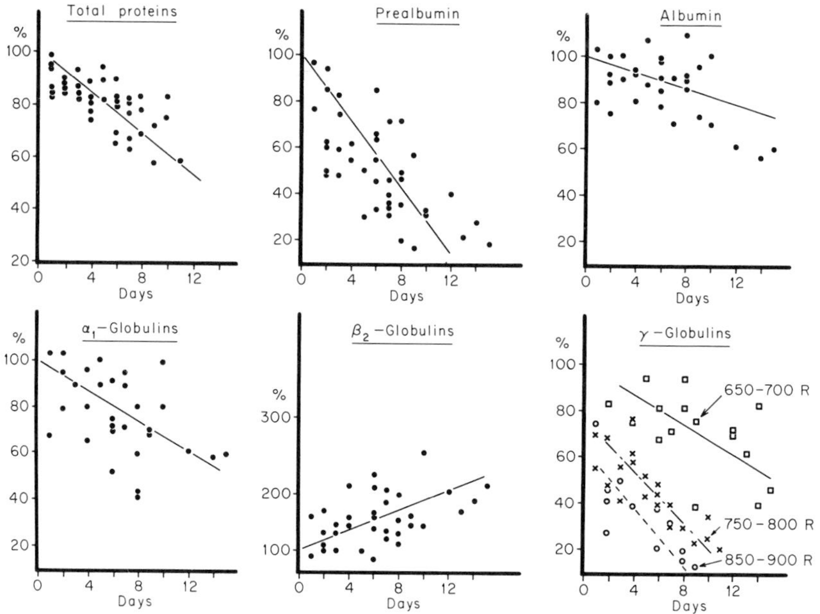

FIG. VIII-12. Changes with time in the concentration of total proteins, prealbumin, albumin, β-globulins, and γ-globulins in the serum of x-irradiated mice (600 R). Note that total proteins, albumin, prealbumin, and α_1- and γ-globulins decrease, whereas β_2-globulins increase. [From A. Sassen, F. Kennes, and J. R. Maisin, *Acta. Radiol.,Therapy, Phys. Biol.* **4**, 97 (1966).]

in lipoproteins may be related to muscle injury after irradiation (*198*, *199*, *272*), but, more likely, it reflects the partial starvation and the inability of the irradiated animal to adjust its metabolism to the starved state. The direction and time course of the changes in lipoproteins differ from one species to another (*60*, *117*) [rabbits (*8*, *111*, *116*) and dogs (*95*)]. In mice, low-density lipoproteins increase and high-density lipoproteins decrease (*220*), but in rabbits both high- and low-density lipoproteins increase (*111*, *117*).

2. Mechanisms of the Changes in Serum Proteins after Irradiation

Changes in the concentration of proteins in the serum could result from alterations in synthesis, catabolism, or distribution between intra- and extravascular space (Fig. VIII-13). Synthesis of serum proteins has been studied

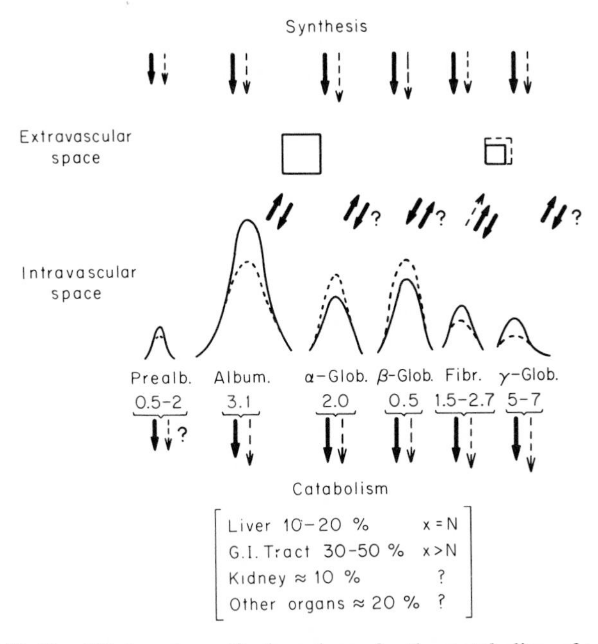

FIG. VIII-13. Simplified semiquantitative scheme for the metabolism of serum proteins in the rat. Proteins in the intravascular space are represented by their electrophoretic peaks, those in the extravascular space (albumin and fibrinogen only) by cubes, but it should be noted that the pool sizes of the extravascular space are uncertain. The rates of synthesis and catabolism are indicated approximately by the length of the respective arrows. The figures above the arrows represent approximate half lifetimes of the proteins in days. Changes after irradiation are shown as dashed lines; controls by solid lines. The figure illustrates the most likely causes for changes in the serum proteins after irradiation, i.e., changes in synthesis, distribution between intra- and extravascular space, and catabolism (loss into gastrointestinal tract).

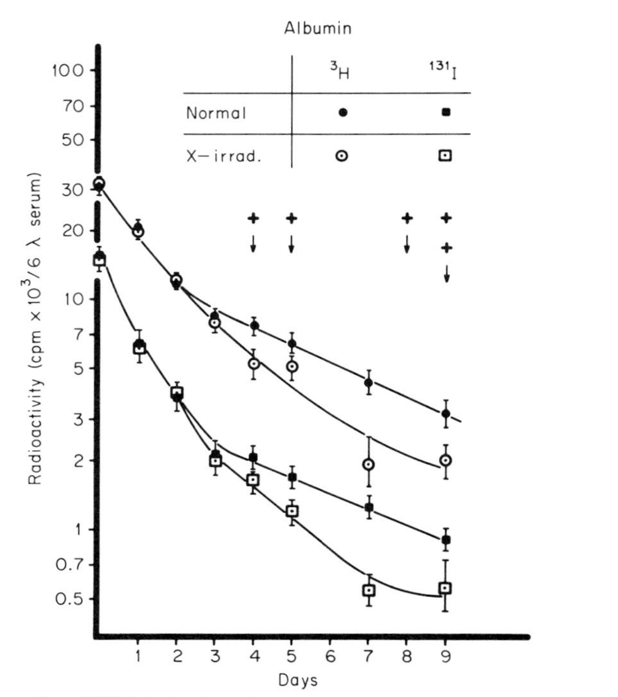

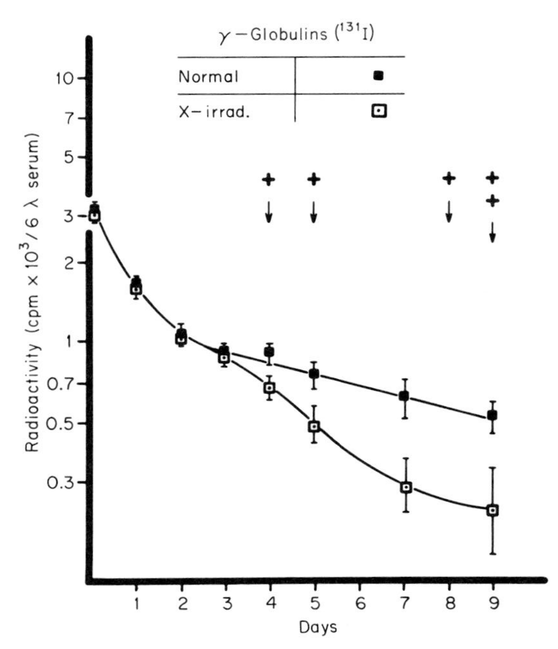

FIG. VIII-14. Radioactivity of albumin and γ-globulins in blood after injection of a mixture of ^{3}H- and ^{131}I-labeled serum proteins into normal and x-irradiated rats (800 R). The proteins were biosynthesized from phenylalanine-^{3}H in a liver perfusion; therefore, albumin, not γ-globulins, is labeled with ^{3}H. The injection of proteins was carried out immediately after exposure. Note that the radioactivity falls rapidly on the first and second days after injection, but at the same rate in normal and x-irradiated rats. After this period of equilibration between the intra- and extravascular spaces, the loss of radioactivity from the serum on days 4 and 5 is more rapid for albumin and γ-globulins in x-irradiated than in normal rats. [From A. M. Reuter *et al.*, *Radiation Res.* **30**, 725 (1967).] (See 84.)

using labeled methionine, lysine, glycine, phenylalanine, tryptophan, etc. as precursors. Incorporation of methionine into total serum proteins and into albumin is depressed after irradiation (*191a*, *228*, *243*, *248*, *278*), but that into α- and β-globulins is normal or even increased (*159*, *225*). On the other hand, incorporation of phenylalanine or tryptophan into albumin is normal in the total animal (*223*) or in the perfused liver (*221*, *230*), and that into α- and β-globulins is elevated. When glycine is utilized as precursor, a higher incorporation is found at 3 days, and a lower incorporation at later times after exposure to 300 rads of neutrons (*12a*). The pools of tryptophan and phenylalanine are not grossly altered in the liver after irradiation (*223*), whereas the metabolism of methionine is probably modified (*206*). Thus, most likely, synthesis of serum proteins is not depressed early after irradiation but, in the case of certain fractions, may even be intensified. Indeed, net synthesis of albumin in perfused liver was found to be unaltered, and that of acid glycoproteins increased (*137a*), whereas replacement of albumin in albumin-depleted rats was delayed (*245a*).

Incorporation of glucosamine into the carbohydrate moiety of β-globulins is elevated and that into comparable fractions of γ-globulins is slightly diminished (*70*). Surprisingly, in view of the severe depression of the antibody levels in irradiated animals (*113*, *260*), the incorporation of tryptophan and phenylalanine into γ-globulins was reported to be increased after irradiation (*233*). Perhaps not all γ-globulins carry immunological information, and the percentage of non-specific γ-globulins might increase after irradiation. The determination of the rate of disappearance of labeled proteins from the serum or from the total organism allows us to assess the amount of proteins catabolized and lost into the extravascular compartments of the organism. Removal and destruction of proteins by the isolated liver is not conspicuously altered in irradiated rats (*222*). On the contrary, labeled proteins, especially albumin (*176a*, *249*, *259*), fibrinogen (*235*), and γ-globulins disappear more rapidly from the circulation of the total animal after irradiation (Fig. VIII-14), although the radioactivity of the total organism does not diminish correspondingly (*84*, *222*) [more rapid albumin turnover (*179a*)]. Loss of proteins into the extravascular space is one explanation for these observations (*235*). Indeed, as discussed elsewhere (p. 312), the permeability of many tissues (e.g., skin, intestine, and brain) to macromolecules increases after irradiation. Direct measurements of the extravascular space with labeled albumin have, however, indicated a decrease rather than an increase (*228a*). In the irradiated animals, ^{131}I-labeled proteins and their breakdown products appear in the gastric contents, from which they are removed, but slowly due to gastric retention (*260a*). Thus, whole-body counting is not very suitable for assessing protein breakdown in irradiated animals.

Irradiated rats also lose more labeled albumin (*19*, *31*, *68*, *69*, *266*, *273a*) or polyvinylpyrrolidone (*68*, *69*, *255*) into the gastrointestinal tract than non-irradiated rats.

TABLE VIII-6

SELECTED BIOCHEMICAL DATA FROM RADIATION ACCIDENTS (MORE THAN 200 R)[a]

Accident	Dose	Outcome	Electrolytes in serum and urine	Amino acids in in urine	Proteins in serum	Nucleic acid metabolites	Other effects
Los Alamos 5–21–46 (*114*)	900 rep neutrons (2044 Rγ)	✝ 9 days	Na^+ S↓ 2days Cl^- S↓ 4 days; K^+ U↑ 6 days	Total AA ↑ 6 days, glycine↑ taurine ↑ aspartate↑	Albumin↓ (γ-globulins↓ α-globulins↑)	S uric acid↑ 9 days	S urea↑ 4 days†, U porphyrin↑ 5–7 days (also urobilirogen) NPN↑ 4 days †, U corticoids↑
Los Alamos 8–21–45 (*114*)	800 rep neutrons (590 Rγ)	✝ 24 days		Total AA↑ 24 days, glycine↑, proline↑, taurine↑		S uric acid↑	S Urea↑ U porphyrins↑ 24 days, U corticoids 24 days
Los Alamos 5–21–46 (*114*)	416 R	—	K^+ U↑	Total AA↑ 6 days, taurine↑ 1–6 days, glycine↑ 1–6 days, aspartate↑ 1–11 days	Albumin↓	—	U Corticoids↑ 5–7 days
Los Alamos 12–30–58 (*238*)	12,000 rem	✝ $1\frac{1}{2}$ days	Na^+ S∅; Cl^- S↑ 1 day K^+ S↑	Taurine↑, BAIBI↑ creatine↑?	Albumin↓, SGOT↑, MDH↑	S uric acid↑, deoxycytidine↑	NPN↑, U catecholamines↑ S urea↑
Vinca 10–18–58 (*135*)	436 rad² (five others received 207–606 rads biochemical data not given separately	✝ 31 days (five others survived)	Na^+ Cl^- S↓ 1–7 days, K^- S↑	Proline↑, phenylalanine↑, tryptophan↑, alanine↑	Total proteins↓, γ-globulins↓, α-globulins↑	—	NPN↑ 1–7 days, U urobilin↑

Oak Ridge 6–16–58 (*9*, *32*)	365 rads (four others received from 236–339 rads	—	No changes	BAIBA↑, taurine↑	SGPT; SPGOT ∅̸, LDH ∅̸	S uric acid ∅̸	NPN ∅̸, urinary nitrogen↑
USSR (*102*) 1953–1954	300 R	—	Cl^- S ∅̸	—	—	—	No changes in urine
USSR 1953–1954	450 R	—	Cl^- S↓	—	—	—	NPN↑
Wood River 7–24–64 (*142*)	12,000 rem (8800 rads)	✝ 2 days	No changes except Cl^- S slight↑	—	Proteins ∅̸, SGOT↑	S uric acid↑	S creatinine↑, S Urea↑
Mol (*136*) 12–30–65	400 rads	—	K^+ U↑ 1 day and 15–21 days, Na^+ U, Cl^- U, Na^+ S↓ 1–3 days	BAIBA↑, taurine↑, serine↓, creatine↑ } 1st week	Proteins↓, albumin↓	—	U 17-ketosteroids↑ 1st day, glycemia 1st day

[a] *Abbreviations* used: S = serum; U = urine; AA = amino acid; NPN = blood nonprotein nitrogen; SGOT = serum glutamate oxalacetate transferase; SGPT = serum glutamate pyruvate transferase; MDH = malate dehydrogenase; LDH = lactate dehydrogenase.

Symbols used: ✝ = death; † = until time of death; ↓ = decrease; ↑ = increase; ∅̸ = no change.

The doses are taken from reference *219c*.

F. Aids in the Biochemical Diagnosis of Radiation Injury

Biological diagnosis of radiation injury in man resulting from accidental exposure to ionizing radiation involves the evaluation of metabolic, hematological, and cytological changes to determine the extent of the damage and to establish the prognosis (*49*, *80*, *81*). Unlike physical dosimetry, it does not depend on the distribution of the dose in the irradiated organism or on the sensitivity of the individual organism.

Many functions are altered in the irradiated organism, but only a few of these changes are suitable for biological diagnosis, since the tests must meet several requirements: (1) they must be specific for radiation damage; (2) they must be a function of the dose in order to permit a distinction between clinically important categories of radiation injury; (3) they must demonstrate radiation damage soon after exposure; (4) they should not harm the patient; (5) they should be simple to perform.

Several types of reactions based on hematogical, cytological, or biochemical techniques are useful in assessing radiation injury. Although usually not as a specific and sensitive as hematological or cytological tests, biochemical tests have definite advantages since they can detect an early response of the body and can be easily carried out. Therefore, biochemical diagnosis should be most useful in situations where many persons have been exposed, when a rapid decision must be made as to who would benefit most and need the most specialized care.

Unfortunately, the dose response is known for only a few biochemical reactions in animals, and no data at all are available on such dose responses in man. On the basis of our present knowledge, the following tests appear to be of the greatest possible value: urinary excretion of deoxycytidine, taurine, creatine, indoxylsulfate, pseudouridine, and 5-OH-indoleacetic acid. Other potentially useful tests are: urinary excretion of kynurenine, kynurenic acid, xanthurenic acid, purines, amino acids (glycine, tryptophan, tyrosine, and β-amino isobutyric acid), several biogenic amines, hormones and their metabolites, radioactive phosphorus [after injection of ^{32}P inorganic phosphate (*160*)], as well as the determination of levels of serum proteins, amino transferases, lactate dehydrogenases, aldolases, and the conductivity (*120*) and density of the blood (*47*).

The further development of biochemical diagnosis is essential but will only result from a large-scale effort in animals and man, for which a few guidelines should be mentioned. The studies should be carried out over a wide dose range and should include groups of sham-irradiated, fed, and starved controls. Ideally several biological tests should be studied simultaneously in the same group of animals, and the investigations should involve several different species. Standardization of techniques of irradiation, animal care, and chemical determination could allow such a large study to be carried out cooperatively in several laboratories.

Studies in man must eventually supplement animal investigations. Patients irradiated locally or totally for the therapy of malignant disease offer certain opportunities for study, but it must be kept in mind that the metabolism of such patients may not react in a normal manner and that the local exposure causes different metabolic disturbances than does whole-body irradiation. Therefore, specimens from persons accidentally exposed should be investigated as exhaustively as possible. In order to facilitate such studies, a synopsis of important biochemical data on accidents in man is given in Table VIII-6.

REFERENCES

1. Abaturova, E. A., Elpatevskaya, G. N., Sviridov, N. K., and Shubina, A. V., *Zh. Khim. Biol. Khim.* **65**, 1050 (1961).
2. Aebi, H., Bernays, L., Fluckiger, H., Schmidli, B., and Zuppinger, A., *Helv. Physiol. Pharmacol. Acta* **13**, c49 (1955).
3. Aebi, H., *Bull. Schweiz. Akad. Med. Wiss.* **12**, 326 (1956).
4. Aebi, H., Lauber, K., Schmidli, B., and Zuppinger, A., *Biochem. Z.* **328**, 391 (1957).
5. Albaum, H. G., *Radiation Res.* **12**, 186 (1960).
6. Almonte, T. J., Barnes, J. H., Bautista, S. C., De La Cruz, B., Lansangan, L., and Pineda, P. A., *Intern. J. Radiation Biol.* **9**, 37 (1965).
7. Anderson, D. R., Joseph, B. J., Krise, G. M., and Williams, C. M., *Am. J. Physiol.* **192**, 247 (1958).
8. Andreoni, O., Dompe, M., and Russo, A. M., *Ric. Sci. Suppl.* **25**, 1393 (1955).
9. Andrews, G. A., Sitterson, B. W., Kretchmar, A. L., and Brucer, M., *Health Phys.* **2**, 134 (1959).
10. Angel, C. R., and Noonan, T. R., *Radiation Res.* **15**, 298 (1961).
11. Arient, M., Dienstbier, Z., and Shejbal, J., *Minerva Fisiconucl.* **7**, 57 (1966).
12. Baker, D. G., and Monkhouse, F. C., *Radiation Res.* **20**, 8 (1963).

12a. Baker, D. G., Carsten, A., Jahn, A., and McFadyen, D., *Radiation Res.* **32**, 265 (1967).

12b. Barnes, J. H., Lowman, D. M. R., Bautista, S. C., De La Cruz, B., and Lansangan, L., *Intern. J. Radiation Biol.* **14**, 417 (1968).

13. Barth, G., Besenbeck, F., and Schön, H., *Strahlentherapie* **119**, 59 (1962).
14. Bates, T. H., Smith, C. I., and Smith, H., *Nature* **203**, 843 (1964).
15. Becker, F. F., Williams, R. B., and Voogd, J. L., *Radiation Res.* **20**, 221 (1963).
16. Beneš, J., Bergsteinova, V., and Brejcha, M., *Neoplasma* **4**, 351 (1957).

16a. Beneš, J., and Zicha, B., *Strahlentherapie* **137**, 612 (1969).

17. Berry, H. K., Saenger, E. L., Perry, H., Friedman, B. I., Kereiakes, J. G., and Sheel, C., *Science* **142**, 396 (1963).
18. Bigwood, E. J., *Strahlenschutz Forsch. Praxis* **4**, 183 (1964).
19. Birke, G., Liljedahl, S. O., Plantin, L. O., and Wetterfors, J., *Nature* **194**, 1243 (1962).
20. Blackwell, L. H., Sinclair, W. K., and Humphrey, R. M., *Am. J. Physiol* **203**, 87 (1962).
21. Bohuon, C., Delarue, J. C. H., and Tubiana, M., *Compt. Rend.* **257**, 4227 (1963).

22. Boquet, P. L., and Fromageot, P., *Intern. J. Radiation Biol.* **6**, 376 (1963).
23. Boquet, P. L., and Fromageot, P., *Biochim. Biophys. Acta* **97**, 222 (1965); **111**, 40 (1965).
23a. Boquet, P., and Fromageot, P., *Intern. J. Radiation Biol.* **12**, 61 (1967); **13**, 343 (1968).
24. Bowers, J. Z., and Scott, K. G., *Proc. Soc. Exptl. Biol. Med.* **78**, 645 and 648 (1951).
25. Bowers, J. Z., Davenport, V., Christensen, N., and Goodner, C. J., *Radiology* **61**, 97 (1953).
26. Braun, H., Hornung, G., and Neumeyer, M., *Strahlentherapie* **126**, 454 (1965).
27. Brent, R. L., Berman, H., McLaughlin, M. M., and Stabile, J. N., *Radiation Res.* **7**, 305 (1957).
28. Brent, R. L., McLaughlin, M. M., and Stabile, J. N., *Radiation Res.* **9**, 24 (1958).
29. Breuer, H., Knuppen, R., and Parchwitz, H. K., *Z. Naturforsch.* **13b**, 741 (1958).
30. Brin, M., and McKee, R. W., *Arch. Biochem. Biophys.* **61**, 384 (1956).
31. Bromfield, A. R., and Dykes, P. W., *Nature* **201**, 633 (1964).
32. Brucer, M., ORINS-25 (1959).
33. Caster, W. O., and Armstrong, W. D., *Proc. Soc. Exptl. Biol. Med.* **90**, 56 (1955).
34. Caster, W. O., and Armstrong, W. D., *Radiation Res.* **5**, 189 (1956).
35. Cavalieri, R. R., van Metre, M. T., Chambers, F. W., Jr., and King, E. R., *J. Nucl. Med.* **1**, 186 (1960).
36. Cavalieri, R. R., van Metre, M. T., and Sivertsen, W. I., *Nature* **191**, 1303 (1961).
37. Chanutin, A., and Ludewig, S., *Am. J. Physiol.* **166**, 380 (1951).
38. Chien, S., Lukin, L., Holt, P., Cherry, S. H., Root, W. S., and Gregersen, M. I., *Radiation Res.* **7**, 277 (1957).
39. Chiriboga, J., *Nature* **198**, 803 (1963); *Experientia* **23**, 903 (1967).
40. Colgan, J., Gates, E., and Miller, L. L., *J. Exptl. Med.* **95**, 531 (1952).
41. Cornatzer, W. E., Engelstad, O. D., and Davison, J. P., *Am. J. Physiol.* **175**, 153 (1953).
42. D'Addabbo, A., Viterbo, F., and Fanfani, G., *Strahlentherapie* **130**, 426 (1966); D'Addabo, A., *ibid.* **133**, 108 (1967).
43. Dalos, B., *Nature* **190**, 817 (1961).
44. Dalrymple, G. V., Lindsay, I. R., and Ghidoni, J. J., *Radiation Res.* **25**, 377 (1965).
45. De Leeuw, N. K. M., Wright, C. S., and Morton, J. L., *J. Lab. Clin. Med.* **42**, 592 (1953).
46. Detrick, L. E., Upham, H. C., Springsteen, R. W., and Haley, T. J., *Intern. J. Radiation Biol.* **7**, 161 (1963).
47. Dienstbier, Z., and Rakovič, M., *Nature* **195**, 1011 (1962).
48. Dienstbier, Z., *Acta Univ. Carolinae, Med.* **11**, 3 (1965).
49. Dienstbier, Z., Arient, M., and Shejbal, J., *Rep. SZS* **12** (1966).
50. Di Luzio, N. R., and Simon, K. A., *Radiation Res.* **7**, 79 (1957).
51. Dimitrov, L., *Strahlentherapie* **124**, 114 (1964).
52. Ditzel, J., *Radiation Res.* **17**, 694 (1962).
53. Drahovsky, D., Winkler, A., and Skoda, J., *Nature* **201**, 411 (1964).
54. Drahovsky, D., Winkler, A., and Skoda, J., *Neoplasma* **12**, 561 (1965).
54a. Eberhagen, D., and Horn, U., *Strahlentherapie* **135**, 364 (1968).
55. Elko, E. E., and Di Luzio, N. R., *Radiation Res.* **11**, 1 (1959).
56. Elko, E. E., and Di Luzio, N. R., *Radiation Res.* **14**, 760 (1961).
57. Elko, E. E., Wooles, W. R., and Di Luzio, N. R., *Radiation Res.* **21**, 493 (1964).
58. Ellinger, F., "Medical Radiation Biology." Thomas, Springfield, Illinois, 1957.

59. Ellis, J. P., Clark, R. T., Rambach, W. A., and Pickering, J. E., *Am. J. Physiol.* **198**, 1245 (1960).
60. Entenman, C., Neve, R. A., Supplee, H., and Olmsted, C. A., *Arch. Biochem. Biophys.* **59**, 97 (1955).
60a. Evans, A. S., Quinn, F. A., Brown, J. A., and Strike, T. A., *Radiation Res.* **36**, 128 (1968).
61. Fedorov, I. V., AEC-tr-3661 (1957); *Conf. Med. Radiol. Exptl. Med., Moscow* p. 202 (1957).
62. Fedorov, N. A., and Kosenko, A. S., *Radiobiologiya* **6**, 534 (1966).
63. Fedorov, N. A., Bogomazov, M. Y., and Kosenko, A. S., *Byul. Eksperim. Biol. i Med.* **62**, 91 (1966).
64. Fedorova, T. A., and Larina, M. A., *Med. Radiol.* **1**, 36 (1956).
65. Fedorova, T. A., Uspenskaja, M. S., Vasileiskii, S. S., and Belyaeva, E. M., *Med. Radiol.* **5**, 42 (1960).
66. Fischer, M. A., Magee, M. Z., and Coulter, E. P., *Arch. Biochem. Biophys.* **56**, 66 (1955).
67. Fox, M., and Woodruff, M. F. A., *Nature* **202**, 918 (1964).
68. Friedberg, W., *Intern. J. Radiation Biol.*, **2**, 185 (1960).
69. Friedberg, W., Jenkins, V. K., and Muderrisoglu, O. V., *Intern. J. Radiation Biol.*, **4**, 465 (1961).
70. Friedberg, W., Faulkner, D. N., Abbott, J. K., and Friedberg, M. H., *Intern. J. Radiation Biol.* **8**, 101 (1964).
70a. Frøholm, L. O., *Intern. J. Radiation Biol.* **12**, 35 (1967).
71. Fromageot, P., and Boquet, P. L., *Strahlenschutz Forsch. Praxis* **4**, 199 (1964).
72. Ganis, F. M., and Howland, J. W., *Intern. J. Radiation Biol.* **7**, 107 (1963).
73. Ganis, F. M., Hendrickson, M. W., and Howland, J. W., *Radiation Res.* **24**, 278 (1965).
74. Gerber, G. B., Gertler, P., Altman, K. I., and Hempelmann, L. H., *Radiation Res.* **15**, 307 (1961).
75. Gerber, G. B., Gerber, G., Kurohara, S., Altman, K. I., and Hempelmann, L. H., *Radiation Res.* **15**, 314 (1961).
76. Gerber, G. B., Gerber, G., Altman, K. I., and Hempelmann, L. H., *Intern. J. Radiation Biol.* **3**, 17 (1961).
77. Gerber, G. B., Gerber, G., Koszalka, T., and Miller, L. L., *Proc. Soc. Exptl. Biol. Med.* **110**, 338 (1962).
78. Gerber, G. B., Gerber, G., and Altman, K. I., *Proc. Soc. Exptl. Biol. Med.* **110**, 797 (1962).
78a. Gerber, G. B., Gerber, G., Altman, K. I., and Hempelmann, L. H., *Intern. J. Radiation Biol.* **6**, 17 (1963).
79. Gerber, G. B., Gerber, G., Koszalka, T. R., and Hempelmann, L. H., *Radiation Res.* **23**, 648 (1964).
80. Gerber, G. B., *Strahlenschutz Forsch. Praxis* **4**, 239 (1964).
81. Gerber, G. B., *Ber. Euratom Symp. Unfallbedingte Bestrahlung Arbeitsplatz, Nice* EUR 3666 d-f-i-n/c p. 141 (1967).
82. Gerber, G. B., and Remy-Defraigne, J., *Intern. J. Radiation Biol.* **10**, 141 (1966).
83. Gerber, G. B., and Remy-Defraigne, J., *Arch. Intern. Physiol. Biochim.* **74**, 785 (1966).
84. Gerber, G. B., Reuter, A. M., Kennes, F., and Remy-Defraigne, J., *Conference on Problems Connected with the Preparation and Use of Labeled Proteins in Tracer Studies, PISA, 1966* Eur 2950 d, t, e, p. 165 (1966).

85. Gerber, G. B., *in* "Stoffwechsel der isoliert perfundierten Leber" (W. Staib and R. Scholz, eds.), p. 168, Springer-Verlag, Heidelberg, 1968.
86. Gerber, G. B., Altman, K. I., Gerber, G., and Deknudt, G., *Strahlentherapie* **137**, 620 (1967).
86a. Gerber, G. B., unpublished data (1968).
86b. Gerber, G. B., and Remy-Defraigne, J., *Studio Biophys.* **12**, 113 (1968); *Radiation Res.* **40**, 105 (1969).
86c. Gerber, G. B., and Deroo, J., *Strahlentherapie* **137**, 724 (1969).
86d. Gerber, G. B., Zicha, B., and Deroo, J., *Intern. J. Radiation Biol.* **15**, 425 (1969).
86e. Gerber, G. B., and Remy-Defraigne, J., *Ztschrft. Ges. Exp. Med.* **150**, 33 (1969).
87. Ghossein, N. A., Azar, H. A., and Williams, J., *Am. J. Pathol.* **43**, 369 (1963).
88. Gilbert, C. W., Paterson, E., and Haigh, M. V., *Intern. J. Radiation Biol.* **5**, 1 (1962).
89. Gilbert, C. W., Paterson, E., Haigh, M. V., and Schofield, R., *Intern. J. Radiation Biol.* **5**, 9 (1962).
90. Gjessing, E. C., and Warren, S., *Radiation Res.* **15**, 276 (1961).
91. Glenn, W. G., *Radiation Res.* **13**, 691 (1960).
92. Glinos, A. D., *Radiation Res.* **19**, 286 (1963).
93. Goetze, T., and Goetze, E., *Acta Biol. Med. Ger.* **16**, 337 (1966).
93a. Goldberg, D. M., MacVicar, J., and Watts, C., *Clin. Sci.* **32**, 99 (1967).
94. Goldman, D., *Am. J. Roentgenol. Radium Therapy* **50**, 381 (1943).
95. Goldwater, W. H., and Entenman, C., *Radiation Res.* **4**, 243 (1956).
96. Goldwater, W. H., and Entenman, C., *Am. J. Physiol.* **188**, 409 (1957).
97. Goodman, R. D., and Vogel, M., *Am. J. Physiol.* **175**, 29 (1953).
97a. Gordon, S. A., and Buess, E., *Ann. N. Y. Acad. Sci.* **144**, 136 (1967).
98. Gorizontov, P. D., Fedorova, T. A., Zharkov, Yu. A., Tereshchenko, O. Y., Khnychev, S. S., and Sbitneva, M. F., *Radiobiologiya* **3**, 514 (1963).
98a. Goyer, R. A., and Yin, M. W., *Radiation Res.* **30**, 301 (1967).
99. Grabar, P., Kashkin, K. P., and Courcon, J., *Rev. Franc. Etudes Clin. Biol.* **8**, 565 (1963).
99a. Greengard, O., Baker, G. T., Horowitz, M. L., and Knox, W. E., *Proc. Natl. Acad. Sci. U.S.* **56**, 1303 (1966).
99b. Greengard, O., *Advan. Enzyme Regulation* **5**, 397 (1967).
100. Grimm, C. A., *Radiology* **78**, 811 (1962).
101. Gros, C., and Mandel, P., *Compt. Rend.* **231**, 631 (1950).
101a. Guri, C. D., Swingle, K. F., and Cole, L. J., *Intern. J. Radiation Biol.* **12**, 355 (1967).
101b. Guri, C. D., and Cole, L. J., *Clin. Chem.* **14**, 383 (1968).
101c. Guri, C. D., Swingle, K. F., and Cole, L. J., *Proc. Soc. Exptl. Biol. Med.* **129**, 31 (1968).
102. Guskova, A. K., and Baisogolov, G. D., *Proc. Intern. Conf. Peaceful Uses At. Energy, Geneva, 1955* Vol. **11**, p. 35. Columbia Univ. Press (I.D.S.), New York, 1956.
103. Gustafson, G. E., and Koletsky, S. *Am. J. Physiol.* **171**, 319 (1952).
104. Haberland, G. L., Schreier, K., Bruns, F., Altman, K. I., and Hempelmann, L. H., *Nature* **175**, 1039 (1955).
105. Haley, T. J., Flesher, A. M., and Komesu, N., *Radiation Res.* **8**, 535 (1958).
106. Haley, T. J., Cartwright, F. D., and Brown, D. G., *Nature* **212**, 820 (1966).
107. Hall, C. C., and Whipple, G. H., *Am. J. Med. Sci.* **157**, 453 (1919).
108. Hartweg, H., and Böwing, G., *Strahlentherapie* **106**, 289 (1958).
109. Hawrylewicz, E. J., *Radiation Res.* **14**, 473 (1961).
110. Hawrylewicz, E. J., and Blair, W. H., *Radiation Res.* **25**, 196 (1965).
111. Hayes, T. L., and Hewitt, J. E., *Am. J. Physiol.* **181**, 280 (1955).

112. Heile, *Z. Klin. Med.* **55**, 508 (1904).
113. Hektoen, J., *J. Infect, Diseases* **17**, 415 (1915).
114. Hempelmann, L. H., Lisco, H., and Hoffman, J. C., *Ann. Internal Med.* **36**, 279 (1952).
115. Herzfeld, E., and Schinz, H. R., *Strahlentherapie* **15**, 84 (1923).
116. Hewitt, J. E., Hayes, T. L., Gofman, J. W., Jones, H. B., and Pierce, F. T., *Am. J. Physiol.* **172**, 579 (1953).
117. Hewitt, J. E., and Hayes, T. L., *Am. J. Physiol.* **185**, 257 (1956).
118. Hewitt, R., and Noonan, T. R., *Radiation Res.* **25**, 9 (1965).
118a. Highman, B., Hansell, J., and White, D., *Proc. Soc. Exptl. Biol. Med.* **125**, 606 (1967).
118b. Highman, B., Stout, D. A., and Hanks, A. R., *Proc. Soc. Exptl. Biol. Med.* **129**, 857 (1968).
119. Himeji, T., *Osaka Shiritsu Daigaku Igaku Zasshi* **9**, 4693 (1960).
120. Hlavaty, V., Žák, M., and Jirounek, P., *Nature* **206**, 1057 (1965).
120a. Hlavaty, V., Žák, M., Jirounek, P., and Bendová, L., *Acta Univ. Carolinae, Med.* **12**, 229 (1966).
121. Hochrein, H., Braun, H., Reinert, M., and Kriegsmann, B., *Strahlentherapie* **120**, 587 (1963).
122. Höhne, G., Künkel, H. A., and Anger, R., *Klin. Wochschr.* **33**, 284 (1955).
123. Höhne, G., Jaster, R., and Künkel, H. A., *Klin. Wochschr.* **30**, 952 (1952).
124. Holle, F., *Z. Ges. Exptl. Med.* **115**, 107 (1949).
125. Holoček, V., Petrášek, J., Schreiber, V., Dienstbier, Z., Kmentová, V., *Physiol. Bohemoslov.* **11**, 119 (1962).
126. Hornung, G., Lehmann, F.-G., and Braun, H., *Strahlentherapie* **128**, 595 (1965).
127. Hornung, G., Braun, H., Bohndorf, W., and Nahm, H., *Strahlentherapie* **133**, 52 (1967).
128. Hornykiewytsch, T., Seydl, G., and Thiele, O. W., *Strahlentherapie* **95**, 523 (1954).
129. Huang, K. C., and Bondurant, J. H., *Am. J. Physiol.* **185**, 446 (1956).
130. Hughes, L. B., Los Alamos Sci. Lab. Rep. LA-2167, 1 (1958).
131. Hunter, C. G., *Proc. Soc. Exptl. Biol. Med.* **96**, 794 (1957).
131a. Hunter, R. L., Bennett, D. C., and Dodge, A. H., *Ann. N.Y. Acad. Sci.* **151**, 594 (1968).
131b. I-Wen Chen, Kereiakes, J. G., Friedman, B. I., and Saenger, E. L., *Radiology* **91**, 343 (1968).
132. Jackson, K., and Entenman, C., *Am. J. Physiol.* **189**, 315 (1957).
133. Jackson, K. L., Rhodes, R., and Entenman, C., *Radiation Res.* **8**, 361 (1958).
134. Jackson, K. L., and Entenman, C., *Proc. Soc. Exptl. Biol. Med.* **97**, 184 (1958).
135. Jammet, H., Mathé, G., Pendic, B., Duplan, J. F., Maupin, B., Latarjet, R., Kalic, J., Schwarzenberg, L., Djukic, Z., and Vigne, J., *Rev. Franc. Etudes Clin. Biol.* **4**, 210 (1959).
136. Jammet, H., Gongora, R., LeGo, R., Marble, G., and Faes, M., *Proc. 1st Intern. Congr. Radiation Protect. Assoc., 1966*, p. 1249.
137. Jehotte, J., *Compt. Rend. Soc. Biol.* **148**, 941 (1954).
137a. John, D. W., and Miller, L. L., *J. Biol. Chem.* **243**, 268 (1968).
138. Jovanović, M., and Voncina, A., *Bull. Inst. Nucl. Sci. "Boris Kidrich"* (Belgrade) **9**, 189 (1959).
139. Jovanović, M., *Bull. Inst. Nucl. Sci. "Boris Kidrich" (Belgrade)* **10**, 137 (1960).
140. Jovanović, M., and Svecenski, N. V., *Acta Med. Iugoslav.* **27**, 46 (1963).
141. Kalashnikova, V. P., *Med. Radiol.* **8**, 58 (1963).

142. Karas, J. S., and Stanbury, J. B., *New Engl. J. Med.* **272**, 755 (1965).
143. Kärcher, K. H., and Kato, H. T., *Strahlentherapie* **125**, 548 (1964).
144. Kärcher, K. H., and Kato, H. T., *Strahlentherapie* **127**, 439 (1965).
145. Kärcher, K. H., *Strahlentherapie* **129**, 80 (1966).
146. Kashkin, K. P., and Alesandrova, S. V., *Byul. Experim. Biol. i Med.* **62**, 57 (1966).
146a. Kato, H., *Strahlentherapie* **133**, 42 (1967).
147. Katz, E. J., and Hasterlik, R. J., *J. Natl. Cancer Inst.* **15**, 1085 (1955).
148. Kay, R. E., and Entenman, C., *Proc. Soc. Exptl. Biol. Med.* **91**, 143 (1956).
149. Kay, R. E., Harris, D. C., and Entenman, C., *Am. J. Physiol.* **186**, 175 (1956).
150. Kay, R. E., Early, J. C., and Entenman, C., *Radiation Res.* **6**, 98 (1957).
151. Kay, R. E., and Entenman, C., *Arch. Biochem. Biophys.* **82**, 362 (1959).
152. Kay, R. E., and Entenman, C., *Radiation Res.* **11**, 357 (1959).
153. Kerova, N. I., *Fiziol. Zh., Akad. Nauk Ukr. RSR* **5**, 99 (1959).
154. Kessler, G., Hermel, M. B., and Gershom-Cohen, J., *Proc. Soc. Exptl. Biol. Med.* **98**, 201 (1958).
155. Khursin, N. E., *Fiziol. Zh., Akad. Nauk Ukr. RSR* **11**, 223 (1965).
156. Kikuchi, Y., Hornung, G., and Braun, H., *Strahlentherapie* **130**, 567 (1966).
156a. Klinger, W., Finck, W., and Leibe, H., *Acta Biol. Med. Ger.* **16**, 415 (1966).
157. Kohn, H. I., *Am. J. Physiol.* **162**, 703 (1950).
158. Kohn, H. I., *Am. J. Physiol.* **165**, 27 (1951).
159. Kollmorgen, G. M., and Osborne, J. W., *Radiation Res.* **22**, 206 (1964).
160. Kolousek, J., and Dienstbier, Z., *Intern. J. Radiation Biol.* **6**, 271 (1963).
161. Konno, K., Traelnes, K. R., and Altman, K. I., *Intern. J. Radiation Biol.* **8**, 367 (1964).
162. Kosyakov, K. S., Libikova, N. I., and Merkina, T. N., *Med. Radiol.* **7** (3), 31 (1962).
163. Koszalka, T. R., Bowerman, C. I., Hempelmann, L. H., and Markham, J., *Proc. Soc. Exptl. Biol. Med.* **89**, 662 (1955).
164. Kowlessar, O. D., Altman, K. I., and Hempelmann, L. H., *Arch. Biochem. Biophys.* **43**, 240 (1953).
165. Kowlessar, O. D., Altman, K. I., and Hempelmann, L. H., *Arch. Biochem. Biophys.* **52**, 362 (1954).
166. Kowlessar, O. D., Altman, K. I., and Hempelmann, L. H., *Arch. Biochem. Biophys.* **54**, 355 (1955).
167. Kretchmar, L. A., *Nature* **183**, 1809 (1959).
168. Krise, G. M., Williams, C. M., and Anderson, D. R., *Proc. Soc. Exptl. Biol. Med.* **95**, 764 (1957).
168a. Kroeger, H., and Greuer, B., *Biochem. Z.* **341**, 190 (1965).
169. Kumta, U. S., Gurnani, S. U., and Sahasrabudhe, M. B., *Experientia* **13**, 372 (1957).
170. Kurohara, S. S., Rubin, P., and Hempelmann, L. H., *Radiology* **77**, 804 (1961).
171. Kurohara, S. S., and Altman, K. I., *Nature* **196**, 285 (1962).
172. Kurohara, S. S., *Acta Radiol.* Suppl. **244**, 125p (1965).
173. Ladner, H. A., and Hahn, K., *Naturwissenschaften* **53**, 226 (1966).
174. Ladner, H. A., and Wollschlaeger, G., *Strahlentherapie* **131**, 461 (1966).
175. Langendorff, H., Melching, H. J., and Streffer, C., *Strahlentherapie* **114**, 525 (1961).
176. Langendorff, H., Scheuerbrandt, G., Streffer, C., and Melching, H. J., *Naturwissenschaften* **51**, 290 (1964).
176a. Lefaure, J., Cier, J., and Cier, A., *Compt. Rend. Soc. Biol.* **160**, 2367 (1966).
177. Leone, C. A., Hartnett, A. R., Crist, R., and McBeth, C., *Radiation Res.* **10**, 357 (1959).

178. Levitt, S. H., Bogardus, C. R., Wood, G. A., and Ficken, V. J., *Radiol. Clin. Biol.* **34**, 347 (1965).
179. Linser, P., and Sick, K., *Deut. Arch. Klin. Med.* **89**, 412 (1906–1907).
179a. Lippincott, S. W., Azzam, N. A., and Rogers, C. C., *Radiology* **92**, 630 (1969).
180. Loiseleur, J., and Petit, M., *Compt. Rend.* **257**, 3235 (1963).
181. Lukin, L., *Radiation Res.* **7**, 150 (1957).
182. Lukin, L., and Gregersen, M. I., *Radiation Res.* **7**, 161 (1957).
183. Lundin, J., Clemedson, C. J., and Nelson, A., *Acta Radiol.* **48**, 52 (1957).
184. Luzzio, A. J., Sutton, G., and Kereiakes, J. G., *Proc. Soc. Exptl. Biol. Med.* **97**, 593 (1958).
185. Luzzio, A. J., Friedman, B. I., Kereiakes, J. G., and Saenger, E. L., *J. Immunol.* **96**, 64 (1966).
186. Marich, L. N., *Dokl. Akad. Nauk SSSR* **120**, 311 (1958).
187. Mathé, G., Pays, M., Bourdon, R., and Maroteaux, P., *Rev. Franc. Etudes Clin. Biol.* **4**, 272 (1959).
188. Mazurik, V. K., *Byull. Eksperin. Biol. Med.* **60** (12), 43 (1965) and **61** (2), 45 (1966).
188a. Mazurik, V. K., *Radiobiologiya* **6**, 482 (1966).
189. McDonald, R. E., Jensen, R. E., Urry, H. C., and Price, P. B., *Am. J. Roentgen Radium Therapy Nucl. Med.* **74**, 701 and 889 (1955).
190. McLaughlin, M. M., Nury, F. S., and Brent, R. L., *Radiation Res.* **7**, 436 (1957).
191. Mefferd, R. B., Jr., and Martens, H. H., *Science* **122**, 829 (1955).
191a. Mehran, A. R., and Loiselle, T. M., *Arch. Intern. Physiol. Biochim.* **76**, 91 (1968).
192. Melching, H. J., *Strahlentherapie* **120**, 34 (1963).
193. Melching, H. J., Streffer, C., and Sauer, H., *Strahlentherapie* **123**, 571 (1964).
194. Melching, H. J., Streffer, C., and Allert, U., *Strahlentherapie* **124**, 288 (1964).
195. Melching, H. J., Abe, M., and Streffer, C., *Strahlentherapie* **125**, 352 (1964).
196. Melching, H. J., Streffer, C., and Mavely, D., *Strahlentherapie* **127**, 268 (1965).
197. Messerschmidt, O., Shibata, K., and Melching, H. J., *Strahlentherapie* **120**, 74 (1963).
198. Milch, L. J., Yarnell, R. A., Stinson, J. V., Collins, J. G., Robinson, L. G., Croye, B. L., and Chang, S. S. M., *Science* **120**, 713 (1954).
199. Milch, L. J., Stinson, J. V., and Albaum, H. G., *Radiation Res.* **4**, 321 (1956).
200. Milch, L. J., and Albaum, H. G., *Proc. Soc. Exptl. Biol. Med.* **93**, 595 (1956).
200a. Mishkin, E. P., and Shore, M. L., *Radiation Res.* **33**, 437 (1968).
201. Monkhouse, F. C., and Baker, D. G., *Can. J. Biochem. Physiol.* **41**, 1901 (1963).
202. Moos, W. S., Mason, H. C., and Counelis, M., *Am. J. Physiol.* **199**, 1101 (1960).
203. Muntz, J. A., Barron, E. S. G., and Prosser, C. L., *Arch. Biochem.* **23**, 434 (1949).
204. Musiyko, V. A., *Ukr. Biokhim. Zh.* **38**, 55 (1966).
205. Nerurkar, M. K., and Sahasrabudhe, M. B., *Intern. J. Radiation Biol.* **2**, 237 (1960).
205a. Nerurkar, M. K., and Sahasrabudhe, M. B., *Intern. J. Radiation Biol.* **2**, 210 (1960).
206. Neuwirt, J., Pokorny, Z., Borova, J., and Sulc, K., *Strahlentherapie* **121**, 312 (1963).
207. Nyffenegger, E., Lauber, K., and Aebi, H., *Biochem. Z.* **333**, 226 (1960).
208. Ogasawara, K., *Wakayama Med. Rept.* **5**, 9 (1960).
209. Oswald, H., Günter, P. G., and Lorenz, W., *Strahlentherapie* **109**, 331 (1959).
210. Pany, J., *Naturwissenschaften* **49**, 164 (1962).
211. Parizek, J., Arient, M., Dienstbier, Z., and Skoda, J., *Nature* **182**, 721 (1958).
212. Parizek, J., Arient, M., Dienstbier, Z., and Skoda, J., *Med. Radiol.* **5**, 31 (1960).
213. Pearson, W. N., Owens, J. P., Hudson, G. W., and Darby, W. J., *Am. J. Physiol.* **173**, 120 (1953).

214. Pentz, E. I., *J. Biol. Chem.* **231**, 165 (1958).
215. Peters, K., and Bauer, R., *Strahlentherapie* **113**, 553 (1960).
216. Petersen, D. F. R., and Gould, R. G., LAMS Rept. 2445, 11 (1959).
217. Pikulyev, A. T., and Smelyanskaya, G. N., *Ukr. Biokhim. Zh.* **38**, 383 (1966).
218. Podil'Chak, M., Kuz'Menko, L., Maker, D., Yurmin, E., Chepovskii, V., and Provezailo, V., *Acta Biol. Med. Ger.* **12**, 137 (1964).
219. Prosser, C. L., Painter, E. E., Lisco, H., Brues, A. M., Jacobson, L. O., and Swift, M. N., *Radiology* **49**, 299 (1947).
219a. Quaife, M. A., Ogborn, R. E., and Odland, L. T., *J. Occupational Med.* **9**, 103 (1967).
219b. Quaife, M. A., and Ogborn, R. E., *Proc. Soc. Exptl. Biol. Med.* **126**, 377 (1967).
219c. Recht, P., *Proc. Symp. Accidental Irradiation at Place of Work, Nice, 1966* EUR 36, d-f-i-n/e, p. 405 (1967).
220. Rehnborg, C. S., Ashikawa, J. K., and Nichols, A. V., *Radiation Res.* **16**, 860 (1962).
221. Reuter, A. M., Gerber, G. B., Altman, K. I., Deknudt, G., and Kennes, F., *Klin. Wochschr.* **44**, 1310 (1966).
222. Reuter, A. M., Gerber, G. B., Kennes, F., and Remy-Defraigne, J., *Radiation Res.* **30**, 725 (1967).
223. Reuter, A. M., Sassen, A., and Kennes, F., *Radiation Res.* **30**, 445 (1967).
224. Robertson, J. E., *Am. J. Roentgenol. Radium Therapy* **50**, 392 (1943).
225. Rodionov, V. M., Chudinovskikh, A. V., Antokol'Skaya, Z. A., and Loboda, L. A., *Byul. Eksperim. Biol. i Med.* **47**, 43 (1959).
225a. Roesler, B., and Reichel, G., *Radiobiol. Radiotherap.* **7**, 765 (1966); **8**, 239 (1967).
226. Rosenthal, R. L., *Science* **110**, 43 (1949).
227. Rubini, J. R., Cronkite, E. P., Bond, V. P., and Fliedner, T. M., *Proc. Soc. Exptl. Biol. Med.* **100**, 130 (1959).
227a. Rust, J. H., Visek, W. J., and Roth, L. J., *Arch. Biochem. Biophys.* **72**, 189 (1957).
227b. Sanna, G., *Minerva Radiol.* **9**, 400 (1964).
228. Saritskin, I. V., and Zelemski, V. B., *Ukr. Biokhim. Zh.* **36**, 14 (1964).
228a. Sassen, A., personal communication (1968).
229. Sassen, A., Kennes, F., and Maisin, J. R., *Compt. Rend. Soc. Biol.* **157**, 1122 (1963).
230. Sassen, A., Gerber, G. B., Kennes, F., and Remy-Defraigne, J., *Radiation Res.* **25**, 158 (1965).
231. Sassen, A., Kennes, F., and Maisin, J. R., *Acta Radiol., Therapy, Phys., Biol.* **4**, 97 (1966).
232. Schön, H., Sitzman, F. S., and Barth, G., *Strahlentherapie* **105**, 585 (1958).
233. Schram, E., *Arch. Intern. Physiol.* **69**, 101 (1961).
234. Schwartz, S., *J. Clin. Invest.* **28**, 809 (1949).
235. Shaber, G. S., and Miller, L. L., *Proc. Soc. Exptl. Biol. Med.* **113**, 346 (1963).
236. Shacter, B., Supplee, H., and Entenman, C., *Am. J. Physiol.* **169**, 508 (1952).
237. Shastin, R., Kucheryavyy, F. Kh., and Krantikova, T. V., *Med. Radiol.* **5** (7), 88 (1960).
237a. Shelley, R. N., *Radiation Res.* **29**, 608 (1966).
238. Shipman, T. L., Lushbaugh, C. C., Petersen, D. F., Langham, W. H., Harris, P. S., and Lawrence, J. N. P., *J. Occupational Med.* **3**, 146 (1961).
239. Skalka, M., *Folia Biol. (Prague)* **7**, 275 (1961).
240. Smith, H., and Martin, J. H., *Health Phys.* **9**, 877 (1963).
241. Smith, H., Bates, T. H., and Smith, C. I., *Intern. J. Radiation Biol.* **8**, 263 (1964).

242. Smith, H., and Bates, T. H., *Symp. Personnel Dosimetry for High Level Exposure, Vienna, March 1965* STI/PUB/99, p. 199 (1965).
242a. Smith, H., and Langlands, A. O., *Intern. J. Radiation Biol.* **11**, 487 (1966).
242b. Smith, H., Wallas, J., and Sturrock, M., *Intern. J. Radiation Biol.* **11**, 383 (1967).
243. Smolichev, E. P., *Tr. Vses. Konf. po Med. Radiol., Moscow* p. 100 (1957).
244. Soberman, R. J., Keating, R. P., and Maxwell, R. D., *Am. J. Physiol.* **164**, 450 (1951).
245. Sörbo, B., *Acta Radiol.* **58**, 186 (1962).
245a. Staehler, F., and Otten, G., *Strahlentherapie* **136**, 718 (1968).
246. Stender, H., and Elbert, O., *Strahlentherapie* **89**, 275 (1952).
247. Stern, D. N., and Stim, E. M., *Proc. Soc. Exptl. Biol. Med.* **101**, 125 (1959).
248. Stevens, K. M., Gray, I., Schwartz, M. S., Lichter, R. J., Highland, G. P., and Dubois, K. G., *Am. J. Physiol.* **175**, 141 (1953).
249. Stevens, K. M., Gray, I., Schwartz, M. S., Lichter, R. J., and Highland, G. P., *Am. J. Physiol.* **175**, 147 (1953).
250. Storey, R. H., Wish, L., and Furth, J., *Proc. Soc. Exptl. Biol. Med.* **74**, 242 (1950).
251. Streffer, C., Melching, H. J., and Mattausch. H., *Strahlentherapie* **130**, 146 (1966).
251a. Streffer, C., and Melching, H. J., *Strahlentherapie* **128**, 406 (1965).
252. Streffer, C., and Messerschmidt, O., *Strahlentherapie* **130**, 285 (1966).
253. Streffer, C., and Langendorff, H. U., *Strahlentherapie* **131**, 453 (1966).
253a. Streffer, C., and Melching, H. J., *Strahlentherapie* **125**, 341 (1964).
254. Streffer, C., *Intern. J. Radiation Biol.* **12**, 487 (1967).
254a. Streffer, C., and Langendorff, H. U., *Intern. J. Radiation Biol.* **11**, 455 (1966).
254b. Streffer, C. *Ger. Med. Monthly* **12**, 35 (1967); *J. Vitaminol.* **14**, 130 (1968).
255. Sullivan, M. F., *Intern. J. Radiation Biol.* **2**, 393 (1960).
256. Sullivan, M. F., *Am. J. Physiol.* **209**, 158 (1965).
257. Supplee, H., Hausschild, J. D., and Entenman, C., *Am. J. Physiol.* **169**, 483 (1952.)
258. Swift, M. N., and Taketa, S. T., *Radiation Res.* **8**, 516 (1958).
259. Szabo, G., Magyar, Z., Kertai, P., and Zadory, E., *Z. Ges. Exptl. Med.* **130**, 452 (1958).
260. Taliaferro, W. H., Taliaferro, L. G., and Jaroslow, B. N., "Radiation and Immune Mechanisms." Academic Press, New York, 1964.
260a. Telia, M., Sassen, A., Reuter, A. M., Kennes, F., and Maisin, J. R., *Arch. Franc. Mol. App. Digest* **2**, Suppl. 2, 191 (1969).
261. Tereshenko, O. Y., *Byul. Eksperim. Biol. i Med.* **62**, 53 (1966).
261a. Tereshenko, O. Y., and Skuikhina, M. M., *Radiobiologiya* **5**, 761 (1965).
261b. Tereshenko, O. Y., *Biokhimiya* **5**, 822 (1965).
262. Thomson, J. F., and Mikuta, E. T., *Proc. Soc. Exptl. Biol. Med.* **85**, 29 (1954).
263. Thomson, J. F., and Klipfel, F. J., *Proc. Soc. Exptl. Biol. Med.* **95**, 107 (1957).
264. Tribukait, B., *Acta Radiol.* **3**, 1 (1965).
265. Uspenskaya, M. S., *Radiobiologiya* **2**, 418 (1962).
265a. Uspenskaya, M. S., *Radiobiologiya* **5**, 822 (1965).
265b. Varga, L., Braun, H., and Hornung, G., *Strahlentherapie* **134**, 284 (1967).
266. Vatistas, S., and Hornsey, S., *Brit. J. Radiol.* **39**, 547 (1966).
267. Volek, V., and Dienstbier, Z., *Strahlentherapie* **123**, 125 (1964).
267a. Waldschmidt, M., *Intern. J. Radiation Biol.* **11**, 429 (1966).
268. Waldschmidt, M., Pollak, H., and Streffer, C., *Strahlentherapie* **131**, 415 (1966).
269. Waldschmidt-Leitz, E., and Keller, L., *Z. Physiol. Chem.* **323**, 88 (1961).
270. Watson, G. M., *Intern. J. Radiation Biol.* **5**, 79 (1962).
271. Watts, C., MacVicar, J., and Goldberg, D. M., *Brit. J. Cancer* **20**, 282 (1966).
271a. Watts, C., and Goldberg, D. M., *Enzymol. Biol. Clin.* **8**, 379 (1967).

272. Weiner, N., Albaum, H. G., Milch, L. J., and the Cardiovascular Group, *A.M.A. Arch. Pathol.* **60**, 621 (1955).
273. Westphal, U., Priest, S. G., Stets, J. F., and Selden, G. L., *Am. J. Physiol.* **175**, 424 (1953).
273a. Wetterfors, J., Liljedahl, S. O., Plantin, L. O., and Birke, G., *Acta Med. Scand.* **177**, 227 (1965).
274. White, J., Burr, B. E., Cool, H. T., David, P. W., and Ally, M. S., *J. Natl. Cancer Inst.* **15**, 1145 (1955).
275. Wiehrmann, D., and Pospisil, M., *Naturwissenschaften* **45**, 92 (1958).
276. Williams, C. M., Krise, G. M., Anderson, D. R., and Dowben, R. M., *Radiation Res.* **7**, 176 (1957).
277. Winkler, C., and Paschke, G., *Radiation Res.* **5**, 156 (1956).
278. Zamyatkina, O. G., *Vopr. Med. Khim.* **10**, 362 (1964).
278a. Zharkov, Yu, A., Fedorova, T. A., and Mikhailova, L. F., *Radiobiologiya* **5**, 675 (1965).
279. Zhulanova, Z. I., and Romantsev, E. F., *Med. Radiol.* **5** (9), 39 (1960).
280. Zícha, B., and Beneš, J., *Proc. 5th European Radiobiol. Symp.*, p. 251 (1967).
281. Zícha, B., Beneš, J., and Hrulovsky, J., *Proc. 6th European Radiobiol. Symp.*, p. 139 (1968).
282. Zicha, B., Gerber, G. B., and Deroo, J., *Experientia* **25**, 1039 (1969).

Chapter IX

HORMONES AND SYSTEMIC EFFECTS

A. Introduction

The ability to respond rapidly to many dangerous situations is vital to the mammalian organism. Irradiation constitutes an insult to the body which elicits a regulatory response but simultaneously damages the systems effecting this regulation. The response to irradiation is determined by many factors which vary among different species and even among individuals of the same species. Thus, it is not surprising that certain observations discussed in this chapter seem contradictory at first sight.

Moreover, the insult resulting from exposure to radiation is neither specific nor the consequence of the irradiation alone. In addition to the irradiation itself, one must consider concomitant results of the exposure, e.g., partial starvation, infection, and even the unavoidable handling of the animals. Since each regulatory function depends upon an equilibrium between at least two forces, it should be kept in mind that one force can be altered only in relation to the other(s). Each regulating action engenders a process which counteracts it and, thus, tends to restrict it, even to the extent that, under certain conditions, the counterresponse may actually overshoot the original regulating action. In the pages that follow, we shall treat, first, the effects of irradiation on the morphology and function of individual endocrine glands, then proceed to more complex systemic responses, and, finally, deal with the questions of to what extent the regulations are specific for radiation injury and whether they benefit or harm the irradiated organism.

B. Diencephalon and Central Nervous System Regulation after Irradiation

The central nervous system (CNS) participates in many systemic responses of the body, either by direct stimulation or by release of neurosecretory hormones, e.g, ACTH-releasing factor. Although it has been known for a long

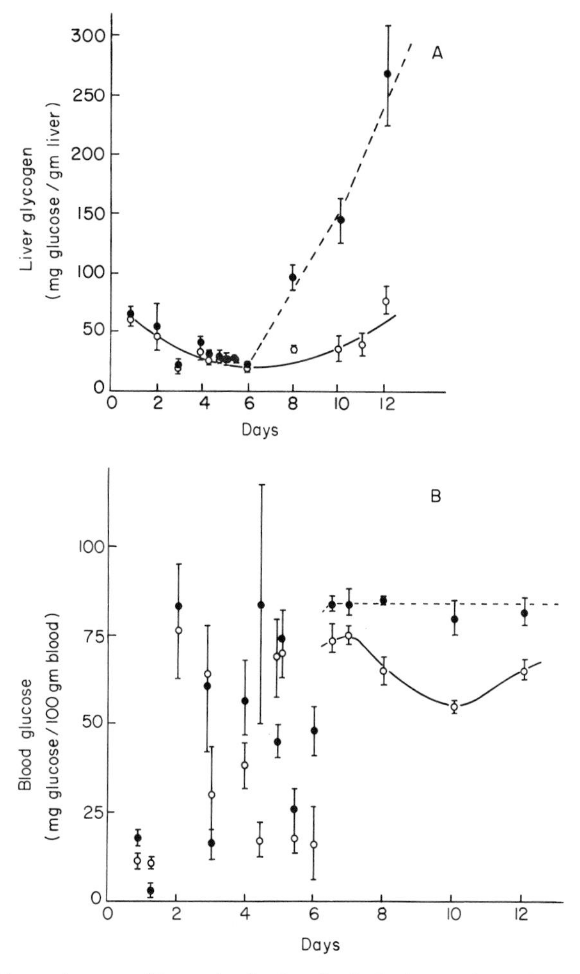

FIG. IX-1. Liver glycogen (determined after hydrolysis to glucose) (A) and blood glucose (B) in newborn rats 24 hr after x-irradiation (1000 R) (closed circles) and/or initiation of fasting (open circles) as a function of the age in days. Note that liver glycogen increases relative to the fasting levels only in rats irradiated after the seventh postnatal date. At the same time, irradiation also commences to raise blood glucose levels, whereas very young irradiated and nonirradiated rats display great variations in blood glucose levels. [From R. E. Thurber, and L. F. Nims, *Endocrinology* **70**, 595 (1962).]

time that damage to the CNS is the determining factor in death after irradiation with very high doses, the study of the response of the CNS and of the diencephalon, particularly to low doses of radiation, has just begun (see also p. 219). Therefore, the data presented in this section appear as isolated observations rather than as a coherent picture.

Extirpation of the hypothalamus abolishes changes in adrenal weight and cholesterol after irradiation (*8*). Even if only the neural connection between the hypothalamus and the pituitary gland is lost, as when the gland is transplanted into the eye chamber, the early response of the adrenals to irradiation no longer takes place (*6*). Drugs, such as morphine, which suppress the activity of the diencephalon prevent the initial stress response but do not influence the second late rise in blood corticoids (*6*). Regulation of the stress response is also essential for the changes in carbohydrate metabolism after irradiation. In newborn animals, the neural control of the pituitary gland is not yet established, and irradiation does not evoke the fall in adrenal ascorbic acid (*7*) or the increase in liver glycogen compared to starved animals (*180*) (Fig. IX-1).

On the other hand, irradiation of the head evokes many "stress" reactions similar to those resulting from whole-body irradiation, e.g, changes in adrenals and liver glycogen (see p. 318). To mention only one more example, an increase in specific activity of splenic RNase was found after irradiation of the head as well as after whole-body exposure (see p. 71), but, as pointed out in more detail below, caution must be exercised before explaining all abscopal effects of head irradiation on the basis of diencephalon or pituitary stimulation.

C. Pituitary Gland

From a morphological point of view, the pituitary gland appears to be rather radioresistant, and in the pituitary, as in the adrenals or the thyroid, it is difficult to draw a line between functional changes and direct damage to the organ. Local irradiation of the pituitary up to kiloroentgen levels leads to hyperemia, occasional pycnosis of β- and γ-cells, and a decrease in number of α-cells. In contrast, after whole-body exposure (1000 R = an LD_{100}), the number of α-cells increases and β-cells enlarge and lose their granules (e.g., *22*). No histological alterations were found after exposure to low doses of whole-body irradiation (300 R) (*37*).

Pituitary function may be altered even in the absence of detectable histological damage. A temporary increase in adrenocorticotrophin (ACTH), thyrotrophin (TSH), and a decrease in gonadotrophins was reported 1 hr after whole-body (*92a*) or local (*117*) irradiation of the hypophysis with β-rays. However, these results have not been confirmed by more recent investigations, in which no change (*107*) or a decrease in ACTH (Fig. IX-2) (*24*), and no change in the gonadotrophins FSH and LSH was detected (*23*). It is interesting that sham irradiation also resulted in a similar, though smaller, decrease in pituitary ACTH (*24*). The discrepancies between these findings may be the result of differences in the radiation dose, the handling of the animals, the method

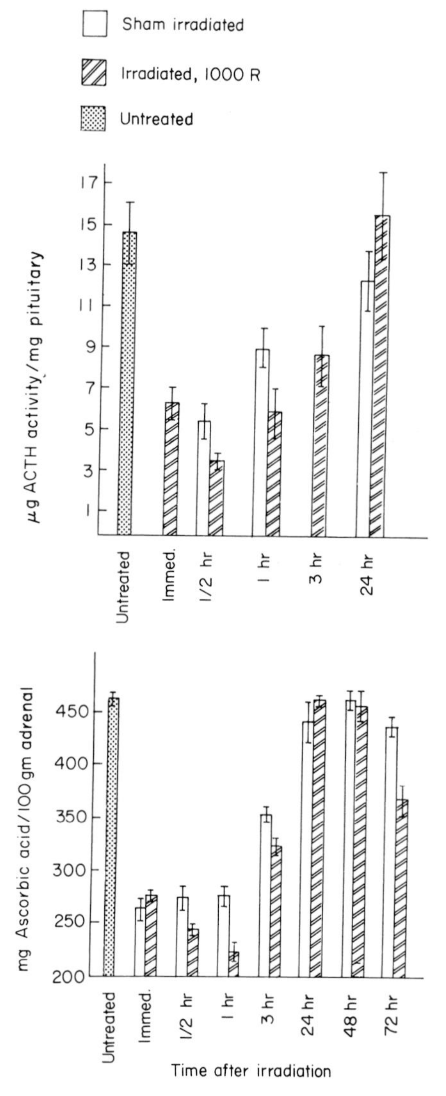

FIG. IX-2. ACTH content of the pituitary and ascorbic acid content of the adrenals after whole-body (1000 R) and sham irradiation. Note that ACTH and adrenal ascorbic acid decrease after x-irradiation but also, to a smaller extent, after sham irradiation. [From R. T. Binhammer, and J. R. Crocker, *Radiation Res.* **18**, 429 (1963).]

of hormone assay, or the time of sampling (circadian rhythm). *In vitro* incorporation of labeled amino acids into ACTH by the pituitary is the same for normal and x-irradiated (800 R) rats (*22a*).

Although ACTH in the blood has been reported to decrease temporarily at an unaltered rate of utilization (*10*), many observations on reactions regulated by ACTH suggest that ACTH secretion is enhanced after irradiation. To mention one example, irradiation of the head as well as the total body stimulates the adrenals and causes (a) a fall in adrenal cholesterol and ascorbic acid, (b) a rise in steroid hormones (see Fig. IX-5) of blood and urine, and (c) eosinopenia. Irradiation of the pituitary also leads to changes in the secretion of somatotrophic hormone (STH) and gonadotrophins. Growth is retarded and spermatogenesis is delayed after irradiation of the pituitary region of young rats with 2000 R (*45*) or less (*161*). Mature animals are more resistant, requiring a dose of 4000 R for inhibition of spermatogenesis (*45*). Although TSH has not yet been measured directly after irradiation, its secretion is probably altered, since the uptake of ^{131}I by the thyroid is increased after exposure of distant parts of the body (*58*). Moreover, urinary excretion of TSH is enhanced after whole-body irradiation of guinea pigs (*37a*).

As discussed elsewhere (p. 312), whole-body irradiation causes polyuria, probably via altered hormonal secretion of the neurohypophysis and the hypothalamus. The neurosecretory granules disappear from the hypothalamus and from the posterior part of the pituitary gland immediately after exposure to 200 R or more (*2*, *47*, *72b*, *76*, *80*), and antidiuretic hormonal activity in blood decreases after whole-body irradiation of rats with 500 R (*190a* [no changes were found in the hypophysis and blood of rats after 600 R (*80*)]. A high dose of irradiation (10 kR) to the head results in an increased oxytocin activity in the posterior lobe of the pituitary (*131*), an effect perhaps produced by impaired release of this hormone from the gland.

D. Adrenal Cortex

The adrenals appear to be relatively radioresistant if studied by morphological methods. Hyperemia and hemorrhages, especially in the zona fasciculata, are observed a few hours after irradiation of the total body, the adrenal region, and even distant parts of the body. Increase in lipofuscin, vacuolization, and similar changes suggesting accelerated aging are found in hamsters, even after an exposure to 100 R or less (*159*).

Adrenals transplanted into the anterior chamber of the eye cease growing after doses as large as 10 kR, but certain structures in the transplanted cortex appear to be damaged after 3 kR (*116*). The function of the adrenals transplanted into adrenalectomized recipients, however, is impaired after an *in vitro* irradiation of the gland with 2000 R; these animals display greater sensitivity to a subsequent stress (*207*). Functional changes in the adrenals, as demonstrated by histochemical methods, become noticeable a few hours after whole-body irradiation. Sudanophil granules representing cholesterol diminish or disappear after whole-body or local irradiation, but abscopal exposure with the adrenals shielded or with only one leg irradiated also leads to a similar decrease in lipids (*137*). Concurrent with the depletion of lipids in the irradiated adrenals (*87*, *88*), esterases diminish but acid phosphatase, alkaline phosphatase, ATPase, oxidative enzymes, and the number of lysosomes in the cell increase.

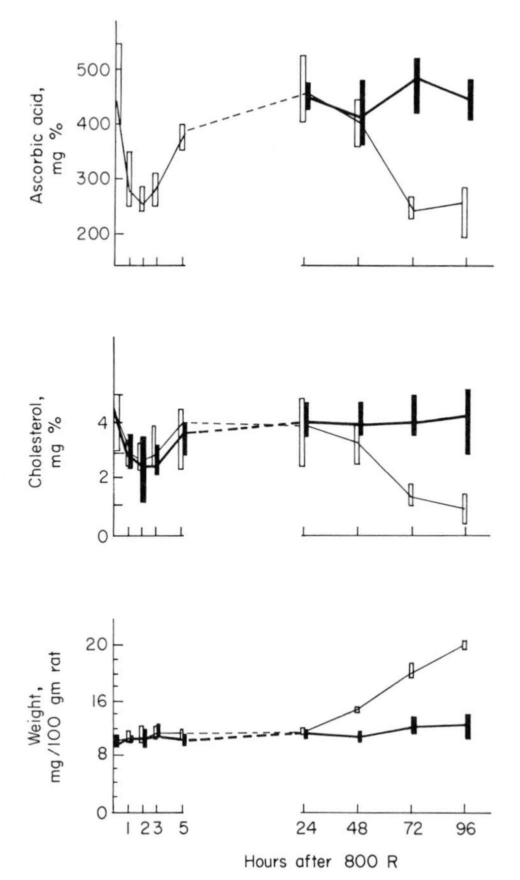

FIG. IX-3. Adrenal weight, adrenal ascorbic acid, and adrenal cholesterol after whole-body irradiation of rats (800 R) with (heavy line) and without (thin line) protection by cysteamine. [From Z. M. Bacq and P. Fisher, *Radiation Res.* 7, 365 (1957).] Note that adrenal ascorbic acid and cholesterol decrease 1–3 hr and again 48–98 hr after exposure. Only the initial decrease is prevented after injection of radioprotectors.

Adrenal activity is enhanced soon after whole-body or local irradiation, as shown by the increase in adrenal weight (*7*, *187*) (but compare Fig. IX-18), dehydroascorbic acid (*149a*), and RNA polymerase (*207a*) and the decrease in adrenal cholesterol and ascorbic acid (*22*, *134*, *187*) (Figs. IX-2 and IX-3). Whereas cholesterol esters and polyunsaturated fatty acids decrease 3 hr after exposure of rats to 1500 R, triglycerides increase (*48*, *49*). Synthesis of cholesterol in adrenal slices increases in response to the decrease in cholesterol content after irradiation (*70*) (Fig. IX-4), and this change depends on the age of the irradiated rats (*181*). Adrenal cholesterol decreases again and triglycerides and polyunsaturated fatty acids increase 2 days after 1500 R (*48*, *49*). Cholesterol

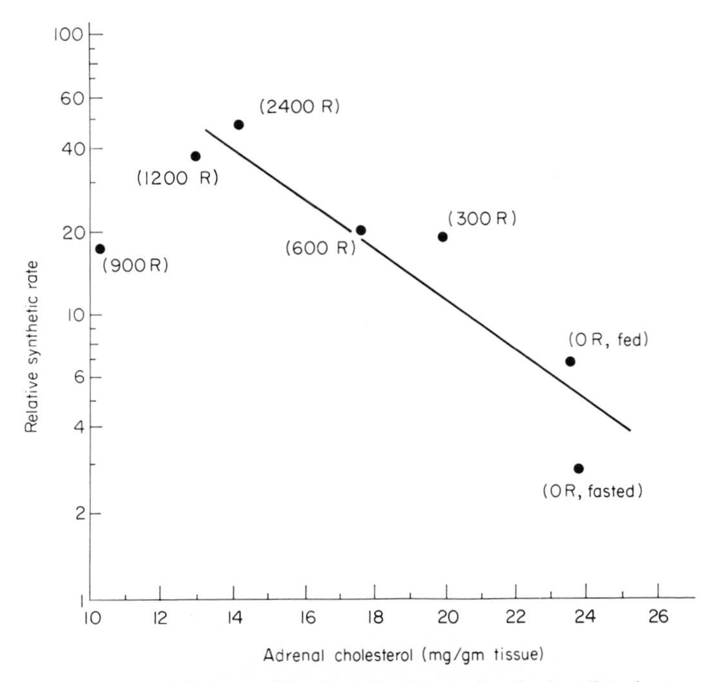

FIG. IX-4. Synthesis of cholesterol in adrenals of normal and x-irradiated rats as a function of cholesterol content. Note that synthesis of cholesterol increases after irradiation in proportion to the fall in cholesterol content of the adrenals. [From R. G. Gould, V. L. Bell, E. H. Lilly, P. Keegan, J. Van Riper, and M. L. Jonnard, *Am. J. Physiol.* **196**, 1231 (1959).]

has, however, also been reported to increase in rat adrenal from 3 days after an 850 R exposure (*163a*). The early decrease of ascorbic acid is independent of the dose over a wide range of radiation (*79*), is more pronounced in males than in females (*197*), and also occurs, although to a smaller extent, after sham irradiation and other types of stress (*24*). Thus, irradiation of a leg results in a significant reduction in the content of cholesterol and ascorbic acid (*137*). On the

TABLE IX-1

METABOLISM OF ADRENAL HORMONES IN IRRADIATED ANIMALS

Factor affected	Early period (0–8 hr)	Intermediate period (about 1st day)	Late period (2nd day and later)	Remarks
Hormone synthesis	Normal synthesis 500 R (*12*), increased synthesis in homogenates 750 R (*41*, *51*)	Low doses (750 R): normal synthesis (*41*, *51*)	Low doses (750 R): homogenates (rat) increased synthesis (*41*, *51*) (Fig. IX-7)	During irradiation of perfused calf adrenals, decreased conversion of progesterone to hydroxycortisone (*156*) direct irradiation of adrenal homogenates alters synthesis only after 500 kR (*68*)
		High doses (2 kR): increased synthesis (*163*, *204*) (not found by *31b*)	High doses (2 kR): progressive rise of synthesis in homogenates of guinea pigs until death (*204*)	
	11-β-Hydroxylation normal (750 R) (*85*); aldosterone synthesis increased (*85*, *209*) or normal (*71*); decreased corticosterone content in venous return from adrenals restored by stimulation with ACTH (*106*)		Perfused calf adrenals (after 600 R) decreased output per gm tissue, but increased total gland due to hypertrophy (*157*, *158*)	
Hormone catabolism	Normal rate of disappearance of cortisol from the blood (*205*)	Decreased catabolism in liver homogenates (*59*, *114*, *163*); decreased formation of glucuronides with 11-ketotetrahydroprogesterone or corticosterone days 1–21 (500 R); partially reversible with UDP–glucose (*172*, *173*)	Decreased uptake of cortisone by irradiated perfused liver (*51*)	Excretion of total glucuronides increases the 1st day, see p. 257

Hormone content—adrenals	Increase after head and whole-body irradiation (rat) (Fig. IX-5) (*75*), especially of glucocorticoids (*62a,b,c*, *208*)	Low doses (650 R): normal (*75*) High doses (1000 R): increased (*61*)	650 R normal (*75*); 1000 R progressive increase until death in rats (*61*, *62*), also in guinea pigs (7000 R) (*68*); increase in hydroxycortisone in guinea pigs only after 1000 R (*31a*)	
Hormone content—blood	Increase in rats 3 hr after head and whole-body exposure (*51*, *75*) (Fig. IX-5) (mostly bound corticoids) (*188*); also increase in monkeys (4–8 hr) (*67*) (Fig. IX-6), burros (*74*)	Normal (*67*)	Often increase (*41*, *51*, *67*) (Figs. IX-5 and IX-6)	Initial rise abolished by hypophysectomy (*8*), morphine (*7*), etc., preterminal rise not eliminated by morphine but mitigated by radioprotectors (*6*); only secondary increase in monkey continuously irradiated (*205*)
Hormone excretion—urine	Mouse: corticoids, later 17-ketosteroids increase after sublethal exposure, decrease after lethal exposure (*199*) Rat: after adrenal irradiation (Fig. IX-8) (*179*), increased 17-ketosteroids, decreased corticoids	Normal	Increased 17-ketosteroids 5–7 days (*113*) dogs; guinea pig corticoids increase until death (2000 R) (*204*); rats dying early after 800 R show higher 17-ketosteroid excretion than those surviving (*178b*)	Increased excretion of 17-ketosteroids and corticoids (especially cortisol) (*171*) during therapy and after accidents reported by many investigators, see p. 274

other hand, the response of the adrenals to stress may become impaired by direct irradiation of the glands; when shielded during the whole-body exposure, the adrenals react more strongly and show a more pronounced fall in cholesterol and ascorbic acid (*198*).

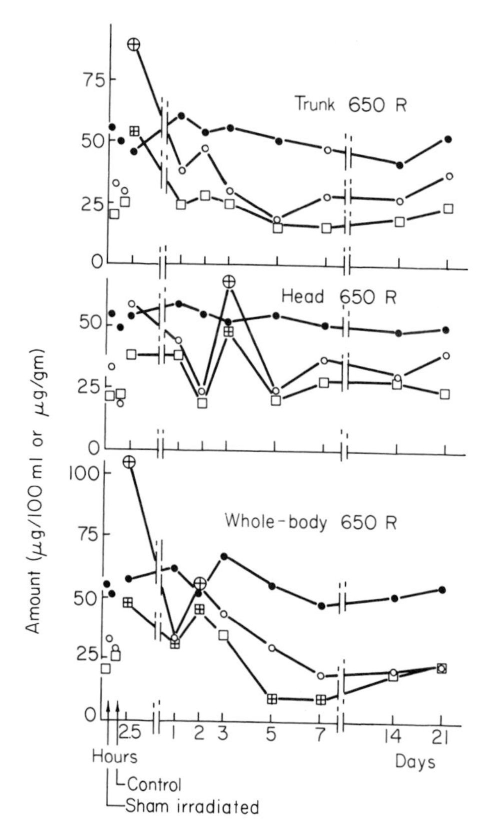

FIG. IX-5. Adrenal weight (solid circles, mg) and deoxycorticosterone content in adrenals (squares, μg/gm) and plasma (open circles, μg/100 ml) after whole-body, head, or trunk irradiation (650 R) of rats. Circles and squares with crosses indicate significantly different values. Note that adrenal and plasma deoxycorticosterones increase 2.5 hr and 2 days after irradiation of the head or the whole body. [From J. M. A. Hameed, and T. J. Haley, *Radiation Res.* **23**, 620 (1964).]

The metabolism of steroid hormones in irradiated animals has been studied extensively. Important observations on synthesis and catabolism, as well as on levels of steroid hormones in adrenals, blood, and urine are summarized in Table IX-1. Although the data reported are not always in agreement with each other, a certain basic pattern can be traced. Irradiation causes a stress response in the body and stimulates synthesis of steroid hormones via the hypothalamic–pituitary system. During the early phase, which in rats develops 3 hr and in

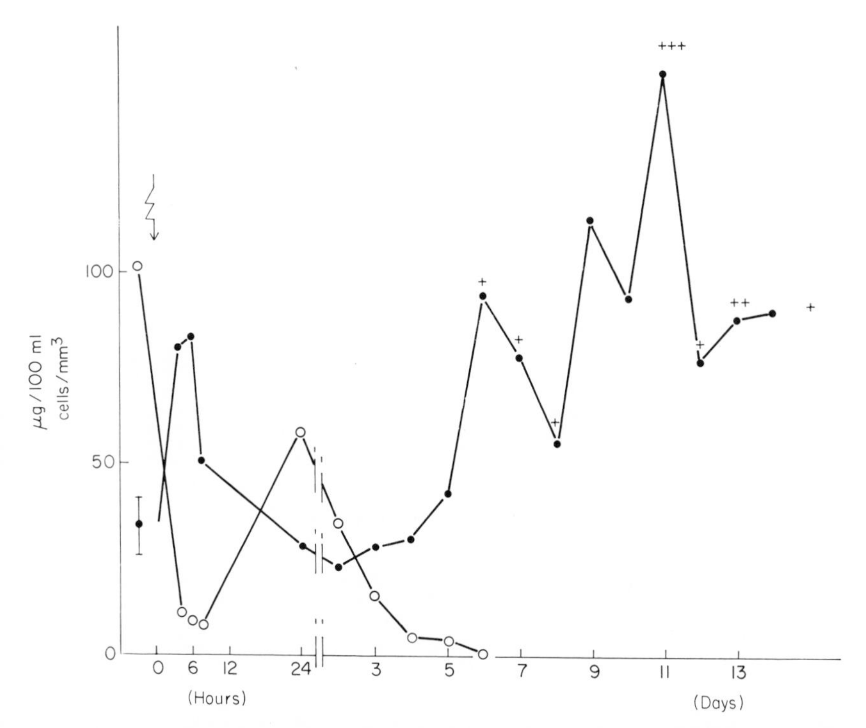

FIG. IX-6. Hydroxycorticoids (solid circles, μg/ml) and eosinophilic cells (open circles, per mm³ plasma) in the plasma of monkeys after whole-body irradiation with 800 R. Crosses represent death of animals. Note that eosinophils decrease and plasma hydroxycorticoids increase during the first day after exposure. Normal values are found one day after irradiation, but hydroxycorticoids later increase progressively until death. [From A. B. French, C. J. Migeon, L. T. Samuels, and J. Z. Bowers, *Am. J. Physiol.* **182**, 469 (1955).]

monkeys about 5 hr after irradiation, hormone synthesis is enhanced. Concurrently, ascorbic acid and cholesterol content in the adrenals and blood decreases. The catabolism of hormones appears to be slightly impaired, but this effect does not interfere seriously with the catabolism, conjugation, and subsequent excretion of excess hormones. As a result, urinary levels of corticoids and ketosteroids may increase during the first days after irradiation.

Adrenal activity returns to normal about 1 day after a sublethal exposure, but usually remains elevated after a lethal exposure until death. A second period of

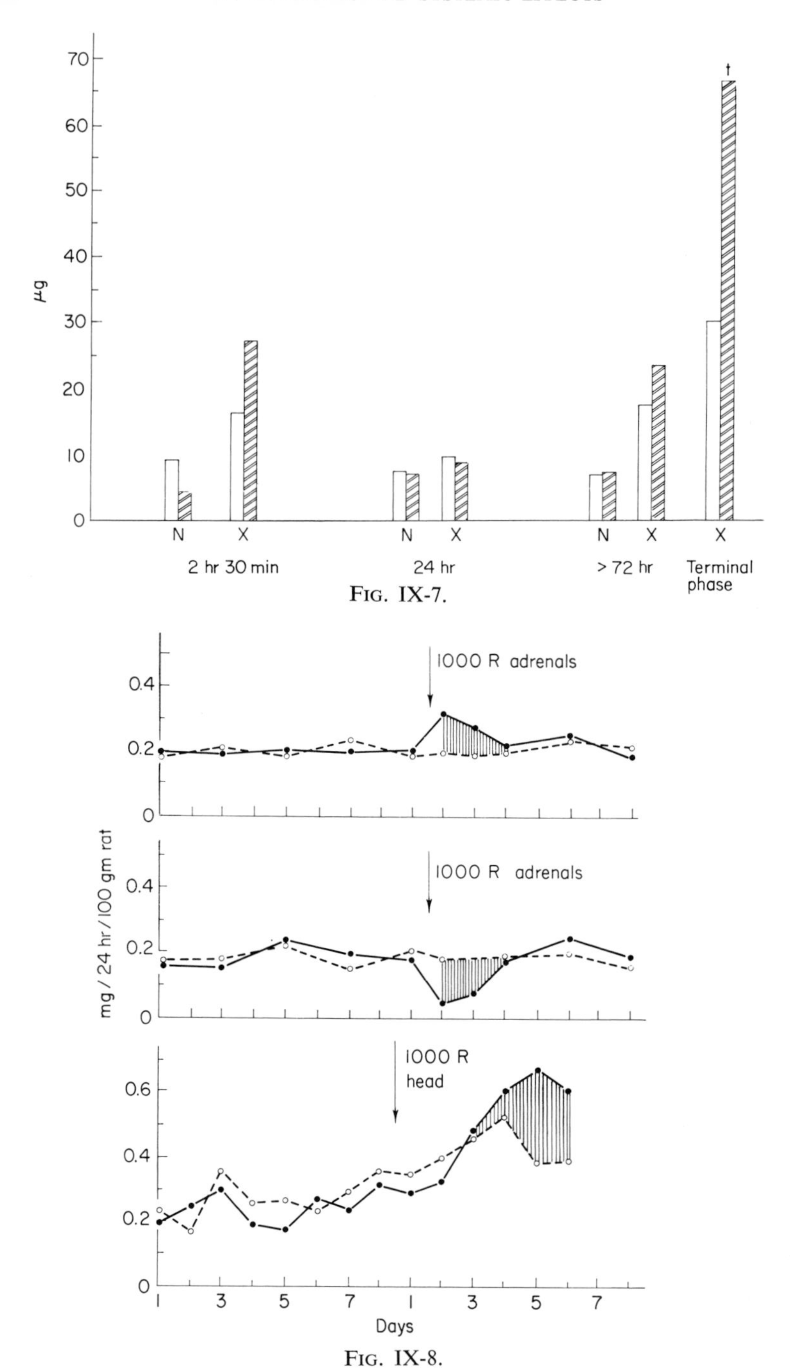

FIG. IX-7.

FIG. IX-8.

adrenal activity often ensues after a few days as a consequence of the radiation illness. This period is characterized by another decrease in adrenal cholesterol and ascorbic acid and a rise in hormone synthesis. Animals exposed to doses of radiation causing gastrointestinal death usually display a progressive increase in adrenal activity until death. Irradiation may directly damage adrenal function to some extent but does not eliminate the response of the adrenals or even lead to their exhaustion.

E. Reproductive Organs

1. Testes

Since the detrimental action of radiation on the testes was recognized by Albers-Schönberg (*3*) in 1903, an immense literature has accumulated which has recently been reviewed (*32*). Spermatogonia (especially those of type B in the mouse, but also those of type A in the rat) are killed after exposure of the testes to relatively low doses of radiation and die in early prophase or late metaphase. As the supply of germ cells derived from spermatogonia becomes exhausted, the testes are progressively depleted of these cells until 2–4 weeks postexposure; mature sperm cells also disappear. If the dose has not been excessive, regeneration from type A spermatogonia spared from death begins, and the testes simultaneously display degeneration and regeneration. In contrast, the constituent Leydig and Sertoli cells of the interstitial tissues are relatively radioresistant; testes atrophied because of radiation damage appear to contain more interstitial tissue in relation to germ cells.

Libido and secondary sexual characteristics usually remain intact after irradiation of the testes, even when azoospermia and infertility result. Indeed, the synthesis of androgens is relatively radioresistant, and testes irradiated with 5–10 kR still respond to stimulation by pituitary gonadotrophins (*1*). Excretion of androgens in man continues after radiation castration (1500 R),

---

Fig. IX-7. Plasma corticosterone (shaded columns, μg/100 ml plasma) and adrenal production of corticosterone (unshaded columns, μg/kg body weight) in rats at different times after whole-body irradiation (750 R). Note that synthesis increases 2 hr 30 min and 72 hr (i.e., before death) after exposure at a time when plasma levels are maximal. [From W. Eechaute, G. Demeester, and I. Lensen, *Arch. Intern. Pharmacodyn.* **135**, 235 1962).] N: nonirradiated; x: irradiated; † = death.

Fig. IX-8. Excretion of 17-ketosteroids (top and bottom figures) and 17-ketogenic steroids (middle figure) after x-irradiation of the head or the adrenal region of rats with 1000 R. Shaded region: period of significant differences between x-irradiated and control rat. Note that excretion of 17-ketosteroids increases and that of 17-ketogenic steroids decreases 1–2 days after exposure of the adrenals. Head irradiation also causes increased excretion of 17-ketosteroids, but not before the fourth day after exposure. [From W. Teller, W. Staib, Haibach, and M. Hohn, *Strahlentherapie* **111**, 107 (1960).] Solid circles: irradiated; open circles: nonirradiated.

but is abolished by orchiectomy (*25*). Nevertheless, synthesis of androgens is probably affected to a certain extent by high doses of radiation (1000 R), since more pituitary gonadotrophins must be produced to stimulate androgen formation (*24a*, *162*).

Exposure of the total body of mice to 540 R or of the testes of rats to 1500 R causes a decrease in conversion of progesterone or 5-pregnenolone to testosterone in testicular homogenates 20 days after exposure (*17*, *54*). Normal

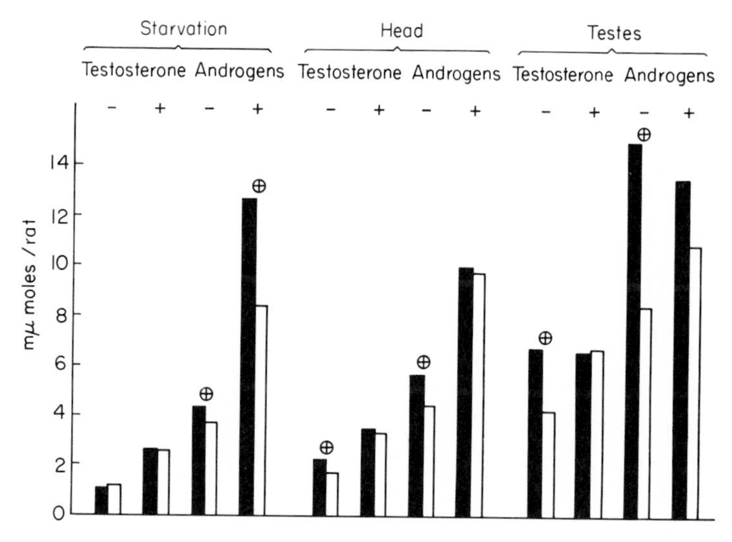

FIG. IX-9. Production of testosterone and of total androgens in homogenates of testes 20 days after irradiation of the head and the testes (1500 R) or after partial starvation with (+) and without (−) addition of reduced nicotinamide adenine dinucleotide (NADPH). Significant differences between nonirradiated and x-irradiated homogenates are marked ⊕. White columns: starved or irradiated; black columns: controls. Note that production of androgens and testosterone decreases after irradiation of the testes and, somewhat less, of the head. The inhibition after irradiation can, at least in part, be reversed by addition of NADPH. On the other hand, androgen synthesis diminished by partial starvation does not respond to NADPH. [From D. L. Berliner *et al.*, *Radiation Res.* **22**, 345 (1964); **23**, 156 (1964); **24**, 572 (1965).]

synthesis can be restored if reduced nicotinamide adenine dinucleotide (NADPH) is added to the assay system (Fig. IX-9) (*17*, *55*). In mice, the response to whole-body irradiation is biphasic, the drop in synthesis of testosterone around the twentieth day being followed by an increase (*54*). Since irradiation of the heads of rats with 1500 R also results in a similar decrease in testosterone production (*55*), it is likely that this effect is, in part, mediated via the pituitary gland. Moreover, an *in vitro* irradiation of testicular homogenates results in different biochemical alterations (*163b*). A drop in hormone production is also observed during partial starvation but, in contrast to the behavior after irradiation, NADPH apparently cannot restore normal conditions (*55*). Isolation of the metabolites formed in homogenates prepared from testes of irradiated rats indicates that the radiation interferes with the cleavage of the side chain of hydroxyprogesterone. Accumulation (*17*) of 20α-dihydroxyprogesterone (*18*), thus, results.

The weight of the testes diminishes as a function of the radiation dose due to the loss of germ cells, and this is accompanied by a simultaneous decrease in content (*4*, *185*) and synthesis (*160*) of nucleic acids. The synthesis of phosphate esters, especially of guanosine phosphates, diminishes in the testes after whole-body irradiation (*153*). Histochemical methods demonstrate that the activity of oxidative enzymes and dehydrogenases diminishes progressively for 1–2 weeks after irradiation (*84*). Irradiated testes contain more stearic and oleic acids and less arachidonic and palmitic acids, suggesting a defect in lipid metabolism (*39*). The synthesis of cholesterol from acetate-^{14}C by rat testes, however, is unaltered after exposure to 800 R whole-body irradiation (*181*).

2. Ovaries

Since Halberstaedter (*73*) and Bergonie (*16*) recognized in 1905 that the ovaries atrophy after irradiation and that temporary or permanent sterility may ensue, considerable information has appeared on the effects of irradiation on the ovaries. We shall only cite two recent reviews (*32*, *102*). In order to evaluate studies of irradiated ovaries, stress must be laid upon the fact that no other organs present such large differences in response after irradiation between species or between individuals within a species. In the developing fetus, oogonia are relatively radioresistant but oocytes in primordial follicles are extremely radiosensitive (D_{37} = 91 R, extrapolation number n = 2–3) (*124*). Sensitivity diminishes as the follicles mature, and is low before ovulation. Thus, female (as well as male) germ cells are most sensitive during the last premiotic divisions, becoming more and more resistant during the premiotic prophase and the development into mature gametes. This pattern of response has been linked to changes in the condensation of chromatin. Nevertheless, it is important not to overlook the crucial differences between male and female germ cells, namely, that male germ cells represent a renewing cell population whereas female germ cells do not. The fertile–sterile–fertile pattern often found in the ovaries after irradiation is not a result of regeneration from a stem-cell pool, as it is in males, but of the higher sensitivity of the intermediate follicular stages compared with that of the primitive and mature stages. Granulosa cells in the developing follicles are damaged even earlier than the oocytes, but in the mature follicles and the corpus luteum they appear more resistant. Also, the germinal epithelium and interstitial glands are relatively radioresistant in rodents.

The function of the ovaries after irradiation is dependent upon the presence of the morphological structure capable of synthesizing the hormones. The pattern of the functional response of the ovaries to irradiation may, therefore, vary as much as does the histological picture. Temporary and permanent sterility after irradiation is usually accompanied by destruction of the follicles, loss of secondary sexual characteristics, and suspension of the cyclical functions. Nevertheless, certain species, such as mice and rats, continue abortive estrus cycles recognizable by the cornification of the vaginal epithelium after

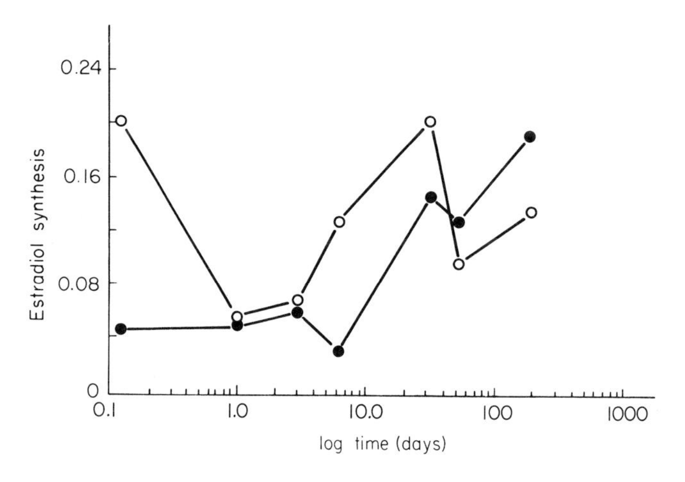

FIG. IX-10. Synthesis of radioactive estradiol (percentage conversion) from testosterone-^{14}C in ovaries of rats at different times after whole-body irradiation. Ovaries from irradiated rats (400 R): solid circles; ovaries from nonirradiated rats: open circles. Note that a significant decrease occurs only 2 hr after exposure, but values from normal and irradiated rats vary widely. [From S. R. Stitch, R. E. Oakey, and S. S. Eccles, *Biochem. J.* **88**, 76 (1963).]

irradiation. Other species, such as rabbits and guinea pigs, sometimes produce abnormal cystic structures in irradiated ovaries which synthesize hormones. Anomalous development of secondary sexual characteristics and continuous estrus can then result.

The immediate effect of irradiation on hormone production by the ovary is probably small. A corpus luteum is still formed upon stimulation by LSH, even after doses of 10 kR (*169*). Synthesis of estrogens from acetate-^{14}C is not much altered directly after exposure of a human ovary to 2 kR (*112*). Conversion of testosterone to estrogens diminishes shortly after whole-body irradiation of rats (400 rads), but then returns to normal (Fig. IX-10) (*174*). On the other hand, catabolism and conjugation of estradiol in kidney and intestine apparently are slightly impaired (*50*). The excretion of estrogens by patients whose ovaries have been irradiated decreases below the level found during menopause but remains higher than that after surgical ovariectomy (*46*). Excretion of gonadotrophins, however, increases as much after irradiation as it does after ovariectomy (*90*).

F. Thyroid Gland

Morphology and function of the normal thyroid appear to be relatively resistant to direct effects of irradiation, whereas the hyperactive thyroid responds more readily, and stimulated cell renewal in the thyroid is as radiosensitive as it is in other organs. High doses of radiation can, however, damage even the normal thyroid permanently and eventually cause myxedema, characterized by a flat epithelium, few follicles, and an increase in connective tissue. Cellular damage at the ultrastructural level becomes noticeable in the irradiated thyroid a few hours after exposure; the microvilli degenerate, and the number of lysosomes and lysosomal enzymes increase (*166*).

Whole-body irradiation appears to modify the normal thyroid more conspicuously than local exposure; the thyroid then shows a histological structure which suggests important functional changes. Most likely, however, these functional changes following whole-body exposure result from abscopal mechanisms mediated by the pituitary. Thus, after a sublethal exposure of rats to 500 R, the epithelium increases in height and colloid disappears, whereas a lethal dose (800 R) is characterized by the development of low epithelium and deposition of colloid (*22*). Recently, the radiosensitivity of thyroid tissue has been determined by grafting the irradiated thyroid tissue of rats into the anterior chamber of the eye. More than 1.2 kR are needed to completely stop the growth of the graft, but somewhat lower doses result in a delay in growth (*146*) and an impaired response to a goitrogenic stimulus (*68a*).

Local external or internal irradiation of the thyroid causes a reduction in the uptake of iodine and in the formation of protein-bound iodine, especially in the hyperactive glands. Relatively high doses are needed to influence the normal thyroid gland, and the basal metabolic rate is not affected immediately after exposure to several kR (*15*). Injection of large doses of radioactive iodine depresses the formation of protein-bound iodine one day later (after a dose of about 150 krep) (*60*). Protein-bound iodine then increases temporarily before decreasing permanently. In contrast, a local dose of only 500 R was found to delay the thyroid hyperplasia induced in opossums by feeding a protein-rich diet (*194*).

Thyroid function is altered after whole-body irradiation, but the direction and extent of the changes depend on the species and strain (*183*), pretreatment of the animals, dose of radiation, caging density (*72a*), and dietary regimen (*184*). Increases in uptake of iodine and protein-bound iodine have been observed 1–3 days after a lethal or sublethal exposure in rats and other species (*5*, *126*, *184*, *191*). Thyroid function increases even after a dose of 2000 R which

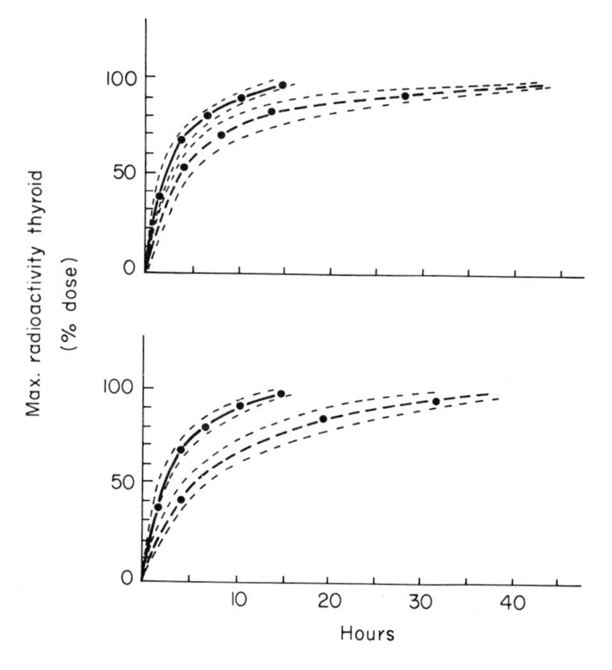

FIG. IX-11. Time course of the uptake of ^{131}I by the thyroids of normal (continuous line) and x-irradiated (dashed line) rats 24 hr (upper figure) and 48 hr (lower figure) after whole-body irradiation (800 R). Note that the rate of iodine uptake decreases after irradiation with a lethal dose. The inhibition is more pronounced 48 hr than 24 hr after exposure. [From E. H. Betz, "Contribution à l'etude du syndrome endocrinien provoqué par l'irradiation totale de l'organisme." Masson, Paris, 1956.]

causes gastrointestinal death (*163*). In one strain of rats, protein-bound iodine decreases after a lethal exposure and increases after a sublethal one, thereby providing a sensitive prognostic sign for survival (*175*). Other investigators have found consistently lowered values for uptake of iodine and for protein-bound iodine after exposure of rats (*21*) or mice (*111*, *167*, *168a*) to a lethal dose (Fig. IX-11), but usually not after a sublethal exposure (*22*, *101*).

The basal metabolic rate, as well as the oxygen consumption, increases after exposure to 900 R, compared with starved animals (*165*), but decreases if compared with fed or pair-fed controls (Fig. IX-12) (*177*). Oxygen consumption is, however, not significantly different in rats exposed to protracted irradiation (80 R/day) and their pair-fed controls (*13*). The respiratory quotient decreases in burros after protracted irradiation (*104*), whereas the oxygen consumption is equal to that of starved controls. These findings emphasize the need for studies on thyroid activity in which animal care and dietary regimen are carefully standardized.

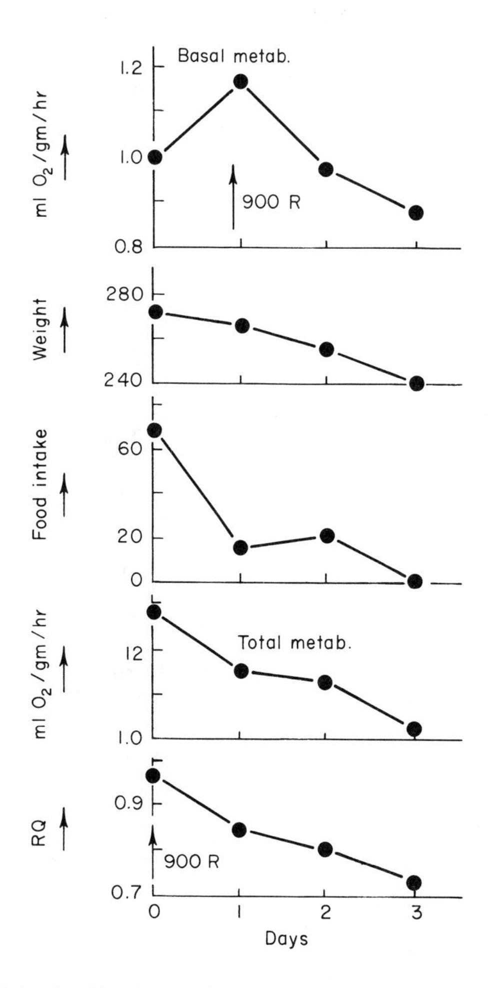

FIG. IX-12. Weight, food intake, total oxygen consumption, and RQ (respiratory quotient) in fed rats, and basal metabolic rate in starved rats after x-irradiation with 900 R. Note that food intake, O_2 consumption, and RQ decrease progressively, whereas the basal metabolic rate increases on the first day after irradiation. [From O. Strubelt, *Strahlentherapie* **126**, 283 (1965).]

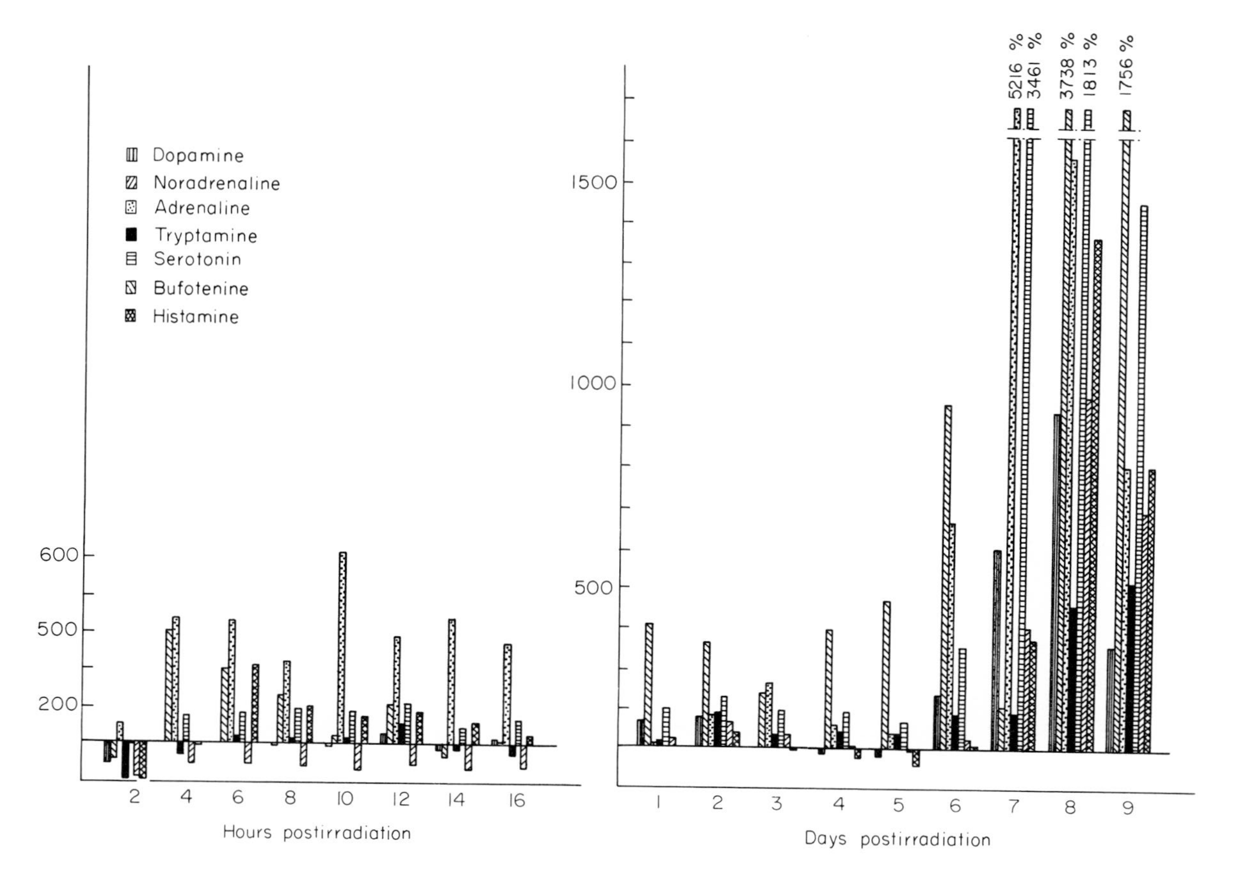
Dopamine
Noradrenaline
Adrenaline
Tryptamine
Serotonin
Bufotenine
Histamine
Excretion / 24 hr / rat
600
500
200
2 4 6 8 10 12 14 16
Hours postirradiation
1500
1000
500
1 2 3 4 5 6 7 8 9
Days postirradiation
5216 %
3461 %
3738 %
1813 %
1756 %

G. Biogenic Amines

As early as 1897 (*193*) i.e., within 1 year of the discovery of x-rays, it was recognized that irradiation not only produces local effects but also causes systemic and abscopal reactions. Reactions, such as nausea, fall in blood pressure, headache, and gastrointestinal disorders, occur as symptoms of a general "intoxication" (radiation sickness), and the search for toxic products causing these reactions occupied much of the early investigators' efforts. Since the symptoms resemble, in several respects, those caused by histamine and related substances, liberation of biogenic amines (histamine) was considered to be responsible for the systemic and many local effects of irradiation (*53*). Although this hypothesis is not tenable as such at this time, it might apply to several specific effects of irradiation.

Radiation-induced changes in the content of biogenic amines in organs, blood, and urine are summarized in Table IX-2. These observations suggest that shortly after irradiation biogenic amines are liberated from subcellular granules. As a result, biogenic amines decrease in organs and increase in blood. Moreover, increased amounts of biogenic amines and their metabolites are excreted in the urine. A second period of excretion of biogenic amines has been observed as the acute phase of radiation illness develops. It should be pointed out, however, that the methods used for the determination of biogenic amines have not been adequate and, in some cases, have not been published in detail (*65*).

The earlier investigations were concerned primarily with the metabolism of histamine and catechol amines. During recent years, interest has focused on serotonin. In this respect, the observations of hydroxyindoles in the brain appear particularly important. Local irradiation with high doses (10 kR) causes an immediate increase in (total) 5-hydroxyindoles in the brain, most notably in the pons, medulla, and brain stem (Table IX-4) (*115*, *128*, *139*, *152*). Most likely, this effect results from a liberation of serotonin from subcellular granules in the brain and its subsequent degradation to hydroxyindoleacetate. Indeed, whereas the total 5-hydroxyindoles increase, serotonin apparently diminishes (*51a*, *154*, *170*). Lower doses of whole-body irradiation (900 R) produce an immediate increase in total 5-hydroxyindoles only in the brain stem (*122*, *139a*, *141–143*) or in the total brain of adrenalectomized rats (*139*). Rats loaded with 5-hydroxytryptophan prior to exposure (900 R) show a decrease in serotonin and an increase in 5-hydroxyindoleacetate (*140*). Serotonin levels in the brain

---

FIG. IX-13. Excretion of biogenic amines (percent of normal values) at different times after whole-body irradiation of rats (800 R). All animals died by day 9. Note that the excretion of some biogenic amines increases during the initial hours and again from days 6–9 after irradiation. [From F. Franzen, H. Gross, and G. Thielicke, *Strahlentherapie* **120**, 598 (1963).]

TABLE IX-2

CONCENTRATION OF BIOGENIC AMINES IN ORGANS, BLOOD, AND URINE AFTER IRRADIATION

Amine	Blood	Organs	Urine	Remarks
Histamine	Increases concurrently with fall in blood pressure in rats (*58a*, *196*) and man (*63*, *95*)	Decreases in lung, intestine, and skin from 3 hr after irradiation (*14*, *52*, *58a*) (minimal values 5th day in rat) (*52*)	Increases 1st day and 8th day (Fig. IX-13) (*65*, *66*, *110*)	Histidine decarboxylase diminished in rat stomach after 2000 R but also after starvation (*93*)
Catechol amines	Unchanged (*72*)	Decrease in adrenals (*97a*, *120*, *203*); no decrease in adrenals (*33*); decreases in heart and brain 24 hr after 850–1000 R (*189a*)	Increases (*27*, *65*, *66*) (Fig. IX-13); also after accidents in man (*120a*); no change after head irradiation (*28*)	No change in synthesis in adrenal slices (*120*)
Serotonin	Decreases (*36*, *57*, *125*, *129*, *130*, *144*)	Decreases in intestine (*34*, *118*, *119b*, *189*) and other organs (*56*, *57*) (Table IX-3); decreases in brain (see text and Table IX-4)	Increase in serotonin (*65*, *66*) and its catabolite 5-hydroxyindoleacetate (*42–44*, *122*, *155*) (Fig. IX-14) in rat, but not in man (*20*)	Tryptophan hydroxylase (*192*), 5-OH-tryptophan decarboxylase (*105*), and monoaminooxidase (*176*) diminished in liver; less 5-OH-tryptophan converted (*35*) to serotonin in total animal; liberation of serotonin occurs also in isolated uterus (*190*) and in amphibians (*31*)
Acetylcholine	—	Decreases, then increases in intestine (*38*, *100*)	—	Cholinesterase decreases in blood, brain, and intestine (*38*) (see p. 89)

TABLE IX-3

SEROTONIN CONTENT IN BLOOD AND VARIOUS ORGANS OF THE RAT AFTER WHOLE-BODY IRRADIATION WITH 450 OR 900 R[a]

Tissue	Dose of radiation (R)	Serotonin (μg/gm)			
		Control	1 day	3 days	6 days
Blood	450	0.73	0.37	0.41	0.43
	900		0.33	0.29	0.26
Spleen	450	2.99	1.42	1.29	1.58
	900		1.33	0.96	1.38
Brain	450	0.54	0.39	0.23	0.24
	900		0.47	0.29	0.20
Intestine	450	5.04	3.03	2.73	3.69
	900		3.14	2.26	3.20

[a] From Ershoff *et al. Proc. Soc. Exptl. Biol. Med.* **110**, 536 (1962).

TABLE IX-4

HYDROXYINDOLES IN DIFFERENT PARTS OF THE BRAIN IMMEDIATELY AFTER WHOLE-BODY IRRADIATION (900 R) OF RATS[a]

Tissue	Total 5-OH-indoles, μg/gm		
	Control	X-irradiated (900 R)	$t(p^*)$[b]
Whole brain	0.958 ± 0.047	1.223 ± 0.036	<0.02
Hemispheres	0.957 ± 0.09	1.149 ± 0.119	>0.05
Brain stem	1.218 ± 0.038	1.736 ± 0.119	<0.01
Cerebellum	0.610 ± 0.118	0.445 ± 0.079	>0.05
Blood	1.130 ± 0.066	0.964 ± 0.046	>0.05

[a] Note that the increase in hydroxyindoles is significant in the whole brain and the brain stem. [From D. Palaić, Ž. Deanović, and Z. Supek, *J. Neurochem.* **11**, 761 (1964).]
[b] Statistical significance (Student *t* test).

remain low for several days after irradiation (*56*, *57*, *128*), but starvation alone has also been reported to reduce serotonin in the brain (*128*). Serotonin thus liberated may be excreted unaltered (*65*, *66*) or converted to 5-hydroxyindoleacetate. The amount of 5-hydroxyindoleacetate excreted by rats during the first day after exposure is a function of the radiation dose and is related to survival time (*42–44*, *178a*) (Fig. IX-14). The level of excretion after irradiation does not depend on the presence of the adrenals (*151*); it differs, however, in male and female rats (*42*).

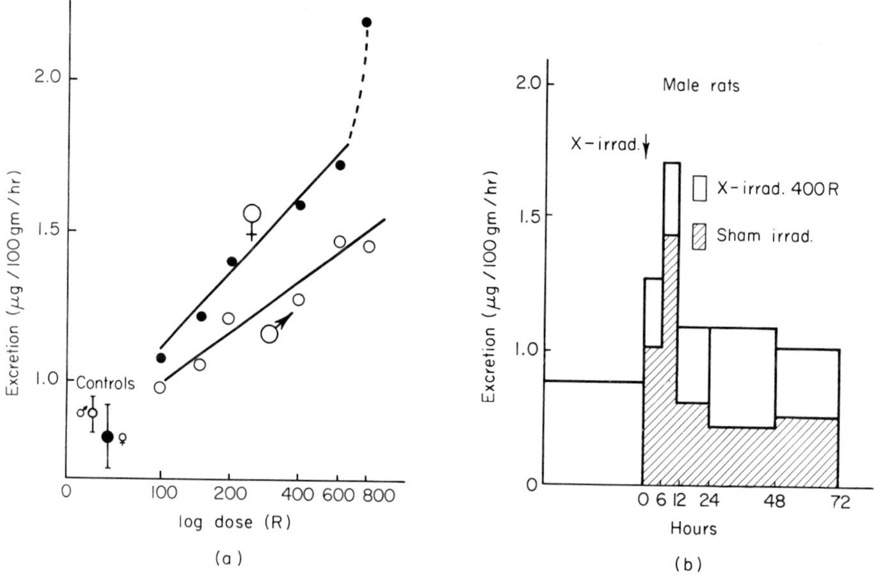

FIG. IX-14. Excretion of 5-hydroxyindoleacetic acid as function of time after irradiation (b) and dose (a, 1 day after irradiation) in male and female rats. Note that excess excretion takes place within 6 hr after exposure and increases with the dose of whole-body radiation. The effect is less pronounced in males than in females. [From Ž. Deanović, Z. Supek, and M. Randic, *Intern. J. Radiation Biol.* 7, 1 (1963).]

Biogenic amines can protect animals if given before whole-body irradiation, but the extent of protection depends on the compound investigated and on the species [see recent reviews (*9*, *123*)]. In 1925, Jolly suggested that the protection of irradiated lymphoid tissue by adrenaline is a result of hypoxia (*86*). Other mechanisms of protective action by biogenic amines have been proposed (*9*, *123*), but it is not yet possible to decide what role, if any, these mechanisms play aside from hypoxia. Since biogenic amines capable of protecting against radiation damage (e.g., serotonin) are liberated after irradiation, it has been proposed that this effect may constitute a physiological reaction serving as a defense against radiation injury. Although it seems unlikely that such a process could respond rapidly enough to protect against the immediate effects of irradiation, it may mitigate certain secondary symptoms of radiation sickness, such as fall in blood pressure. Indeed, a certain correlation has been observed between the serotonin content in the intestine and the radiosensitivity of an animal (*118*).

H. Systemic Effects

1. Introduction

Irradiation causes several systemic effects such as changes in gastrointestinal motility, excretion of water, blood pressure, and permeability. These responses can often also be elicited as abscopal effects, and may be modified by biogenic amines and hormones.

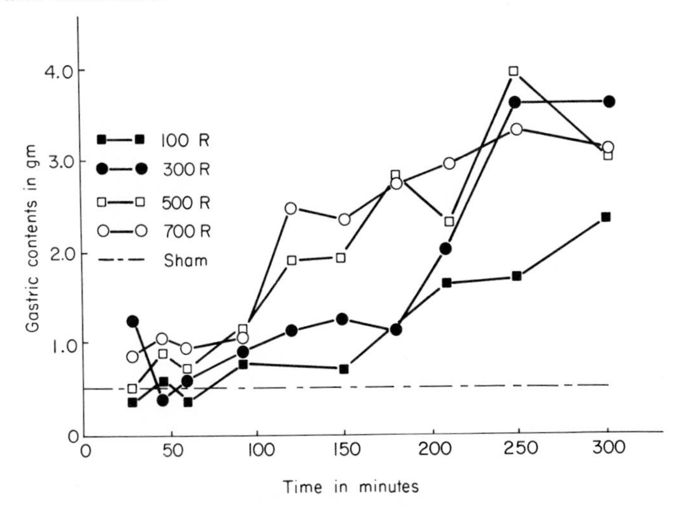

FIG. IX-15. Gastric content of rats at different times after whole-body irradiation with doses from 100 to 700 R. Note that the gastric content increases after irradiation as a function of dose and time. [From D. G. Baker and C. G. Hunter, *Radiation Res.* **9**, 660 (1958).] Recent experiments (M. B. Yatvin, and G. B. Gerber, *Intern. J. Radiation Biol.* show that the increase in gastric content is not the result of antiperistalsis but of desire for abnormal food (eating of woodshavings, feces, etc.).

2. Effects of Irradiation on Gastrointestinal Motility

If substantial parts of the body are irradiated, animals suffer from anorexia and experience a weight loss for several days, depending on the radiation dose. If radiation illness ensues, a second period of weight loss usually follows. The main causes of early anorexia are delay in gastric emptying, increase in stomach weight (*82*), and, in certain species, vomiting. The stomach loses its tonus within 30–60 min after exposure to 100–200 R, and this may persist for 1–2 days after higher doses of radiation (*11*, *36a*, *38*, *103*, *178*, *206*). If autopsied during this period, the stomach is atonic and distended with food, even though the animals have eaten little or nothing after irradiation. The weight of the gastric content can actually increase during the postirradiation period, since antiperistalsis in the intestine and regurgitation of intestinal contents into the stomach occur (*38*) (Fig. IX-15). Irradiated adrenalectomized rats also show a delay in gastric emptying (*81*), but it differs from that in nonadrenalectomized irradiated animals (*89*). Since delay in gastric emptying occurs only after abdominal irradiation (*83*), this might represent a direct effect of radiation due to release of biogenic amines from subcellular particles (*103*). Aversion to feeding is strongly correlated to the delay in gastric emptying but, in addition, may be elicited by head exposure (*82, 83a*). The cat, monkey, dog, and man can vomit, and vomiting occurs in these species following irradiation; emesis appears to result from stimulation of the brain stem, mediated via the vagus. The rat and mouse cannot vomit (*25a, 31a*). Intestinal motility in the rat is enhanced after local irradiation (*38*) but later diminishes. However, decreased motility in the rabbit early after local exposure has been observed (*119, 119a*). After whole-body irradiation, the tonus and the motility of the intestinal tract are abnormal and antiperistalsis is seen (*11, 100*). Total transit time in the intestine, however, is apparently unaltered (*89* but compare *38, 100*).

3. Polyuria after Irradiation

Most, but not all, animals excrete considerably more urine and drink more water during the initial period after a whole-body irradiation (1–2 days); not all individuals react in the same way (*2, 80, 133, 147, 200*, and many others) (Fig. VIII-2, p. 251), and the reaction depends on the age of the animals irradiated (*138*) and their ability to survive (*44a*). Polyuria is most conspicuous if free access to water is provided, but it is apparently not dependent on polydipsia (*147*). Polyuria is associated with stimulation of the hypothalamus and of the pituitary gland. No relationship exists between the histological appearance of neurosecretory granules and the incidence of polyuria (*2*) (see p. 291). Shielding of the liver or kidney (*92*), adrenalectomy (*50a, 97, 147*), hypophysectomy (*147*), thyroidectomy (*97*), or injection of pitressin (*200, 201*) abolish or diminish polyuria.

4. Changes in Blood Pressure and Permeability after Irradiation

A fall in blood pressure is an early result of therapeutic irradiation, especially after irradiation of the abdominal region. Irradiation of the total body or of the head of the rat also causes a drop in blood pressure, perhaps as a result of an increased level of histamine in the plasma. A substantial decline in aortic pressure is observed only after a whole-body exposure to 1000 R or more (*98*, *127*, *149*). A fall in blood pressure at the periphery (tail vessels), however, can occur after exposure to lower doses of irradiation (*149*) (Fig. IX-16).

The blood flow of the liver, spleen, and, most notably, skin, changes in rate and volume after local and whole-body irradiation. After local exposure, vessels in the dermis first constrict, then dilate. These phenomena are probably related to erythema occurring in the early postirradiation period and to altered permeability of the skin following exposure to radiation. Capillaries in irradiated skin, as well as in other tissues, are more permeable after exposure to

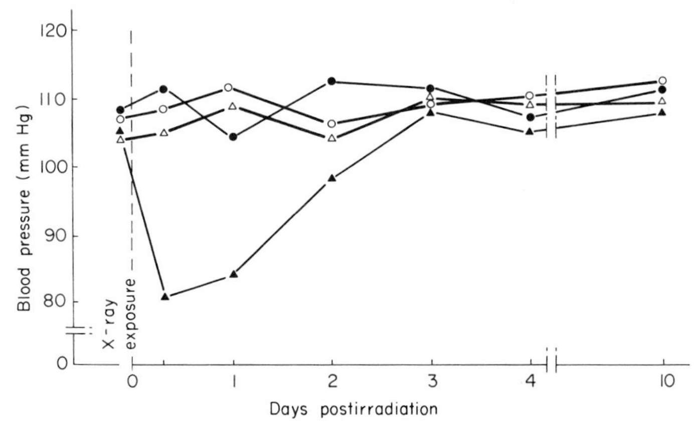

FIG. IX-16. Simultaneous measurements of aortic and tail arterial blood pressure during 10 days after x-irradiation. Aortic blood pressure, controls: open circles; aortic blood pressure, irradiated: solid circles; tail blood pressure, controls: open triangles; tail blood pressure, irradiated: solid triangles. Note that the marked fall in tail blood pressure is not matched by a similar decrease in aortic pressure. Peripheral resistance must therefore be diminished after irradiation. [From R. D. Phillips and D. J. Kimeldorf, *Radiation Res.* **18**, 86 (1963).]

doses of 100 R and more (*132*). Connective tissue offers less resistance to an injection of water after irradiation (*29*). Permeability to water increases immediately, even after irradiation with low doses (*186*), perhaps as a result of an immediate depolymerization of mucopolysaccharides (*30*). The transfer rate of labeled albumin, as well as the plasma volume of skin increases in a biphasic manner after local exposure to 3000 rep of β-rays (*168*) (Fig. IX-17). Changes in permeability also occur in other tissues, and may be detected by determining the uptake of labeled proteins, labeled phosphate, or various dyes. Changes in permeability are particularly marked in the intestinal tract, and these changes can be partially reversed by the administration of antihistamines (*202*) or monoamine oxidase inhibitors (*203*).

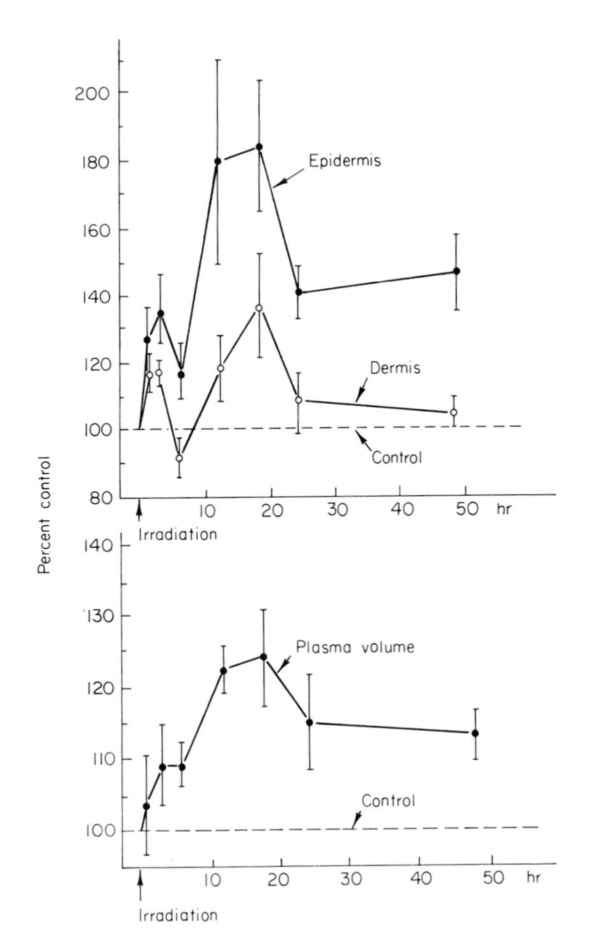

FIG. IX-17. Changes in intradermal plasma volume and in transfer rates of iodine-labeled plasma proteins into the extravascular space of the dermis and epidermis after a surface dose of 3000 rep β-rays. Mean $\pm$ standard error of the mean for 3 to 6 animals. Note that the transfer rate, especially into the epidermis, increases and the plasma volume expands immediately after exposure until the second day. [From C. W. Song, R. S. Anderson, and J. Tabachnick, *Radiation Res.* **27**, 604 (1966).]

I. Irradiation, Stress, and Systemic Responses

1. Introduction

The preceding paragraphs present many, often confusing, data. In order to understand these data, and to integrate them into an overall picture, one must consider the *modus operandi* by which hormonal and systemic responses modify

TABLE IX-5

EFFECTS OF HORMONES ON RADIOSENSITIVITY

Endocrine gland	Effect of removal	Restorative hormone therapy after gland removal	Protection in normal animals	Secretion of hormones after irradiation
Pituitary gland	Greater radiosensitivity; dose-dependent survival for doses producing gastrointestinal death	ACTH?	ACTH protects before irradiation sensitizes afterwards; STH protects (?); Pitressin protects	ACTH and FSH increase 1st day
Adrenal cortex	Greater radiosensitivity	Corticosterone (good); cortisone (moderate)	Bone marrow death: corticosterone protects, cortisone sensitizes Gastrointestinal death: no effect CNS death: corticosterone protects	Biphasic secretion in many species (a few hours after exposure and from day 2)
Adrenal medulla	Not known	Not known	Adrenaline and other biogenic amines protect	Increased secretion 1st day
Thyroid gland	Little or no effect	—	Thyroxine sensitizes, especially if combined with antithyroid drugs	Secretion increases 1st day after sublethal exposure, decreases after lethal exposure
Testis	Protects slightly	—	Testosterone given several days before exposure protects, given immediately before or after exposure sensitizes	No immediate change, later (21st day) decreased synthesis
Ovary	Sensitizes slightly	—	Estrogens given several days before exposure protect; immediately before or after exposure they sensitize	No immediate changes, later changes depend on morphological structure

and possibly determine the behavior of the irradiated organism. Yet, in view of the fact that certain reactions only occur under special circumstances, it is difficult to generalize, except by simply stating the obvious, namely, that radiation elicits systemic responses which somehow influence the behavior of the irradiated organism.

We propose, therefore, to select only the following few important topics for further discussion: (1) What are the effects of the removal of endocrine glands on biochemical responses, and how does the administration of hormone affect the irradiated organism? (2) What is the nature of and the most important mechanism underlying abscopal effects? (3) To what extent does the action of radiation resemble other agents producing stress? (4) To what extent do symptoms of irradiation depend on the capacity of the organism to react effectively to the insult of irradiation? (5) Are the responses of the body beneficial or detrimental to the irradiated organism?

2. Effects of Removal of Endocrine Glands and Action of Hormones on Irradiated Individuals

TABLE IX-6

Selected Data on the Influence of Removal of Endocrine Glands on Radiation-Induced Biochemical Changes

Gland removed	Effect abolished in part or in full	Effect not altered or doubtful alterations
Pituitary	Polyuria and polydipsia (*147*), glycemia (*94*), deposition of hepatic glycogen (*121*, *134*), decrease in adrenal cholesterol and ascorbic acid (*134*) [perhaps only in part (*145*)], changes in hepatic 6-phosphogluconate dehydrogenase and phosphohexose isomerase (*195*), increased incorporation of acetate into liver lipids (in part) (*69*, *91*)	Change in hepatic glucokinese (*26*)
Adrenals	Polyuria and polydipsia (*147*), early effects on intestinal absorption of Na^+ and glucose (*77*), changes in glucose and hepatic glycogen metabolism (*121*), increased synthesis of hepatic lipids (*91*), radiation-induced changes in DNA metabolism mitigated (*78*, *164*) (Table IX-7)	Delayed gastric emptying (*81*) (effect found by *89*), abscopal effect on lymphatic tissue, creatinuria (*96*) (effect found by *97*), electrolyte excretion (*148*), decrease in blood pressure (*127*), changes in permeability (*203*), changes in thyroid activity (*58*)
Thyroid	Changes in hepatic cholesterol synthesis (in part) (*181*), delays onset of creatinuria (*97*, *99*)	

Information on the influence of removal of endocrine glands and treatment with hormones on the survival of irradiated animals is presented in Table IX-5. In view of the large number of papers on this subject, no attempt has been made to document these data here. Selected information on the influence of the removal of endocrine glands on the biochemical effects of irradiation is summarized in Table IX-6 (data included in Table IX-7 show the influence of the adrenal on DNA synthesis in irradiated animals).

TABLE IX-7

INCORPORATION OF ^{32}P INTO DUODENAL DNA AT DIFFERENT TIMES AFTER 600 R WHOLE-BODY IRRADIATION (PERCENT OF UNIRRADIATED UNTREATED CONTROLS)[a]

Subjects	Before irradiation	Time after irradiation or treatment			
		2 hr	4 hr	12 hr	24 hr
X-irradiated mice, sham operated	100.0 ±10.8	33.8 ±3.3	37.5 ±4.6	22.7 ±2.9	55.2 ±4.4
X-irradiated mice, adrenalectomized	84.3 ±8.6	54.8 ±5.2	59.3 ±4.2	44.7 ±6.3	44.0 ±4.1
X-irradiated mice + deoxycorticosterone (1 mg)	100.0 ±6.7	54.8 ±5.3	52.3 ±2.7	—	60.1 ±5.0
Normal mice + cortisone (2.5 mg)	100.0 ±6.9	90.3 ±5.2	76.4 ±8.7	—	87.7 ±7.3
X-irradiated mice + cortisone (2.5 mg)	100.0 ±6.9	20.0 ±5.3	21.1 ±2.5	—	23.4 ±3.7

[a] Note that the decrease in incorporation after exposure is less marked after adrenalectomy but is intensified if cortisone, but not deoxycorticosterone, is administered. [From M. Skalka and J. Soska, *Compt. Rend. Soc. Biol.* **157**, 455 (1963).]

Studies of irradiated animals from which an endocrine organ has been removed are of great value in elucidating the nature of the hormone response, but it should be noted that such animals are in a nonphysiological and often debilitated state. Under such conditions, reactions normally not regulated by the excised gland may also be affected by its removal. One must, therefore, guard against drawing far-reaching conclusions solely on the basis of experiments on animals from which the adrenal or pituitary glands have been removed. Furthermore, it should be recalled that supernumerary adrenal tissue is often present in rodents and higher vertebrates, including man, and may be stimulated to grow by ACTH. No studies on adrenalectomized rats are, therefore, complete, unless the absence of such adrenal tissue has been verified microscopically.

3. Abscopal Effects of Irradiation

Many biochemical changes in irradiated animals occur in parts of the body not directly exposed to the radiation beam (abscopal). [Formerly, the term "indirect effect" was often used to refer to changes occurring in nonirradiated parts of the body, but this term should now be reserved for effects transmitted by short-lived substances capable of causing chemical reactions close to the original site of absorption of radiation (Vol. I, p. 17).] We speak of a "systemic" reaction when we describe effects elicited by neural or humoral transmitters (including toxic materials); such reactions can take place inside as well as outside (abscopal) the irradiated tissue mass. Abscopal effects were

TABLE IX-8

Possible Mechanisms Underlying Abscopal Effects of Irradiation

Mechanism mediated by	Examples (where known)
Hypothalamus–pituitary system	Polyuria (?); vomiting; increased activity of splenic RNase (see p. 71); inhibition of development of liver enzymes (*128a*)
Adrenals (and pituitary gland)	Changes in hepatic glycogen (p. 125)
Thyroid (and pituitary gland)	Changes in basal metabolism (p. 305)
Liberation of biogenic amines	Fall in blood pressure (?); delayed gastric emptying (?)
Nervous regulation	Ca uptake by the bone (p. 201)
Toxic factors	Lipid peroxides, O-chinones (p. 144)
Infection	—
Partial starvation	Decrease in thymus weight
Unknown factors	Decreased uptake of creatine by muscle

recognized early in the history of radiation biology, but opinions of their importance have varied. Some have denied that abscopal effects are important, whereas others have used them to explain almost all actions of irradiation. Several different mechanisms of action may be responsible for abscopal effects. The most important ones together with a few pertinent examples are summarized in Table IX-8.

In searching for the causal factors leading to abscopal effects, one should not overlook the possible role of such factors as partial starvation after irradiation. Thus, damage to the thymus after abdominal irradiation has long been cited as a classic example of an abscopal effect, mediated via the adrenals (*109*). Recent careful studies have, however, cast doubt on these conclusions, and suggest that partial starvation, rather than mechanisms involving the adrenal glands, is responsible for the abscopal effect in the thymus (*40*, *108*) (Table IX-9) as well as in other organs (see p. 216).

TABLE IX-9

THYMUS WEIGHT AND DNA CONTENT 48 HR AFTER IRRADIATION OF NORMAL AND ADRENALECTOMIZED RATS[a]

Area of body irradiated	Dose	Postirradiated regime	Thymus weight (%)	Thymus DNA phosphorus (%)
Cranial	50 R	Starved	68	60
		Fed	71	71
Cranial	100 R	Starved	43	38
		Fed	45	34
Whole body	50 R	Fed	66	67
	100 R	Fed	41	36
Caudal	5000 R	Starved, sham operated	57 ± 4	68 ± 2
		Starved, adrenalectomized	58 ± 2	72 ± 4
Caudal	5000 R	Fed, sham operated	50 ± 5	50 ± 5
		Fed, adrenalectomized	61 ± 8	62 ± 15

[a] All data are expressed in percent of respective litter-mate controls. For presentation we have combined the different groups in Table IV of M. J. Corp, A. W. Law, and R. H. Mole, *Am. J. Roentgenol. Radium Therapy Nucl. Med.* **85**, 550 (1961). Note that exposure of the caudal portion is much less effective than that of the total body or the cranial portion. Adrenalectomy has no significant influence on the abscopal effect (*40*, *108*).

4. SIMILARITY BETWEEN RESPONSE TO IRRADIATION AND TO OTHER STRESS-PRODUCING AGENTS

Certain postirradiation symptoms have much in common with the general phenomena of stress. In discriminating between specific and nonspecific effects of irradiation, one must, of course, take into consideration that sham-irradiation alone causes a stress response in animals due to the handling and restraint involved in this procedure. Thus, a fall in pituitary ACTH and adrenal ascorbic acid (*24*) (see Fig. IX-2), a release of leucocytes (*64*), a mild degree of creatinuria (p. 260), and many other reactions take place after sham irradiation, though to a milder degree than after x-irradiation.

Aside from the handling of animals, starvation is an important factor in modifying the biochemical response of the irradiated animal. Fasting alone causes the weight of the adrenals to increase to almost the same extent as fasting combined with irradiation (*135*, *136*) (Fig. IX-18). On the other hand, fasting

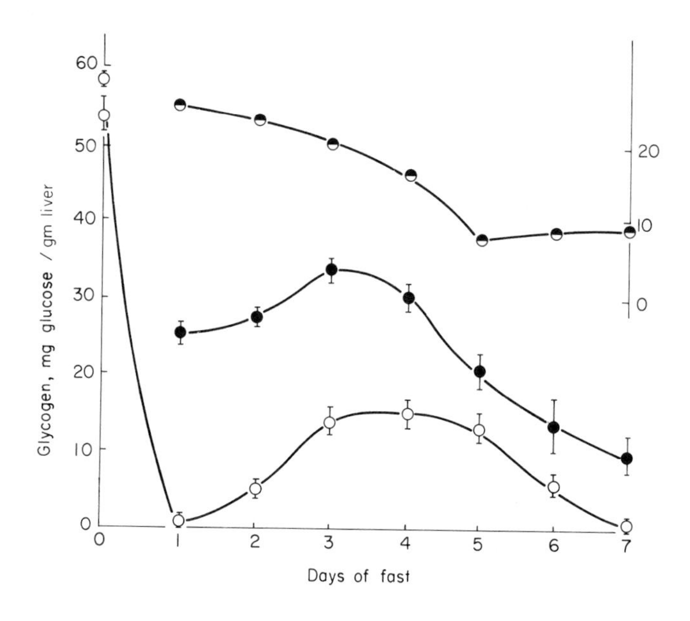

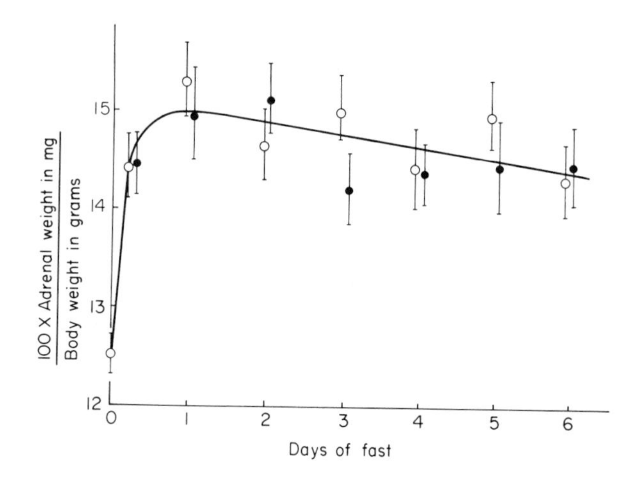

FIG. IX-18. Adrenal weight (lower figure) and glycogen in liver (upper figure) as a function of the duration of fasting. Fasting rats: open circles; fasting rats that received 2000 R of whole-body irradiation 5–6 hr before sampling: solid circles; difference between glycogen levels in fasted and in fasted x-irradiated rats: half-solid circles. Note that x-irradiation does not change the increase in adrenal weight of fasted rats. Hepatic glycogen, which diminishes during fasting, increases if the rats are x-irradiated. [From L. F. Nims and J. L. Geisselsoder, *Radiation Res.* **5**, 58 (1956).]

does not appreciably alter the decrease in adrenal cholesterol and ascorbic acid (*135*) or the radiosensitivity (*46a*) after irradiation. Hepatic glycogen diminishes markedly during fasting, but does so to a lesser extent if the fasted animals have been irradiated (see p. 122) (Fig. IX-18). The changes in various hepatic enzymes involved in carbohydrate metabolism suggest that irradiation prevents the normal adaptation of the organism to the conditions of starvation

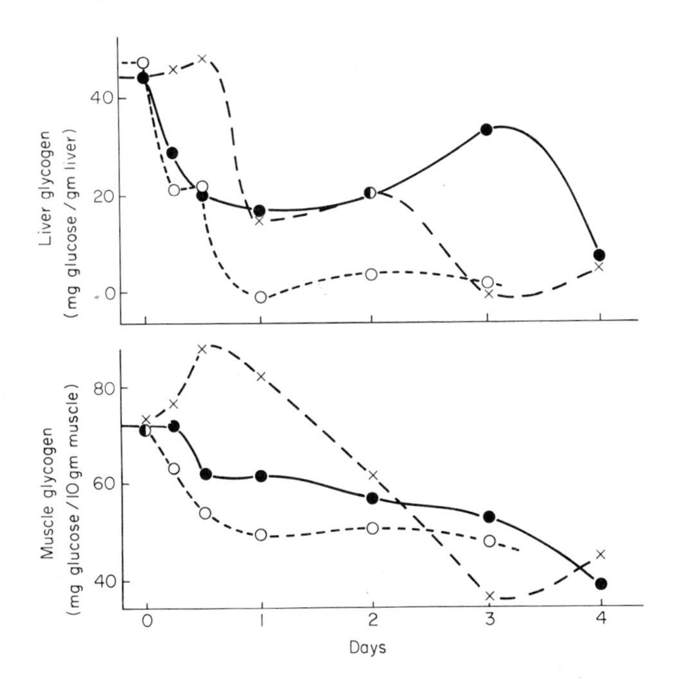

FIG. IX-19. Changes with time in liver and muscle glycogen in starved rats after whole-body irradiation (1000 R) (solid circles), during exposure to a cold environment (between 2°–5°C) (open circles), or to hypoxia equivalent to that produced by an altitude of 20,000 feet (crosses). Note that each type of stress causes a characteristic response in the glycogen level of liver or muscle. [From L. F. Nims and J. L. Geisselsoder, *Radiation Res.* **12**, 251 (1960).]

(see p. 140). These observations may help to explain why the response of muscle and liver glycogen to hypoxia or cold differs from that to irradiation (Fig. IX-19). These examples may suffice to illustrate how difficult it is to separate certain systemic responses due to stress and starvation from the effects of irradiation. Similar conditions probably prevail with respect to other metabolic reactions, most notably those related to lipid and amino acid metabolism; these reactions are also affected by radiation-induced starvation, but have not yet been studied in a more detailed fashion.

5. Capacity of the Organism to React to Radiation Injury

The ability of the organism to react to radiation injury plays an important role in the outcome of radiation illness, as has been shown in previous sections, e.g., the dependence of radiosensitivity on the activity of the thyroid and adrenals. In general, it appears that animals whose endocrine responses are not excessive also have the best chances for survival after irradiation. Thus, if rats are tested as to their electrolyte excretion in response to a starvation stress

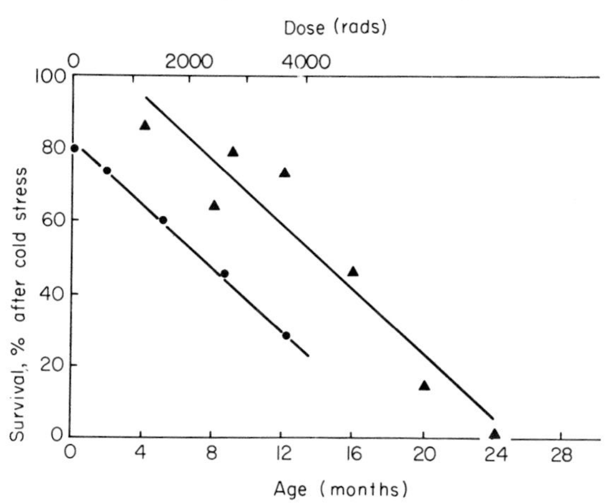

Fig. IX-20. Comparison between the effects of protracted γ-irradiation (circles) (2.2 rads/day) and age (triangles) on ability of RF female mice to survive a cold stress at 6°–7°C for 14 days. Note that after γ-irradiation, and with advancing age, the sensitivity to stress increases. [From T. T. Trujillo, J. F. Spalding, and W. H. Langham, *Radiation Res.* **16**, 144 (1962).]

and then irradiated, it is observed that animals which excreted only a moderate excess of electrolytes are the ones most likely to recover from radiation (*150*). Furthermore, animals subjected to additional stress after exposure, such as strenuous exercise (*92b*) or a combination of irradiation and skin injury (*123a*), suffer a reduction in survival. The ability to react to injury also may be impaired by irradiation. To mention one example, rats exposed to protracted irradiation display a diminished ability to survive a prolonged cold stress, just as they would with advancing age (*182*). On this basis, one can calculate that the relative increase in age due to 1 rad of radiation exposure is equivalent to 0.093 days (Fig. IX-20).

6. Are the Systemic Responses of the Body Beneficial or Harmful to the Irradiated Organism?

It has been argued that the adrenal response and the liberation of biogenic amines may actually harm the individuals and impair their recovery after irradiation. Most of the experiments and arguments which appear to favor such an interpretation have already been mentioned elsewhere and need only be recapitulated (*22*). They are: (1) large doses of glucocorticoids can intensify the effects of irradiation on cell renewal and DNA synthesis; (2) removal of adrenals, although deleterious to survival, promotes the recovery of the bone marrow; (3) adrenal (glucocorticoids), thyroid, pituitary, and sex hormones, if given at a certain time with respect to exposure, can harm the irradiated animal, whereas when given at another time may protect it (see also Table IX-5); (4) similarities exist between the toxic effects of histamine and certain sequelae of irradiation [Ellinger's hypothesis (*53*)].

On the other hand, many observations also indicate that the systemic response is necessary for survival: (a) the increased radiosensitivity after removal of the adrenals or the pituitary gland; (b) the protection afforded by release of biogenic amines. Reviewing these arguments, it appears to us that an overwhelming number of systemic factors tend to be beneficial to the irradiated organism and harm only under specific circumstances.

Finally, we must consider whether the terminal phase of radiation disease constitutes an "adaptation syndrome," in the sense that Selye thought of it, with death intervening as a result of the exhaustion and failure of the adrenals. Although it is likely that irradiated animals "need" more adrenal hormone and depend on the adrenals far more during the postirradiation period than do normal animals, almost all biochemical data support the conclusion that adrenal activity is relatively high before death and that a "dysgenesis" of hormones (i.e., abnormal synthesis of hormones) does not occur. Indeed, infection, hemorrhage, and fluid loss due to cell depletion of bone marrow and intestine are the immediate causes of death.

REFERENCES

1. Abbott, C. R., *J. Endocrinol.* **19**, 33 (1959).
2. Afifi, A. K., Osborne, J. W., and Mitchell, C. T., *Radiation Res.* **23**, 1 (1964).
3. Albers-Schönberg, H. E., *Münch, Med. Wochschr.* **50**, 1859 (1903).
4. Alekseeva, A. M., and Timofeeva, N. M., *Med. Radiol.* **6**, 54 (1961).
5. Anbar, M., and Inbar, M., *Nature* **197**, 302 (1963).
6. Bacq, Z. M., and Fischer, P., *Radiation Res.* **7**, 365 (1957).
7. Bacq, Z. M., Martinovitch, P., Fischer, P., Pavlovitch, M., Sladitch, D. S., and Radivojevitch, D. V., *Radiation Res.* **7**, 373 (1957).
8. Bacq, Z. M., Smelik, P. G., Goutier-Pirotte, M., and Renson, J., *Brit. J. Radiol.* **33**, 618 (1960).
9. Bacq, Z. M., "Chemical Protection Against Ionizing Radiation." Thomas, Springfield, Illinois, 1965.

10. Bagramyan, E. R., *Vopr. Endokrinol. i. Gormonoterap.* **6**, 27 (1960); *Vopr. Med. Khim.* **10**, 265 (1964).
11. Baker, D. G., and Hunter, C. G., *Radiation Res.* **9**, 660 (1958).
12. Baker, D. G., Hunter, C. G., and Schönbaum, E., *Radiation Res.* **12**, 409 (1960).
13. Baker, D. G., and Hunter, C. G., *Radiation Res.* **15**, 573 (1961).
14. Beaumariage, M. L., and Lecomte, J., *Compt. Rend. Soc. Biol.* **151**, 1976 (1957).
15. Bender, A. E., *Brit. J. Radiol.* **21**, 244 (1948).
16. Bergonie, J., Tribondeau, L., and Recamier, D., *Compt. Rend. Soc. Biol.* **57**, 284 (1905).
17. Berliner, D. L., Ellis, L. C., and Taylor, G. N., *Radiation Res.* **22**, 345 (1964).
18. Berliner, D. L., and Ellis, L. C., *Radiation Res.* **24**, 368 (1965).
19. Berliner, D. L., and Ellis, L. C., *Radiation Res.* **24**, 572 (1965).
20. Bettendorf, G., Moje, A. P., and Schmerund, H. J., *Strahlentherapie* **117**, 370 (1962).
21. Betz, H., *Ann. Endocrinol.* (*Paris*) **15**, 391 (1954).
22. Betz, E. H., "Contribution à l'étude du syndrome endocrinien provoqué par l'irradiation totale de l'organisme." Masson, Paris, 1956.
22a. Bezdrobnyi, Yu. V., and Grodzenskii, D. E., *Byull. Eksperim. Biol. Med.* **63**, 49 (1967) and **64**, 49 (1967).
23. Binhammer, R. T., and Herman, I., *Radiation Res.* **18**, 277 (1963).
24. Binhammer, R. T., and Crocker, J. R., *Radiation Res.* **18**, 429 (1963).
24a. Binhammer, R. T., *Radiation Res.* **30**, 676 (1967).
25. Birke, G., Franksson, C., Hultborn, K. A., and Plantin, L. O., *Acta Chir. Scand.* **110**, 469 (1956).
25a. Borison, H. L., *J. Comp. Neurol.* **107**, 439 (1957).
26. Borrebaek, B., Abraham, S., and Chaikoff, I. L., *Biochim. Biophys. Acta* **90**, 451 (1964).
27. Braun, H., and Kuschke, H. J., *Fortschr. Gebiete Roentgenstrahlen Nuklearmed.* **94**, 827 (1961).
28. Braun, H., and Kuschke, H. J., *Fortschr. Gebiete Roentgenstrahlen Nuklearmed.* **96**, 683 (1962).
29. Brinkman, R., Lamberts, H. B., Wadel, J., and Zuideveld, J., *Intern. J. Radiation Biol.* **3**, 205 (1961).
30. Brinkman, R., Lamberts, H. B., and Zuideveld, J., *Intern. J. Radiation Biol.* **3**, 509 (1961).
31. Brinkman, R., and Veninga, T. S., *Intern. J. Radiation Biol.* **4**, 249 (1962).
31a. Brizzee, K. R., Calton, F. M., and Vitale, D. E., *Anat. Record* **130**, 533 (1958).
31b. Burstein, S., and Pincus, G., *Endocrinology* **80**, 947 (1967).
32. Carlson, W. D., and Gassner, F. X., "Effects of Radiation on the Reproductive System." Pergamon Press, Oxford, 1964.
33. Cession-Fossion, A., Beaumariage, M. L., and Vandermeulen, R., *Intern. J. Radiation Biol.* **10**, 277 (1966).
34. Chernov, G. A., and Raushenbakh, M. O., *Probl. Hematol. Blood Transfusion* (*USSR*) (*English Transl.*) **5**, 577 (1960).
35. Chernov, G. A., and Morozoskaya, L. M., *Med. Radiol.* **6**, 10, 59 (1961).
36. Chernov, G. A., Sheremet, Z. I., and Lenskaya, R. V., *Med. Radiol.* **9**, 12, 58 (1964).
36a. Chmelar, V., Grossmann, V., Hais, I. A., and Deml, F., *Radiation Res.* **37**, 627 (1969).
37. Comsa, J., *Strahlentherapie* **126**, 366 and 541 (1965).
37a. Comsa, J., *Ann. Endocrinol.* **13**, 931 (1952).

38. Conard, R. A., *Radiation Res.* **5**, 167 (1956).
39. Coniglio, J. G., Culp, F. B., Davis, J., Ford, W., and Windler, F., *Radiation Res.* **20**, 372 (1963).
40. Corp, M. J., Law, A. W., and Mole, R. H., *Am. J. Roentgenol. Radium Therapy Nucl. Med.* **85**, 550 (1961).
41. David, G., Faradi, L., and Tanka, D., *Radiobiol. Radiotherap.* **3**, 91 (1962).
42. Deanović, Ž., Supek, Z., and Randić, M., *Intern. J. Radiation Biol.* **7**, 1 (1963).
43. Deanović, Z., and Supek, Z., *Intern. J. Radiation Biol.* **7**, 569 (1963).
44. Deanović, Ž., and Supek, Z., *Intern. J. Radiation Biol.* **10**, 199 (1966).
44a. Deanović, Ž., and Supek, Z., *Strahlentherapie* **137**, 375 (1969).
45. Denniston, R. H., *J. Exptl. Zool.* **91**, 237 (1942).
46. Diczfalusy, E., Notter, G., Edsmyr, F., and Westman, A., *J. Clin. Endocrinol. Metab.* **19**, 1230 (1959).
46a. Dragic, M., Ninkov, V., Hajdukovic, S., and Karanovic, J., *Strahlentherapie* **136**, 617 (1968).
47. Duchesne, P. Y., and Beaumariage, M. L., *Bull. Acad. Roy. Med. Belg.* [7] **11**, 1340 (1965).
48. Eberhagen, D., and Zöllner, N., *Z. Ges. Exptl. Med.* **138**, 581 (1965).
49. Eberhagen, D., and Plette, G., *Z. Physiol. Chem.* **343**, 17 (1965).
50. Eccles, S. S., Oakey, R. E., and Stitch, S. R., *Biochem. J.* **85**, 38p (1962).
50a. Eddy, H. A., and Casarett, G. W., *Radiation Res.* **37**, 287 (1969).
51. Eechaute, W., Demeester, G., and Lensen, I., *Arch. Intern. Pharmacodyn.* **135**, 235 (1962).
51a. Egana, E., and Velarde, M. I., *Experientia* **23**, 526 (1967).
52. Eisen, V. D., Ellis, R. E., and Wilson, C. W. M., *J. Physiol.* (*London*) **133**, 506 (1956).
53. Ellinger, F., *Schweiz. Med. Wochschr.* **81**, 55 (1951).
54. Ellis, L. C., and Berliner, D. L., *Radiation Res.* **20**, 549 (1963).
55. Ellis, L. C., and Berliner, D. L., *Radiation Res.* **23**, 156 (1964); **32**, 520 (1967).
56. Ershoff, B. H., and Gal, E. M., *Proc. Soc. Exptl. Biol. Med.* **108**, 160 (1961).
57. Ershoff, B. H., Hellmers, R., and Wells, A. F., *Proc. Soc. Exptl. Biol. Med.* **110**, 536 (1962).
58. Evans, T. C., Clarke, G., and Sobel, E., *Anat. Record* **99**, 577 (1947).
58a. Fedorova, A. V., and Mirzoyan, A. A., *Vopr. Radiobiol. Akad. Nauk Arm. SSR Sektor Radiol. S.B.* **3–4**, 127 (1963).
59. Fedotov, V. P., and Rynova, N. N., *Byull. Eksperim. Biol. Med.* **61**, 150 (1966).
60. Feller, D. D., Chaikoff, I. L., Taurog, A., and Jones, H. B., *Endocrinology* **45**, 464 (1949).
61. Flemming, K., and Yegin, M., *Z. Naturforsch.* **19b**, 745 (1964).
62. Flemming, K., *Naturwissenschaften* **52**, 515 (1965); *Strahlentherapie* **134**, 565 (1967) and **137**, 62 (1969).
62a. Flemming, K., and Geierhaas, B., *Intern. J. Radiation Biol.* **13**, 13 (1967).
62b. Flemming, K., Hemsing, W., and Geierhaas, B., *Z. Naturforsch.* **22b**, 85 (1967).
62c. Flemming, K., Geierhaas, B., Hemsing, W., and Mannigel, U., *Intern. J. Radiation Biol.* **14**, 93 (1968).
63. Fochem, K., Formanek, K., and Smetana, W., *Wien. Med. Wochschr.* **106**, 543 (1956).
64. Forssberg, A., Tribukait, B., and Vikterlöf, K. J., *Acta Physiol. Scand.* **52**, 1 (1961).
65. Franzen, F., Gross, H., and Thielicke, G., *Strahlentherapie* **120**, 598 (1963).

66. Franzen, F., Gross, H., and Friedrich, G., *Strahlentherapie* **122**, 591 (1963) and **124**, 270 (1969).
67. French, A. B., Migeon, C. J., Samuels, L. T., and Bowers, J. Z., *Am. J. Physiol.* **182**, 469 (1955).
68. Gerber, G. B., Unpublished data (1955).
68a. Gibson, J. M., and Doniach, I., *Brit. J. Cancer* **21**, 519 (1963).
69. Gould, R. G., Bell, V. L., and Lilly, E. H., *Radiation Res.* **5**, 609 (1956).
70. Gould, R. G., Bell, V. L., Lilly, E. H., Keegan, P., van Riper, J., and Jonnard, M. L., *Am. J. Physiol.* **196**, 1231 (1959).
71. Griffith, J. Q., and Griffith, M. G., *J. Albert Einstein Med. Center* **11**, 254 (1963).
72. Griffith, M. G., Griffith, J. Q., Hermel, M. B., and Gershon-Cohen, J., *Radiology* **77**, 91 (1961).
72a. Hahn, E. W., *Radiation Res.* **31**, 846 (1967).
72b. Hajduković, S., and Duchesne, P., *Compt. Rend. Soc. Biol.* **160**, 1971 (1966).
73. Halberstaedter, L., *Berlin. Klin. Wochschr.* **42**, 64 (1905).
74. Haley, T. J., Komesu, N., and Brown, D. G., *Radiation Res.* **27**, 566 (1966).
75. Hameed, J. M. A., and Haley, T. J., *Radiation Res.* **23**, 620 (1964).
76. Heckmann, U., Kleinert, H., and Künkel, H. A., *Naturwissenschaften* **50**, 732 (1963).
77. Hehl, R., and Wilbrandt, W., *Helv. Physiol. Acta* **22**, 319 (1964).
78. Hemingway, J. T., *Experientia* **15**, 189 (1959).
79. Hochman, A., and Block-Frankenthal, L., *Brit. J. Radiol.* **26**, 599 (1953).
80. Holoček, V., Petrášek, J., Schreiber, V., Dienstbier, Z., and Kmentová, V., *Physiol. Bohemoslov.* **11**, 119 (1962).
81. Hulse, E. V., *Brit. J. Exptl. Pathol.* **38**, 498 (1957).
82. Hulse, E. V., and Dempsey, B. C., *Intern. J. Radiation Biol.* **8**, 97 (1964).
83. Hulse, E. V., *Intern. J. Radiation Biol.* **10**, 521 (1966).
83a. Hulse, E. V., and Mizon, L. G., *Intern. J. Radiation Biol.* **12**, 515 (1967).
84. Ito, M., *Radiation Res.* **28**, 266 (1966).
85. Ivanenko, T. J., *Radiobiologiya* **5**, 338 (1965).
86. Jolly, J., and Ferroux, R., *Compt. Rend. Soc. Biol.* **92**, 125 (1925).
87. Jonek, J., Grzybek, H., and Jonek, T., *Protoplasma* **58**, 391 (1964).
88. Jonek, J., Stoklosa, E., and Konecki, J., *Strahlentherapie* **123**, 226 (1964).
89. Jones, D. C., and Kimeldorf, D. J., *Radiation Res.* **11**, 832 (1959).
90. Juret, P., Henry, R., Lalanne, C. M., Hourtoulle, F. G., and Fermanian, J., *Rev. Franc. Etudes Clin. Biol.* **11**, 176 (1966).
91. Kay, R. E., Warner, W. L., and Entenman, C., *Radiation Res.* **9**, 137 (1958).
92. Kay, R. E., and Entenman, C., *Am. J. Physiol.* **197**, 169 (1959).
92a. Kharitonova, L. V., *Dokl. Akad. Nauk SSSR* **176**, 946 (1967).
92b. Kimeldorf, D. J., and Baum, S. J., *Growth* **18**, 79 (1954).
93. Kobayashi, Y., *Radiation Res.* **24**, 503 (1965).
94. Kohn, H. I., Swingley, N., Robertson, W. J., and Kirslis, M., *Am. J. Physiol.* **165**, 43 (1951).
95. Koslowski, L., Poppe, H., and Walther, E., *Strahlentherapie* **97**, 266 (1955).
96. Koszalka, T. R., Kurohara, S. S., Gerber, G. B., and Hempelmann, L. H., *Radiation Res.* **20**, 309 (1963).
97. Krise, G. M., and Williams, C. M., *Am. J. Physiol.* **196**, 1352 (1959).
97a. Kulinskii, V. I., and Semenov, L. F., *Radiobiologiya* **5**, 494 (1965).
98. Kundel, H. L., *Radiation Res.* **27**, 406 (1966).
99. Kurohara, S. S., and Altman, K. I., *Radiation Res.* **21**, 473 (1964).

100. Kurtsin, I. T., "Effects of Ionizing Radiation on the Digestive System." Elsevier, Amsterdam, 1963.
101. Laborde, M., Commanay, L., Meyniel, G., and Blanquet, P., *Compt. Rend. Soc. Biol.* **160**, 321 (1966).
102. Lacassagne, A., Duplan, J. F., Marcovich, H., and Raynaud, A., *in* "The Ovary" (S. Zuckerman, ed.), p. 463, Academic Press, New York, 1962.
103. Lamberts, H. B., and Dijken, G. B., *Intern. J. Radiation Biol.* **4**, 43 (1962).
104. Lane, J. L., Wilding, J. L., Rust, J. H., Trum, B. F., and Schoolar, J. C., *Radiation Res.* **2**, 64 (1955).
105. Langendorff, H., Scheuerbrandt, G., Streffer, C., and Melching, H. J., *Naturwissenschaften* **51**, 290 (1964).
106. Larina, M. A., and Sakhatskaya, T. S., *Probl. J. Endokrinol. i Gormonoterap.* **4**, 21 (1958).
107. Larina, M. A., *Probl. Endokrinol i Gormonoterap.* **6**, 18 (1960).
108. Law, A. W., and Mole, R. H., *Intern. J. Radiation Biol.* **3**, 233 (1961).
109. Leblond, C. P., and Segal, G., *Am. J. Roentgenol Radium Therapy* **47**, 302 (1942).
110. Leitch, J. L., Debley, V. G., and Haley, T. J., *Am. J. Physiol.* **187**, 307 (1956).
111. Lemaire, M., and Colson, J., *Compt. Rend. Soc. Biol.* **157**, 1513 (1963).
112. Lemon, H. M., Wotiz, H. H., Sommers, S. C., Parson, L., and Davis, J. W., *Proc. 2nd Intern. Symp. Mammary Cancer, Univ. Perugia, 1957* (L. Severi, ed.) p. 111. Div. Cancer Res., Perugia, 1958.
113. Leusen, J., Claeys, E., and de Schrijver, A., *Verhandel. Koninkl. Vlaam. Acad. Geneeskunde Belg.* **20**, 240 (1958).
114. Lott, J. R., and Pryor, N. W., *Radiation Res.* **11**, 452 (1959).
115. Lott, J. R., and Hines, J. F., *Radiation Res.* **25**, 213 (1965); *Texas J. Sci.* **20**, 91 (1968).
116. Martinovitch, P., Pavitch, D., Pavlovitch-Hournac, M., and Zivkovitch, N., *Intern. J. Radiation Biol.* **5**, 255 (1962).
117. Mateyko, G. M., and Edelmann, A., *Radiation Res.* **1**, 470 (1954).
118. Matsuoka, O., Tsuchiya, T., and Furukawa, Y., *J. Radiation Res.* (*Tokyo*) **3**, 104 (1962).
119. Matsuoka, O., Tsuchiya, T., and Eto, H., *Nippon Acta Radiol.* **24**, 157 (1964).
119a. Matsuoka, O., Tsuchiya, T., and Eto, H., *Nippon Acta Radiol.* **24**, 157 and 910 (1964).
119b. Matsuoka, O., Tsuchiya, T., Kashima, M., and Eto, H., *Proc. Symp. Gastrointestinal Radiation Injury*, Richland (M. F. Sullivan, ed.) p. 310. Excerpta. Med. Found., Amsterdam, 1968.
120. McGoodall, C., and Long, M., *Am. J. Physiol.* **197**, 1265 (1959).
120a. McGoodall, C., *Health Phys.* **14**, 199 (1968).
121. McKee, R. W., and Brin, M., *Arch. Biochem. Biophys.* **61**, 390 (1956).
122. Melching, H. J., Ernst, H., and Rösler, H., *Strahlentherapie* **113**, 394 (1960).
123. Melching, H. J., "Current Topics in Radiation Research," Vol. 1, p. 93. North-Holland Publ., Amsterdam, 1965.
123a. Messerschmidt, O., *Strahlentherapie* **131**, 298 (1966).
124. Mole, R. H., and Papworth, D. G., *Intern. J. Radiation Biol.* **10**, 609 (1966).
125. Momma, H., Nakamura, W., and Eto, H., *Compt. Rend. Soc. Biol.* **159**, 2094 (1965).
126. Monroe, R. A., Rust, J. H., and Trum, B. F., *Science* **119**, 65 (1954).
127. Montgomery, P. O'B., and Warren, S., *Proc. Soc. Exptl. Biol. Med.* **77**, 803 (1951).

128. Nair, V., *Nature* **208**, 1293 (1965).
128a. Nair, V., and Bau, D., *Proc. Soc. Exptl. Biol. Med.* **126**, 853 (1967).
129. Nakamura, W., Momma, H., and Eto, H., *Compt. Rend. Soc. Biol.* **159**, 1629 (1965).
130. Nakamura, W., Shibata, K., Momma, H., and Eto, H., *Compt. Rend. Soc. Biol.* **158**, 1426 (1964).
131. Natarajan, S., and Dykes, P. W., *Intern. J. Radiation Biol.* **7**, 315 (1963).
132. Neumayr, A., and Thurnher, B., *Strahlentherapie* **84**, 297 (1951); **86**, 207 (1952).
133. Nims, L. F., and Sutton, E., *Am. J. Physiol.* **171**, 17 (1952).
134. Nims, L. F., and Sutton, E., *Am. J. Physiol.* **177**, 51 (1954).
135. Nims, L. F., and Geisselsoder, J. L., *Radiation Res.* **5**, 58 (1956).
136. Nims, L. F., and Geisselsoder, J. L., *Radiation Res.* **12**, 251 (1960).
137. Nizet, A., Heusghem, C., and Herve, A., *Compt. Rend. Soc. Biol.* **143**, 876 (1949).
138. Osborn, G. K., Jones, D. C., and Kimeldorf, D. J., *Radiation Res.* **23**, 119 (1964).
139. Palaić, D. J., Randić, M., and Supek, Z., *Intern. J. Radiation Biol.* **6**, 241 (1963).
139a. Palaić, D. J., Deanović, Z., and Supek, Z., *J. Neurochem.* **11**, 761 (1964).
140. Palaić, D., and Supek, Z., *J. Neurochem.* **12**, 329 (1965).
141. Palaić, D., and Supek, Z., *Intern. J. Radiation Biol.* **9**, 601 (1965).
142. Palaić, D., and Supek, Z., *Nature* **210**, 970 (1966).
143. Palaić, D., and Supek, Z., *J. Neurochem* **13**, 705 (1966).
144. Parin, V. V., Antipov, V. V., Raushenbakh, M. O., Saksonov, P. P., Shashkov, V. S., and Chernov, G. A., *Izv. Akad. Nauk SSSR, Ser. Biol.* **1**, 3 (1965).
145. Patalova, N. V., *Bull. Exptl. Biol. Med.* (*USSR*) (*English Transl.*) **62**, 68 (1966).
146. Pavlović-Hournac, M. R., Radivojević, D. V., and Pantić, V., *Intern. J. Radiation Biol.* **10**, 51 (1966).
147. Pentz, E. I., and Hasterlik, R. J., *Am. J. Physiol.* **189**, 11 (1957).
148. Pentz, E. I., *Exptl. Cell Res.* **32**, 170 (1963); *Proc. Soc. Exptl. Biol. Med.* **96**, 829 (1957).
149. Phillips, R. D., and Kimeldorf, D. J., *Radiation Res.* **18**, 86 (1963).
149a. Polikarpova, L. I., *Radiobiologiya* **5**, 896 (1965).
150. Pospíšil, M., and Šikulová, J., *Intern. J. Radiation Biol.* **9**, 597 (1965).
151. Randić, M., and Supek, Z., *Intern. J. Radiation Biol.* **4**, 151 (1961).
152. Randić, M., and Supek, Z., *Intern. J. Radiation Biol.* **4**, 637 (1962).
153. Rappoport, D. A., *Radiation Res.* **11**, 229 (1959).
154. Renson, J., and Fischer, P., *Arch. Intern. Physiol.* **67**, 142 (1959).
155. Renson, J., *J. Physiol.* (*Paris*) **52**, 208 (1960).
156. Rosenfeld, G., Ungar, F., Dorfman, R. I., and Pincus, G., *Endocrinology* **56**, 24 (1955).
157. Rosenfeld, G., *Am. J. Physiol.* **192**, 232 (1958).
158. Rosenfeld, G., *J. Lab. Clin. Med.* **51**, 198 (1958).
159. Sahinen, F. M., and Soderwall, A. L., *Radiation Res.* **24**, 412 (1965).
160. Sapsford, C. S., *Australian J. Biol. Sci.* **18**, 653 (1965).
161. Savkovic, N., Kacaki, J., Andjus, R., and Malcic, K., *Strahlentherapie* **130**, 432 (1966).
162. Schoen, E. J., *Endocrinology* **7**, 56 (1964).
163. Schorn, H., Doctoral thesis, Bonn (1967).
163a. Shelley, R. N., *Radiation Res.* **29**, 608 (1966).
163b. Simpson, C. G., and Ellis, L. C., *Radiation Res.* **31**, 139 (1967).
164. Skalka, M., and Soška, J., *Compt. Rend. Soc. Biol.* **157**, 455 (1963).

165. Smith, D. E., Tyree, E. B., Patt, H. M., and Jackson, E., *Proc. Soc. Exptl. Biol. Med.* **78**, 774 (1951).
166. Sobel, H. J., *Arch. Pathol.* **78**, 53 (1964); *Am. J. Pathol.* **50**, 39 (1967).
167. Song, C. W., and Evans, T. C., *Radiation Res.* **22**, 238 (1964).
168. Song, C. W., Anderson, R. S., and Tabachnick, J., *Radiation Res.* **27**, 604 (1966).
168a. Song, C. W., and Evans, T. C., *Radiation Res.* **33**, 480 (1968).
169. Spalding, J. F., Hawkins, S. B., and Strang, V. G., *Radiation Res.* **11**, 67 (1959).
170. Speck, L. B., *J. Neurochem.* **9**, 573 (1962).
171. Stark, G., and Altinli, A., *Arch. Gynaekol.* **196**, 13 (1961).
172. Stevens, W., Berliner, D. L., and Dougherty, T. F., *Radiation Res.* **16**, 556 (1962).
173. Stevens, W., and Berliner, D. L., *Radiation Res.* **20**, 510 (1963).
174. Stitch, S. R., Oakey, R. E., and Eccles, S. S., *Biochem. J.* **88**, 76 (1963).
175. Storer, J. B., and Simonson, H. C., *Proc. Soc. Exptl. Biol. Med.* **90**, 369 (1955).
176. Strubelt, O., *Strahlentherapie* **124**, 570 (1964).
177. Strubelt, O., *Strahlentherapie* **126**, 283 (1965).
178. Swift, M. N., Taketa, S. T., and Bond, V. P., *Am. J. Physiol.* **182**, 479 (1955).
178a. Tačeva, H., and Pospíšil, M., *Folia Biol.* (*Prague*) **13**, 164 (1967).
178b. Tačeva, H., and Pospíšil, M., *Intern. J. Radiation Biol.* **13**, 289 (1967).
179. Teller, W., Staib, W., Haibach, H., and Hohn, M., *Strahlentherapie* **111**, 107 (1960).
180. Thurber, R. E., and Nims, L. F., *Endocrinology* **70**, 595 (1962).
181. Tretyakova, K. A., and Grodzenskii, D. E., *Biokhimiya* **25**, 399 (1960); *Vopr. Med. Khim.* **6**, 611 (1960).
182. Trujillo, T. T., Spalding, J. F., and Langham, W. H., *Radiation Res.* **16**, 144 (1962).
183. Tsuchiya, T., Hayakawa, J. L., Muramatsu, S., and Eto, H., *Radiation Res.* **19**, 316 (1963).
184. Ulmer, D. D., Perkins, L. B., and Kereiakes, J. G., *Radiation Res.* **11**, 810 (1959).
185. Valenta, M., Kolousek, J., and Fulka, J., *Intern. J. Radiation Biol.* **6**, 81 (1963).
186. Van Caneghem, P., and Lachapelle, J. M., *Intern. J. Radiation Biol.* **9**, 115 (1965).
187. Van Cauwenberge, H., *Arch. Intern. Pharmacodyn* **106**, 473 (1956).
188. Van Cauwenberge, H., Fischer, P., Vliers, M., and Bacq, Z. M., *Comp. Rend. Soc. Biol.* **151**, 198 (1957).
189. Varagic, V., Krstic, M., Stepanovic, S., and Hajdukovic, S., *Intern. J. Radiation Biol.* **9**, 153 (1965).
189a. Varagic, V., Stepanovic, S., Svecenski, N., and Hajdukovic, S., *Intern. J. Radiation Biol.* **12**, 113 (1967).
190. Veninga, T. S., *in* "Biogenous Amines in Radiobiology," Jansen en Zoon, Groningen, Netherlands, 1965.
190a. Vierling, A. F., Radford, E. P., and Little, J. B., *Radiation Res.* **36**, 441 (1968).
191. Vittorio, P. V., Allen, M. J., and Small, D. L., *Radiation Res.* **15**, 625 (1961).
192. Waldschmidt, M., Pollak, H., and Streffer, C., *Strahlentherapie* **131**, 415 (1966).
193. Walsh, D., *Brit. Med. J. II*, 272 (1897).
194. Walter, O. M., Anson, B. J., and Ivy, A. C., *Radiology* **18**, 583 (1932).
195. Weber, G., and Cantero, A., *Am. J. Physiol.* **197**, 1284 (1959).
196. Weber, R. P., and Steggerda, F. R., *Proc. Soc. Exptl. Biol. Med.* **70**, 261 (1949).
197. Wexler, B. C., Pencharz, R., and Thomas, S. F., *Proc. Soc. Exptl. Biol. Med.* **79**, 183 (1952).
198. Wexler, B. C., Pencharz, R., and Thomas, S. F., *Am. J. Physiol.* **183**, 71 (1955).
199. Wilhelm, G., *Strahlentherapie* **97**, 75 (1955).
200. Williams, C. M., and Krise, G. M., *Science* **128**, 475 (1958).

201. Williams, C. M., and Krise, G. M., *Radiation Res.* **10**, 410 (1959).
202. Willoughby, D. A., *Nature* **184**, 1156 (1959).
203. Willoughby, D. A., *J. Pathol. Bacteriol.* **83**, 389 (1962).
204. Winkler, C., and Schorn, H., *Strahlentherapie* **127**, 121 (1965).
205. Wolf, R. C., and Bowman, R. E., *Radiation Res.* **23**, 232 (1964).
206. Woodward, K. T., and Rothermel, S. M., *Radiation Res.* **5**, 441 (1956).
207. Wyman, L. C., Witney, R., Goldman, R. F., Griffin, P. L., and Patt, D. I., *A. J. Physiol.* **192**, 51 (1958).
207a. Yago, N., Omato, S., Kobayashi, S., and Ichii, S., *J. Biochem.* (*Tokyo*), **62**, 339 (1967).
208. Yegin, M. Z., *Endokrinologie* **49**, 205 (1966).
209. Zsebök, B., Petranyi, G., Baumgartner, E., and Fachet, J., *Radiobiol. Radiotherap.* **8**, 201 (1967).

AUTHOR INDEX

Numbers in parentheses are reference numbers and indicate that an author's work is referred to although his name is not cited in the text. Numbers in italics show the page on which the complete reference is listed.

A

Abaturova, E. A., 254(1), 260(1), *277*
Abbadessa, S., 223(1, 10, 11), *225*
Abbei, L., 243(112), *247*
Abbott, C. R., 299(1), *323*
Abbott, J. K., 273(70), *279*
Abdullaev, V. M., 218(2), *225*
Abe, M., 266(195), *283*
Abraham, S., 123(31), 125(31), 126(31), 127(31), 130(31), 131(31), 141(31), 156(31), *186*, 316(26), *324*
Abrams, H. L., 5(1), 6(1), *10*
Abrams, R., 15(1, 2), 21(1), 24(1), 25(1), 26(2), *41*, 84(1), 89(1), *99*
Ackermann, I. B., 6(24), *11*
Acland, J. D., 95(1a), *99*
Adachi, K., 206(3), *225*
Adachi, R., 206(111), 229
Adams, K., 19(15, 123), *41*, *44*
Adelstein, S. J., 53(100), 54(1a), 66(1), *73*, *76*, 138(1), 159(1), *185*, 216(4), *225*
Aebi, H., 15(3), 39(76), 41(76), *41*(3), *43*, 142(226), *192*, 255(2, 3, 4), 258(207), *277*, *283*
Afifi, A. K., 291(2), 312(2), *323*
Agarwal, S. K., 118(1a), 122(1a), *185*
Agostini, C., 117(3), 118(2), 119(3), 128(3), *185*
Agracheva, N. D., 221(242), *232*
Aguirre, G., 22(178), 25(178), *46*
Ahlström, L. H., 52(2), 56(2), *73*
Akaboshi, M., 238(111), *247*
Åkeson, Å., 211(28a), *225*, *226*
Albaum, H. G., 212(189, 283), *231*, *234*, 268(5), 269(5, 200), 271(199, 272), *277*, *283*, *286*
Albers-Schönberg, H. E., 299(3), *323*
Albert, M. D., 162(4), 166(4), 171(4), *185*
Aldridge, W. G., 62(52), 70(3), *73*, *75*
Alekseeva, A. M., 301(4), *323*
Alesandrova, S. V., 268(146), 270(146), *282*
Alexander, P., 59(146), 60(146), 68(9), 71(102), *73*, *76*, *77*, 144(230), *192*, 236(1), 237(7), 244(48), *244*, *245*
Alexandrov, S. N., 221(5), *225*
Alexanian, R., 15(3a), 29(3a), 30(3a), *41*
Alfin-Slater, R. B., 199(195), *231*
Allen, M. J., 304(91), *329*
Allert, U., 255(194), 264(194), 265(194), 266(194), *283*

Ally, M. S., 146(292), *194*, 253(274), 256 (274), *286*
Almand, J. R., 214(116), *229*
Almeida, A. B., 111(244), 173(244), *193*
Almonte, T. J., 269(6), *277*
Alpen, E. L., 19(70), 24(70), *41*, *43*, 49(4), 52(4, 24), *73*, *74*, 96(97b), *102*
Alper, C., 185(224), *192*
Alper, T., 4(11), *11*
Alt, H. L., 18(186), 19(186), *46*
Altenbrunn, H.-J., 49(5), 52(5), 66(5), *73*, 142(5), 144(5), *185*, 240(2), 241(2), *244*
Altinli, A., 295(171), *329*
Altman, K. I., 15(4, 5), 16(5), 17(180), 19(4, 35, 206), 20(35, 191, 192, 206), 27(192), 29(192), 31(180), 39(180), *41*, *42*, *46*, 62(48, 49, 50, 51, 52), *75*, 87(25, 26), *100*, 140(224a), 141 (7, 86, 148, 224a), 146 (86, 87, 148), 148(85), 153 (6), 155(216), 169(88), *185*, *188*, *190*, *192*, 198(87, 89, 90, 102, 141, 142, 155, 240), 199(89, 91), 202(90, 91), 212(89), 213(87, 88, 90, 155), 221(5), *225*, *228*, *230*, *232*, 254(171), 257(161), 260(74, 75, 76, 78, 104, 171), 263(78a), 267(86), 269(164, 165, 166), 273(221), *279*, *280*, *282*, *284*, 316(99), *326*
Ammon, K., 214(107), *229*
Amore, G., 221(28), 224(28), *226*
Amoroso, C., 212(7), *225*
Anbar, M., 304(5), *323*
Anderegg, J. W., 23(19), *42*, 168(29), 179 (29), 180(29), 182(29), 183(29), *186*
Anderson, C. F., 146(8), *185*
Anderson, D. R., 260(7, 168, 276), 262(7), *277*, *282*, *286*
Anderson, E. P., 169(298), 175(298), *194*
Anderson, R. E., 198(283), *234*
Anderson, R. S., 313(168), 314(168), *329*
Andjus, R., 291(161), *328*
Andreoni, O., 270(8), 271(8), *277*
Andrews, F., 198(26), *226*
Andrews, G. A., 275(9), *277*
Andrews, J. R., 202(50), *227*
Angel, C. R., 255(10), *277*
Anger, R., 268(12), *281*
Ansari, P. M., 90(2), *99*
Ansfield, F. J., 237(23), *245*
Anson, B. J., 303(194), *329*
Anton, E., *244*(12)
Anton, E. J., 244(86a), *246*
Antipov, V. V., 308(144), *328*
Antokol'Skaya, Z. A., 225, 273(225), *284*
Appelman, A. W. M., 63(83), 64(83, 85, 86), *76*, 243(78), *246*
Applegate, R. L., 180(16), 182(16), *186*
Arabei, L., 223(8), *225*
Araki, K., 54(109), 55(109), 61(109), 64(6), 65(6), *73*, *76*
Archambeau, J. O., 3(3), *10*, 80(7), 81(7), 91(4), 93(3), *99*
Archuleta, R. F., 5(25), *11*
Arends, I. A., 29(109), *44*
Arient, M., 256(211, 212), 262(11, 211, 212), 263(11), 276(49), *277*, *278*, *283*
Arkinstall, W. W., 54(109), 55(109), 61 (109), *76*
Armstrong, W. D., 7(5), *10*, 96(9a), *99*, 196(39, 40), 220(41), 221(41), *226*, 252 (33, 34), *278*
Arnold, R., 25(194), *46*
Artom, C., 199(9), *225*
Ashikawa, J. K., 270(220), 271(220), *284*
Ashwell, G., 58(7), 62(104), 63(7, 104), 64 (74), *73*, *75*, *76*
Ashwood-Smith, M. J., 58(8), 63(8), *73*
Asnen, J., 31, 35(213), *46*
Astaldi, G., 243(3), *244*
Atkins, H. L., 31(202), *46*
Aurand, K., 2(17a), *11*
Auricchio, F., 23, 24, 31(23, 24), 33(23, 24), 35(24), 38(24), *42*
Avellone, S., 222(28), 223(1, 10, 11), 224 (28), *225*, *226*
Ayre, H. A., 243(39), 244(39), *245*
Azar, H. A., 270(87), *280*
Azevedo, N. L. C., 31(60), 36(60), *43*
Azzam, N. A., 273(179a), *283*

B

Babicky, A., 201(13), 202(12, 13), 203(13, 14, 14a, 140), *225*, *226*, *230*
Bachofer, C. S., 221(15), 225(15), *226*
Backmann, R., 238(4, 49), *244*, *245*
Bacq, Z. M., 19(169), 20(170), 21(171), *45*, 68(9), *73*, 221(16), 225(16), *226*, 259 (6), 289(6, 7, 8), 292(6), 293(7), 295(8, 188), 310(9), *323*, *329*
Badalyan, G. E., 222(193), *231*
Baev, I. A., 30(266, 267), *48*
Baeyens, W., 153(8a), 154(8a), *185*
Bagramyan, E. R., 291(10), *324*

Bailey, J., 198(26), *226*
Baisogalov, G. D., 274(102), *280*
Baker, D. G., 7(2), *10*, 90(60), 96(5), 97(5), *99*, *101*, 269(12, 201), 273(12a), *277*, *283*, 294(12), 305(13), 311(11), 312(11), *324*
Baker, G. T., 267(99a), *280*
Barnes, J. H., 269(6), *277*
Ballin, J. C., 70(43), *74*, 90(22), *100*
Balmukhanov, S. B., 236(5), *244*
Banerjee, R., 218(55, 160), *227*, *230*
Bankowski, Z., 206(17), *226*
Barbiroli, B., 90(5a), *99*, 151(173), *191*
Barboglio, F., 206(18), *226*
Barkley, H., 49(4), 52(4), *73*
Barnabei, O., 151(9), 159(9, 10), *186*
Barnum, C. P., 111(129), 117(129), 165 (129), *189*, 237(6), 238(115), *244*, *247*
Barrionuevo, M., 118(238), 119(239), *193*
Barron, E. S. G., 15(2), 26(2), *41*, 91(6), *99*, 142(11), *186*, 236(203), *244*, 268(203), *283*
Barth, G., 254(13, 232), *277*, *284*
Bartsch, G. G., 107(12), 110(12), 112(12), 115(12), 116(12), *186*
Bates, N., 220(97), *228*
Bates, T. H., 256(241), 257(242), 263(14, 241, 242), *277*, *284*, *285*
Bau, D., 160(189), *191*, 318(128a), *328*
Bauer, R., 268(215), *284*
Baugnet-Mahieu, L., 177(100), 180(13, 100), 181(13, 100), *186*, *188*
Baum, S. J., *41*, 322(92b), *326*
Baumgartner, E., 294(209), *330*
Baur, R. D., 60(10), *73*
Bautista, S. C., 269(6), *277*
Bavetta, L. A., 58(11), *73*, 197(200, 201), *231*
Baxi, A. J., 212(236), *232*
Baxter, C. F., 17(6), 18(6), 19(6), 26(6), 29(6), *41*
Bezděk, M., 184(251), 185(251), *193*
Beach, G., 165(135a), 168(135a), 169(135a), *189*, 237(74), *246*
Bealmear, P., *102*
Beaumariage, M. L., 208(19, 271, 273), 221 (15), 225(15), *226*, *233*, 291(47), 308 (14, 33), *324*, *325*
Beck, K., 148(126), *186*
Beck, L. H., 142(299, 300), *194*
Becker, F. F., 269(15), *277*
Beer, J. Z., 237(7), *244*
Bekhor, I. J., 58(11), *73*
Bekkum, D. W. Van, 142(14), *186*
Belcher, E. H., 17(6), 18(6, 7, 8, 9), 19(6, 7, 9), 20(6, 7, 8, 9), 26(6), 29(6), *41*
Bell, M. C., 202(171), *230*
Bell, V. L., 105(97), 106(97), 107(97), 109 (97), 113(97), 114(97), 115(97), *188*, 293(70), 316(69), *326*
Beltz, R. E., 117(15), 166(17), 167(15), 173 (17), 179(17), 180(15, 16), 181(15), 182 (16), *186*
Belyaeva, E. M., 256(65), *279*
Bender, A. E., 303(15), 324
Bendová, L., 253(120a), *281*
Beneš, J., 39(265), *48*, 129(305), 130(305), 131(305), 132(305), 134(306), 135(305), 138(306), 142(303), 144(304, 306), *194*, 254(16), 257(280), 268(280, 281), *277*, *286*
Beneš, L., 22(228), 23(228), 24(10, 45), 25 (10, 228), 26(228), *41*, *42*, *47*, 54(12), *74*
Benjamin, T. L., 71(13), *74*, 142(18), *186*
Bennett, D. C., 269(131a), *281*
Bennett, E. L., 168(19), 174(19), *186*
Bennett, L. R., 95(6a, b), *99*
Bennett, V. C., 95(6a), *99*
Ben-Parath, M., 58(62), *75*
Benson, B., 105(51), 114(51), 117(51), 137 (51), *187*
Berech, J. Jr., 214(20), *226*
Berg, T. L., 173(20, 21), 176(21), 178(20, 21), 182(21), *186*
Bergeder, H. D., 208(21, 22, 23), *226*
Bergen, J. R., 220(24), *226*
Bergonie, J., 301(16), *324*
Bergsteinova, V., 254(16), *277*
Bergstrand, H., 236(91, 92), *247*
Bergström, R. M., 208(25), 210(25), *226*
Berlin, B. S., 52(14), *74*
Berliner, D. L., 294(172, 173), 300(17, 19, 54, 55), 301(18, 55), *324*, *325*, *329*
Berman, H., 269(27), *278*
Bernays, L., 255(2), *277*
Berndt, H., 238(60), 241(60), *246*
Berndt, J., 117(23, 24, 24a, 25a), 123(23, 25a), 128(24a), 131(24a), 132(25a), 133 (25, 25a), 135(24, 25), 137(24a), 138 (22, 23), 142(22), *186*
Bernhard, K., 206(282), *234*
Bernheim, F., 14(11), 17(11), *41*, 190(72a, b) *101*

Bernick, S., 199(195), *231*
Berry, H. K., 256(17), *277*
Berry, R. J., 24(12), 25(12), *41*, 236(9), 237 (8), 239(9), *244*
Bertazzoni, U., 66(19), *74*
Bethard, W. F., 19(85), *43*
Bertram, B., 240(59), 241(59), *246*
Besenbeck, F., 254(13), *277*
Betel, I., 63(83), 64(14a, 83, 86), 72(14a), *74*, *76*
Bettendorf, G., 62(15), *74*, 153(237), *192*, 239(10, 101), 241(101), *244*, *247*, 308 (20), *324*
Betz, E.-H., 124(156), 141(156), *190*, 289 (22), 293(22), 303(22), 304(21, 22), 323 (22), *324*
Beyl, G. E., 31(215), *47*
Bezdrobnyi, Yu. V., 291(28a), *324*
Bhaskaran, S., 236(10a), *244*
Bianchi, M. R., 35(13, 14), *41*
Bide, R. W., 31(162, 163), 33, 34(163), 36 (162, 163), 37(163), 38(162), 39(162), *45*, 61(112), *76*
Biggs, S. L., 66(1), *73*
Bigwood, E. J., 255(18), 259(18), *277*
Binhammer, R. T., 289(24), 290(23, 24a), 293(24), 300(24a), 319(24), *324*
Birke, G., 268(273a), 273(19), *277*, *286*, 300 (25), *324*
Birzle, H., 148(26), 149(26a), *186*
Bishop, C. W., 54(16), *74*
Bishop, R. C., 55(47), *74*
Bitonto, G. D., 159(10), *186*
Bjorkstein, J., 198(26), *226*
Blackburn, J., 203(27), *226*
Blackwell, L. H., 253(20), *277*
Blackett, N. M., 19(15, 123), *41*, *44*
Blafield, R., 208(25), 210(25), *226*
Blair, W. H., 268(110), 269(110), *280*
Blais, R., 95(62), *101*
Blandino, G., 38(16), *41*
Blanquet, P., 304(101), *327*
Blaschko, H., 146(8), *185*
Blasi, F., 128(36), 133(36, 37), 136(36, 37), 141(37), *186*
Block-Frankenthal, L., 293(79), *326*
Bloedorn, F., *244*(12)
Blokhina, V. D., 118(27), 119(27), 120(27), 153(28), *186*
Blomgren, H., 49(15a), *74*
Blum, K. U., 38(17), *42*
Boag, J. W., 218(216), *232*
Bocqacci, M., 35(13, 4), *41*
Bogardus, C. R., 250(178), *283*
Bogomazov, M. Y., 256(63), *279*
Bogotavlenskaya, M. P., 19(102), 24(102), 25(102,) *44*
Bohländer, A., 124(138, 139), *190*
Bohndorf, W., 256(127), 269(127), *281*
Bohuon, C., 254(21), *277*
Bollinger, J. N., 14(18), 16(18), *42*
Bollum, F. J., 23(19), *42*, 168(29), 179(29), 180(29), 182(29), 183(29), *186*
Bölöni, E., 153(122), 154(122), *189*
Bonavita, V., 222(28), 224(28), *226*
Bond, V. P., 3(3), *10*, 18(174), 20(174), 22(54, 250), 24(152, 250), 25(152, 250), 31 (68), 35(68), 42, 43, *45*, *47*, 80(7), 81(7), 98(3), *99*, 221(294), *234*, 256(227), 263 (227), *284*, 311(178), *329*
Bondurant, J. H., 250(129), *281*
Bonnichsen, R., 18(20, 81), 19(20, 81), *42*, *43*, 211(28a), *226*
Bonoppis, G., 198(250), *233*
Bonsignore, D., 122(30), 128(30), 130(30), 131(30), 132(30), 313(30), *186*
Boquet, P. L., 258(23, 71), 259(22, 23, 23a, 71), 260(23a), *278*, *279*
Bordner, L. F., 143(267), 144(267), *193*
Borgers, M., 90(36), *100*
Borison, H. L., 312(25a), *324*
Borova, J., 39(265), *48*, 258(206), 273(206), *283*
Borrebaek, B., 316(26), *324*
Borrebaek, D., 123(31), 125(31), 126(31), 127(31), 130(31), 131(31), 141(31), 156(31), *186*
Bose, A., 19(21), *42*
Botteghi, F., 19(51), 26(51), *42*
Bottero, M., 197(29, 232), *226*, *232*
Boudnitskaya, E. V., 238(11), *244*
Bourdon, R., 268(187), 270(187), *283*
Bourne, G. H., 208(219), *232*
Bouronele, B. A., 31(261), *48*
Bowerman, C. I., 254(163), 256(163), *282*
Bowers, J. Z., 252(24, 25), 253(24, 25), *278*, 295(67), 297(67), *326*
Böwing, G., 264(108), *280*
Bowman, R. E., 294(205), 295(205), *330*
Boyd, M. J., 185(224), *192*
Brandes, D., 208(219), *232*, *244*
Brandes, P., 244(86a), *246*

Braselmann, H., 199(146, 147), *230*
Braun, H., 40(260), *48*, 51(7), *74*, 141(32, 33), 148(291), 150(34, 35), 153(34, 35), *186*, *194*, 211(111a), 212(30, 114), 214 (107), 215(30, 114, 131), *226*, *229*, 250 (121), 252(121), 254(265a), 256(127), 268(126), 269(26, 126, 127, 156, 265a), *278*, *281*, *282*, *285*, 308(27, 28), *324*
Brawner, H. P., 37(22), *42*
Brecher, G., 31(231), 37(22, 231), *42*
Brejcha, M., 254(16), *277*
Brennan, P., 98(24), *100*
Brent, R. L., 269(27, 28), 270(190), *278*, *283*
Bresciani, F., 31(23, 24), 33(23, 24), 35(24), 38(24), *42*, 128(36), 133(36, 37), 136 (36, 37), 141(37), *186*, 208(31, 59), 209(31, 59), *226*, *227*, 243(12a), *244*
Bresnick, E., 31(204), 36(204), 38(127, 204), *44*, *46*
Breuer, H., 39(35), *42*, 54(18), 58(19), *74*, 91(100), 92(100), 94(100), *102*, 211(32), 214(32, 33), *226*, 253(29), *278*
Brin, M., 122(170), 125(170), 150(38), 151 (38), 153(38), 159(38), *186*, *190*, 269 (30), *278*, 316(121), *327*
Brinkman, R., 308(31), 313(29, 30), *324*
Britun, A. I., 203(95), *228*
Brizzee, K. R., 295(31a), 312(31a), *324*
Brodskaya, N. I., 221(34), *226*
Brodskii, V. Y., 22(26), 24(26), 25(26), *42*
Brohee, H., 55(155a), 59(155a), 61(155a, 181, 182, 183), 64(185), 65(185), 66 (184), 69(190), 72(155a, 180, 187), *78*, *79*
Bromfield, A. R., 273(31), *278*
Brooks, P. M., 220(93), *228*
Brown, D. G., 253(106), *280*, 295(74), *326*
Brown, J. A., 270(60a), *279*
Brownell, L. W., 184(236), *192*
Brownson, R. H., 222(35, 122a), *226*, *229*
Bruce, W. R., 52(20), *74*
Brucer, M., 275(9, 32), *277*, *278*
Brues, A. M., 268(219), *284*
Brückner, J., 89(25), *100*, 184(84), *188*
Brunfaut, M., 238(11), *244*
Bruni, L., 206(36), 207(37), *226*
Brunkhorst, W. K., 57(73a), *75*
Bruns, F., 260(104), *280*
Bucher, N. L. R., 114(40), 162(4, 39), 163 (39, 39a), 166(4), 171(4), 177(39, 39a), *185*, *186*
Buchsbaum, R., *42*
Buhlmann, A., 31(28), *42*
Buck, A. T., 216(268), *233*
Buess, E., 267(97a), *280*
Bugliani, D., 206(18), *226*
Bührer, G., 142(193), 144(82), *188*, *191*
Bukovsky, J., 69(141), *77*
Burdin, K. S., 214(37a), *226*
Burg, C., 117(47), 119(47), *187*
Burger, D. O., 24(157), 38(157), 139(157), *45*
Burk, D., 239(13), *245*
Burke, W. T., 112(41), *186*
Burn, J. H., 90(8), *99*, 146(8), *185*
Burr, B. E., 146(292), *194*, 253(274), 256 (274), *286*
Burstein, S., 294(31b), *324*
Busch, S., 167(45a), 169(45a), 182(45a), *187*
Bush, R. S., 52(20), *74*
Butler, C. L., 84(21), 87(21), *99*
Butler, J. A. V., 153(42), 184(42), *186*

C

Caceres, E., 214(187), *231*
Cain, J., 221(188b), *231*
Cairnie, A. B., 216(268), *233*
Caldavera, C. M., 151(173), *191*
Calton, F. M., 295(31a), 312(31a), *324*
Calzavara, F., 90(9), *99*
Cammarano, P., 239(14, 15), *245*
Campagnari, F., 66(21), *74*
Canonica, C., 122(30), 128(30), 130(30), 131 (30), 132(30), 133(30), *186*
Cantero, A., 123(287), 124(287), 128(287), 130(287), 131(287), *194*, 316(195), *329*
Cantor, L. N., 19(205), 29(205), 30(205), 36(205), *46*
Capalin, S., 223(8), *225*
Caponetti, R., 38(16), *41*
Caputo, A., 239(16), 240(15), *245*
Cardiovascular Group, 271(272), *286*
Carlough, M., 85(86), 88(86), *101*
Carlson, L. D., 6(4), *10*
Carlson, W. D., 299(32), 301(32), *324*
Carr, S., 153(119), *189*, 211(109), *229*
Carrol, H. W., 123(43), *186*
Carsten, A., 273(12a), *277*
Carter, R. R., 52(22), *74*
Carttar, M. S., 24(240), 27(240), 29(240), *47*, 63(170), *78*, 165(263), 172(263), *193*
Cartwright, F. D., 253(106), *280*
Caruso, P., 38(16), *41*

Casarett, G. W., 153(6), *185*, 198(38), 211 (6), *225*, *226*, 312(50a), *325*
Cascarano, J., 63(80, 81), *75*, *76*, 129(135d, 136), *190*
Casper, R., 224(275a), *233*
Caspersson, O., 236(18), 238(17), 239(17), *245*
Caster, W. O., 7(5), *10*, 96(9a), *99*, 196(39, 40), 220(41), 221(41), *226*, 252(33, 34), *278*
Cate, D. L., 105(51), 114(51), 117(50, 51), 137(50, 51), *187*
Cater, D. B., 161(44), 162(44), 164(44), 172 (44), 178(44), *187*
Cathie, J. A. B., 235(19), *245*
Cattaneo, S. M., 205(41a), *226*
Cavalieri, R. R., 18(95), 19(95), *43*, 255(35), 257(36), *278*
Cavina Pratesi, A., 91(10), 99
Celada, F., 52(22), *74*
Cenciolti, L., 221(41b), *226*
Cepicka, I., 64(68), 66(68), *75*
Cescon, I., 91(10), *99*
Cession-Fossion, A., 308(33), *324*
Cha, H. S., 157(45), *187*
Chadwick, C., 52(99a), *76*
Chaika, Ya. P., 89(10a, 87), *99*, *101*, 211 (259), 215(259), *233*
Chaikoff, I. L., 107(123), 123(31, 123), 125 (31), 126(31), 127(31, 123), 129(79), 130(31, 79), 131(31), 132(79), 134(79), 137(123), 140(79), 141(31), 150(79), 156(31), *186*, *188*, *189*, 303(60), 316(26), *324*, *325*
Chally, C. H., 222(49), *227*
Chambers, F. W., Jr., 255(35), *278*
Chambers, H., 36(29), *42*
Chambon, P., 25(154), 26(154), 38(30), 39 (30), *42*, *45*, 55(78), 57(101), *75*, *76*, 147 (45a), 169(45a), 182(45a), *187*
Chan, H., *229*
Chanana, A. D., 31(202, 203), *46*
Chang, L. F., 205(266), 206(266), 207(266), *233*
Chang, L. O., 162, *187*, 237(19a), *245*
Chang, S. S. M., 271(198), *283*
Chanutin, A., 31(31), *42*, 215(173), *230*, 253(37), *278*
Chapman, W. H., 2(6), 6(6), *10*
Chase, H. B., 208(83), *228*
Chastain, S. M., 95(6b), *99*
Chepovskii, V., 254(218), *284*
Cherkasova, L. S., 212(42), 221(42), 221 (42), *226*
Chernavskaya, N. M., *234*
Chernevskaya, Kh. I., 221(42a), *226*
Chernov, G. A., 308(34, 35, 36, 144), *324*, *328*
Cherry, S. H., 251(38), *278*
Chetverikov, D. A., 222(86), *228*
Chevallier, A., 117(47), 119(47), *187*
Chien, S., 220(97), *228*, 251(38), *278*
Chiriboga, J., 148(48, 49), *187*, 257(39), *278*
Chirpaz, B., 24(193), *46*
Chlumeckà, V., 22(219), *47*, 59(159), *78*
Chmelar, V., 311(36a), *324*
Chow, D. C., 206(3), *225*
Christensen, G. M., 49(77a), 55(77), *75*, 98 (22a), *100*
Christensen, N., 252(25), 253(25), *278*
Christie, S., 96(5), 97(5), *99*
Chudinovskikh, A. V., 273(225), *284*
Church, M. L., 31(164), 33(164), 36(164), 38(164), *45*
Ciccarone, P., 117(3), 119(3), 128(3), 185 (101), *185*, *188*
Cier, A., 273(176a), *282*
Cier, J., 273(176a), *282*
Cividalli, G., 31(32, 33), 36(32), 37(33), 38 (32), *42*
Claeys, E., 295(113), *327*
Clark, J. B., 135(49a), 142(49a), 144(49a), *187*
Clark, T. W., 18(186), 19(186), *46*
Clark, R. T., 253(59), 254(59), *279*
Clarke, G., 291(58), 316(58), *325*
Clarkson, T. B., 199(9), *225*
Clemendson, C. J., 28(146, 147), 38(147), *45*, 269(182), *283*
Click, G., 252(158), 254(158), 268(158), *282*
Clifton, K. H., 237(20, 21, 22, 23), *245*
Cloutier, P. F., 155(178), *191*
Clugston, H., 6(199), *11*
Crosfill, M. L., 5(6a), *11*
Coburn, J. W., 214(43), *226*
Cochran, D. W., 222(65), *227*
Cochran, K. W., 63(31), *74*, 91(19), *99*, 135 (63), 140(63), *187*
Cohan, S., 223(43a), *226*
Cohen, J. A., 172(289), 178(289), *194*
Cohn, P., 153(42), 184(42), *186*
Cohn, S. H., 202(44, 45), *227*

Cole, L. J., 23(48), 24(50, 152), 25(50, 152), 26(39), *42*, *45*, 52(105a), 59(23, 61, 168a), 60(99, 168, 168a), 61(168a), 69 (60, 61), 70(60, 61), *74*, *75*, *76*, *38*, 216 (46, 176, 279, 280, 281), 217(176), *227*, *231*, *234*, 243(25), *245*, 256(101a,b,c), *280*
Coles, D., 96(5), 97(5), *99*
Colgan, J., 269(40), *278*
Collins, J. G., 271(198), *283*
Collins, V. P., 19(138), 21(138), *44*, 58(135), *77*, *192*, 216(226), *232*
Colson, J., 304(111), *327*
Comar, C. L., 95(50), *100*
Commanay, L., 304(101), *327*
Comsa, J., 289(37), 291(37a), *324*
Conard, R. A., 84(11), 89(11), 90(11), *99*, 203(47), *227*, 308(38), 311(38), 312 (38), *325*
Congdon, C. C., 141(168), 144(168), *190*
Coniglio, J. G., 91(12), *99*, 105(51), 106 (53, 54), 114(51), 117(50, 51), 118(52), 119(52), 120(52), 123(53, 54, 54a), 128 (53), 137(50, 51), *187*, 301(39), *325*
Conti, F., 38(16), *41*
Conzelman, G. M., Jr., 169(55), 175(55), *187*, 238(25), *245*
Cool, H. T., 146(292), *194*, 253(274), 256 (274), *286*
Coon, J. M., 211(71a), *227*
Cooper, E. H., 52(24), 70(3), *73*, *74*
Cooper, J. A. D., 18(186), 19(186), *46*
Cordic, A., 223(120b), *229*
Cornatzer, W. E., 58(24a), *74*, 110(56), *187*, 216(48), 222(49), *227*, 254(41), 268(41), *278*
Corp, M. J., 319(40), *325*
Correa, J. N., 202(50), *227*
Costello, M., 63(45), *74*
Cottier, H., 18(174), 20(174), *45*
Coulter, E. P., 63(45), *74*, 268(66), *279*
Counelis, M., 253(202), *283*
Courcon, J., 268(99), 270(99), *280*
Coursaget, J., 158(108), *189*
Courtenay, V. D., 18(7), 19(7), 20(7), *41*
Crabtree, H. C., 239(26), *245*
Cramer, W., 94(66), 95(66), *101*
Crathorn, A. R., 153(42), 184(42), *186*, 238 (27), *245*
Creasey, W. A., 54(26), 63(25), 68(27), *74*
Cress, E. A., 14, 15(224), 17(224), *47*
Crist, R., 268(177), *282*
Crocco, E. M., 212(196), 214(196), *231*
Crocker, J. R., 289(24), 290(24), 293(24), 319(24), *324*
Cromray, H. L., 55(193), 63(193), 64(193), *79*
Cronkite, E. P., 31(202, 203, 231), 37(231), *46*, *47*, 256(227), 263(227), *284*
Croye, B. L., 271(198), *283*
Culp, F. B., 301(39), *325*
Cultrera, G., 95(12a), *99*
Cumming, J. D., 30(33a), *42*
Curran, P. F., 96(13), *99*
Curtis, H. J., 214(20), *226*

D

D'Addabbo, A., 257(42), *278*
Daiseley, K. W., 218(51), *227*
dalla Palma, L., 94(14), *99*
Dalos, B., 269(43), *278*
Dalrymple, G. V., 55(145), 62(144), *77*, 269(44), *278*
Dal Santo, G., 19(82), 21(82), *43*
D'Amore, A., 34(34), *42*
Dancewicz, A. M., 19(35), 20(35, 36, 153, 155), *42*, *45*, 128(57), 131(57a), *187*
Danysz, A., 223(52), *227*
Darby, W. J., 118(52), 119(52), 120(52), *187*, 257(213), *283*
Darden, E. B., 198(53a), 208(53), *227*
Datskiv, M. Z., 211(259), 215(259, 259a), *233*
Davenport, V., 252(25), 253(25), *278*
David, G., 294(41), *325*
David, P. W., 146(292), *194*, 253(274), 256 (274), *286*
Davidová, E., 221(270b), *233*
Davidson, J. N., 24(221), 25(221), *47*, 54 (16), *74*
Davis, A. K., *41*
Davis, G. K., 212(68), *227*
Davis, J., 301(39), *325*
Davis, J. W., 302(112), *327*
Davis, R. W., 20(38), 31(38), 36(38), *42*
Davis, W. E., 26(39), *42*
Davison, J. P., 58(24a), *74*, 110(56), *187*, 216 (48), *227*, 254(41), 268(41), *278*
Davydova, I. N., 222(54), *227*
Dawson, K. B., 225(53b, 230, 241), *227*, *232*

Day, P. L., 27(41), 29(41), *42*, 69(30), *74*, 185(74), *187*
De, N., 236(28), *245*
Deakin, H., 54(28), 61(28), 67(28), *74*
Deanovic, D., 307(139a), 309(139a), *328*
Deanović, Z., 308(42, 43, 44), 310(42, 43, 44), 312(44a), *325*
Debenedetti, A., 211(190), *231*
Debley, V. J., 308(110), *327*
de Boer, W. G. R. M., 199(154), *230*
Decker, A. B., 95(6b), *99*
Deckers, C., 21(264), 22(264), *48*
DeGowin, E. L., 16(214), 31(214), *46*
Deknudt, G., 148(85), 155(216), *188*, *192*, 267(86), 273(221), *280*, *284*
De La Cruz, B., 269(6), *277*
Delarue, J. C. H., 254(21), *277*
De Leeuw, N. K. M., 269(45), *278*
Del Favero, A., 118(58), 119(58), *187*
Dellenback, R. J., 220(231), 221(231), 222 (231), *232*
Deme, I., 206(85), *228*
Demeester, G., 294(51), 295(51),29 9(51), *325*
Demin, N. N., 90(15), *99*, 118(27), 119(27), 120(27), *186*
Deml, F., 311(36a), *324*
Dempsey, B. C., 311(82), *326*
Denniston, R. H., 291(45), *325*
Denny, W. F., 215(167), *230*
Denson, J. R., 123(59), 125(59), *187*
Deroo, J., 94(19a, 55), *99*, *101*, 257(86c), 263(282), 268(86c, d), *280*, *286*
Desai, I. D., 70(29), *74*, 143(60), *187*
de Schrijver, A., 295(113), *327*
Detrick, L. E., 95(16, 17, 18), *99*, 257(46), *278*
De Vellis, J., 223(54a), 224(239), *227*, *232*
Devi, A., 153(227), *192*, 218(55, 160), *227*, *230*
Devik, F., 204(56), 207(118), *227*, *229*
de Vries, M. J., 64(85), 71(176), *76*, *78*, 243 (78), *246*
De Wolfe-Slade, D., 54(109), 55(109), 61 (108, 109, 110), 72(108, 110), *76*
Diczfalusy, E., 303(46), *325*
Dienstbier, Z., 31(209, 210), 39(227, 265), *46*, *47*, *48*, 66(75), *75*, 94(96a), *102*, 129 (305), 130(305), 131(305), 132(305), 134(306), 135(306), 138(306), 142(305), 144(306), *187*, *194*, 251(125), 252(125), 253(48), 256(160, 211, 212), 262(11, 48, 211, 212), 263(11), 269(267), 276(47, 49, 160), *277*, *278*, *281*, *282*, *283*, *285*, 291 (80), 312(80), *326*
Dieterich, E., 148(26), *186*
Dietz, A. A., 14(40), 15(40), *42*
Di Ferrante, N., *192*, 198(240), *232*
Di Giacomo, R., 31(203), *46*
Dijken, G. B., 311(103), 312(103), *327*
Dikovenko, E. A., 221(57), *227*
Di Luzio, N. R., 15(47), 16(47), *42*, 112(66), 119(65), 120(65, 66), *187*, 199(71), *227*, 253(56, 57), 254(50, 55, 56, 57), 255 (50), *278*
Di Marco, N., 118(195), *191*
Dimitrov, L., 269(51), *278*
Dinning, J. S., 27(41), 29(41), *42*
Di Pietrantorij, F., 31(42), 37(42), *42*
Dische, Z., 218(58), *227*
Dittrich, W., 236(10a), *244*
Ditzel, J., 268(52), *278*
Djukic, Z., 274(135), *281*
Dodd, M. C., 31(261), *48*
Dodge, A. H., 269(131a), *281*
Doida, Y., 236(69), *246*
Dole, N., 20(38), 31(38), 36(38), *42*
Dompe, M., 270(8), 271(8), *277*
Doneda, G., 118(195), 153(197), *191*
Doniach, I., 303(68a), *326*
Dorfman, R. I., 294(156), *328*
Dose, K., 208(31, 59), 209(31, 59), *226*, *227*, 239(29, 30, 31), 240(29), 243(12a), 244 (31), *244*, *245*
Dose, U., 239(29), 240(29, 31), 244(31), *245*
Doty, S. B., 203(60), *227*
Douglass, C. D., 69(30), *74*, 185(74), *187*
Doull, J., 63(31), *74*, 91(19), *99*, 135(63), 140(63), *187*, 222(65), *227*
Dowben, R. M., 39(43), *42*, 211(61), 213 (61), *227*, 260(276), *286*
Dragic, M., 321(46a), *325*
Dragoni, G., 215(62, 63), *227*, 244(32), *245*
Drahovsky, D., 256(53, 54), *278*
Drapers, N. R., 237(22), *245*
Drašíl, V., 22(228), 23(228), 24(44, 45, 229), 25(44, 45, 228), 26(228), *42*, *47*
Dreisbach, L., 111(240), 118(238, 239), 119 (238, 239), *193*
Drew, A. H., 94(66), 95(66), *101*
Dreyfus, G., 221(110), *229*
Dubinskii, A. M., 221(64), *227*

Dubois, K. G., 273(248), *285*
Dubois, K. P., 58(32), 63(31, 167), *74*, *78*, 91(19), *99*, 135(63), 140(63), 160(62, 122a), 168(122a), *187*, *189*, 215(66), 222 (65, 66), *227*
Duchesne, P. Y., 291(47, 72b), *325*, *326*
Duhig, J. T., 199(66a), *227*
Dulcino, J., 94(19a, 55), *99*, *101*
Dunjic, A., 21(264), 22(264), *48*
Dunlap, A. K., 95(16), *99*
Duplan, J. F., 274(135), *281*, 301(102), *327*
Durst, F., 215(71b), *227*
Dux, A., 206(278), *234*
Dwyer, F. M., 123(245), *193*
Dykes, P. W., 273(31), *278*, 291(131), *328*
Dziewiatkowski, D. D., 202(67), *227*

E

Early, J. C., 146(132), *189*, 252(150), 253 (150), 254(150), 255(150), *282*
Easly, J. F., 212(68), *227*
Eberhagen, D., 119(64), *187*, 254(54a), *278*, 293(48, 49), *325*
Ebert, M., 237(8), *244*
Ebina, T., 239(33), *245*
Eccles, S. S., 302(174), 303(50, 174), *325*, *329*
Eckfeldt, G., 206(111), *229*
Eddy, H. A., 312(50a), *325*
Edelmann, A., 289(117), *327*
Eder, H., 90(2), *99*
Edsmyr, F., 303(46), *325*
Edwards, C. H., 21(46), *42*, 89(20), 95(20), *99*, 215(69), *227*
Edwards, G. A., 21(46), *42*, 89(20), 95(20), *99*, 203(69), *227*
Eechaute, W., 294(51), 295(51), 299(51), *325*
Efimov, M. L., 236(5), *244*
Egana, E., 307(51a), *325*
Egineta, M., 212(196), 214(196), *231*
Ehrenstein, V. G., *225*
Ehrhardt, E., 66(5), *73*, 240(2), 241(2), *244*
Eichel, H. J., 58(33), 63(34), 69(36, 139, 140, 141), *74*, *77*, 138(64a), 185(224), *187*, *192*
Eisen, V. D., 308(52), *325*
Elbert, O., 268(246), *285*
Eliashof, P. A., 218(105), *229*
Elkina, N. L., 203(70), *227*
Elkind, M. M., 84(101a), *102*
Elko, E. E., 15(47), 16(47), *42*, 112(66), 119 (65), 120(65, 66), *187*, 199(71), *227*, 253 (54, 57), 254(55, 56, 57), *278*
Ellinger, F., 103(67, 68), *187*, 252(58), 253 (58), 254(58), *278*, 307(53), 323(53), *325*
Ellinwood, L. E., 211(71a), *227*
Elliot, J., 218(58), *227*
Ellis F., 17(234), 18(234), 19(234), 20(234), 24(118, 119, 120), 25(118, 119), *44*, *47*
Ellis, J. P., 253(59), 254(59), *279*
Ellis, L. C., 300(17, 19, 54, 55, 153b), 301 (18, 55), *324*, *325*, *328*
Ellis, M. E., 23(48), *42*, 59(23), 60(99), *74*, *76*
Ellis, R. E., 308(52), *325*
Ellis, R. F., 84(97a), *102*
Elpatevskaya, G. N., 254(1), 260(1), *277*
Eltgen, D., 18(49), 19(49), *42*
Ely, J. O., 122(222), *192*
Emmel, V. M., 70(3), *73*
Endicott, K. M., 37(22), *42*
Engelstad, O. D., 58(24a), *74*, 110(56,) *187*, 216(48), *227*, 254(41), 268(41), *278*
Entenman, C., 84(43), 85(43), 91(44), 96 (39), 97(39), *100*, 106(158), 108(158), 110(288), 111(288), 119(126), 120(126), 122(133), 145(126), 146(132), 153(134), 168(209), 169(209), 173(209), 174(209), *189*, *190*, *192*, *194*, 250(257), 252(150), 253(133, 150), 254(66, 148, 149, 150, 152, 236), 255(150, 152), 256(132, 134), 258(157), 260(152), 268(96), 270(60), 271(60, 95), *279*, *280*, *281*, *282*, *284*, *285*, 312(92), 316(91), *326*
Erdman, K., 193(76), 94(76), *101*
Erede, P., 215(71b), *227*
Erickson, C. A., 24(50), 25(50), *42*
Ericson, U., 31(136), 37(136), *44*
Ernst, H., 60(37, 38, 40), 61(39, 40), 64 (68), 66(68), *74*, *75*, 184(69, 70, 71), *187*, 307 (122), 308(122), *327*
Errera, M., 57(41, 95, 96), *74*, *76*, 170(161), 174(161), 184(161), *190*, 238(11), *244*
Ershoff, B. H., 199(195), *231*, 308(56, 57), 309(57), 310(57), *325*
Eto, H., 304(183), 308(119b, 125, 129, 130), 312(119, 119a), *327*, *328*, *329*
Evans, A. S., 270(60a), *279*
Evans, N. T. S., 63(40b), *74*
Evans, R. G., 53(40a), *74*
Evans, T. C., 244(47), *245*, 291(58), 304 (167, 168a), 316(58), *325*, *329*

F

Faber, M., 105(117), 106(117), 145(93, 94), *188*, *189*
Fachet, J., 294(209), *330*
Faes, M., 275(136), *281*
Fahmy, A. R., 242(34), 243(34), *245*
Fain, J. N., 199(71c), *227*
Fanfani, G., 257(42), *278*
Fantoni, A., 19(51, 52), 26(51, 52), *42*
Faradi, L., 294(41), 295(41), *325*
Farkas, T., 112(208), 119(208), 120(208), *192*
Faulkner, D. N., 273(70), *279*
Faures, A., 57(41), *74*
Fausto, N., 55(42), 69(55), *74*, 162(72), 168 (73, 278), 172(63, 279), 173(63, 278), 174(279), 177(63, 278), 178(63, 278, 279), 179(278), 181(63, 278), 182(63, 278), *187*, *193*
Fedarov, I. V., 253(61), *279*
Fedorov, N. A., 18(53), 31(53), *42*, 256(62, 63), *279*
Fedorova, A. V., 308(58a), *325*
Fedorova, T. A., 256(64, 65, 98, 278a), 260 (64), *279*, *280*, *286*
Fedorovskii, L. L., 98(30), *100*
Fedotov, V. P., 294(59), *325*
Feine, U., 214(73), 215(72), *227*, 237(35), *245*
Feinendegen, L. E., 18(174), 20(174), 22 (54, 250), 24(250), 25(250), *42*, *45*, *47*, 221(294), *234*
Feinstein, R. N., 69(44), 70(43), *74*, 84(21), 87(21), 90(22), *99*, *100*
Fellas, V. M., 185(74), *187*
Feller, D. D., 303(60), *325*
Fenaroli, A., 117(3), 119(3), 128(3), *185*
Ferbel, B., 85(86), 88(86), *101*
Fermanian, J., 303(90), *326*
Ferroux, R., 310(86), *326*
Fershing, J. L., 214(94), *228*
Fiam, B., 215(127), *229*
Ficken, V. J., 250(178), *283*
Ficq, A., 57(95, 96), *76*, 170(161), 174(161), 184(161), *190*
Filippova, Z., 123(76), 131(76), 132(76), 149(75), *187*
Finck, W., 257(156a), *282*
Finnegan, C., 220(188a), *231*
Finston, R. A., 203(74), *227*
Fiore, C., 31(23, 24), 33(23, 24), 35(24), 38 (24), *42*
Firfarova, K. F., 218(207), *231*
Fischel, E., 122(77), *188*
Fischer, M. A., 63(45), *74*, 268(66), *279*
Fischer, P., 123(78), *188*, 259(6), 289(6, 7), 292(6), 293(7), 295(7, 188), 304(154), *323*, *328*, *329*
Fisher, J. W., 30(55), *42*, 214(75), 216(75), *227*
Fishler, M. C., 31(68), 36(68), *43*
Fitch, M. F., 129(79), 130(79), 132(79), 134(79), 140(79), 150(79), 151(79), *188*
Flemming, K., 5(6b), *11*, 36(56, 57, 58), *42*, *43*, 244(35a), *245*, 295(61, 62, 62a, b, c), *325*
Flesher, A. M., 213(103a), *228*, 253(105), *280*
Fliedner, T. M., 3(3), *10*, 22(250), 24(250), 25(250), *47*, 52(46), *74*, 80(7), 81(7), *99*, 256(227), 263(227), *284*
Florsheim, W. H., 111(80), *188*
Floyd, K. W., 165(96a), *188*
Fluckiger, H., 255(2), *277*
Fochem, K., 6(7, 8), *11*, 308(63), *325*
Follis, R. H., 202(185), 203(185), *231*
Fomichenko, K. V., 212(42), 221(42), 222 (42, 76), *226*, *227*
Ford, D., 223(43a), *226*
Ford, W., 301(39), *325*
Forkey, D. J., 49(77a), *75*, 98(22a), *100*
Formanek, K., 308(63), *325*
Forssberg, A., 105(121a), *189*, 202(77), 211 (77, 78), *228*, 237(36), 238(36, 77), 243 (77), *245*, *246*, 319(64), *325*
Fowler, C. F., 215(167), *230*
Fowler, J. F., 5(8a), *11*
Fox, M., 253(67), *279*
Fragomele, F., 128(36), 133(36), 136(36), *186*
Franguelli, R., 218(181, 182), *231*
Frank, S., 239(68), 240(68), 241(68), *246*
Franke, H. D., 221(79, 293a), 222(293a), *228*, *234*
Frankenberg, D., 211(224), *232*
Franklin, T. J., 206(80), 216(81), *228*
Franksson, C., 300(25), *324*
Frantz, I. D., 153(119), *189*
Franzen, F., 307(65), 308(65, 66), 310(65, 66), *325*, *326*
Franzius, E., 149(26a), *186*

Frautz, J. D., 211(109), *229*
Frawley, H. E., 243(109), *247*
Freed, R. M., 205(265, 266, 267), 206(266, 267), 207(266), *233*
Freinkel, N., 5(12), 6(12), *11*
French, A. B., 90(23), *100*, 295(67), 297 (67), *326*
Frenkel, E. P., 47(55), 55(166), 56(166), *74*, *78*
Frick, K., 239(37), *245*
Friedberg, M. H., 273(70), *279*
Friedberg, W., 273(68, 69, 70), *279*
Friedman, B. I., 256(17), 263(131b), 170 (185), *277*, *281*, *283*
Friedman, N. B., 103(81), 146(81), *188*
Friedrich, J., 308(66), 310(66), *326*
Fritz, R. R., 39(187), 40, *46*
Fritz, T., 98(24), *100*, 144(82), *188*
Frøholm, L. O., 256(70a), *279*
Fromageot, P., 258(23, 71), 259(22, 23, 23a, 71), 260(23a), *278*, *279*
Fry, R. J. M., 98(24), *100*
Fujii, M., 19(59), *43*
Fulka, J., 301(185), *329*
Furchner, J. E., 19(232), *47*
Furlan, M., 61(166a), 69(47a, b), *74*, *78*, 244 (48), *245*
Furukawa, Y., 308(118), 311(118), *327*
Furth, J., 19(88), 20(88), 31(88), 37(88), *43*, 250(250), 251(250), *285*

G

Gabar, M., 148(291), *194*
Gabay, S., 198(250), *233*
Gadsen, E. L., 21(46), *42*, 89(20), 95(20), *99*, 215(69), *227*
Gaffney, G. W., *192*, 198(240), *232*
Gahlen, W., 206(82), *228*
Gajos, S., 212(120), *229*
Gal, E. M., 308(56), *325*
Galbraith, D. B., 208(83), *228*
Galucci, V., 197(29, 232), *226*, *232*
Galvao, J. P., 31(60), 36(60), *43*
Gangloff, H., 219(84), *228*
Ganis, F. M., 256(72, 73), *279*
Garattini, S., 108(83), 115(83), 116(83), *188*
Garazsi, M., 206(85), *228*
Garcia, E. N., 243(24), *245*
Garcia, J. F., 19(85), *43*
Gardella, J. W., 238(38), 239(38), *245*
Gassner, F. X., 299(32), 301(32), *324*
Gasteva, S. V., 222(86), *228*
Gates, E., 269(40), *278*
Gattung, H. W., 41(258), *48*
Gaumert, R., 117(23, 24, 24a), 123(23), 128 (24a), 131(24a), 133(25), 135(24, 25), 137(24a), *186*
Gautereaux, M. E., 221(15), 225(15), *226*
Gawehn, K., 239(119, 120, 121), *247*
Geel, S., 224(269), *233*
Geierhaas, B., 295(62a, 62b), *325*
Geissler, A. W., 239(119, 120, 121, 122), *247*
Geissler, G., 208(23), *226*
Geisselsoder, J. L., 122(190), *191*, 212(202), *231*, 319(135, 136), 320(135), 321(135, 136), *328*
Genazzani, E., 215(86a), *228*
Geoffrey, H. B., 215(184), *231*
Gerasimova, G. K., 31(61), 38(61), *43*
Gerbaulet, K., 89(25), *100*, 184(84), *188*
Gerber, G., 2(9), *11*, 62(48, 49, 50, 51, 52), 63(134), *75*, *77*, 87(25a, 26), *100*, 141 (86), 146(86, 87), 148(85), 149(85a, 89), 155(91), 162(86a), 169(88), *188*, 198 (87, 90), 199(89, 91), 202(90, 91), 212 (89), 213(87, 88, 90), 215(72), *227*, *228*, 243(87), *246*, 256(75), 257(175), 260 (76, 77, 78, 79), 263(78a), 267(86), *279*, *280*
Gerber, G. B., 2(10), *11*, 61(169a), 62(48, 49, 50, 51, 52), *75*, *78*, 87(25a, 26, 28), 89 (28a), 92(27), 93(28a), 98(58a), *100*, 101, 107(12), 110(12), 115(12), 116 (12), 141(7, 86, 89), 146(86, 87, 89), 148 (85), 149(85a, 89), 155(91, 216, 217, 228), 162(84a, 86, 90), 167(190), 169 (190), *185*, *186*, *188*, *192*, 198(87, 89, 90), 199(89, 91), 202(90, 91), 212(89), 213 (87, 88, 89, 90), *228*, 253(86e), 257(75, 86c), 259(82, 86a), 260(74, 75, 76, 77, 78, 79), 261(80), 262(86b), 263(78, 78a, 83, 85, 282), 267(86), 268(86c, d), 267 (86), 272(84, 222), 273(84, 221, 222, 230), 276(80, 81), *279*, *280*, *284*, *286*, 294(68), 295(68), 316(96), *326*
Gerhardt, P., 39(62), *43*
Gershom-Cohen, J., 269(154), *282*, 294(72), 308(72), *326*
Gerstner, H. B., 208(92a), 220(93), 225(92, 92b), *228*
Gerth, N., 197(200), *231*

Gertler, P., 260(74), *279*
Gewitz, H. S., 239(119, 120, 121), *247*
Ghidoni, J. J., 269(44), *278*
Ghizari, E., 223(8), *225*
Ghossein, N. A., 270(87), *280*
Gibson, J. M., 303(68a), *326*
Gilbert, C., W., 19(63, 64, 65), 31(63, 65), 37, (64, 65), *43*, 250(89), 253(88), *280*
Gilbert, I. G. F., 18(8), 19(8), *41*
Giovanella, B., 239(16), 240(16), *245*
Gitelzon, I. I., 31(66), *43*
Giudice, G., *188*
Gjessing, E. C., 256(90), *280*
Glagov, S., 215(228), *232*
Glas, U., 236(90), *247*
Glaubitt, D., *100*
Glavind, J., 145(93, 94), *188*
Glenn, W. G., 270(91), *280*
Glickman, R. M., 142(299), *194*
Glicksman, A. S., 197(136), *229*
Glinos, A. D., 268(92), 270(92), *280*
Glynn, I. M., 31(67). *43*
Godiniaux, F., 19(156), *45*
Godlewski, H., 206(278), *233*
Goessner, W., 90(29), *100*
Goetze, E., 107(95), 115(95), *188*, 269(93), *280*
Goetze, T., 107(95), 115(95), *188*, 269(93), *280*
Gofman, J. W., 270(116), 271(116), *281*
Goldberg, D. M., 243(39), 244(39), *245*, 269 (93a, 271, 271a), *280*, *285*
Goldfeder, A., 142(96), *188*, 214(94, 194), *228*, *231*, 236(42, 43), 239(40, 41), 244 (44), *245*, *246*
Goldinger, J. M., 15(2), 26(2), *41*
Goldman, D., 253(94), *280*
Goldman, R. F., 292(207), *330*
Goldmann, E., 5(18a, b), *11*
Goldschmidt, L., 31(68), 36(68), *43*
Goldstein, A. L., 142(114), 143(114), *189*
Goldwasser, E., 25(103), 29(103), 30(103), *44*
Goldwater, W. H., 268(96), 271(95), *280*
Gollin, F. F., 237(23), *245*
Golub, F. M., 203(95), *228*
Golubentsev, D. A., 212(96), *228*
Gong, J. K., 202(44), *227*
Gongora, R., 275(136), *281*
Goodman, R. D., 252(97), 253(97), *280*
Goodner, C. J., 252(25), 253(25), *278*
Gordon, E. R., 59(120), 60(120), 69(53, 120), 71(54), *75*, *77*
Gordon, S. A., 267(97a), *280*
Gorinzontov, P. D., 98(30), *100*, 256(98), *280*
Gostmzyk, J. G., 25(194), *46*
Gould, D. M., 165(96a), *188*
Gould, E., 114(40), *186*
Gould, R. G., 105(97, 98), 106(97), 107(97), 109(97), 113(97), 114(97, 98), 115(97), *188*, 259(216), *284*, 293(70), 316(69), *326*
Goutier, R., 59(57, 58), 69(55, 56, 57, 58), 70(57), *75*, 127(281), 145(102), 153 (8a), 154(8a), 162(98a), 163(99, 99a), 173(20, 21, 99a, 281), 176(21, 99a), 177 (100), 178(20, 21, 281), 180(13, 100), 181(13, 100), 182(21, 99a, 281), 184 (99a), 185(101, 102), *186*, *188*, *194*
Goutier-Pirotte, M., 59(57), 69(57, 59), *75*, 145(102), 185(101, 102), *188*, 289(8), 295(8), *323*
Goyer, R. A., 259(98a), 260(98a), *280*
Grabar, P., 268(99), 270(99), *280*
Grachus, T., 95(6a), *99*
Graevskaya, B. M., 141(103), *188*
Graevskii, E. Y., 18(69), 22(69), 24(69), *43*, 236(45), *245*
Granick, S., 20(89), *43*
Gray, E. J., 123(59), 125(59), 138(59), *187*
Gray, I., 273(248, 249), *285*
Gray, J. L., 123(59), 125(59), 138(59), *187*
Greenberg, L. J., 59(61), 69(60, 61), 71(60, 61), *75*
Greenberg, M. L., 31(202), *46*
Greengard, O., 140(106), 157(105), 158 (105), 161(104, 105), *188*, *189*, 267(99a, b), *280*
Gregersen, M. I., 220(97), *228*, 251(38, 182), *278*, *283*
Gregolin, C., 222(223), *232*
Gregora, V., 97(79), *101*
Gresham, P. A., 90(30a), 98(30a), *100*
Greuer, B., 157(140), 161(140), *190*, 267 (168a), *282*
Griem, M. L., 208(98), *228*
Griem, N., 91(4), *99*
Griffin, E. D., 19(70), 24(70), *43*
Griffin, P. L., 292(207), *330*
Griffith, J. Q., 294(71, 72), 308(72), *326*
Griffith, M. J., 294(71, 72), 308(72), *326*
Grigorovitch, N. A., 59(61a), *75*

Grimm, C. A., 269(100), *280*
Grodzenskii, D. E., 106(276), 113(276), 118 (276), 120(276), 155(276), *193*, 291 (28a), 293(181), 301(181), 316(181), *324*, *329*
Gromov, V. A., 144(107), *189*
Gronow, M., 236(46), *245*
Gros, C., 207(180b), *231*
Gros, C. M., 25(154), 26(154), *45*, 253(101), *280*
Gros, P., 158(108), *189*
Grossi, E., 105(109), 107(110), 108(109, 110), 114(110), 115(109, 110), *189*, 222 ((99, 100), *228*
Grossmann, V., 149(75), *187*
Grzybek, H., 212(143), *230*, 292(87), *326*
Guarneri, R., 222(28), 224(28), *226*
Guggenheim, K., 58(62), *75*
Gump, H., 37(22), *42*
Günter, R. G., 269(209), *283*
Guri, C. D., 256(101a, b, c), *280*
Gurnani, S. U., 257(169), 260(169), *282*
Gurney, C. W., 17(119a), 18(72), 19(72), 29(72), 30(70a, 71, 72), *43*, *44*
Guskova, A. K., 274(102), *280*
Gustafson, G. E., 253(103), *280*
Gutniak, O., 20(36), *42*
Gutnitsky, A., 19(73), 29(73), 30(73), *43*, 216(101), *228*
Györgyi, S., 18(99), 19(99), 31(99), *44*

H

Haack, K., 224(238), *232*
Haberland, G. L., 141(148), 146(148), *190*, 198(102, 155), 213(155), *228*, *230*, 260 (104), *280*
Hagemann, R. F., 244(47), *245*
Hagen, U., 59(66), 60(65), 61(66), 64(68), 66 (68), 69(63, 64, 67), *75*, 145(111), 149 (113), 184(112), *189*
Hahn, B., 31(176), *45*
Hahn, E. W., 304(72a), *326*
Hahn, K., 254(173), *282*
Haibach, H., 295(179), 299(179), *329*
Haigh, M. V., 19(63, 64, 65), 31(63, 65), 37 (64, 65), *43*, 250(89), 251(89), 253(88), *280*
Hais, I. A., 311(36a), *324*
Hajduković, S., 28(74), *43*, 215(103), *228*, 291(72b), 308(189, 189a), 321(46a), *325*, *326*, *329*
Halberstaedter, L., 31(75), *43*, 301(73), *326*
Hale, D. B., 206(174, 175), 207(174, 175), *230*
Halevy, S., 58(62), *75*
Haley, T. J., 95(16, 17), *99*, 213(103a), 219 (104, 104a), *228*, 253(105, 106), 257(46), *278*, *280*, 295(74, 75), 296(75), 308(110), *326*, *327*
Hall, C. C., 253(107), 254(107), *280*
Hall, J. C., 142(114), 143(114), *189*
Hallén, O., 220(104b), 222(104b), 223 (104b), *228*
Hamberger, A., 220(104b), 222(104b), 223 (104b), *228*
Hameed, J. M. A., 295(75), 296(75), *326*
Hamilton, H. E., 16(214), 31(214), *46*
Hamilton, L. D. G., 244(48), *245*
Hanel, H. K., 106(116), 113(115), 115(115), 149(116), *189*
Hanks, A. R., 269(118b), *281*
Hanks, G. E., 52(69), *75*
Hansell, J., 269(118a), *281*
Hansen, H. J. M., 105(117), 106(117), *189*
Hansen, L., 105(117), 106(117), 117(118), 122(118), 145(94), 146(118), *188*, *189*
Harbers, E., 238(4, 49, 50, 51, 52, 53, 116), 241(50, 51, 52, 53), *244*, *245*, *247*
Harding, C. V., 218(105), *229*
Hargan, L. A., 214(116), *229*
Harper, P., 91(4), *99*
Harrington, H., 52(69), 59(70a), *75*, 165 (118a), *189*, 237(54, 56), 238(55), 241 (54, 55, 56), *245*, *246*
Harris, D. C., 254(149), 256(149), *281*
Harris, P. S., 19(232), *47*, 256(238), 274 (238), *284*
Harrison, R. G., 204(119), *229*
Harriss, E. B., 17(6), 18(6, 9), 19(6, 9), *41*
Hartiala, K., 90(31), 91(31), *100*, 148(118b), *189*
Hartnett, A. R., 268(177), *282*
Hartung, R., 119(64), *187*
Hartweg, H., 264(108), *280*
Hasl, G., 241(57), 244(57), *246*
Hassler, O., 221(106), *229*
Hasterlik, R. J., 255(147), 256(147), *282*, 312(147), 316(147), *328*
Hausschild, J. D., 250(257), 268(257), *285*
Havens, V. W., 19(205), 29(205), 30(205), 36(205), *46*
Hawkins, S. B., 302(169), *329*

Hawrywicz, E. J., 268(110), 269(109, 110), *280*
Hawtrey, A. O., 240(107), *247*
Hayakawa, J. L., 304(183), *329*
Hayes, J. M., 19(205), 29(205), 30(205), 36 (205), *46*
Hayes, T. L., 270(111, 117), 271(111, 116, 117), *280*, *281*
Haymaker, W., 221(106a, 291), *229*, *234*
Heckmann, U., 291(76), *326*
Hedges, M. J., 95(33a), *100*
Hehl, R., 94(32), *100*, 316(77), *326*
Heidelberger, C., 237(23), 238(50), 241(50), *245*
Heidland, A., 214(107), *229*
Heile, Z., 254(112), 255(112), *281*
Heineke, H., 49(71), *75*
Heiniger, J. P., 39(76), 41(76), *41*(3)
Heise, E. R., 150(34), 153(34), *186*
Heistracher, P., 208(108), *229*
Hektoen, J., 273(113), *281*
Hell, E. A., 24(12), 25(12), *41*, 237(8), *244*
Hemingway, J. T., 316(78), *326*
Hempelmann, L. H., 59(120), 60(120), 62 (49, 50, 51, 52), 69(53, 54, 120, 121), 70 (3, 122), 71(53, 54, 122), *73*, *75*, *77*, 87 (26), *100*, 141(7, 148), 146(87, 148), 153 (119, 227), 169(88), 180(200), 184(236a), *185*, *188*, *189*, *190*, *191*, 198(87, 90, 102, 141, 155, 240), 199(89, 91), 202(90, 91), 204(139), 211(109), 212(89, 149), 213 (87, 88, 90, 149, 155), *228*, *229*, *230*, *232*, 254(163), 255(114), 256(75, 163), 257 (75), 260(74, 75, 76, 79, 104, 170), 263 (75, 78a), 269(164, 165, 166), 274 (114), *279*, *280*, *281*, *282*, 316(96), *326*
Hemphill, C. A., 181(188), *191*
Hemsing, W. 295 (62b, c), *325*
Hendrickson, M. W., 256(73), *279*
Hennessy, T. G., 18(78), 19(77, 78), *43*
Henri, V., 36(79), *43*
Henry, R., 303(90), *326*
Herbert, F. J., 123(59), 125(59), 138(59), *187*
Herdan, A., 84(97a), *102*
Herman, I., 290(23), *324*
Herman, N. P., 237(6), *244*
Hermel, M. B., 269(154), *282*, 294(72), 308 (72), *326*
Herranen, A., 57(73, 73a), 59(72), *75*
Herrera, F. M., 34(126), *44*
Herrick, S. E., 19(77), *43*
Herve, A., 19(169), 20(170), 21(171), *45*, 292(137), 293(137), *328*
Herzfeld, E., 268(115), *281*
Heuse, O., 2(17a), *11*
Heusghem, C., 292(137), 293(137), *328*
Heuss, K., 4(10a), *11*
Hevesy, G., 18(20, 81), 19(20, 80, 81, 82), 21(80, 82), 31(136), 37(136), *42*, *43*, *44*, 105(121a), 110(120, 121), 137(121a), 153(121a), 169(120a), *189*, 202(77), 211(28a, 77), 221(110), *225*, *226*, *228*, *229*
Hewitt, J. E., 270(111, 116, 117), 271(111, 116, 117), *280*, *281*
Hewitt, R., 258(118), *281*
Hiatt, N., 92(32a), *100*, *101*
Hickman, J., 58(7), 63(7), 64(74), *73*, *75*
Hidvégi, E. J., 153(122), 154(122), *189*
Hietsbrink, B. E., 160(122a), 168(122a), *189*
Highland, G. P., 273(248, 249), *285*
Highman, B., 269(118a, b), *281*
Hilding, G., 236(90), *247*
Hildinger, M., 21(182), *46*
Hilf, R., 206(111), *229*
Hilgertovà, J., 66(75), *75*
Hill, B., 62(147), *77*, 142(231, 232), *192*
Hill, R., 107(123), 123(123), 127(123), 129 (79), 130(79), 132(79), 134(79), 137 (123), 140(79), 150(79), 151(79), *188*, *189*
Hilton, S., 69(142), 70(142), *77*
Hilz, H., 238(60), 240(58, 59, 60), 241(58, 58a, 59, 60, 96), 242(58a), *246*, *247*
Himeji, T., 264(119), *281*
Hines, J. F., 221(169), *230*, 307(115), *327*
Hines, M., 208(219, 220, 221, 221a), *232*
Hirashima, K., 29(185), 30(185), *46*
Hirsch, J. D., 165(135a), 168(135a), 169 (135a), *189*, 237(74), 238(74), *246*
Hirsch-Hoffman, A., 241(58a), 242(58a), *246*
Hjart, G., 113(115), 115(115), *189*
Hlavaty, V., 253(120a), 276(120), *281*
Hlavica, P., 240(59), 241(59), *246*
Ho, G. B., 105(225), *192*
Hochman, A., 58(62), *75*, 293(79), *326*
Hochrein, H., 211(111a), *229*, 250(121), 252 (121), *281*
Hockert, H., 220(104b), 222(104b), 223 (104b), *228*

Hockwin, O., 208(21), 218(112, 112a, 139a), *226*, *229*, *230*
Hodgson, G. S., 19(83), 22(178), 25(178), 29(83), *43*, *46*
Hoffman, J. C., 255(114), 274(114), *281*
Hofstra, D., 30(71), *43*
Hohn, M., 295(179), 299(179), *329*
Höhne, G., 239(61, 62), 240(89), 243(82), *246*, *247*, 268(122, 123), *281*
Hokin, L. E., 91(33), *100*
Hokin, M. R., 91(33), *100*
Holland, J., 153(122), 154(122), *189*
Holle, F., *281*
Hollingshead, S., 244(65), *246*
Holmes, B. E., 161(44, 125), 162(44, 124, 125), 164(44, 124, 125), 165(125), 171 (124), 172(44), 177(125), 178(44), *187*, *189*, 239(66, 67), *246*
Holoček, V., 251(125), 252(125), *281*, 291 (80), 312(80), *326*
Holt, P., 251(38), *278*
Holthusen, H., 30(84), 31(84), *43*
Hölzel, F., 153(166), *190*, 239(63), 240(63), 241(58a, 64), 242(58a, 64), *246*
Holzer, H., 239(68), 240(68), 241(68), *246*
Honjo, I., 238(111), *247*
Horikawa, M., 236(69), *246*
Horn, U., 254(54a), *278*
Horn, Y., 31(33), 37(33), *42*
Hornsey, S., 4(11), *11*, 84(34), 95(33a, 35, 97), *100*, *102*, 204(113), *229*, 236(106), *247*, 273(266), *285*
Hornung, G., 150(34, 35), 153(34, 35), *186*, 212(30, 114), 215(30, 114, 131), *226*, *229*, 254(265a), 256(127), 268(126), 269(26, 126, 127, 156, 265a), *278*, *281*, *282*, *285*
Hornykiewytsch, T., 254(128), *281*
Horowitz, M. L., 267(99a), *280*
Hosek, B., 221(270b), *233*
Høst, H., 30(84a), *43*, 216(115), *229*
Hourtoulle, F. G., 303(90), *326*
Hovsepian, J. A., 96(13), *99*
Howard Flanders, P., 218(217, 218), *232*
Howland, J. W., 4(26), *11*, 256(72, 73), *279*
Hrulovsky, J., 268(281), *286*
Huang, K. C., 214(116), *229*, 254(129), *281*
Hubmann, B., 240(58), 241(58), *246*
Hudson, G. W., 91(12), *99*, 105(51), 106 (53, 54), 114(51), 117(50, 51), 118(52), 119(52), 120(52), 123(53, 54, 54a), 128 (53), 137(51), *187*, 257(213), *283*
Huff, R. L., 18(78), 19(78, 85), *43*
Hug, O., 208(117), 219(84), *228*, *229*
Hughes, L. B., 269(130), *281*
Hughes, W. L., 22(54), *42*
Hugon, J., 90(36, 37), *100*
Hulse, E. V., 98(92), *102*, 311(82), 312(81, 83, 83a), 316(81), *326*
Hultborn, K. A., 300(25), *324*
Humphrey, G. F., 24(221), 25(221), *41*
Humphrey, R. M., 253(20), *277*
Hunt, E. L., 219(132), *229*
Hunt, H. B., 31(255), *48*
Hunter, C. G., 7(2), *10*, 256(131), *281*, 294 (12), 305(13), 311(11), 312(11), *324*
Hunter, R. L., 269(131a), *281*
Huntley, G. H., 52(76), 54(76), *75*
Huseby, R. A., 238(115), *247*
Hyman, G. A., 19(138), 21(138), *44*

I

Ibsen, K. H., 243(24), *245*
Ichii, S., 159(202a), *191*, 293(207a), *330*
Il'Ina, L. I., 89(38), *100*
Inbar, M., 304(5), *323*
Ingbar, S. H., 5(12), 6(12), *11*
Ingram, M., 18(86), 19(86), *43*
Inouye, K., *70*
Isley, J. K., 94(80), *101*
Ito, M., 301(84), *326*
Ito, T., 244(71), *246*
Ivanenko, T. J., 294(85), *326*
Iverson, O. H., 207(118), *229*
Ivy, A. C., 303(194), *329*
I-Wen Chen, 263(131b), *281*
Izzo, M. J., 20(38), 31(38), 36(38), *42*

J

Jackson, B. H., 6(4), *10*
Jackson, E., 6(21), *11*, 303(165), 304(165), *329*
Jackson, K. L., 49(77a), 55(77), *75*, 96(39), 97(39), 98(22a, 40), *100*, 119(126), 120 (126), 145(126), *189*, 253(133), 256 (132, 134), *281*
Jacob, M., 55(78), *75*
Jacobson, L. O., 19(85), *43*, 268(219), *284*
Jaffe, J. J., 164(127), 184(127), *189*

Jahn, A., 273(12a), *277*
Jamieson, D., 66(79), *75*, 142(128), *189*
Jammet, H., 274(135), 275(136), *281*
Janík, B., 60(79a), *75*
Janney, C. D., 16(214), 31(214), *46*
Janossy, G., 94(103), 96(103), *102*
Jardetzky, C. D., 111(129), 117(129), 165 (129), *189*
Jaroslaw, B. N., 273(260), *285*
Jasinska, J., 22(87), 24(87), 25(87), *43*
Jaster, R., 268(123), *281*
Jaudel, C., 24(193), 25(154), 26(154), *45*, *46*
Jehotte, J., 253(137), *281*
Jelinek, J., 31(209, 210), *46*
Jenkins, J. O., 221(192), 222(192), *231*
Jenkins, V. K., 273(69), *279*
Jenner, E., 153(166), *190*
Jensen, H., 123(59), 125(59), 138(59), *187*
Jensen, R. E., 250(189), 252(189), 253(189), *283*
Jericijo, M., 69(47b), *74*
Jervis, E. L., 91(74), 93(74), 94(74), 96(74), *101*
Jesdinsky, 5(18b), *11*
Jesseph, J. E., 98(3), *99*
Jirounek, P., 253(120, 120a), 276(120, 120a), *281*
Joel, D. D., 31(203), *46*
Johansen, G., 113(115), 115(115), *189*
John, D. W., 155(178), *191*, 273(137a), *281*
Johnson, D. S., 5(25), *11*
Johnson, H. A., 31(202), *46*
Johnston, J., *194*
Jolles, B., 204(119), *229*
Jolly, J., 310(86), *326*
Jonek, J., 90(41, 46), *100*, 212(120, 143), 222(120a), *229*, *230*, 292(87, 88), *326*
Jonek, T., 222(120a), *229*, 292(87), *326*
Jones, D. C., 312(89, 138), 316(89), *326*, *328*
Jones, H. B., 271(116), *281*, 303(60), *325*
Jonnard, M. L., 293(70), *326*
Jordan, D. L., 18(186), 19(186), *46*
Joseph, B. J., 260(7), 262(7), *277*
Jovanovic, M., 223(120b, 121, 122), *229*, 256(140), 269(138, 139), *281*
Juret, P., 303(90), *326*

K

Kacaki, J., 291(161), *328*
Kagan, E. H., 222(122a), *229*
Kahn, J. B., 19(88), 20(88), 31(88), 37(88), *43*
Kainova, A. S., 110(130), 119(130), *189*
Kaiser, J., 90(41, 46), *100*
Kalashnikova, V. P., 142(131), *189*, *281*
Kalendo, G. S., 238(80), 246
Kalic, J., 274(135), *281*
Kallee, F., 70(129), *77*
Kapis, M. V., 211(259), 215(259), *233*
Kaplan, H. S., 52(69), *75*
Kappas, A., 20(89), *43*
Karanovic, J., 31(90), 36(90), *43*, 321(46a), *325*
Karas, J. S., 275(142), *282*
Kärcher, K. H., 90(42), *100*, 199(125), 207 (122b), 215(123, 124), *229*, 243(72, 73), *246*, 254(145), 268(144), 269(143, 144), *282*
Karon, H., 38(30), 39(30), *42*
Karpfel, Z., 22(228, 229), 23(228), 24(229), 25(228), 26(228), *47*
Karpova, E. V., 204(135), *229*
Kashima, M., 308(119b), *327*
Kashkin, K. P., 268(99, 146), 270(99, 146), *280*, *282*
Kato, H. T., 90(42), *100*, 215(123), *229*, 268 (144), 269(143, 144, 146a), *282*
Katz, E. J., 255(147), 256(147), *282*
Kay, E. R. M., 24(221), 25(221), *47*
Kay, R. E., 84(43), 85(43), 91(44), *100*, 122 (133), 146(132), 153(134), *189*, *229*, 252 (150), 253(150), 254(148, 149, 152), 255 (150, 152), 256(149), 258(151), 260 (152), *282*, 312(92), 316(91), *326*
Keating, R. P., 250(244), 252(244), 268 (244), *285*
Kedrova, E. M., 218(207), *231*
Keegan, P., 293(70), *326*
Keim, O., 199(146), *230*
Keith, C. K., 27(41), 29(41), *42*
Keller, M., 215(127), *229*, 268(269), *285*
Kelly, L., 168(19), 174(19), *186*
Kelly, L. S., 165(135a), 168(135b, 209), 169 (135b, 209), 173(209), 174(209), 177 (135), *189*, *192*, 237(74), 238(74), *246*
Kember, N. F., 202(128), *229*
Kennes, F., 58(136), 61(169a), *77*, *78*, 155 (91, 216, 217, 228), *188*, *192*, 268(229, 231, 260a), 272(84, 222, 229), 273(84, 221, 222, 223, 230, 231, 260a), *279*, *284*, *285*
Kent, S. P., 218(129), *229*
Kent, T. H., 92(44a), *100*

Kereiakes, J. G., 256(17, 184), 263(131b), 270(185), *277*, *281*, *283*, 304(184), *329*
Kerova, N. I., 269(153), *282*
Kertai, P., 273(259), *285*
Kessler, G., 269(154), *282*
Khanson, K.P., 142(135c), *189*
Kharitonova, L. V., 239(92a), *326*
Khnychev, S. S., 256(98), *280*
Khoyi, M. A., 30(33a), *42*
Khursin, N. E., 269(155), *282*
Kihn, L., 206(130), *229*
Kikuchi, Y., 215(131), *229*, 269(156), *282*
Killander, D., 239(75, 76), *246*
Killar, M., 223(52), *227*
Kimeldorf, D. J., 7(14a), *11*, 203(211), 219 (132), *229*, *232*, 312(89, 138), 313(149), 316(89), 322(92b), *326*, *328*
King, E. R., 255(35), *278*
King, R., 59(120), 60(120), 69(120), *77*
Kinnamon, K. E., 198(133), *229*
Kinoshita, J. H., 218(153), *230*
Kirschman, J. C., 123(53), 128(53), *187*
Kirslis, M., 252(158), 254(158), 268(158), *282*, 316(94), *326*
Kirsten, E., 62(15), *74*
Kirzon, M. V., 225(134), *229*
Kisecleski, W., 203(60), *227*
Kiselev, P. N., 204(135), *229*
Kisielski, W. E., 98(24), *100*
Kitagawa, T., 197(136), *229*
Kivy-Rosenberg, E., 63(80, 81), *75*, *76*, 129 (135d, 136), *190*
Kiyasu, J., 107(123), 123(123), 127(123), 137(123), *189*
Klapheck, U., 29(112), 30(112), *44*
Klatzo, I., 221(291), *234*
Klebanoff, S. J., 39(92), 41(91, 92), *43*
Kleeman, C. R., 214(43), *226*
Klein, E., 238(17), *245*
Klein, G., 237(36), 238(36, 77), 243(77), *245*, *246*
Klein, J. R., 18(93, 95), 19(93, 94, 95), *43*
Kleinbock, T. S., 236(5), *244*
Kleinest, H., 291(76), *326*
Klemm, J., 31(176), *45*
Klemm, S., 240(79), 241(79), *246*
Klener, V., 19(97), 30(96, 97), *43*, *44*
Kleschick, A., 118(238, 239), 119(238, 239), *193*
Klimovskaya, L. D., 223(137, 138), *230*
Klinger, W., 257(156a), *282*
Klipfel, F. J., 157(264), 161(264), *193*, 267 (263), *285*
Klouwen, H. M., 56(82), 63(83), 64(83, 84, 85, 86), 71(176), *76*, *78*, 173(137), *190*, 243(78), *246*
Klütsch, K., 214(107), *229*
Kmentova, V., 251(125), 252(125), *281*, 291 (80), 312(80), *326*
Knowlton, N. P., 204(139), *230*
Knox, W. E., 267(99a), *280*
Knuppen, R., 54(18), *74*, 253(29), *278*
Kobayashi, Y., 90(45), *100*, 293(207a), 308 (93), *326*, *330*
Kobert, E., 142(5), 144(5), *185*
Koch, H. R., 6(13), *11*, 218(139a), *230*
Koch, R., 18(49), 19(49, 98), *42*, *44*, 124 (138, 139), 149(113), *189*, *190*, 240(79), 241(79), *246*
Koch, W., 40(260), *48*
Kocmierska-Grodzka, D., 223(52), *227*
Koella, W. P., 220(24), *226*
Koenig, V. L., 218(233), *232*
Kohen, C., 239(79a), *246*
Kohen, E., 239(79a), *246*
Kohn, H. I., 17(145), *45*, 138(1), 159(1), *185*, 216(4), *225*, 252(157, 158), 254(157, 158), 268(157, 158), *282*, 316(94), *326*
Kolár, J., 201(13), 202(12, 13), 203(13, 14, 14a, 140), *225*, *226*, *230*
Koldobskaya, F. D., 212(42), 221(42), 222 (42), *226*
Koletsky, S., 253(103), *280*
Kollmann, G., 31(211, 212), 35(212, 213), 36(211), 38(211), *46*
Kollmorgen, G. M., 273(159), *282*
Koloušek, J., 18(99), 19(99), 31(99), *44*, 256 (160), 276(160), *282*, 301(185), *329*
Komei, U., 221(198), *231*
Komesu, N., 253(105), *280*, 295(74), *326*
Komesu, U. M., 213(103a), *228*
Konecki, J., 292(88), *326*
Konno, K., 198(141, 142), *230*, 257(161), *282*
Konstantinova, M. M., 236(45), *245*
Konstantinova, M. S., 58(87), *76*
Konstantinova, V. V., 24(130), 28(130), *44*
Kopylov, V. A., 23(107), 24(107), *44*
Kordik, P., 90(8), *99*
Korneeva, N. V., 90(15), *99*
Kosenko, A. S., 256(62, 63), *279*
Koslowski, L., 308(95), *326*

Kosmider, S., 90(41, 46), *100*, 212(120, 143), 222(120a), *229*, *230*
Kostyk, I., 118(238), 119(239), *193*
Kosyakov, K. S., 256(162), *282*
Koszalka, T. R., 254(163), 256(163), 260 (77, 79), *279*, *282*, 316(96), *326*
Koths, M., 55(143), *77*
Kovacs, L., 94(103), 96(104), *102*
Kover, G., 31(100), 38(100), *44*
Kowalski, E., 20(36, 239), 25(239), *42*, *47*
Kowlessar, O. D., 269(164, 165, 166), *282*
Kozinets, G. I., 19(101, 102), 21(101), 24 (101, 102), 25(101, 102), *44*
Krantikova, T. V., 269(237), *284*
Krantz, S. B., 25(103), 29(103), 30(103), *44*
Kraupp, O., 208(108), *229*
Krause, P., 80(47), *100*
Krause, R. F., 17(104), *44*
Krebs, J. S., 109(157), 162(157), 171(157), *190*
Kretchmar, A. L., 256(167), 275(9), *277*, *282*
Kreuckel, B., 168(19), 174(19), *186*
Kriegsmann, B., 211(111a),*2 29*, 250 (121), 252(121), *281*
Krise, G. M., 260(7, 168, 276), 262(7), *277*, *282*, *286*, 312(97, 200, 201), 313(201), 316(97), *326*, *329*, *330*
Kritskii, G. A., 22(106, 108), 23(105, 107, 108), 24(107), 29(108, 109), *44*
Kroeger, H., 267(168a), *282*
Kröger, H., 157(140), 161(140), *190*
Krohm, G., 225(277a), *233*
Krstic, M., 308(189), *329*
Kubena, K., 218(144), *230*
Kucherenko, M. E., 221(145, 253, 253a), *230*, *233*
Kucheryavyy, F. Kh., 269(237), *284*
Kukushkina, V. A., 212(42), 221(42), 222 (42), *226*
Kulinskii, V. I., 308(97a), *326*
Kumatori, T., 24(120), *44*, 64(118a), *77*
Kumta, U. S., 257(169), 260(169), *282*
Kundel, H. L., 312(98), *326*
Künkel, H. A., 19(111), 92(112), 30(112), 31(113), 37(113), *44*, *45*, 62(15), *74*, 153 (237), *192*, 231(61, 62), 239(83, 101), 240(83, 89), 241(101), 243(82, 83), *246*, *247*, 268(122, 123), *281*, 291(76), *326*
Kunz, J., 199(146, 147), *230*
Kurata, N., 211(148), 212(148), *230*
Kurnick, N. B., 28(114, 115), 29(114, 115), *44*, 60(10), 69(88, 89, 90), 70(88), 71 (89, 91), *73*, *76*, 185(140a), *190*
Kurohara, S. S., 212(149), 213(149), *230*, 254(171), 260(75, 170, 171, 172), *279* *282*, 316(96, 99), *326*
Kurosu, M., 239(33), *245*
Kurtsin, I. T., 91(48), 92(48), *100*, 308(100), 312(100), *327*
Kuschke, H. J., 308(27, 28), *324*
Kusnets, E., 142(141), *190*
Kuss, B., 91(100), 92(100), 94(100), *102*
Kuzin, A. M., 60(92), 72(93), *76*, 212(150), *230*, 238(80, 113), *246*, *247*
Kuz'Menko, L., 254(218), *284*
Kyo, S. K., 202(151), *230*

L

Labanova, N. M., 212(152), *230*
Laborde, M., 304(101), *327*
Lacassagne, A., 301(102), *327*
Lachapelle, J. M., 208(271, 272), *233*, 313 (186), *329*
Ladner, H. A., 212(239a), *232*, 254(173), 264(174), *282*
La Grutta, G., 223(1, 10, 11), *225*
Lajtha, L. G., 15(3a), 17(116, 116a, 119, 121, 234), 18(21, 234), 19(116, 234), 20 (121, 234), 24(12, 117, 118, 119, 120), 25(12, 118, 119), 29(3a, 116a, 185), 30 (3a, 70a, 116a, 184, 185), *41*, *43*, *44*, *46*, *47*, 52(76), 54(76), *75*, 164(127), 184 (127), *189*, 237(8), *244*
Lalanne, C. M., 303(90), *326*
Lamar, C., Jr., 156(213, 214), 158(213), 159 (213), 160(213), 161(214), *192*
Lambert, B. W., 218(153), *230*
Lambert, S., 20(170), 21(171), *45*
Lamberts, H. B., 199(154), *230*, 311(103), 312(103), 313(29, 30), *324*, *327*
Lambiet, M., 94(55), *101*
Lamdin, E., 153(119), *189*, 211(109), *229*
Lamerton, L. F., 17(6), 18(6, 8, 9), 19(6, 8, 9, 123, 183), *41*, *44*, *46*, 84(39), *100*
Lancker, J. L. Van, 55(42, 177), 69(42), *74*, *78*, 162(72), 163(145b), 166(17), 167 (142, 143, 144, 145), 168(73, 144, 145, 278), 172(73, 279), 173(17, 73, 278), 174(279), 177(72, 73, 278), 178 (73, 278, 279), 180(142, 143, 144, 145), 181 (144, 278), 182(278), *186*, *187*, *190*, *193*

Lane, J., 252(158), 254(158), 268(158), *282*
Lane, J. L., 105(146), *190*, 305(104), *327*
Langendorff, H., 6(13), *11*, 18(49), 19(49), *42*, 58(94), *76*, 146(147), 149(113), *189*, *190*, 255(175), 264(175), 265(175), 266 (176), 267(176), *282*, 308(105), *327*
Langendorff, H.-U., 151(258), *193*, 264 (254a), 265(254a), 266(253, 254a), 267 (253, 254a), *285*
Langham, W. H., 19(232), *47*, 256(238), 274 (238), *284*, 322(182), *329*
Langlands, A. O., 257(242a), 264(242a), *285*
Lansangan, L., 269(6), *277*
Larina, M. A., 256(64), 260(64), *279*, 294 (106, 107), *327*
Larsen, K., 215(228), *232*
Lartigue, O., 124(163), *190*
Lascelles, J., 164(127), 184(127), *189*
Latarjet, R., 274(135), *281*
Lathrop, T. P., 154(297a), *194*
Lauber, K., 255(4), 258(207), *277*, *283*
Lauenstein, K., 141(148), 146(148), *190*, 198(155), 213(155), *230*
Laughlin, J. S., 200(293), 202(293), *234*
Lavik, P. S., 52(70), *75*, 165(118a), *189*, 237 (54, 56), 238(55), 241(54, 55, 56), *245*, *246*
Law, A. W., 319(40, 108), *325*, *327*
Lawrence, J. H., 19(85), *43*
Lawrence, J. N. P., 256(238), 274(238), *284*
Lebedeva, G. A., 98(30), *100*
Lebedinskii, A. U., 219(156), *230*
Leblond, C. P., 118(149), *190*, 318(109), *327*
Lecomte, J., 308(14), *324*
Ledford, E. A., 252(158), 254(158), 268 (158), *282*
Ledney, G. D., 6(14), *11*
Lee, T. C., 58(94a), *76*, 111(150), *190*
Lefaure, J., 273(176a), *282*
Legakis, N. J., 58(105b), *76*
Le Go, R., 275(136), *281*
Lehmann, F., 31(124), *44*
Lehmann, F.-G., 212(114), 215(114), *229*, 268(126), 269(126), *281*
Lehnert, S. M., 167(152, 152a, 153), 168 (152a), 169(153), 180(153), 183(152), 184(151, 152, 153), *190*
Leibe, H., 257(156a), *282*
Leibowitz, J., 31(75), *43*
Leitch, J. L., 308(110), *327*
Leites, F. L., 199(157), *230*
Lejsek, K., 134(306), 138(306), 144(304, 306), *194*
Lelièvre, P., 124(154, 155, 156), 141(154, 156), *190*
Lelievre, P., 211(158), 222(158), *230*
Lemaire, M., 304(111), *327*
Le May, M., 215(159), *230*
Lemon, H. M., 302(112), *327*
Lengemann, F. W., 95(50), *100*
Lensen, I., 294(51), 295(51), 299(51), *325*
Lenskaya, R. V., 308(36), *324*
Lenzinini, L., 122(30), 128(30), 130(30), 131(30), 132(30), 133(30), *186*
Leoncini, G., *42*
Leone, C. A., 268(177), *282*
Leong, G. F., 109(157), 162(157), 171(157), *190*
Lerman, S., 218(160, 161, 162), *230*
Lerner, S. R., 106(158), 108(158), 110(288), 111(288), *190*, *194*
Le Roy, G. V., 105(225), *192*
Lesher, S., 156(214), 161(214), *192*
Lessler, M. A., 34(125, 126), *44*
Lett, J. T., 237(7), *244*
Leuchtenberger, C., 22(251), 24(251), *47*
Leusen, J., 295(113), *327*
Levdikova, G. A., 218(207), *231*
Levia, G. M., 58(105b), *76*
Levin, W. C., 24(157), 31(204), 36(204), 38 (127, 157, 204), 39(157), *44*, *45*, *46*
Levine, D. L., *101*
Levis, G. M., 16(158), 17(158), *45*
Levitt, S. H., 250(178), *283*
Levy, B., 109(159), 124(159), 141(159), *190*
Levy, C. K., 218(163), 219(163), 220(24), *226*, *230*
Levy, L., 33(165), 36(165), *45*
Lewis, R. B., 208(92a), *228*
Li, S.-G., 142(296, 297), *194*
Li, S.-K., 142(296, 297), *194*
Liberman, A. N., 202(259b), *233*
Libikova, N. I., 256(162), *282*
Libinzon, R. E., 22(131), 24(128, 130, 132), 28(129, 130), 29(128, 129), *44*
Lichter, R. J., 273(248, 249), *285*
Lichtler, E. J., 238(38), 239(38), *245*
Lieberman, I., 177(277), *193*
Liechti, A., 31(28, 133), *42*, *44*

Lierse, W., 221(79, 293a,) 222(293a), *228*, *234*
Likhotkin, I. P., 221(64), *227*
Liljedahl, S. O., 273(19, 273a), *277*, *286*
Lilly, E. H., 105(97), 106(97), 107(97), 109 (97), 113(97), 114(97), 115(97), *188*, 293(70), 316(69), *326*
Lim, J. K., 15, 16(134), *44*, 119(159a), *190*
Lin, C.-H., 142(296, 297), *194*
Lin, I-San R., 224(239), *232*
Lindemann, B., 31(135), 34(135), *44*
Lindop, P. J., 5(6a), *11*
Lindsay, I. R., 269(44), *278*
Linser, P., 255(179), 256(179), *283*
Lipiński, B., *42*
Lipkan, N. F., 197(164), 205(164), 206(164), *230*
Lipkin, M., 89(51), *100*
Lippincott, S. W., 273(179a), *283*
Lisco, H., 255(114), 268(219), 274(114), *281*, *284*
Little, J. B., *329*
Liu, C. T., 213(165), 214(165), *230*
Lluch, M., *101*
Loboda, L. A., 273(225), *284*
Locke, B., 218(277b), *233*
Lockner, D., 31(136), 37(136), *44*
Loeffler, R. K., 19(138), 21(137, 138), 24 (137), *44*
Loew-Beer, A., 239(81), *246*
Lofland, H. B., 199(9), *225*
Logan, R., 57(95, 96), *76*, 169(160), 170(160, 161), 174(160, 161), 175(160), 184(160, 161), *190*
Lohmann, W., 215(167), *230*
Löhr, E., 27(140, 141), 28(140), 39(139), *45*
Loiseleur, J., 39(142), *45*, 254(180), *283*
Loiselle, T. M., 271(191a), *283*
Loken, M. K., 58(94a), *76*, 111(150), *190*, 243(97), *247*
Lombardi, S., 212(196), 214(196), *231*
Lomova, M. A., 199(168), *230*
Lónai, P., 153(122), 154(122), *189*
Long, M., 308(120), *327*
Looney, W. B., 84(52), 85(52), 87(52), 89 (52), *100*, 162, *187*, 237(19a), *245*
Lord, B. I., 17(143, 144), *45*
Lorenson, M. G., 177(171), *191*
Lorenz, W., 212(7), *225*, 239(122), *247*, 269 (209), *283*
Loring, W., 60(10), *73*
Lossen, H., 212(7), *225*
Lott, J. R., 221(169), 225(170), *230*, 294 (114), 307(115), *327*
Lotz, W. E., 203(60), *227*
Loud, A. V., 114(40), *186*
Lourau, M., 123(162), 124(162, 163), *190*
Louraw-Pitres, M., 104(164), 108(164), 109 (164), 117(164), 137(164), *190*
Low-Beer, B. V. A., 58(163), *78*, 162(249), *193*, 222(248), *233*
Lowman, D. M. R., *277*
Lowrey, R. S., 202(171), *230*
Lucas, D. R., 217(172), *230*
Ludwig, F. C., 17(145), *45*
Ludewig, S., 215(173), *230*, 253(37), *278*
Luk'ašová, E., 54(12), *74*
Lukin, L., 250(181), 251(38, 182), *278*, *283*
Lundgren, P. R., 221(192), *231*
Lundin, J., 28(146, 147), 38(147), *45*, 269 (183), *283*
Lushbaugh, C. C., 97(153), *101*, 206(174, 175), *230*, 256(238), 274(238), *284*
Lutwak-Mann, C., 15(149), 16(149), 22 (148), 23, 24(148, 149), *45*
Luz, A., 90(29), *100*
Luzzio, A. J., 256(184), 270(185), *283*
Lyman, C. P., 53(100), *76*
Lyons, C., 197(201), *231*
Lyons, E. A., 6(17), *11*

M

Maass, H., *44*, 56(97), 62(15, 98), *74*, *76*, 135 (165, 167), 138(167), 153(166, 237), *190*, *192*, 239(10, 62, 63, 88, 89, 101), 240(63, 83), 241(101), 242(58a), 243 (82, 83), *244*, *246*, *247*
McBeth, C., 268(177), *282*
McCandless, R. G., 95(18), *99*
MacCardle, R. C., 141(168), 144(168), *190*
McCay, P. B., 138(168), 167(168), *190*
McCormick, D. B., 91(12), *99*, 106(53, 54), 123(53, 54), *187*
McCormick, W. G., 213(103a), *228*
McCullah, E. F., 213(103a), *228*
McCulloch, E. A., 24(150), 26(241, 242), *45*, *47*
McDonald, R. E., 250(189), 252(189), 253 (189), *283*
McElya, A. B., 23(19), *42*, 168(29), 179(29), 180(29), 182(29), 183(29), *186*
Macfarlane, M. G., 138(169), 167(169), *190*

McFayden, D., 90(60), *101*, 273(12a), *277*
McGarrahan, K., 114(40), *186*
McGoodall, C., 308(120), *327*
MacIntosh, I. J. C., 246(107), *247*
McKee, R. W., 122(170), 123(170), 150(38), 151(38), *186*, *190*, 243(24), *245*, 269(30), *278*, 316(121), *327*
McKelvey, R. W., 215(184), *231*
McKenney, J. R., 96(61), *101*
McLaughlin, M. M., 269(27, 28), 270(190), *278*, *283*
MacVicar, J., 269(93a, 271), *280*, *285*
Maetz, M., 98(3), *99*
Magalhaes, E. M., 31(60), 36(60), *43*
Magdon, E., 39(151), 41(151), *45*, 236(83a), *246*
Magee, M. Z., 268(66), *279*
Magyar, Z., 273(259), *285*
Main, R. K., 24(50, 152), 25(50, 152), *42*, *45*, 62(99), *76*, 217(176), *231*
Maisin, J. R., 84(56), 86(56), 89(56), 91(56), 94(19a, 55, 56), 95(54), 96(54, 56), *99*, *101*, 268(229, 231), 270(229, 231), 273 (260a), *284*, *285*
Maker, D., 254(218), *284*
Makhlina, A. M., 213(177), *231*
Makinodan, T., 52(99a), *76*
Malcic, K., 291(161), *328*
Maletta, G. J., 223(178, 179), *231*
Maley, F., 177(171), *191*
Maley, G. F., 177(171), *191*
Malina, L., 39(227), *47*
Malinowska, T., 20(36, 153, 239), 25(239), *42*, *45*, *47*
Malkinson, F. D., 208(98), *228*
Manasek, F. J., 53(100), *76*
Manasek, G. B., 54(1a), *73*
Mandel, H. G., 169(55), 175(55), *187*, 238 (25), *245*
Mandel, P., 24(193), 25, 26(154), 38(30), 39(30), *42*, *45*, *46*, 55(78), 57(101), *75*, *76*, 138(172), 141(172), 167(45a), 169 (45a), 182(45a), *187*, *191*, 207(179a, 180, 180b), 218(180a), 220(180a), *231*, 253(101), *280*
Mannigel, U., 295(62c), *325*
Mantyeva, V. L., 239(84), *246*
Mantzos, J. D., 16(158), 17(158), *45*, 58 (105b), *76*
Maor, D., 71(102), *76*
Maraini, G., 218(181), *231*
Marble, G., 275(136), *281*
Marcato, M., 218(181, 182), *231*
Marcovich, H., 301(102), *327*
Marchetti, M., 151(173), *191*
Marcus, C. S., 95(57), *101*
Marich, L. N., 253(186), *283*
Maros, T., 118(174), *191*
Maroteaux, P., 268(187), 270(187), *283*
Marotta, U., 34(34), *42*
Marsella, A., 34(34), *42*
Marsiti, G., 91(58), *101*
Martens, H. H., 256(191), *283*
Martin, J. H., 263(240), *284*
Martin, R. L., 69(60), 71(60), *75*
Martinovitch, P., 289(7), 291(116), 293(7), 295(7), *323*, *327*
Marvin, J. F., 243(97), *247*
Maslova, A. F., 223(137, 138), *230*
Mason, H. C., 211(183), *231*, 253(202), *283*
Mason, W. B., 18(86), 19(86), *43*
Masoro, E. J., 117(175), *191*
Massey, B. W., 28(114), 29(114), *44*, 69(88, 89), 70(88), 71(89), 76, 185(140a), *190*
Masters, R., 153(119), *189*, 211(109), *229*
Mateyko, G. M., 297(117), *327*
Mathé, G., 268(187), 270(187), 274(135), *281*, *283*
Mathieu, O., 98(58a), *101*
Matsudaira, H., 236(84a), *246*
Matsuoka, O., 308(118, 119b), 311(118), 312(119, 119a), *327*
Matsuzawa, T., 81(59), 84(59), *101*, *102*
Mattausch, H., 255(251), *285*
Matyášová, J., 22(217, 218, 219, 220), 24 (218), 25(218), *47*, 59(103, 158, 159), *76*, *78*
Maupin, B., 274(135), *281*
Maurer, W., 89(25), *100*, 184(84), *188*
Mauss, H.-J., 19(111), 29(112), 30(112), 31 (113), 37(113), *44*, *45*
Mavely, D., 260(196), 265(196), *283*
Maxwell, E., 62(104), 63(104), *76*
Maxwell, R. D., 250(244), 252(244), 268 (244), *285*
Mayer, A., 36(79), *43*
Mayr, H.-J., *45*
Mazanowska, A., 20(36, 155), *42*, *45*, 128 (57), *187*
Mazia, D., 72(104a), *76*
Mazurik, V. K., 256(188, 188a), 263(188), *283*

Mazza, A., 207(37), *226*
Mazzone, A., 151(9), 159(9), 170(9), *186*
Mead, J. F., 95(6b), *99*
Meardi, G., 31(90), 36(90), *43*, 243(3), *244*
Mechalli, D., 19(156), *45*
Mee, L. K., 161(44, 125), 162(44, 125, 175a), 164(44, 125), 165(125), 172(44), 177 (125), 178(44), *187*, *189*, *191*, 239(67), *246*
Mefferd, R. B. Jr., 256(191), *283*
Mehran, A. R., 95(62), *101*, 271(191a), *283*
Mehrisch, J. N., 244(35a, 108), *245*, *247*
Mehrotra, R. M. L., 118(1a), *185*
Melching, H.-J., 58(94a, 105, 165), *76*, *78*, 146(147, 259), 150(260), 151(260), 152 (260), 153(260), *190*, *193*
Melanotte, P. L., 202(185), 203(185), *231*
Melching, H. J., 215(255), *233*, 255(175, 192, 193, 194, 195, 251), 258(192), 260(192, 193, 196, 197), 264(175, 192, 193, 194, 253a), 265(175, 192, 193, 194, 195, 196), 266(176, 192, 193, 194, 195), 267(176), *282*, *283*, *285*, 307(122), 308(105, 122), 310(123), *327*
Melik-Musyan, A. B., 218(186), *231*
Mellmann, J., 5(18b), *11*
Mellmann, T., 5(18a), *11*
Mendelsohn, M. L., 214(187, 188), *231*
Meredith, O. M., 220(188a), *231*
Merenov, V. V., 21(235), *47*
Meriam, G. R., 218(58), *227*
Merits, I., 221(188b), *231*
Merkina, T. N., 256(162), *282*
Meschan, I., 27(41), 29(41), *42*, 185(74), *187*
Messerschmidt, O., 256(252), 260(197), *283*, *285*, 322(123a), *327*
Meyniel, J., 301(101), *327*
Michaelis, M., 224(229), *232*
Michaelson, S. M., 4(26), *11*
Michalika, W., 6(7, 8), *11*
Michalowski, A., 22(87), 24(87), 25(87), *43*
Migion, C. J., 295(67), 297(67), *326*
Mihm, K., 31(113), 37(113), *44*
Mikhailova, L. F., 256(278a), *286*
Mikuta, E. T. ,157(265, 266), 161(265, 266), *193*, 267(262), *285*
Milch, L. J., 212(189, 284), *231*, *234*, 269 (198, 199, 200), 271(198, 199, 200, 272), *283*, *286*
Miller, C. J., 52(105a), *76*
Miller, H. M., 223(234), *232*
Miller, L. A., 236(43), 244(44), *245*
Miller, L. L., 112(41), 147(177), 155(178, 243), *186*, *191*, *193*, 260(77), 269(40), 273(137a, 235), *278*, *279*, *281*, *284*
Millo, A., 211(190), *231*
Mills, G. C., 24(157), 38(157), 39(157), 45
Minayev, P. F., 223(191), *231*
Minder, W., 202(298), *234*
Miquel, J., 221(192, 291), 222(192), *231*, *234*
Miras, C. J., 16, 17(158), *45*, 58(105b), *76*
Mironova, T. M., 212(42), 221(42), 222(42), *226*
Mirzoyan, A. A., 308(58a), *325*
Mishkin, E. P., 267(200a), *283*
Mishkin, P. E., 158(179), 161(179), *191*
Misiti-Dorello, P., 35(13), *41*
Misurale, F., 124(202), *191*
Mitchell, C. T., 291(2), 312(2), *323*
Mitchell, R. I., 96(5), 97(5), *99*
Mizon, L. G., 312(83a), *326*
Mkhitaryan, V. G., 222(193), *231*
Modig, H., 236(93), 243(94), 244(93), *247*
Moguilevsky, J. A., 224(269), *233*
Moje, A. P., 308(20), *324*
Mole, R. H., 84(63), 90(8), 98(92), *99*, *101*, *102*, 122(180), 125(180), 148(8), *185*, *191*, 301(124), 319(40, 108), *325*, *327*
Mollura, J. L., 214(194), *231*
Momma, H., 308(125, 129, 130), *327*, *328*
Monacelli, R., 34(34), *42*
Monkhouse, F. C., 269(12, 201), *277*, *283*
Monroe, R. A., 304(126), *327*
Monsul, R. W., 218(233), *232*
Montano, A., 69(89), 71(89), *76*
Montemagno, U., 215(86a), *228*
Montgomery, P. O. B., 313(127), 316(127), *327*
Moorehead, W. R., 237(85), 239(85, 86), *246*
Moos, W. S., 211(183), *231*, 253(202), *283*
Morcek, A., 31(159), 38(159), *45*
Morczek, A., 31(160), *45*,
Morehouse, M. G., 94(64), *101*, 107(182), 109(184), 110(181, 184), 111(181, 183, 184), 112(183, 184), 122(180), 123(182), 125(180), 128(182), 136(185), *191*
Morgenstern, L., *101*
Mori, K. J., 56(106), *76*
Morita, T., 52(99a), 56(106), 76
Morozoskaya, L. M., 308(35), *324*
Morris, H. P., 237(19a), *245*
Morrison, L. M., 199(195), *231*

Morton, J. L., 269(45), *278*
Morton, M. E., 111(80), *188*
Morton, T. L., 31(261), *48*
Moss, A. J., 215(167), *230*
Moss, W. T., 94(65), *101*
Mosser, D. G., 58(94a), *76*, 111(150), *190*, 243(97), *247*
Mottram, J. C., 94(66), 95(66), *101*
Mozurri, M. S., 90(5a), *99*
Muderrisoglu, O. V., 273(69), *279*
Muggia, F., 89(51), *100*
Mukerjee, H., *246*
Muntz, J. A., 91(6), *99*, 268(203), *283*
Muramatsu, S., 304(183), *329*
Musiyko, V. A., 269(204), *283*
Muto, V., 212(196), 214(196), *231*
Mutton, D. E., 225(53b), *227*
Myers, D. K., 31(162, 163, 164, 179), 33 (164, 165), 34(163), 35(166), 36(162, 163, 164, 165), 37(163), 38(162, 164, 165), 39(161, 162), *45*, *46*, 52(113), 53 (113), 54(109, 113), 55(109), 56(113), 61 (108, 109, 110, 112), 64(6), 65(6), 66 (107, 110), *73*, *76*, *77*, 142(186), 179 (187), 181(187, 188), 182(187), *191*

N

Naegele, W., 90(2), 99
Nagasawa, H., 236(84a), *246*
Nahm, H., 256(127), 269(127), *281*
Nair, V., 160(189), *191*, 220(197), *231*, 307 (128), 310(128a), 318(128a), *328*
Nakamura, W., 308(125, 129, 130), *327*, *328*
Nakazawa, S. Y., 221(198), *231*
Nakhil'nitskaya, Z. N., 31(167), *45*, 219 (156), *230*
Nakken, K. F., 31(168), 35(168), *45*
Nam, S. B., 202(151), *230*
Nance, S. L., 143(267), 144(267), *193*
Näntö, V., 90(31), 91(31), *100*, 148(118b), *189*
Napier, J. A. F., 244(35a), *245*
Natarajan, S., 308(131), *328*
Neumayr, A., 313(132), *328*
Nejtek, V., 94(96a), *102*
Nekrasova, I. V., 236(45), *245*
Nelson, A., 28(147), 38(147), *45*, 269(183), *283*
Nelson, L., 208(220), *232*
Nerurkar, M. K., 212(236), *232*, 257(205a), 262(205), *283*
Nettesheim, P., 52(99a), *76*
Neumeyer, M., 150(35), 153(35), *186*, 212 (30), 215(30), *226*, 269(26), *278*
Neuwirt, J., 18(247), 19(247), 39(265), *47*, *48*, 258(206), 273(206), *283*
Neve, R. A., 254(60), 270(60), 271(60), *279*
Newson, B. D., 7(14a), *11*
Nguyen, H. C., 223(199), *231*
Nichols, A. V., 270(220), 271(220), *284*
Nickolai, D. J., 223(234), *232*
Nickson, J. J., 197(136), *229*
Nicolaev, J. V., 29(109), *44*
Nicrosini, F., 118(58), 119(58), *187*
Niggli, H., 144(82), *188*
Nikólaeva, N. D., 221(244), *232*
Nimmi, M. E., 197(200, 201), *231*
Nims, L. F., 6(15), *11*, 122(190, 191, 194), 124(191), 125(192, 269), 127(269), 141 (191), *191*, *193*, 212(202), *231*, 288(180), 289(190), 293(134), 312(133), 316(134), 319(135, 136), 320(135), 321(135, 136), *328*, *329*
Ninkov, V., 321(46a), *325*
Nitz-Litsow, D., 142(193), *191*
Nizet, A., 19(169), 20(170), 21(170), *45*, 292 (137), 293(137), *328*
Nohr, M. L., 19(73), 29(73), 30(73), *43*, 216 (101), *228*
Nokay, N., 28(115), 29(115), *44*, 69(90), 71(91), *76*
Noonan, T. R., 4(26), *11*, 15(4), 19(4), *41*, 153(6), *185*, 211(6), 255(10), *225*, 258 (118), *277*, *281*
Norman, A., 53(40a), *74*
Noronha, J. M., 58(114), *77*
North, N., 122(194), *191*
Notario, A., 21(172), 22(172), *45*, 116(220), 118(195), 129(196, 196a), 132(196a), 134(196a), 137(196, 196a), 153(197, 220), *191*, *192*
Notter, G., 303(46), *325*
Novelli, G. D., *188*
Noyes, P. P., 141(198), 143(198), *191*
Nury, F. S., 270(190), *283*
Nyffenegger, E., 258(207), *283*
Nygaard, O. F., 25(173), *45*, 52(115, 118), 53(115), 55(115, 116), 62(118), *77*, 84 (67, 68, 70), 86(57), 88(69), *101*, 162 (199), *191*

O

Oakey, R. E., 302(174), 303(50, 174), *325* *329*
O'Bourke, F., 240(100), *247*
O'Brien, E. M., 243(109), *247*
Odartchenko, N., 18(174), 20(174), *45*
Odland, L. T., 269(219a), *284*
Oganesjan, S. S., 212(203), *231*
Ogasawara, K., 264(208), 267(208), *283*
Ogborn, R. E., 253(219b), 269(219a), *284*
Ohyama, H., 64(118a, 189a), *77*, *79*
Ojak, M., 222(204), *231*
Okada, S., 59(120), 60(120), 69(53, 54, 120, 121), 70(119, 122), 71(53, 54, 122), *75*, *77*, 145(201), 167(152, 152a, 153), 168 (152a), 169(153), 180(153, 200), 183 (152), 184(151, 152, 153), *190*, *191*
Okulov, N. M., 95(71), *101*
Okunewick, J. P., 19(77), *43*
Oldekop, M., 240(58), 241(58, 96), 242(58a), *246*, *247*
Oliva, L., 124(202), *191*
Oliver, R., 17(119a, 234), 18(234), 19(234), 20(234), 24(117, 118, 119, 119a, 120), 25(118, 119), 30(70a), *43*, *44*
Olmstead, C. A., 254(60), 270(60), 271(60), *279*
Omata, S., 159(202a), *191*
Omato, S., 293(207a), *330*
Onkelinx, C., 202(206), 211(205), *231*
Ontko, J. A., 237(85), 239(85, 86), *246*
Ord, M. G., 52(125, 138), 54(28, 124, 125, 128b), 56(138), 59(126), 61(28, 128, 128a, b), 62(123), 63(123, 127), 64 (127), 67(28, 128), 72(128b), *74*, *77*, 142 (203, 205), 153(219), 164(127), 165 (204), 168(219), 169(219), 172(204), 174(204), 184(127, 219), *189*, *191*, *192*
Orekhovich, V. N., 218(207), *231*
Orin, H., 236(84a), *246*
Orlova, L. V., 31(175, 248), *45*, *47*
O'Rourke, F., 142(235), *192*
Orth, J. S., 225(92b), *228*
Ortolani, E. A., 223(10), *225*
Osborn, G. K., 312(138), *328*
Osborne, J. W., 92(44a), 95(77), 98(72), *100*, *101*, 273(159), *282*, 291(2), 312 (2), *323*
Ostadalova, I., 203(14a), *226*
Oswald, H., 212(7), *225*, 269(209), *283*
Otten, G., 273(245a), *285*
Ottolenghi, A., 14(11), 17(11), *41*, 90(72a, b), *101*
Overman, R. R., 213(165), 214(166), *230*
Owens, J. P., 257(213), *283*

P

Pabst, H. W., 31(176), *45*
Painter, E. E., 268(219), *284*
Palaić, D., 307(139, 139a, 140, 141, 142, 143), 309(139a), *328*
Paleček, E., 22(228), 23(228), 25(228), 26 (228), *47*
Pallavicini, C., 220(97), *228*
Palmer, W., 165(135a), *189*
Pande, S. V., 107(206), 114(206), *192*
Panfil, B., 128(57), 131(57a), *187*
Pantić, V., 303(146), *328*
Pany, J., 24(177), *45*, 257(210), *283*
Paoletti, M., 91(58), *101*
Paoletti, P., 105(109), 108(83, 109, 207), 115 (83, 109, 207), 116(83), *188*, *189*, *192*, 222(99), *228*
Paoletti, R., 105(109), 107(110), 108(83, 109, 110, 207), 112(208), 114(110), 115 (83, 109, 110, 208), 116(83), 119(208), 120(208), *188*, *189*, *192*, 222(99, 100), *228*
Paolini, F., 215(63), *227*
Papworth, D. G., 301(124), *327*
Parchwitz, H. K., 39(25), *42*, 54(18, 128a), 58(19), *74*, *77*, 91(100), 92(100), 94 (100), *102*, 211(32), 214(32), *226*, 243 (87), *246*, 253(29), *278*
Parin, V. V., 308(144), *328*
Paris, J. E., 244(86a), *246*
Parizek, J., 203(14a), *226*, 256(211), 262 (212), *283*
Parkhomenko, I. M., 214(37a), *226*
Parretta, M., 22(178), 25(178), *46*
Perris, A. D., 31(179), *46*
Parson, L., 302(112), *327*
Paschke, G., 268(277), *286*
Pasotti, C., 118(58), 119(58), *187*
Patalova, N. V., 316(145), *328*
Patek, P. R., 199(195), *231*
Paterson, E., 19(63, 64, 65), 31(63, 65), 37 (64, 65), *43*, 250(89), 251(89), 253(88), *280*
Patt, D. I., 192(207), *330*

Patt, H. M., 6(21), *11*, 81(73), *101*, 204(208, 209), 216(254), 218(209), *231*, *233*, 303 (165), 304(165), *329*
Pauly, H., 63(134), *77*, 142(215), 144(215), *192*, 214(210), *231*, 243(87), 244(57), *246*
Pavitch, D., 291(116), *327*
Pavlovitch, M., 289(7), 293(7), 295(7), *323*
Pavlovitch-Hournac, M., 291(116), 303 (146), *327*, *328*
Payne, A. H., 168(135b, 209), 169(135b, 209), 173(209), 174(209), *189*, *192*
Pays, M., 268(187), 270(187), *283*
Peachy, L. D., 145(201), *191*
Pearse, A. G. F., 90(85), *101*
Pearson, E., 218(58), *227*
Pearson, W. N., 257(213), *283*
Pencharz, R., 293(197), 296(198), *329*
Pendic, B., 274(135), *281*
Pentz, E. I., 260(214), *284*, 312(147), 316 (147, 148), *328*
Peraino, C., 156(212, 213, 214), 158(213), 159(213), 160(213), 161(214), *192*
Perkins, L. B., 304(184), *329*
Perkins, W. H., 215(167), *230*
Perez, J. C., 208(219, 220, 221, 221a), 210 (221), *232*
Perlish, J. S., 205(265, 266, 267), 206(266, 267), 207(266), *233*
Perris, A. D., 53(128g), 59(128d, e), 61 (128f, g, 188a), 72(128d, g), *77*, *79*, 91 (74, 75a), 93(74), 94(74, 75, 75a), 96 (74), *101*
Perrotta, C. A., 85(86), 88(86), *101*
Perry, H., 256(17), *277*
Pesotti, R. L., 109(157), 162(157), 171(157), *190*
Peters, K., 6(16), *11*, 268(215), *284*
Petersen, D. F., 58(32), *74*, 256(238), *284*
Petersen, D. F. R., 259(216), *284*
Petit, M., 39(142), *45*, 254(180), *282*
Petrakis, N. L., 237(74), 238(74), *246*
Petranyi, G., 94(103), 96(103, 104), *102*, 294 (209), *330*
Petrášek, J., 251(125), 252(125), *281*, 291 (80), 312(80), *326*
Petropoulous, M. C., 17(180), 31(180), 39 (180), *46*
Petrov, R. V., 89(38), *100*
Petrović, S., 174(210), *192*
Pfleger, R. C., 16(181), *46*
Philipps, G. R., 21(182), *46*
Phillips, R. D., 203(211), *232*, 313(149), *328*
Phillips, S. J., 211(212), *232*
Phillips, T. L., 216(213, 281), *232*, *234*
Philpot, J. S. L., 144(210a), *192*
Piantadosi, C., 15(134), 16(134), *44*, 119 (159a), *190*
Picha, E., 6(7, 8), *11*
Pickering, J. E., 253(59), 254(59), *279*
Pierce, F. T., 271(116), *281*
Pierucci, O., 53(128h), 69(129), 70(129), 71 (129), *77*
Pihl, A., 31(237), 33(237), 34(237), 35(237), 36(237), *47*
Pikulev, A. T., 150(211), 151(211), *192*, 212 (214, 215), *232*, *284*
Pillat, B., 208(108), *229*
Pilot, H. C., 156(212, 213, 214), 158(213), 159(213), 160(213), 161(214), *192*
Pincus, G., 294(31b, 156), *324*, *328*
Pineda, P. A., 269(6), *277*
Pirie, A., 218(216, 217, 218, 274), *232*, *233*
Pitts, J. F., 243(39), 244(39), *245*
Pizarello, D. J., 6(17), *11*
Plantin, L. O., 273(19, 273a), *277*, *286*, 300 (25), *324*
Plette, G. Z., 293(49), *325*
Podesta, A. M., 215(71b), *227*
Podil'Chak, M., 254(218), *284*
Poggi, M., 107(110), 108(110), 112(208), 114(110), 115(110), 119(208), 120(208), *189*, *192*, 222(100), *228*
Pokorny, Z., 258(206), 273(206), *283*
Polikarpova, L. I., 293(149a), *328*
Polis, B. D., 111(240), *193*
Polis, E., 111(240), *193*
Pollak, H., 266(268), 267(268), *285*, 308 (192), *329*
Pomini, P., 124(202), *191*
Poncelet, J., 196(39), *226*
Pontifex, A. H., 19(123, 183), *44*, *46*
Ponz, F., *101*
Popjack, G., 105(98), 114(98), *188*
Popp, R. A., 95(54), 96(54), *101*
Poppe, H., 308(95), *326*
Poppei, H., 93(76), 94(76), *101*
Portela, A., 208(219, 220, 221, 221a), 210 (221), *232*
Porteous, D. D., 15(3a), 29(3a, 185), 30 (3a, 184, 185), *41*, *46*

Posener, K., 239(37), *245*
Pospišil, J., 39(227), *47*, 66(75), *75*
Pospišil, M., 270(275), *286*, 295(178b), 310 (178a), 322(150), *328*, *329*
Potter, R. L., 23(233), 25(173), *45*, *47*, 52 (115, 118), 53(115), 55(47, 115, 116, 166), 56(166), 62(118), *74*, *77*, *78*, 84 (67, 68, 70), 86(67), 88(69), *101*, 162 (199), *191*
Potter, V. R., 23(19), *42*, 166(17), 168(29), 173(17), 179(17, 29), 180(29), 182(29), 183(29), *186*
Pover, W. F. R., 90(30a, 101b), 98(30a), *100*, *102*
Pozza, F., 90(9), *99*
Prasad, K. N., 96(77), *101*
Price, P. B., 250(189), 252(189), 253(189), *283*
Priest, S. G., 268(273), *286*
Prosser, C. L., 268(203, 219), *283*, *284*
Provezailo, V., 254(218), *284*
Pryor, N. W., 294(114), *327*
Pshennikova, M. C., 225(222), *232*
Pshennikova, M. G., 225(134), *229*
Puca, G. A., 133(37), 136(37), 141(37), *186*
Puck, T. T., 49(130), *77*
Purser, P. R., 113(115), 115(115), *189*

Q

Quaife, M. A., 253(219b), 269(219a), *284*
Quastler, H., 80(78), 81(73), 84(84), 85(86), 88(86), 89(51), *100*, *101*, 204(208, 209), 205(41a), 218(209), *226*, *231*
Quinn, F. A., 270(60a), *279*
Quintilliani, M., 35(13, 14), *41*

R

Rabassini, A., 222(223), *232*
Radford, E. P., *329*
Radivajević, D. V., 303(146), *328*
Radivojevitch, D. V., 289(7), 293(7), 295 (7), *323*
Raffi, A., 59(57), 69(57), *75*
Rahman, Y. E., 69(131, 132), 70(132), 71 (133), *77*, 142(268), *193*
Raimondi, L., 244(32), *245*
Rajewsky, B., 2(17a), *11*, 63(134), *77*, 142 (215), 144 (215), *192*, 208(31), 209(31), 211(224), 214(210), *226*, *231*, *232*, 241 (57), 243(87), 244(57), *246*
Rakovič, M., 276(47), *278*
Rakovich, M., 97(79), *101*
Ramaiah, A., 107(206), 114(206), *192*
Rambach, W. A., 18, 19(186), *46*, 253(59), 254(59), *279*
Randić, M., 308(42), 307(139, 152), 310(42, 151), *325*, 328
Rapoport, E. A., 239(84), *246*
Rappoport, D. A., 39(187, 188), 40(188), *46*, 58(135), *77*, *191*, 216(226), 222(225), *232*, 301(153), *328*
Rasković, D., 215(103), *228*
Rathgen, G. H., 69(135a), *77*, 153(237), *192*, 239(61, 62, 88, 101), 240(89), 241(101), 243(82), *246*, *247*
Rauschkolb, D., 237(54), 238(55), 241(54, 55), *245*, *246*
Raushenbakh, M. O., 31(254), *48*, 308(34, 144), *324*, *328*
Rava, G., 151(9), 159(9), 170(9), *186*
Raymund, A. B., 215(66), 222(66), *227*
Raynaud, A., 301(102), *327*
Recamier, D., 301(16), *324*
Recht, P., 275(219c), *284*
Reddy, V., 208(221), 210(221), *232*
Redetzki, H. M., 142(235), *192*, 240(100), *247*
Redgate, E. S., 220(41), 221(41), *226*
Reeves, R. J., 94(80), *101*
Regelson, W., 69(129), 70(129), 71(129), *77*
Rehnborg, C. S., 270(220), 271(220), *284*
Reichel, G., 268(225a), *284*
Reid, J. A., 211(212), *232*
Reid, J. C., 89(96), *102*
Reincke, U., 5(18a, b), 6(18), *11*
Reinert, M., 211(111a), *229*, 250(121), 252 (121), *281*
Reiss, M., 239(81), *246*
Remberger, V. G., 212(42), 221(42), 222 (42), *226*
Remy-Defraigne, J., 87(28), 89(28a), 92 (27), 93(28a), *100*, 141(89), 146(89), 149(85a, 89), 155(91, 217, 228), 162 (90), 167(90), 169(90), *188*, *192*, 253 (86e), 259(82), 262(86b), 263(83), 272 (84, 222), 273(84, 222, 230), *279*, *280*, *284*
Renson, J., 289(81), 295(8), 307(154), 308 (155), *323*, *328*
Rettegi, C., 118(174), *191*

Reuter, A. M., 58(136), 61(169a), *77*, *78*, 155(91, 216, 217), *188*, *192*, 272(84, 222), 273(84, 221, 222, 223, 260a), *284*, *285*
Reuvers, D. C., 63(83), 64(83), *76*
Révész, L., 49(15a), *74*, 236(18, 90, 91, 92, 93), 243(94), 244(94), *245*, *247*
Revis, V. A., 21(189, 190), *46*, 56(137), 57 (137), *77*, 215(227), *232*
Rhoades, R. P., 103(218), *192*
Rhodes, R., 96(39), 97(39), *100*, 253(133), *281*
Ribbing, C., 239(75), *246*
Rich, J. G., 215(218), *232*
Richards, B. M., 239(75, 76), *246*
Richards, R. D., *232*
Richey, E. O., 208(92a), 225(92b), *228*
Richmond, C. R., 97(53), *101*
Richmond, J. E., 15(5), 16(5), 20(191, 192), 27(192), 29(192), *41*, *46*, 52(138), 56 (138), *77*, 153(219), 168(219), 169(219), 184(219), *192*
Richmond, S. S., 142(300), *194*
Ricotti, V., 116(220), 153(220), *192*
Ringertz, N. R., 238(17), 239(17, 75, 76, 95), *245*, *246*, *247*
Rinne, U. K., 90(31), 91(31), *100*, 148(118b), *189*
Rixon, R. H., 49(138a), 53(128g), 72(128g), *77*
Robbiani, M. G., 31(90), 36(90), *43*
Roberts, B. M., 19(85), *43*
Robertson, J. E., 252(224), 253(224), *284*
Robertson, J. S., 31(202), *46*
Robertson, W. J., 252(158), 254(158), 268 (158), *282*, 316(94), *326*
Robinette, C. D., 55(145), 62(144), *77*
Robinson, L. G., 271(198), *283*
Robson, H. H., 62(191, 192), 63(191, 192), *79*, 142(302, 303), 144(301), *194*
Robustelli della Cuna, G., 95(12a), *99*
Rodesch, J., 24(193), 25(154), 26(154), *45*, *46*, 207(179a, 180, 180b), *231*
Rodgers, C. H., 224(229), *232*
Rodionov, V. M., 273(225), *284*
Roesler, B., 268(225a), *284*
Rogers, C. C., 273(179a), *283*
Roh, B. L., 30(55), *42*, 214(75), 216(75), *227*
Roitman, F. I., 22(108), 23(108), 29(108), *44*
Romano, B., 159(10), *186*
Romantsev, E. F., 144(107), 149(221), 153 (28), *186*, *189*, *192*, 256(279), *286*
Ronyaeva, M. P., 212(214), *232*
Root, W. S., 251(38), *278*
Rosen, D., 225(230), *232*
Rosen, V. J., Jr., 216(46, 280), *227*, *234*
Rosenberg, M., 34(263), *48*
Rosenberg, R., 34(262), *48*
Rosenfeld, G., 294(156, 157, 158), *328*
Rosengren, B., 220(104b), 222(104b), 223 (104b), *228*
Rosenthal, R. L., 31(68), 36(68), *43*, 270 (226), *284*
Rösler, H., 307(122), 308(122), *327*
Ross, M. H., 122(222), *192*
Rossi, C. S., 222(223), *232*
Rossoni, C., 90(5a), *99*
Rotblat, J., 5(6a), *11*
Roth, J. S., 55(143), 69(36, 139, 140, 141, 142), 70(143), *74*, *77*, 185(223, 224), *192*
Roth, L. J., 105(225), *192*, 253(227a), *284*
Rothermel, S. M., 311(206), *330*
Rothman, S., 206(3), *225*
Rothstein, A., 31(252), 35(252), *47*
Rotreklová, E., 24(10), 25(10), *41*
Rowley, G. R., 220(231), 221(231), 222 (231), *232*
Roylance, P. J., 19(15), *41*
Rubin, P., 212(149), 213(149), *230*, 260 (170), *282*
Rubini, J. R., 256(227), 263(227), *284*
Rubini, M. E., 214(43), *226*
Rudolph, W., 22(178), 25(178), *46*
Ruffato, C., 197(29, 232), *226*, *232*
Rugh, R., 6(19, 19a), 98(81), *101*, 109(159), 124(159), 141(159), *190*, 211(212), 218 (105), *229*, *232*
Ruhenstroth-Bauer, G., 25(194), *46*
Ruppe, C. O., 218(233), *232*
Rushmer, H. N., 140(224a), 141(224a), *192*
Russ, S., 36(29), *42*
Russo, A. M., 270(8), 271(8), *277*
Rust, J. H., 105(146, 225), *190*, *192*, 253 (227a), *284*, 304(126), 305(104,) *327*
Rüter, J., 241(96), *247*
Rynova, N. N., 294(59), *325*
Ryser, H., 142(226), *192*
Rysina, T. N., 22(195, 196), 23(195), 24 (195, 197), 29(195), 41(197), *46*, 89 (92), *101*

S

Sabine, J. C., 38(198), 40(198), *46*, 223(234), *232*
Sacher, G. A., 5(20), *11*
Sacukevič, V. B., 223(235), *232*
Saenger, E. L., 256(17), 263(131b), 270(185), *277*, *281*, *283*
Sahasrabudhe, M. B., 212(236), *232*, 257 (169, 205a), 260(169), 262(205), *282*, *283*
Sahinen, F. M., 291(159), *328*
Sakhatskaya, T. S., 294(106), *327*
Saksonov, P. P., 308(144), *328*
Salas, J., 127(250), *193*
Salerno, P. R., 22(251), 24(251), *47*
Salmi, A., 208(25), 210(25), *226*
Salmon, R. J., 58(94a), *76*, 111(150), *190*, 243(97), *247*
Salomon, K., 15(4, 5), 16(4, 5), 17(180), 19 (4, 5, 35, 206), 20(35, 191, 192, 206), 27 (192), 29(192), 31(180), 39(180), *41*, *42*, *46*, 153(6), *185*, 211(6), *225*
Sams, A., 202(237), *232*
Samuels, L. T., 295(67), 297(67), *326*
Sandeen, G., 28(114), 29(114), *44*, 69(88), 70(88), *76*, 185(140a), *190*
Sanders, A. P., 94(80), *101*
Sanders, J. L., 55(145), 62(144), *77*, 165 (96a), *188*
Sanna, G., 221(41b), *226*, 269(227b), *284*
Santagati, G., 95(12a), *99*, 129(196, 196a), 132(196a), 134(196a), 137(196, 196a), 153(197), *191*
Santori, M., 218(181, 182), *231*
Sapsford, C. S., 301(160), *328*
Saritskin, I. V., 273(228), *284*
Sarkar, N. K., 153(227), *192*
Sassen, A., 58(136), *77*, 155(228), *192*, 268 (229, 231), 270(229, 231), 273(223, 228a, 230, 260a), *284*, *285*
Sauer, H., 255(193), 260(193), 264(193), 265(193), 266(193), *283*
Sauerbier, W., 31(199), *46*
Savkovic, N., 291(161), *328*
Savoie, J. C., 18(200), 20(200), *46*
Sawrant, P. L., 70(29), *74*, 143(60), *187*
Sbitneva, M. F., 256(98), *280*
Scaife, J. F., 55(155a), 56(153), 58(152), 59 (146, 155, 155a), 60(146), 61(151, 152, 155a), 62(147, 154), 63(149, 150, 154), 66(148), 67(148, 152), 71(151), 72 (155, 155a), *77*, *78*, 142(231, 232), 143 (229), 144(230), *192*, 238(99), 244(98), *247*
Schall, H., 21(182, 201), *46*
Scheller, M. S., 237(6), *244*
Scherer, E., 141(233, 234), 142(233), *192*
Scheuerbrandt, G., 146(147), *190*, 266(176), 267(176), *282*, 308(105), *327*
Scheyer, W. J., 6(4), *10*
Schian, S., 223(8), *225*
Schiffer, L. M., 31(202), *46*
Schinz, H. R., 268(115), *281*
Schjeide, O. A., 223(54a), 224(238, 239), *227*, *232*
Schlegel, B., 59(120), 60(120), 69(120, 121), *77*
Schleich, A., 243(73), *246*
Schliep, G. I., 208(117), *229*
Schmack, W., 239(63), 240(63), *246*
Schmaglowski, S., 66(5), *73*, 240(2), 241(2), *244*
Schmerund, H. J., 308(20), *324*
Schmidli, B., 142(226), *192*, 255(2, 4), *277*
Schmidt, K. J., 212(239a), *232*
Schmitt, M. L., 218(180a), 220(180a), *231*
Schnappauf, H., 31(203), *46*
Schneider, M., 24(157), 31(204), 36(204), 38(127, 157, 204), 39(157), *44*, *45*, *46*, 142(235), *192*, 240(100), *247*
Schneider, W. C., 184(236), *192*
Scheiderman, M., 31(231), 37(231), *47*
Schocket, S. S., *232*
Schoen, E. J., 300(162), *328*
Schoen, H. D., 18(256), 39(256), *48*
Schofield, R., 19(65), 31(65), 37(65), *43*, 250 (89), 251(89), *280*
Schoffenfiels, E., 31(100), 38(100), *44*
Scholz, M., 240(58), 241(58), *246*
Schön, H., 254(13, 232), *277*, *284*
Schönbaum, E., 294(12), *324*
Schoolar, J. C., 105(146), *190*, 305(104), *327*
Schooley, J. C., 19(205), 29(205), 30(205), 36(205), *46*
Schorn, H., 294(163, 204), 295(204), 304 (163), *328*, *330*
Schram, E., 259(233), *284*
Schreiber, V., 251(125), 252(125), *281*, 291(80), 312(80), *326*
Schreier, K., 198(102, 240), *228*, *232*, 260 (104), *280*
Schrek, R., 52(156, 156a), *78*

Schreler, K., *192*
Schroeder, W., 41(258), *48*, 239(119, 120, 121), *247*
Schubert, G., 153(237), *192*, 239(101, 102), 241(101, 102), *247*
Schulman, M., 19(206), 20(206), *46*
Schulz, G., 215(124), *229*
Schwartz, E. E., 94(83), *101*
Schwartz, M. S., 273(248, 249), *285*
Schwartz, S., 257(234), *284*
Schwarz, H. P., 111(240), 118(238, 239), *193*
Schwarzenberg, L., 274(135), *281*
Scott, C. M., 142(241), *193*
Scott, K. G., 252(24), *278*
Seaman, G. V. F., 244(108), *247*
Searcy, R. L., 94(64), *101*, 107(182), 109 (184), 110(181, 184), 111(181, 183, 184), 112(183, 184), 122(180), 123 (182), 125(180), 128(182), 136(185), *191*
Seay, H. D., 220(24), *226*
Šebestik, V., 31(207, 208, 209, 210), 36(207, 208), *46*
Seed, J., 238(103), 239(104), *247*
Segal, G., 118(150), *190*, 318(109), *327*
Segni, P., *42*
Seibert, R. A., 58(135), *77*, *192*, 216(226), *232*
Seiter, I., 19(98), *44*
Seits, I. F., 123(76), 131(76), 132(76), *187*
Sekiguchi, E., 236(84a), *246*
Sekiguchi, T., 236(84a), 244(105), *246*, *247*
Selden, G. L., 268(273), *286*
Selig, J. N., 244(44), *245*
Semal, M., 177(100), 180(13, 100), 181(13, 100), *186*, *188*
Semenov, L. F., 222(256), *233*, 308(97a), *326*
Sereni, F., 159(10), *186*
Seres-Sturm, L., 118(174), *191*
Sessa, A., 117(3), 118(2), 119(3), 128(3), *185*
Šestan, N., 153(242), *193*
Sewell, B. W., 39(188), 40(188), *46*
Seydl, G., 254(128), *281*
Seymour, R., 225(53b, 241), *227*, *232*
Shabadash, A. L., 221(242), *232*
Shaber, G. S., 155(243), *193*, 273(235), *284*
Shacter, B., 254(236), *284*
Shah, N. S., 58(114), *77*
Shah, P., 58(11), *73*
Shanklin, W. M., 222(35), *226*
Shapiro, B., 31(211, 212), 35 (212, 213), 36(211), 38(211), *46*, 94(83), *101*
Sharpe, K. W., 94(80), *101*
Sharobaiko, V. I., 221(243), *232*
Shashkov, V. S., 308(144), *328*
Shastin, R., 269(237), *284*
Shaternikov, V. A., 90(15), *99*
Shaver, A., 95(6a), *99*
Shchedrina, R. N., 141(103), *188*
Sheel, C., 256(17), *277*
Sheets, R. F., 16(214), 31(214), *46*
Shejbal, J., 262(11), 263(11), 276(49), *277*
Shelley, R. N., 254(237a), *284*, 293(163a), *328*
Shenep, K., 30(55), *42*
Sheppard, C. W., 31(215, 216), 39(216), *47*, 208(287), 211(287), *234*
Sheremet, Z. I., 31(254), *48*, 308(36), *324*
Sherman, F. G., 84(84), *101*, 111(244), 123 (245), 173(244), *193*, 205(41a), *226*
Shevchuk, Ya. G., 211(259), 215(259), *233*
Shibata, K., 260(197), *283*, 308(130), *328*,
Shipman, T. L., 256(238), 274(238), *284*
Shirley, R. L., 212(68), *227*
Shooter, K. V., 238(27), *245*
Shore, M. L., 158(179), 161(179), *191*, 267 (200a), *283*
Shubina, A. V., 254(1), 260(1), *277*
Shungskava, V. E., 221(244), *232*
Sibatani, A., 52(157), 57(169), *78*
Sick, K., 255(179), 256(179), *283*
Sigdestad, C. P., *101*
Šikulová, J., 322(150), *328*
Silaev, M., 212(150), *230*
Silini, G., 236(106), *247*
Silk, M. H., 240(107), *247*
Sillero, A., 127(250), *193*
Silyayeva, M. F., 212(245), 222(245), *232*
Simek, J., 144(304), *194*
Simon, A. B., 196(39), *226*
Simon, K. A., 254(50), 255(50), *278*
Simon-Reuss, T., 244(108), *247*
Simonov, V. V., 119(245), *193*
Simonson, C., 58(24a), *74*, 110(56), *187*, 216(48), *227*
Simonson, H. C., 304(175), *329*
Simpson, C. G., 300(163b), *328*
Sinclair, W. K., 253(20), *277*
Sitterson, B. W., 275(9), *277*
Sitzman, F. S., 254(232), *284*

Sivertsen, W. I., 257(36), *278*
Sjaarda, J. R., 198(283), *234*
Skalka, M., 22(217, 218, 219, 220, 228), 23 (228), 24(218), 25(218, 228), 26(228), *47*, 59(103, 158, 159), *76*, *78*, 110(248), 118(247), 119(247), 184(251), 185(251), *193*, 269(239), *284*, 316(164), 317(164), *328*
Skaugaard, M. R., 85(86), 88(86), *101*
Skjaelaaen, P., 30(84a), *43*, 216(115), *229*
Skoda, J., 256(53, 54, 211, 212), 262(211, 212), *278*, *283*
Skov, K., 52(113), 53(113), 54(113), 56(113), *77*
Skuikhina, M. M., 263(261, 261a), *285*
Skvortsova, R. I., 221(246), 222(246), 223 (191), *231*, *233*
Slade, D. E., 35(166), *45*
Sladitch, D. S., 289(7), 293(7), 295(7), *323*
Slaviero, G. E., 21(172), 22(172), *45*, 118 (195), 129(196, 196a), 132(196a), 134 (196a), 137(196, 196a), 153(197), *191*
Sloan, K., *244*(12)
Small, D. L., 304(191), *329*
Smelik, P. G., 289(8), 295(8), *323*
Smellie, R. M. S., 24(221), 25(221), *47*
Smelyanskaya, G. N., 212(215), *232*, *284*
Smetana, W., 308(63), *325*
Smets, L. A., 54(160), 57(160), *78*
Smit, J. A., 57(161, 162), 58(161, 162), *78*
Smith, C. I., 251(241), 263(14, 241), *277*, *284*
Smith, D. E., 6(21, 22, 23), *11*, 221(247), 222 (247), *233*, 303(165), 304(165), *329*
Smith, F., 6(24), *11*
Smith, H., 256(240), 257(242, 242a), 263 (14, 240, 241, 242a), 264(242, 242a), *277*, *284*, *285*
Smith, K. C., 58(163), *78*, 162(249), *193*, 222(248), *233*
Smith, L. H., 52(164), *78*
Smith, P. K., 169(55, 298), 175(55, 298), *187*, *194*, 238(25), *245*
Smith, R. E., 141(198), 143(198), *191*
Smith, S. A., 220(93), *228*
Smith, W. W., 6(24), *11*
Smolichev, E. P., 273(243), *285*
Smolin, D. D., 144(107), *189*
Smoot, A. O., 55(42), 69(42), *74*
Smyth, D. H., 91(74), 93(74), 94(74), 96 (74), *101*
Snezhko, A. D., 221(249), *233*
Snider, R. S., 213(103a), 219(104, 104a), *228*
Snyder, F., 14(181, 223, 225, 226), 15(134, 223, 224, 226), 16(134, 181, 222, 226), 17(222, 224), 36(223), *44*, *46*, *47*, 119 (159a), *190*
Sobel, E., 291(58), 316(58), *325*
Sobel, H., 198(250), *233*
Sobel, H. J., 303(166), *329*
Soberman, R. J., 250(244), 252(244), 268 (244), *285*
Soderwall, A. L., 291(159), *328*
Soffer, E., 111(240), *193*
Sokolov, I. I., 211(251), *233*
Sokolova, G. N., 144(107), *189*
Sokolova, O. M., 236(45), *245*
Soldatenkova, L. I., 123(76), 131(76), 132 (76), *187*
Sols, A., 127(250), *193*
Sommers, S. C., 302(112), *327*
Song, C. W., 204(252), 205(266), 206(266), 207(266), *233*, 304(167, 168a), 313(168), 314(168), *329*
Sonka, J., 39(227), *47*, 66(75), *75*
Sonnenblick, B. P., 142(114), 143(114), *189*
Sopin, E. F., 221(145, 253, 253a), *230*, *233*
Sörbo, B., 258(245), *285*
Soška, J., 22(220, 228, 230), 24(44, 45, 229), 25(44, 228), 26(228), *42*, *47*, 54(12), 59 (103), *74*, *76*, 184(251), 185(251), *193*, 316(164), 317(164), *328*
Soškova, L., 22(230), 23(230), *47*
Spalding, J. F., 5(25), *11*, 302(169), 322 (182), *329*
Spargo, B., 215(228), *232*
Speck, L. B., 307(170), *329*
Spence, W. P., *194*
Spiers, F. W., 202(292), *234*
Spiro, H. M., 90(85), *101*
Spirtes, M. A., 138(64a), *187*
Spraragen, S. C., 15(4), 19(4), *41*
Spratt, J. L., 105(225), *192*
Springsteen, R. W., 95(17, 18), *99*, 257(46), *278*
Stabile, J. N., 269(27, 28), *278*
Stacey, K. A., 184(252), *193*
Staehler, F., 273(245a), *285*
Staib, W., 295(179), 299(179), *329*
Stanbury, J. B., 275(142), *282*
Stannard, J. N., 31(238), 32(238), 34(238), *47*

Stark, G., 295(171), *329*
Steenbeck, L., 66(5), *73*, 240(2), 241(2), *244*
Steggerda, F. R., 308(196), *329*
Stefanescu, P., 223(8), *225*
Stein, G., 244(108), *247*
Steinberg, B., 14(40), 15(40), *42*
Stender, H., 268(246), *285*
Stenstrom, K. W., 238(115), *247*
Stepanovic, S., 308(189, 189a), *329*
Stern, D. N., 259(247), *285*
Stets, J. F., 268(273), *286*
Stevens, K. M., 273(248, 249), *285*
Stevens, L., 181(253), *193*
Stevens, W., 294(172, 173), *329*
Stewart, M., 31(216), 39(216), *47*
Stewart, P., 85(86), 88(86), *101*, 208(219, 220, 221, 221a), 210(221), *232*
Stewart, R., 118(52), 119(52), 120(52), *187*
Stich, H. F., 162(286), 171(286), *194*
Stim, E. M., 259(247), *285*
Stinson, J. V., 212(189), *231*, 271(198, 199), *283*
Stitch, S. R., 302(174), 303(50, 174), *325*, *329*
Stockell, A., 118(52), 119(52), 120(52), *187*
Stocken, L. A., 52(125, 138), 54(28, 124, 125, 128b), 56(138), 57(160, 161), 58 (160, 161), 59(126), 61(28, 128, 128a, b), 62(123), 63(25, 123, 127), 64(127), 67(28, 128), 72(128b), *74*, *77*, *78*, 142 (203, 205), 153(219), 164(127), 165 (204), 168(219), 169(219), 172(204), 174(204), 181(253), 184(127, 219), *189*, *191*, *192*, *193*
Stoenescu, R., 223(8), *225*
Stohlman, F., Jr., 31(231), 37(231), *47*
Stoklasová, H., 18(247), 19(247), *47*
Stoklosa, E., 292(88), *326*
Storer, J. B., 5(27), *11*, 19(232), 24(240), 27 (240), 29(240), *47*, 206(174), 207(174), *230*, 304(175), *329*
Storey, R. H., 250(250), 251(250), *285*
Stout, D. A., 269(118b), *281*
Strang, V. G., 302(169), *329*
Straube, R. L., 216(254), *233*
Streffer, C., 58(94, 105, 165), *76*, *78*, 117 (255), 121(255), 123(255), 124(255), 133(255), 138(255, 257), 139(255), 142 (254, 256, 257), 146(147, 259), 150(260), 151(260), 152(260), 153(260), *190*, *193*, 215(255), *233*, 255(175, 193, 251), 256 (252), 260(193, 196), 264(175, 193, 194, 253a, 254, 254a), 265(175, 193, 194, 254a, b), 266(193, 194, 195, 196, 251a, 253, 254a, b, 268), 267(253, 253a, 254a, 268), 268(254b), *282*, *283*, *285*, 308 (105, 192), *327*, *329*
Strelkov, R. B., 222(256, 257), *233*
Strike, T. A., 270(60a), *279*
Strom, E., 35(14), *41*
Strubelt, O., 223(258), *233*, 304(177), 305 (177), 308(176), *329*
Stryckmans, P. A., 31(202), *46*
Sturrock, M., 263(242a), *285*
Stüttgen, G., 206(82), *228*
Suetina, I. A., 22(26), 24(26), 25(26), *42*
Sugakawa, T., 236(69), *246*
Sugino, Y., 23(233), *47*, 55(47, 166), 56 (166), *74*, *78*
Suhar, A., 61(166a), *78*
Suit, H. D., 17(121, 234), 18(121, 234), 19 (121, 234), 20(121, 234), *44*, *47*
Sukhomlonov, B. F., 21(235), *47*, 89(87), *101*, 211(259), 215(259, 259a), *233*
Sulc, K., 258(206), 273(206), *283*
Sullivan, M. F., 63(167), *78*, 94(89), 95(88, 91, 93), 98(24, 90, 92, 93), *100*, *102*, 259 (256), 273(255), *285*
Sullivan, W. D., 243(109), *247*
Sundharagiati, B., 34(236), *47*
Sung, S. C., 238(110), *247*
Supek, Z., 307(139, 139a, 140, 141, 142, 143, 152), 308(42, 43, 44), 309(139a), 310 (42, 43, 44, 151), 312(44a), *325*, *328*
Supplee, H., 250(257), 254(60, 236), 268 (257), 270(60), 271(60), *279*, *284*, *285*
Suter, D. B., 222(35, 122a), *226*, *229*
Suter, H., *41*(3)
Sutherland, R. M., 31(237, 238), 32, 33(237), 34(237), 36(237), *47*
Sutter, J. L., 198(133), *229*
Sutton, E., 6(15), *11*, 122(191), 124(191), 141(191), *191*, 293(134), 312(133), 316(134), *328*
Sutton, G., 256(184), *283*
Sutton, J., 97(53), *101*
Svecenski, N., 223(121, 122), *229*, 256(140), *281*, 308(189a), *329*
Sviderskaya, T. A., 202(259b), *233*
Sviridov, N. K., 254(1), 260(1), *277*
Swanson, A. A., 218(260), *233*

Swarm, R. L., 202(50), *227*
Swift, M. N., 98(94), *102*, 251(258), 268 (219), *284*, *285*, 311(178), *329*
Swingle, K. F., 59(168a), 60(168, 168a), 61 (168a), *78*, 256(101a, c), *280*
Swingley, N., 252(158), 254(158), 268(158), *282*, 316(94), *326*
Syromyatnikova, N. V., 146(261), *193*
Sytinskii, I. A., 223(199), *231*
Szabo, G., 273(259), *285*
Szot, Z., 20(239), 25(239), *47*
Szybalski, W., 237(23), *245*

T

Tabachnick, J., 204(252), 205(262, 265, 266, 267), 206(261, 263, 264, 265, 266), 207 (261, 262, 266), *233*
Tábrosky, J., 18(247), 19(247), *47*
Tačeu, T., 24(10), 25(10), *41*
Tačeva, H., 295(178b), 310(178a), *329*
Taits, M. Yu., 212(42), 221(42), 222(42), *226*
Takamori, Y., 238(111), *247*
Takao, H., 221(198), *231*
Taketa, S. T., 92(95), 98(94, 95), *102*, 251 (258), *286*, 311(178), *329*
Taliaferro, L. G., 273(260), *285*
Taliaferro, W. H., 273(260), *285*
Talmage, R. V., 203(60), *227*
Talwar, G. P., 158(108), *189*
Tamanoi, I., 21(249), 22(249), *47*
Tanka, D., 215(127), *229*, 294(41), 295(41), *325*
Tappel, A. L., 70(29), *74*, 143(60), *187*
Taranger, M., 19(156), *45*
Tarasenko, A. G., 236(45), *245*
Taurog, A., 303(60), *325*
Taylor, D. M., 216(268), *233*
Taylor, G. N., 300(17), *324*
Tchoe, Y. T., 57(169), *78*
Teixeira, F. P., 31(60), 36(60), *43*
Telia, M., 61(169a), *78*, 273(260a), *285*
Teller, W., 295(179), 299(179), *329*
Temnikov, V., 220(97, 231), 221(231), 222 (231), *228*, *232*
Temple, D. M., 84(63), *101*
Tereshchenko, O. Y., 256(98), 262(261b), 263(261, 261a, b), *280*, *285*
Terskov, I. A., 31(66), *43*
Tew, J. T., 123(59), 125(59), 138(59), *187*
Thayer, M. N., 218(105), *229*
Theorell, H., *225*
Thiele, O. W., 254(128), *281*
Thielicke, G., 307(65), 308(65), 310(65), *325*
Thoai, N.-V., 118(149), *190*
Thomas, D. W., 58(11), *73*
Thomas, S. F., 293(197), 296(197), *329*
Thomason, D., 244(65), *246*
Thomson, J. F., 24(240), 27(240), 29(240), *47*, 58(171), 62(171), 63(170), *78*, 142 (262, 268), 143(262, 267), 144(242, 267), 157(264, 265, 266), 161(264, 265, 266), 165(263), 172(263), *193*, 221(247), 222 (247), *233*, 267(262, 263), *285*
Thonnard, A., 69(59), *75*
Thorne, W. A., 94(80), *101*
Thorell, B., 239(79a), *246*
Threlfall, G., 216(268), *233*
Thurber, R. E., 125(192, 269), 127(269), *193*, 288(180), 289(180), *329*
Thurnher, B., 313(132), *328*
Tibe, T. A., 31(163), 34(163), 36(163), 37 (163), *45*
Till, J. E., 24(150), 26(241, 242), *45*, *47*
Timiras, P. S., 223(178, 179), 224(269, 275a), *231*, *233*
Timm, M., 56(97), 62(98), *76*, 135(167), 138 (167), *190*
Timofeeva, N. M., 301(4), *323*
Ting, T. P., 31(243, 244, 245), 36(244), *47*
Titov, V. N., 215(227), *232*
Toal, J. N., 89(96), *102*
Tobias, C., 221(291), *234*
Tom, S. S., 123(270), 125(270), *193*
Tomana, M., 223(270), *233*
Tomasi, V., 159(10), *186*
Tonazy, N. E., *193*
Toropova, G. P., 123(273), 138(272), 142 (272), *193*
Toshi, H., 243(73), *246*
Totskii, V. N., 222(270a), *233*
Totter, J. R., 24(246), 25(246), *47*
Tourtellotte, W. W., 24(240), 27(240), 29 (240), *47*, 63(170), *78*, 165(263), 172 (263), *193*
Townsend, C. M., 181(188), *191*
Traelnes, K. R., 198(142), *230*, 257(161), *282*
Trandafirescu, M., 243(112), *247*
Traveníček, T., 18(247), 19(247), *47*
Trenk, B., 198(26), *226*

Tretyakova, K. A., 106(274, 276), 113(274, 275, 276), 118(276), 120(274, 275, 276), 155(275, 276), *193*, 293(181), 301(181), 316(181), *329*
Tribondeau, L., 301(16), *324*
Tribukait, B., 251(264), *285*, 319(64), *325*
Trincher, K. S., 31(175, 248), *45*, *47*
Trujillo, T. T., 322(182), *329*
Trum, B. F., 105(146), *190*, 304(126), 305 (104), *327*
Tsessarskaya, T. P., 19(101, 102), 21(101), 24(101, 102), 25(101, 102), *44*
Tseveleva, I. A., 22(131), 24(132), *44*
Tso, S. C., 29(185), 30(185), *46*
Tsuchiya, T., 21(249), 22(249), *47*, 304(183), 308(118, 119b), 311(118), 312(119, 119a), *327*, *329*
Tsukada, K., 177(277), *193*
Tsuya, A., 22(250), 24(250), 25(250), *47*
Tubiana, M., 254(21), *277*
Tulykes, S. G., 239(84), *246*
Turba, F., 21(182, 201), *46*
Tyree, E. B., 6(21, 22, 23), *11*, 197(136), *229*, 303(165), 304(165), *329*

U

Uchiyama, T., 168(73, 278), 172(73, 279), 173(73, 278), 174(279), 177(278), 178 (73, 278, 279), 179(278), 181(73, 278), 182(73, 278), *187*, *193*
Uhlmann, G., 239(61, 62), *246*
Ulbrich, O., 117(25a), 123(25a), 132(25a), 133(25, 25a), 135(25), *186*
Ulmer, D. D., 304(184), *329*
Umbreit, W. W., *193*
Ung, X. B., 202(151), *230*
Ungar, F., 294(156), *328*
Upham, H. C., 95(16, 17, 18), *99*, 257(46), *278*
Upton, A. C., 198(53a), *227*
Urakami, H., 142(280), 143(280), *194*
Urry, H. C., 250(189), 252(189), 253(189), *283*
Uspenskaya, M. S., 256(65, 265), *279*, *285*
Uyeki, E. M., 22(251), 24(251), *47*, 62(172), *78*

V

Vaček, A., 221(270b), *233*
Vachek, H., 241(96), *247*
Vaeth, J. M., 216(213), *232*
Vainson, A. A., 238(113), *247*
Valenta, M., 301(185), *329*
Valli, P., 124(202), *191*
van Bekkum, D. W., 62(173, 174, 175), 63 (173, 174), *78*
Van Caneghem, P., 208(19, 271, 272, 273), *226*, *233*, 313(186), *329*
Van Cauwenberge, H., 293(187), 295(188), *329*
Vandergoten, R., 127(281), 173(281), 178 (281), 182(281), *194*
Vandermeulen, R., 308(33), *324*
Van Duyet, P., 94(96a), *102*
Van Dyke, D., 19(73), 29(73), 30(73), *43*, 216(101), *228*
Vane, J. R., 22(122), 25(122), *44*
van Heyningen, R., 218(216, 217, 274), *232*, *233*
van Metre, M. T., 255(35), 257(36), *278*
Van Riper, J., 293(70), *326*
Van Steveninck, J., 31(252), 35(252), *47*
Varagic, V., 308(189, 189a), *329*
Varga, L., 254(265a), 269(265a), *285*
Várterész, V., 153(122), 154(122), *189*
Vasileiskii, S. S., 254(65), *279*
Vatistas, S., 84(34, 97a), 95(35, 97), *100*, *102*, 273(266), *285*
Vaughan, B. E., 96(97b), *102*
Vecchioni, R., 197(29, 232), *226*, *232*
Veksler, Ya. I., 222(275), *233*
Velarde, M. I., 307(51a), *325*
Velikii, N. N., 211(259), 215(259), *233*
Veninga, T. S., 308(31, 190), *324*, *329*
Venkitasubramanian, T. A., 107(206), 114 (206), 140(206), *192*
Verhagen, A., 237(114), 243(114), *247*
Verly, W., 94(55), *101*
Vermund, H., 111(129), 117(129), 165(129), *189*, 237(20, 21, 23), 238(115), *245*, *247*
Vernadakis, A., 223(179), 224(275a), *231*, *233*
Vertua, R., 112(208), 119(208), 120(208), *192*
Veržar, F., 198(276), *233*
Vierling, A. F., *329*
Vigne, J., 274(135), *281*
Vikterlöf, K. J., 319(64), *325*
Viktora, L., 31(210), *46*
Vinetski, Y. P., 34(253), 37(253), *48*
Vinogradova, M. F., 105(282), 106(282), 107(282), 113(282), 114(282), 120(282), *194*

Visek, W. J., 253(227a), *284*
Vitale, D. E., 295(31a), 312(31a), *324*
Viterbo, F., 257(42), *278*
Vittorio, P. V., *194*, 304(191), *329*
Vliers, M., 295(188), *329*
Vodicka, I., 215(277d), *234*
Vodop'Yanova, L. G., 221(64), *227*
Voelker, W., 239(120, 121), *247*
Vogel, H. H., 18(186), 19(186), *46*
Vogel, M., 252(97), 253(97), *280*
Vogel, T. S., 220(93), *228*
Vogt, M., 238(116), *247*
Voigt, E. D., 38(17), *42*
Volek, V., 269(267), *285*
Volk, B. W., 91(98), *102*
Voncina, A., 269(138), *281*
von Euler, H., 52(2), 56(2), *73*, 237, *247*
von Gossler, M., 240(58), 241(58), *246*
von Hevesy, G., 52(2), 56(2), *73*, 237, *247*
von Kroebel, W., 225(277a), *233*
von Sallmann, L., 218(277b, c), *233*
Voogd, J. L., 269(15), *277*
Vorbrodt, A., 206(17), *226*
Vorbrodt, K., 206(278), *234*
Vorobyev, V. N., 31(259), *48*
Vos, O., 52(164), 62(173), 63(173), *78*, 95 (57), *101*
Vyhnanek, L., 201(13), 202(13), 203(13), *225*

W

Wachtel, L. W., 216(279, 280, 281), *234*
Wackman, N., 18(72), 19(72), 29(72), *43*
Wadel, J., 313(29), *324*
Waggener, R. E., 31(255), *47*
Wagner, H., 206(282), *234*
Wagner, M., 55(143), *77*
Wagner, O., 206(282), *234*
Waldschmidt, M., 63(178), *78*, 130(285), 135(285a), 136(285a), 138(285a), 142 (283), 146(284), 147(284), 153(284), *194*, 255(267a), 266(268), 267(268), *285*, 308(192), *329*
Waldschmidt-Leitz, E., 268(269), *285*
Walford, R. L., 198(283), *234*
Waligora, Z., 224(285), *234*
Walker, C. D., 202(50), *227*
Walker, J. K., 39(43), *42*
Wall, P. E., 90(23), *100*
Wallas, J., 263(242b), *285*
Waller, H. I., 18(256), 39(256), 41(257, 258), *48*
Walsh, D., 307(193), *329*
Walter, O. M., 303(194), *329*
Walther, E., 308(95), *326*
Walther, G., 212(239a), *232*
Walwick, E. R., 216(176), 217(176), *231*
Warburg, O., 41(268), *48*, 239(119, 120, 121, 122), *247*
Warner, N. E., 92(32a), *100*
Warner, W. L., 106(158), 108(158), *190*, 316(91), *326*
Warren, S., 256(90), *280*, 313(127), 316 (127), *327*
Warren, S. L., 80(99), 90(99), 91(99), *102*
Wase, A., 185(224), *192*
Watras, J., 41(259), *48*
Watson, G. M., 260(270), *285*
Watts, C., 269(93a, 271, 271a), *280*, *285*
Webber, M. M., 162(286), 171(286), *194*
Weber, G., 123(287), 124(287), 128(287), 130(287), 131(287), *194*, 316(195), *329*
Weber, R. P., 308(196), *329*
Webster, E. W., 96(13), *99*
Weed, R. I., 31(238, 252), 32(238), 34(238), 35(252), *47*
Weill, J. D., 167(45a), 169(45a), 182(45a), *187*
Weimer, E. Q., 214(188), *231*
Weiner, N., 212(284), *234*, 271(272), *286*
Weinman, E. O., 110(288), 111(288), *194*
Weiss, C., 206(261), *233*
Welling, W., 172(289), 178(289), *194*
Wellmann, K. F., 91(98), *102*
Wells, A. B., 198(203), *226*
Wells, A. F., 308(57), 309(57), 310(57), *325*
Wells, P., 31(124), *44*
Wende, C. M., 92(44a), *100*
Wender, M., 224(285, 286), *234*
Wen-Mei, S., 142(290), 179(290), *194*
Wernze, H., 40(260), *48*, 148(291), *194*
Wesemann, I., 91(100), 92(100), 94(100), *102*
Westman, A., 303(46), *325*
Westphal, U., 268(273), *286*
Wetterfors, J., 273(19, 273a), *277*, *286*
Weymouth, P. P., 68(179), 69(179), 70(179), *78*
Wexler, B. C., 296(197), *329*
Whipple, G. H., 80(99), 90(99), 91(99), *102*, 253(107), 254(107), *280*

White, D., 269(118a), *281*
White, J., 89(96), *102*, 146(292), *194*, 253 (274), 256(274), *286*
White, N. G., *193*
Whitehead, R. W., 165(96a), *188*
Whitfield, J. F., 53(128g), 59(128d, e), 61 (128f, 181, 182, 183, 186, 188a), 63(184), 64(185), 65(185), 66(21, 184), 69(180), 72(128d, g, 180, 187, 188), *74*, *77*, *78*, *79*
Wiehrmann, D., 270(275), *286*
Wilbrandt, W., 31(28, 133), *42*, *44*, 94(32), *100*, 316(77), *326*
Wilbur, K. M., 14(11), 17(11), *41*
Wilde, W. S., 208(287), 211(287), *234*
Wilding, J. L., 105(146), *190*, 305(104), *327*
Wilhelm, G., 211(224), *232*, 295(199), *329*
Wilian-Ulrich, I., 106(116), 149(116), *189*
Wilkerson, M. C., 118(52), 119(52), 120(52), *187*
Wilkinson, A. E., 70(189), *79*, 144(294), 145 (294), *194*
Williams, C. M., 260(7, 168, 276), 262(7), *277*, *282*, *286*, 312(97, 200, 201), 316 (97, 201), *326*, *329*, *330*
Williams, J., 270(87), *280*
Williams, R. B., 89(96), *102*, 269(15), *277*
Williams, W. J., 242(34), 243(34), *245*
Willoughby, D. A., 308(203), 313(202), 316(203), *330*
Wills, E. D., 70(189), *79*, 144(293, 294), 145 (293, 294), *194*
Wilson, C. W., 84(101a), *102*, 199(289), 202 (288), *234*
Wilson, C. W. M., 308(52), *325*
Wilson, J. E., 211(71a), *227*
Wilson, R., 6(14), *11*, 81(59), 84(59), *101*, *102*
Wimber, D. R., 85(86), 88(86), *101*
Windler, F., 301(39), *325*
Winkler, A., 256(53, 54), *278*
Winkler, C., 211(32), 214(32), *226*, 268(277), *286*, 294(204), 295(204), *330*
Wish, L., 250(250), 251(250), *285*
Wishiewski, K., 223(52), *227*
Witcotski, A. L., 6(17), *11*
Withers, H. R., 84(101a), *102*, 204(290), *234*
Witney, R., 292(207), *330*
Wittekindt, E., 54(128c), *77*
Wittekindt, W., 54(128c), *77*
Wolf, R. C., 294(205), 295(205), *330*
Wolfe, L. S., 221(291), *234*
Wolfsberg, M. F., 85(86), 88(86), *101*
Wolkowitz, W., 91(6), *99*
Wollschlaeger, G., 264(174), *282*
Wood, G. A., 250(178), *283*
Wood, R., 14(225), 15(225), 16(225), *47*
Woodard, H. Q., 200(293), 202(67, 292, 293), 203(74), *227*, *234*
Woodruff, M. F. A., 253(67), *279*
Woods, M., 239(13), *245*
Woodward, K. T., 4(26), *11*, 311(206), *330*
Wooles, W. R., 112(66), 113(295), 120(66, 295), *187*, *194*, 253(57), 254(57), *278*
Word, E. L., 31(31), *42*
Wotiz, H. H., 302(112), *327*
Wrage, D., 221(293a), 222(293a), *234*
Wrigglesworth, J. M., 90(101b), *102*
Wright, C. S., 31(261), 34(236), *47*, *48*, 269 (45), *278*
Wright, R., 14(226), *47*
Wueppen, I., 241(96), *247*
Wyman, L. C., 292(207), *330*

Y

Yago, N., 159(202a), *191*, 293(207a), *330*
Yakubovick, L. S., 150(211), 151(211), *192*
Yamada, T., 64(118a), *77*
Yamada, T. J., 240(124), 241(124), *247*
Yamaguchi, 49(190), *79*
Yamamoto, Y. L., 221(294), *234*
Yamoda, T., 64(189a), *79*
Yang, C. H., 225(170), *230*
Yang, F.-Y., 142(296, 297), *194*
Yarnell, R. A., 271(198), *283*
Yates, C. W., 203(60), *227*
Yatvin, M. B., 154(297a), *194*
Yegin, M., 295(61, 208), *325*, *330*
Yen, C. Y., 169(298), 175(298), *194*
Yin, W. A., 259(98a), 260(98a), *280*
Yoshikawa, H., 244(106), *247*
Yost, H. T., Jr., 62(191, 192), 63(191, 192), 71(13), *74*, *79*, 142(18, 299, 300, 302, 303), 144(301), *186*, *194*
Yost, M. T., 62(191, 192), 63(191, 192), *79*, 142(302, 303), *194*
Youdale, T., 61(128f, 181, 182, 183, 186, 188a), 63(184), 64(185), 65(185), 69 (180), 72(180, 188), *77*, *78*, *79*

Young, L. E., 20(38), 31(38), 36(38), *42*
Yuhas, J. M., 5(27), *11*
Yurmin, E., 254(218), *284*

Z

Začek, J., 34(262, 263), *48*
Zadory, E., 273(259), *285*
Žak, M., 253(120a), 276(120), *281*
Zakharov, S. V., 222(295), *234*
Zaminjan, T. S., 212(203), *231*
Zampaglione, G., 21(172), 22(172), *45*
Zamyatkina, O. G., 273(278), *286*
Zanetti, A., 118(195), *191*
Zanolla, W., 95(12a), *99*
Zaorska, B., 21(264), 22(264), *48*
Zarkh, E. N., 221(244), *232*
Zaruba, K., 214(296), *234*
Zbarsky, I. B., 239(84), *246*
Zehnder, H., 94(102), *102*
Zelemski, V. B., 273(228), *284*
Zelikina, T. I., 221(242), *232*
Zharkov, Yu. A., 256(98, 278a), *280*, *286*
Zhulanova, Z. I., 144(107), 149(221), *189*, *192*, 256(279), *286*
Zhurbin, G. I., 211(297), 213(297), *234*
Zicha, B., 39(265), *48*, 129(305), 130(305), 131(305), 132(305), 134(306), 135(305), 138(306), 142(305), 144(304, 306), *194*, 257(280), 263(282), 268(86d, 280, 281), *277*, *280*, *286*
Ziegler, K., 80(47), *100*
Zigman, S., 218(161), *230*
Zimmerman, D. H., 55(193), 63(193), 64 (193), *79*
Zirkle, R. E., 31(243, 244, 245), 36(244), *42*, *47*
Zivkovitch, N., 291(116), *327*
Zographov, D. G., 30(266, 267), *48*
Zöllner, N., 119(64), *187*, 293(48), *325*
Zsebök, Z. B., 94(103), 96(103, 104), *102*, 294(209), *330*
Zubkova, S. R., *234*
Zucker, M., 80(78), *101*
Zuckerman, L., 211(61), 213(61), *227*
Zuideveld, J., 313(29, 30), *324*
Zuppinger, A., 142(226), *192*, 202(298), *234*, 254(4), 255(2), *277*
Zweifach, B. W., 63(80, 81), *75*, *76*, 129 (135d, 136), *190*

SUBJECT INDEX

A

Acetate
incorporation of
into cholesterol, in liver, 105, 113–115
into DNA, in liver, 169
into estrogens, in ovaries, 302
into heme, in bone marrow, 25
into lipids
in bone marrow, 16, 17
in brain, 222
in gastrointestinal tract, 91
in liver, after hypophysectomy, 316
in thymus, 58
into protein, in brain, 221
utilization of
by bone marrow, 26
by muscle, 209, 211
Acetylcholine
in brain, 223
in small intestine, 82, 89
Acetylcholinesterase
in bone marrow, 29
in brain, 223
intestine, 90
in red blood cells, 39, 40
Acetyl-CoA, in liver, 135
Acetyl-CoA carboxylase, in liver, 133, 135
α_1-Acid glycoprotein, biosynthesis of, in liver, 155
Actinomycin, inhibition of adaptive enzyme biosynthesis by, 161
Acyl-CoA dehydrogenase, in lymphoid tissue, 63
Adaptation syndrome, 323
Adenine, incorporation of
into DNA
in bone marrow, 25
in liver, 170, 174
in lymphoid tissue, 52
into RNA
in liver, 174
in thymus nuclei, 57
Adenosine deaminase (aminohydrolase), in spleen, 58
Adenosinediphosphate, in red blood cells, 39
Adenosinetriphosphatase
activity of, in lymphoid tissue, 62, 63, 68
in adrenal cortex, 292
in brain, 222
in intestine, 90
in lens, 218
in muscle, 208, 212
in spleen, 58
ouabain-resistant, in bone marrow, 24, 26, 29
ouabain-sensitive
in cation transport in red blood cells, 33
in red blood cells, 33, 39

Adenosinetriphosphate
in anaerobic glycolysis, in lymphoid tissue, 64
in bone marrow, 22, 24, 25
in brain, 222
formation of
in lymphocyte nuclei, 50, 54
in mitochondria of thymus and myeloid leukemia cells, and lymphocytes, 50, 62, 64
incorporation of inorganic phosphate into, in muscle, 208, 209, 211
incorporation of, into nRNA, in liver, 173
in lens, 218
in liver, 135
in lymphocytes, 62
in lymphoid tissue, 62
in lymphoid tissue cell nuclei, 50, 62–64, 72
in muscle, 208, 209, 211, 212
in red blood cells, 37, 39
in tumors, 240, 243
Adenylate
in bone marrow, 23
incorporation of inorganic phosphate into
in liver, 173
in muscle, 208, 209
phosphorylation of
in bone marrow, 24
in isolated thymus nuclei, 64
in lymphoid tissue nuclei, 62–64
in mouse lymphosarcoma cell nuclei, 64
in myeloid leukemia cell nuclei, 64
in tumor cell nuclei, 243
in red blood cells 39
Adipose tissue
lipase in, 199
lipolytic activity in fat cells of, stimulation of, by growth hormones, 199
mobilization of lipids from, 253
radioinsensitivity of depot fat metabolism in, 199
Adrenal gland
ascorbic acid in, 289–293, 297, 299, 316, 319
cholesterol in, 291–293, 296, 297, 299, 317
steroid hormones, 291, 294–297, 299, 315
in blood and urine, 291
Adrenal hormones, catabolism of, in liver, 294
Adrenaline
in brain, 223
radiation protection by, in lymphoid tissue, 310
Adrenocorticotrophin, in blood and pituitary, 291
Aging, acceleration of, after irradiation, 198
Alanine
conversion to CO_2, in liver, 136
incorporation of
into glycogen, in liver, 136
into protein
in liver, 153
in muscle, 211
in tumors, 237
in urine, 256
Alanine aminotransferase (glutamic–pyruvic transferase)
in liver, 150, 159
in muscle, 212
Albumin
in blood, 254, 268, 270
transfer of, in skin, 313, 314
Albumin/globulin ratio, in blood, 254
Albuminuria, 214
Aldose
in brain, 222
in muscle, 212
in plasma, 213, 276
Aldosterone, biosynthesis of, in adrenal gland, 294
Allantoin, in urine, 256
Amino acid(s)
activation of
in isolated thymus nuclei, 57, 58
in lymphoid tissue, 56–58
in blood, 254, 255
free
in kidney, 215
in liver, 146
in skin, 206
leakage of, from lens into aqueous humor, 218
in urine, 255, 256, 276
Amino acid decarboxylases, activity of, in vitamin B_6 deficiency, 266
L-Amino acid dehydrogenase, in lymphoid tissue, 63
p-Aminobenzoate, effect of, on red blood cells, 35

γ-Aminobutyric acid (GABA), in brain, 223
2-Aminoethanol, in urine, 257
5-Amino-4-imidazole carboxamide, in liver, incorporation of
  into DNA, 169
  into RNA, 175
α-Aminoisobutyrate
  in blood, 263
  in liver, 256
    from thymidine, 263
  in urine, 256, 262, 263, 276
5-Aminolevulinate, incorporation of, into heme, in bone marrow, 19, 20
5-Aminolevulinate synthase
  in bone marrow, 20, 26
  in red blood cells (avian), 20
Aminopeptidase, in lymphoid tissue, 69
Aminotransferases
  activity of, in Vitamin B_6 deficiency, 266
  comparative radiosensitivity of
    in kidney, 215
    in liver, 153
    in muscle, 147
    in serum, 269
    in spleen, 58
Ammonia
  in brain, 222
  formation of, in bone marrow, 26
Amylase (α), biosynthesis of, in pancreas, 91
Androgens
  biosynthesis of, in testis, 299, 300
  conversion of, to testosterone, in testis, 300
Anthranilate
  hydroxylation of, 264, 266
  in urine, 265
Antibodies, biosynthesis of, in lymphocytes, 52
Antibody levels, depression of, 273
Antidiuretic hormone, in polyuria, 291
Antioxidants
  in intestine, 90
  in red blood cells, 36
Aorta, 199
Arachidonate
  in lecithin, in liver, 119
  in triglycerides, in bone marrow, 16
Arginine, incorporation of, into glycocyamine, 262
Aryl sulfatase
  in intestine, 90
  in lymphoid organs, 68
  release of, from lysosomes of lymphoid tissue, 70
  in tumors, 244
Ascorbate
  in adrenal gland, 289–293, 297, 299, 316, 319
  in urine, 257
Aspartate aminotransferase (glutamic–oxalacetic transferase)
  in liver, 150
  in muscle, 212
Aspartate carbamoyltransferase, in liver, 151
Autolysis in lymphoid tissue, 145

B

Bacteria, intestinal, 147, 258, 259
Basal metabolic rate, regulation of, by thyroid gland, 304, 305, 318
Behenic acid, in total lipids, in liver, 118
Bile, 98
Bile acids
  conjugation of, in liver, 149
  enterohepatic pathway of, 92
  intestinal reabsorption of, 259
Bilirubin
  in bile, 98
  in blood, 254
  conjugation of, in liver, 148
  enterohepatic pathway of, 92
Biogenic amines, 255, 256, 307, 308, 310, 312, 318, 322
Bladder, collagen in, 194
Blood
  albumin in, 254
  alkaline phosphatase in, 202
  alkali reserve of, 253
  amino acids in, 254, 255
  BAIBA in, 263
  cholinesterase in, 308
  conductivity of, 276
  corticoids in, 289
  creatine in, 254, 260
  density of, 276
  deoxycytidine in, 256
  electrolytes in, 251–253
  fatty acids in, 253

Blood (*cont.*)
globulins in, 254
glucose in, 253, 254, 288
glycosaminoglycans in, 199, 254
hydroxyindoles in, 309
K^+ in, 211
lactate in, 254
lipids in, 254
lipoproteins in, 253, 254
malate dehydrogenase in, 269
nonprotein nitrogen in, 254
organic acids in, 254
palmitate in, 253
pH of, 253
phospholipids in, 253
plasmalogen in, 254
polysaccharides in plasma of, 254
prealbumin in, 254
protein in, 254
protein-bound hexosamines in, 254
purines in, 254
serotonin in, 309
steroid hormones in, 291
surface tension of, 253
taurine in, 258
tryptophan in, 254
tyrosine in, 254
urea in, 254
uric acid in, 255
viscosity of, 253
volume of, 251
Blood-brain barrier, movement of substances across, 220
Bone
alkaline phosphatase in, 200, 202
Ca^{2+} uptake by, 201–203, 318
effect of denervation on, 203
collagen in, 202
conversion of soluble to insoluble in, 202
degradation of, 202
deposition of ^{89}Sr in, 203
inhibition of mineralization in, 202
metabolic replacement of glycosaminoglycans in, 202
uptake of inorganic sulfate by, 201
effect of denervation on, 203
Bone marrow
accumulation of deoxypolynucleotides in, 22
acetate utilization in, 26
acetylcholinesterase in, 29
acyl-CoA transferase in, 12–30
adenylate in, 23
alkaline phosphatase in, 24, 26
5-aminolevulinate synthase in, 20, 26
ammonia formation in, 26
ATP in, 22, 24
ATPase, ouabain resistant, in, 24, 26, 29
biosynthesis
of DNA in, 21, 24–26
of fatty acids in, 13, 16, 17
of globin in, 17
of hemoglobin in 17–22, 29, 30
of lipids in (cultures of human), 17
of phospholipids in, 16
of proteins in, 21, 22
of thymidine in, 23, 25, 29
Ca^{2+} in, 29
catabolism of RNA in, 23, 26, 29
cholesterol in, 15
choline dehydrogenase in, 27
cholinesterase in, 28
damage to template DNA in, 29
dCMP restoration of normal base composition in DNA of, 24
dCMP aminohydrolase in, 23
deoxyribonucleotides in, 22, 24
dihydroxyacetone phosphate in, 27
DNA in, 21, 24–26
DNase I in, 26, 28
DNase II in, 26, 28, 29
effect of erythropoietin on, 30, 31
on thymidine incorporation into DNA in, 29
erythroblasts in, 13, 18–20, 29
fatty acid oxidation in, 17
ferrochelatase in, 20
fructose-1,6-diphosphate in, 27
fructose-6-phosphate in, 27
glucose-6-phosphate in, 27
glucose-6-phosphate dehydrogenase in, 27
glutathione reductase in, 27
glyceraldehydephosphate dehydrogenase in, 27
glycolysis in, 26, 27
hexokinase in, 27
incorporation
of acetate, into heme in, 25
into lipids of, 16, 17
of adenine into DNA in, 25

Bone marrow (*cont.*)
incorporation (*cont.*)
of fatty acids into lipids of, 13–17
of formate into DNA of, 25, 26
of glycine
into DNA of, 25
into heme of, 20, 21, 25
into protein of, 21
of inorganic phosphate
into nucleic acids of, 24–26
into ribonucleotides of, 24
of methionine into protein of, 21
of oleate and palmitate into phospholipids in, 15
of porphobilinogen into heme in, 19, 20
of stearate and palmitate into triglycerides in, 15
of thymidine into DNA of, 23, 25, 29
of uridine into RNA of, 29
K^+ ion in, 29
lactate in, 27
lipids in, 14
lipocytes in, 13–15, 17
lipogensis in, 15–17
methionine-methyl groups, in dTMP methylation in, 21
Na^+ ions in, 29
nucleic acid catabolism in, 22
nucleic acids in, 13, 23–26
nucleoproteins in, 22
5′-nucleotidase in, 28, 29
nucleotides and nucleosides in, 13, 22–27
oxidation of fatty acids in, 17
oxygen consumption by, 13, 27, 29
peroxide formation in, 17
phospholipids in, 14–16, 23
porphobilinogen synthase in, 20
protein composition of, 21, 22
pyrimidine deoxyribonucleotides in, 22
pyruvate in, 27
reticulocyte release from, 19
reutilization of pyrimidine bases in, 22
stem cells in, 3, 13, 30
succinate dehydrogenase in, 27
dTMP kinase in, 23
tricarboxylic acid cycle intermediates in, 26
triglycerides in, 14–16
uridylate in, 23, 24
water in, 14

Brain
acetylcholine in, 223
acetylcholinesterase in, 223
acid phosphatase in, 222
aldolase in, 222
alkaline phosphatase in, 222
biosynthesis
of cerebrosides in, 224
of DNA in, 221, 223
of protein in, 221
of RNA in, 221, 223
of sphingomyelin in, 224
of unsaturated fatty acids in, 224
CO_2 production by neonatal, 224
in central nervous system (CNS) syndrome, 219
cholesterol in, 224
degradation of cytoplasmic RNA in, 221
DNA in, 220, 221
gangliosides in, 222
glutamate in, 223
glutamate decarboxylase in, 223
glutaminase in, 223
glutathione in, 223
glycogen in, 221, 222
glycolysis in, 222
glycosaminoglycans in, 221
incorporation
of acetate into lipids in, 222
of inorganic phosphate, into DNA and RNA in, 221
of mevalonate into cholesterol in, 222
K^+ in, 220
lactate accumulation in, 222
lactate dehydrogenase isozymes in, 224
loss
of DNA by, 221
of K^+ by, 225
metabolism of glutamate in, 223
Na^+ in, 220
oxidation of glucose in, 222
oxygen consumption by, 221, 223
phospholipids in, 224
phosphoproteins in, 221
RNA in, 220, 221
succinate oxidase in, 222
total lipids in, 222
utilization of citrate in, 222
water in, 220
Brain stem
biosynthesis of myelin in, 224

Brain stem (*cont.*)
protein in, 221
Butyrate
incorporation of, into fatty acids of lipids, in liver, 105, 106
oxidation to CO_2, in liver, 116
Butyryl-CoA dehydrogenase, in liver, 130

C

Calcium
incorporation of, into bone, 203
in serum, 253
uptake of, by bone, 201, 318
in urine, 202, 253
Calcium ion
in bone marrow, 29
prevention of nuclear structural changes in lymphocytes by, 72
protection from mitotic arrest by, in lymphocytes, 72
in red blood cells, 38
uptake of, by hepatic mitochondria, 144
Carbohydrate, metabolism of, serum enzyme in, 268
Carbon dioxide
bicarbonate, in liver, 136
from kynurenine, 267
production of, by neonatal brain, 224
Carboxypeptidase, in intestine, 90
Cardiolipin, in liver, 119
Catalase
in kidney, 215
leakage from lymphocyte nuclei, 68
in red blood cells, 39, 41
in tumors, 236, 239, 244
Catecholamine, 308
Cathepsins
liberation of, from lysosomes, in liver, 145
in lymphoid tissue, 68, 70
in spleen, 70
Cation transport, in red blood cells, 33
Cephalin, biosynthesis of, in bone marrow, 15
Cerebrosides, biosynthesis of, in brain, 224
Chloride ion
blood levels of, 252
depletion of, 253
in muscle, 213
in organs, 252
in urine, 251, 252
p-Chloromercuribenzene sulfonate, effect on transport phenomena in red blood cells, 32, 35
Cholesterol
in adrenal gland, 291–293, 296, 297, 299, 317
biosynthesis of
in adrenal gland, 293
in liver, 105, 113–115
in blood, 253, 254
in bone marrow, 15
in brain, 224
deposition of, in aorta, 199
esters of, in adrenal gland, 293
incorporation of mevalonate into, in brain, 222
metabolism of, in liver, after thyroidectomy, 316
in serum, 254, 255
Choline dehydrogenase, in bone marrow, 27
Cholinesterase (acylcholine acyl-hydrolase)
in blood, 308
in bone marrow, 28
in brain, 308
in intestine, 308
in serum, 269
in small intestine, 89
Chondroblast, loss of glycogen by, 202
Chondroblastoma, incorporation of inorganic sulfate into, 202
Chondroitin sulfate, in intestine, 89
Citrate
accumulation of, in presence of fluoroacetate, in liver, 135, 140
formation of
in small intestine, 82, 91
in spleen, 63
utilization of, in brain, 222
Collagen
in aorta, 199
biosynthesis of, in connective tissue, of healing wounds, 197
in bladder, 194
in bone, 202
conversion of soluble to insoluble, 197, 198
degradation of
in bone, 202
in muscle, 213
in granulomas, 197
metabolism of, abscopal effects of irradiation on, 198

Collagen (*cont.*)
in muscle, loss of amino acids from, 198
ratio of soluble to insoluble, in relation to aging, 198
in skin, 197, 198
soluble, 198
in tendon, 198
thermal contraction of fibers of, in connective tissue, 198
degradation of, 198
in irradiated bone, 198
in skin, 198
Connective tissue
collagen in, in relation to age, 198
cysteine, free, in, 198
growth of, 50
methionine, free, in, 198
permeability of, 312, 313
Copper, in serum, 253
Corticoids
in adrenal gland, 294–297, 299
in blood, 289
Corticosterone, in adrenal gland, 294
Cortisol, in adrenal gland, 294
Cortisone
in adrenal gland, 294, 315
effect of, on incorporation of inorganic phosphate into DNA, in duodenum, 317
Creatine
biosynthesis of, in liver, 262
in blood, 254, 260–262
in brain, 251, 262
incorporation of, into muscle, 213
metabolism of, 261
in muscle, 213
in muscle, 213, 261, 262, 319
pools of, 261, 262
in red blood cells, 254, 260
in urine, 256, 260, 262, 276
Creatine/creatinine ratio, 260
Creatine kinase (phosphokinase), in serum, 269
Creatine phosphate, 262
Creatine, excretion of, 252, 262
Crypts, of small intestine, biosynthesis of DNA by stem cells in, 81
Cysteic acid, free, in spleen, 58
Cysteine
conversion to taurine, 258
free
in connective tissue, 198
in spleen, 58
Cytidine, incorporation of, into RNA, in brain, 221
Cytidine triphosphate, incorporation of, into nRNA, in liver, 173
Cytidylate
in bone marrow, 13
incorporation
of inorganic phosphate into, in liver, 173
of orotate into, in liver, 174
Cystine, in urine, 256
Cytochrome c, binding of, by mitochondria, 143
Cytochrome oxidase
in intestine, 90
in lymphoid tissue, 63

D

Dehydration, during radiation illness, 250, 251
Deoxyadenylate, in bone marrow, 26
Deoxycorticosterone
in adrenal gland, 296
effect of, on incorporation of inorganic phosphate into DNA, in intestinal mucosa, 317
in plasma, 296
Deoxycytidine
accumulation of, in liver, 164
in blood, 256
catabolism of, 263
incorporation of, into DNA
in liver, 162
in lymphoid tissue, 53, 55
in thymus, 55, 56
in urine, 184, 256, 262, 263, 276
Deoxycytidine deaminase (aminohydrolase), 263
Deoxycytidylate (dCMP)
in bone marrow, 13, 22, 26
incorporation of, into DNA, in liver, 168
restoration of normal base composition on administration of, in bone marrow, 26

Deoxycytidylate deaminase (dCMP aminohydrolase)
in bone marrow, 23
in liver, 163, 177, 181
in lymphoid tissue, 55
in thymocyte suspensions, 55
in tumors, 236
Deoxyribonuclease(s)
in bone marrow lysosomes, 13, 24, 26, 28, 29
in lymphoid tissue lysosomes, 50
Deoxyribonuclease I (DNase I)
in bone marrow, 26, 28
in hepatic mitochondria, 185
in skin, 205, 206
Deoxyribonuclease II (DNase II)
in bone marrow, 26, 28, 29
in lysosomes of liver, 184
in plasma, 269
release from lysosomes in spleen, 70
in skin, 205, 206
in spleen, 60, 70
in thymus, 59, 68–70
in tumors, 238
in urine, 269
Deoxyribonucleic acid (DNA)
biosynthesis of
in bone marrow, 24–26
in brain, 221, 223
in hair follicles, 205
in intestine, 81, 82, 84–88
in lens, 218
in liver, 162–164, 170, 171, 176–179, 183–185
in lymphoid tissue, 52–56, 58, 62, 72, 73
in regenerating crypt epithelium, 87
in skin, 204
in tumors, 236–238, 242
in bone marrow, 21, 24–26
in brain, 220, 221
catabolism of
in bone marrow, 24
in lymphoid tissue, 50
in small intestine, 87
in spleen, 54
in thymus, 54
incorporation
of dTTP into, in tumors, 238
of formate into, in tumors, 237
of glycine into
in bone marrow, 25
in lymphoid tissue, 52
in tumors, 237
of inorganic phosphate into
in brain, 221
in intestine, 86
in tumors, 237
of iododeoxyuridine into, in tumors, 237
of thymidine into
in brain, 221
in intestine, 85–87
in kidney, 215–217
in skin, 204
in tumors, 237, 242
in kidney, 216
in lens, 218
in lymphoid tissue, 52, 262
metabolism of, in lymphocytes, 52, 54, 55
mitochondrial (mDNA)
biosynthesis of
in liver, 167, 169
in tumors, 237
incorporation of thymidine into, in liver, 167
turnover in kidney, 215
nuclear (nDNA), incorporation of thymidine into, in liver, 167
radiation damage to template activity of
in bone marrow, 29
in lymphoid tissue, 54
radiation-induced enlargement of dTTP pool in biosynthesis of, in intestine, 85, 88
radiation-induced structural changes in, in tumors, 238
in skin, 206
in small intestine, 82, 85
specific activity/time plot for small intestine, 87
in tissue, 262
in tumors, 236, 242
Deoxyribonucleic acid nucleotidyltransferase (DNA polymerase)
in kidney, 217
in liver, 177–183
in lymphoid tissue, 55
Deoxyribonucleotides
in bone marrow, 22, 24
incorporation of inorganic phosphate into, in bone marrow, 24

Deoxyribonucleotides (pyrimidine), pool sizes, in bone marrow, 22
Deoxyribose derivatives, in urine, 256
Deoxythymidylate, incorporation of, into DNA, in liver, 168
Deoxyuridine, in urine, 256, 262
Deoxyuridylate, in bone marrow, 13, 22
Dioleylpalmitin, in liver, incorporation of
  into phospholipids, 109–111
  into triglycerides, 111, 112
Diarrhea
  in gastrointestinal syndrome, 92
  prevention of by ligating pancreatic duct, 98
Diencephalon, role in hormonal regulation, 288, 289
Dihydroxyacetone phosphate, in bone marrow, 27

E

Elastin, replacement of
  in muscle, 212, 213
  in skin, 199
Electrolytes
  active transport of, in small intestine, 82, 94
  in blood, 251, 252
  depletion of, in body, 253
  loss of
    by intestine, and survival, 97
    by small intestine, 82
  metabolism of, 252
  in urine, 251
Electron transport
  defect in lymphoid tissue, 63
  in mitochondria
    of lymphoid cells, 50
    of lymphoid tissue, differences in radiosensitivity at various sites of, 62
Enzymes
  adaptive, biosynthesis of, in liver, 161
  glycolytic, loss of, by nuclei of thymus cells, 64, 66
  lysosomal, increased activity of, in irradiated liver, 184, 185
Epidermis
  acid phosphatase in, 205
  mitotic arrest in, 204
Epithelium, of intestinal crypts and villi, loss of cells from, 81
Erythroblast, in bone marrow, 18–20, 29
Erythropoietin
  biosynthesis of, in kidney, 29, 30
  in bone marrow metabolism, 30
  effect of
    on erythroblast differentiation, 29
    on hemoglobin biosynthesis, 29, 30
    on messenger RNA biosynthesis, in bone marrow, 29
    on normoblasts, 29
    on reticulocyte release from bone marrow, 19
    on stem cells, 29
    on thymidine incorporation into DNA, 29
  enhancement of efficacy of isologous bone marrow transfusion by, 30
  inhibition by actinomycin D, 30
  in plasma, 30
  release from kidney, 30
Esterase(s)
  in adrenal cortex, 292
  (E600) resistant, in intestine, 90
  in kidney, 215
  in skin, 207
Estradiol
  biosynthesis of, in ovaries, 302
  catabolism of
    in intestine, 303
    in kidney, 313
Estrogens, in ovaries, 301–303, 315
Ethanolamine, in kidney, 215
N-Ethylmaleimide, effect of, on red blood cells, 35
Extravascular space
  expansion of, 250
  proteins in, 271
    lost into, 273

F

Fatty acids
  biosynthesis of, in bone marrow, 13, 16, 17
  in blood, 253
  oxidation of, in bone marrow, 17
  polyunsaturated, in adrenal cortex, 293
  unsaturated, biosynthesis of, in brain, 224
Feces, Na^+ and K^+ in, in gastrointestinal syndrome, 97
Ferrochelatase, in bone marrow, 20

Fibrinogen
biosynthesis of, in liver, 154
disappearance of, from circulation, 273
Folic acid, in spleen, 58
Follicle-stimulating hormone, 290, 315
Formaldehyde, incorporation of, into DNA, in liver, 167
Formate
incorporation of, into acid-soluble fraction of liver, 170
into DNA
in bone marrow, 25, 26
in liver, 167, 170
in lymphoid tissue, 52
in spleen, 58
in tumors, 237
into nucleoproteins, in liver nuclei, 170
oxidation to CO_2, in liver, 136
Fragility, of red blood cells, 36–38
Fructose
conversion
to fatty acids, in liver, 105, 106, 137
to lactate, in liver, 136
incorporation of, into glycogen, in liver, 136, 137
lipogenesis from, in liver, 117
oxidation to CO_2, in liver, 136, 137
Fructosediphosphate aldolase, in liver, 134
Fructose-1,6-diphosphate
accumulation of
in thymocytes, 64
in tumors, 241
in bone marrow, 27
Fructose-6-phosphate, in bone marrow, 27

G

β-Galactosidase, in lymphoid tissue, 69
Ganglioside, in brain, 222
Gastrointestinal tract
biosynthesis of DNA in, 82, 84–88
diarrhea of, in postirradiation syndrome, 92
incorporation of acetate into lipids in, 91
isozymes in, 90
loss of water by irradiated segments of, 98
motility of, 82, 91, 92
oxygen consumption in, by gastric and intestinal mucosa, 91
proteases in, 82, 90, 91
succinate dehydrogenase in, 90
syndrome of, K^+ and Na^+ levels in urine and feces in, 97
Gland, salivary, 91
Globin
biosynthesis of, in bone marrow, 21
of myoglobin, incorporation of glycine into, 211, 213
Globulin (α, β, γ)
autoimmune response to, 270
in blood, 254, 268, 270, 273
in cell nuclei
of lymphoid tissue, 50
of spleen, 51
of thymus, 51
Glucocorticoids, in adrenal gland, 295
Glucokinase, in liver, 117, 127, 130, 140, 156, 316
Gluconeogenesis, in liver, 105, 117, 135
Glucosamine
biosynthesis of, in intestine, 91
incorporation of, into globulins, 273
metabolic replacement of, in glycosaminoglycans, in skin, 199
Glucose
absorption of, in intestine, 316
active transport in small intestine, 82
in blood, 117, 137, 253, 254, 288
conversion of
to fatty acids, in liver, 105, 106, 108, 109, 137
to glycogen, in liver, 136, 137
incorporation of
into "free" and "bound" glycogen, in brain, 221
into glycosaminoglycans, in connective tissue, 198
lipogenesis from, in liver, 104, 105, 117, 135, 137
in liver, 121
metabolism of, via the HMP shunt, in liver, 117
oxidation of
in brain, 222
in intestine, 91
in liver, 136, 137, 141
phosphorylation of, in liver, 117, 137
transport in red blood cells, 35
utilization of
by muscle, 211
by red blood cells, 33–36, 38–40
in small intestine, 82

Glucose-6-phosphatase, in liver, 128
Glucose-6-phosphate
  in bone marrow, 27
  in liver, 127
Glucose-6-phosphate dehydrogenase
  in bone marrow, 27
  in intestine, 90
  in lens, 218
  in liver, 129
  in red blood cells, 38, 40
Glucosephosphate (phosphohexose) isomerase, in liver, 130
β-Glucuronidase
  after hypophysectomy, 316
  in intestine, 90
  in lymphoid tissue, 68, 69
  in tumors, 244
  in urine, 269
Glucuronides
  biosynthesis of
    in intestine, 90, 91
    in liver, 147, 148
  in urine, 257
Glutamate
  in brain, 223
    metabolism of, 223
  in urine, 256
Glutamate decarboxylase, in brain, 223
L-Glutamate dehydrogenase
  in liver, 151, 153
  in tumors, 243
Glutamate oxalacetate transaminase (SGOT), in serum, 269
Glutamate pyruvate transaminase (SGPT), in serum, 269
Glutaminase, in brain, 223
Glutaryl CoA, in kynurenine degradation, 267
Glutathione
  in brain, 223
  conjugation of, with bromsulphthalein, in liver, 148
  in lens, 218
  loss of by lymphoid tissue, 50
  in lymphocytes, 66
  metabolism of, in red blood cells, 35, 39–41
  in red blood cells, with glutathione reductase deficiency, 40, 41
  in spleen, 58
Glutathione peroxidase, in lymphocytes, 68
Glutathione reductase
  in bone marrow, 27
  in lens, 218
  in lymphocytes, 66
  in red blood cells, 40, 41
Glyceraldehydephosphate dehydrogenase
  in bone marrow, 27
  in liver, 129
Glycerolphosphate dehydrogenase
  in liver, 130
  in serum, 269
Glycine
  expansion of metabolic pool of, in liver, 146
  hippuric acid formation from, in liver, 146, 148
  incorporation of
    into ATP, in tumors, 243
    into bone marrow protein, 21
    into DNA
      of bone marrow, 25
      in liver, 169
      of lymphoid tissue, 52
      in tumors, 237, 238
    into globin of hemoglobin, in bone marrow, 21
    into glycocyamide, in kidney, 262
    into glycogen, in liver, 136
    into heme,
      in bone marrow, 20, 21, 25
      of myoglobin, 211
    into histones, in tumors, 239
    into intestinal proteins, 86
    into lipids, in tumors, 238
    into muscle protein, 211
    into protein, in liver, 146, 273
    into RNA, in tumors, 238
  in urine, 256, 276
Glycocyamine
  biosynthesis of, in kidney, 262
  in biosynthesis of creatine, 262
Glycogen
  accumulation of, in liver, 105, 122
  biosynthesis of, in liver, 117, 141
  in brain, 221, 222
  CO_2-fixation in, in liver, 136
  incorporation of glucose into, in brain, 221
  in liver
    after adrenalectomy, 125
    after hypophysectomy, 124, 125, 127
    of newborn, 125

Glycogen (*cont.*)
  loss of, by chondroblasts, 202
  metabolism of, 253
  in muscle, 212, 213, 321
  in retina, 217, 218
  in skin, 206
  in tumors, 243
Glycogen phosphorylase, in liver, 133
Glycogenolysis, in liver, 117, 141
Glycolysis
  anaerobic
    ATP in,
      in lymphocytes, 64
      in thymocytes, 64, 65
  in bone marrow, 26, 27
  in brain, 222
  effect of hydrogen peroxide on, in tumors, 239, 240
    of loss of NAD on, in tumors, 242
  in kidney, 214
  in lens, 218
  in liver, 117, 137
  in lymphoid tissue, 50
  in muscle, 208, 209, 211, 212
  in red blood cells, 30, 36, 38, 40
  in retina, 217
  in tumors, 237, 239–243
Glycoproteins, biosynthesis of, in serum, 273
Glycosaminoglycans (mucopolysaccharides)
  in aorta, 199
  in blood, 199, 254
  in brain, 221
  incorporation of inorganic sulfate into, 199
    epiphysis of bone, 202
    of glucose into, in connective tissue, 198
  metabolic replacement of
    glucosamine in, in skin, 199
    in bone, 202
    in skin, 199
  in urine, 199
Glycyl-glycine peptidase, in lymphoid tissue, 69
Glyoxalase, in red blood cells, 39, 41
Gonadotrophins
  of pituitary gland, 289–291, 303
  response of testis to, 299
  in urine, 303
Granuloma, 197
Ground substance
  depolymerization in, in skin, 204
  incorporation of inorganic sulfate into, 199
  surrounding tumors, 199
Guanase, in spleen, 58
Guanylate
  biosynthesis of, in testis, 301
  in bone marrow, 23
  incorporation of inorganic phosphate into, in liver, 173

H

Hair follicle, biosynthesis of DNA in, 205
Heinz bodies, in red blood cells, 40
Heme
  biosynthesis of, in bone marrow, 17–21
  incorporation
    of 5-aminolevulinate into, in bone marrow, 20
    of glycine into, in bone marrow, 20
    of iron into, in bone marrow, 20
  of myoglobin, incorporation of glycine into, 211
Hemoglobin
  abnormal, in chick embryo, 21
  biosynthesis of, in bone marrow, 17, 20, 21, 29, 30, 253
  loss of, from red blood cells, 32
  oxidation of to methemoglobin
    associated with methemoglobinemia, role of hydrogen peroxide and anoxia in, 41
    in presence of azide or thiols, 41
    in red blood cells, 36, 39, 41
  oxygen saturation of, 21
  oxygenation of, 21
  in red blood cells, 33
Hemolysis
  associated with oxidation of hemoglobin to methemoglobin, 36, 39, 41
  in red blood cells, 32, 34–38
    incubated without glucose, 36
Heptocyte, DNA in, 171
Hexokinase
  in bone marrow, 27
  in brain, 222

Hexokinase (*cont.*)
in lens, 218
in liver, 127, 131
in muscle, 212
in red blood cells, 38
in skin, 206
Hexosamine
in granulomas, 197
protein-bound, in blood, 254
Hexosediphosphatase, in liver, 128
Hippuric acid, in urine, 146
Histamine, 307, 308, 312
in intestine, 308
in lung, 308
in skin, 308
Histidine
free, in spleen, 58
in tumors, 236
Histidine decarboxylase, in stomach, 90, 308
Histones
biosynthesis of, in liver, 184
in lymphocytes, 61
in lymphoid tissue, 50, 51, 60, 61
phosphorylation of, in thymus, 61
in spleen, 51
in thymus, 51, 61
Hyaluronidase, in skin, 204
Hydrogen peroxide, detoxification of, in lymphocytes, 68
3-Hydroxyanthranilate
in liver, 267
in urine, 264, 265
β-Hydroxybutyrate, ratio of, to acetoacetate, in liver, 135, 139
5-Hydroxyindoleacetate
after adrenalectomy, 308
in brain and brain stem, 307
in urine, 264, 265, 276, 308
3-Hydroxykynurenine
in liver, 267
in urine, 265, 266
Hydroxymethylglutarate, incorporation of into cholesterol, in liver, 105
Hydroxyproline, in urine, 256
5-Hydroxytryptamine, in urine, 265
5-Hydroxytryptophan, 307
in urine, 265
Hyperpotassiemia, 253
Hypothalamus, effect of, on adrenal weight and cholesterol, 289

I

Indoleacetate
biosynthesis of, in liver, 267
in urine, 265
Indoxyl sulfate
biosynthesis of, 92
in urine, 257, 264, 276
Inosinate, in bone marrow, 23
Intestine
accumulation of precursors of RNA in, 89
acetylcholine in, 82, 89, 308
acetylcholinesterase in, 90
acid phosphatase in, 90
alkaline RNase in, 90
arylsulfatase in, 90
ATPase (Mg^{2+}) activity in, 90
bacterial flora in, 92
biosynthesis
of glucosamine in, 91
of glucuronide in, 90, 91
of proteins in, 86, 89
of triglyceride in, 91
catabolism of
estradiol in, 303
tryptophan and metabolites in, by bacteria, 92
urea in, 93, 147
cholinesterase in, 308
chondroitin sulfate in, 89
cytochrome oxidase in, 90
effect of bile on Na^+ leakage from, 98
esterases (E600 resistant) in, 90
formation of lactate in, 91
glucose absorption in, 316
glucose-6-phosphate dehydrogenase in, 90
β-glucuronidase in, 90
histamine in, 308
incorporation
of fatty acids into phospholipids in, 91
of glycine into protein in, 86
lactate dehydrogenase in, 90
leakage of protein from, 82, 95
lipoamide dehydrogenase (NAD-diaphorase) in, 90
loss of electrolytes by, and survival, 97
of *N*-methyl nicotinamide through the wall of, 95
motility in, 311
Na^+ absorption of, 316

Intestine (*cont.*)
oxidation of glucose in, 91
permeability of, 313
phosphodiesterase in, 90
production of juice by, 91
release of taurine from, 260
serotonin in, 82, 98, 308, 309, 311
small
acid ribonuclease in, 82, 90
active transport, of electrolytes in, 82, 94
of glucose in, 82
biosynthesis
of proteins in, 82, 86, 89
of RNA in, 82, 86, 89
catabolism of DNA in, 87
cholinesterase in, 89
DNA in, 82, 85
formation of citrate in, 82, 91
K^+ and Na^+ in, 82, 95
loss of electrolytes by, 82
of water by, 82, 97, 98
lysosomal enzymes in, 82, 90, 91
monoamine oxidase (monooxidase) in, 82, 90, 91
phosphatases in, 82
reactivity to acetylcholine, 98
utilization of glucose in, 82
Iodine, uptake of, by thyroid gland, 291, 303, 304
Iodoacetamide, radiosensitization of red blood cells to, 35
Iodoacetate, radiosensitization of red blood cells to, 35
Iododeoxyuridine, incorporation of, into DNA, in tumors, 237
Iron
incorporation of, into heme, in bone marrow, 17–19
in red blood cells, 250
release of, from siderophilin, 20, 21
in serum, 253
transport of, by siderophilin, 20, 21
uptake of
by bone marrow cells, 17–19
by circulating, nonnucleated red blood cells, 18, 19
by red blood cells, 19, 30
by reticulocytes, 18, 19
Isozyme(s)
in intestine, 90
of lactate dehydrogenase
in brain, 224
in serum, 268
in serum (mice), 269
Isozyme pattern, in serum, 269

J

Juice, production of
gastric, by stomach, 91
intestinal, by intestine, 91
pancreatic, by pancreas, 91

K

Ketohexokinase, in liver, 131
Ketose-1-phosphate aldolase, in serum, 269
Kidney
biosynthesis
of DNA in, after unilateral nephrectomy, 216, 217
of DNA-synthesizing enzymes in, after unilateral nephrectomy, 216
compensatory hypertrophy after unilateral nephrectomy, 216
DNA in, 216
after unilateral nephrectomy, 216
estradiol catabolism in, 313
glycolysis in, 214
isozymes in, 215
malate dehydrogenase in, 215
mitotic activity in, after unilateral nephrectomy, 216
Na^+ and K^+ in, 214
oxygen consumption by, 214
release of endogenous erythropoietin from, 30
thymidine kinase in, 216, 217
turnover of mitochondrial DNA in, 215
Knee joint
incorporation
of inorganic phosphate into, 202
of inorganic sulfate into glycosaminoglycans of, 199
permeability of, to inorganic sulfate, 199
Kynurenate, in urine, 265
Kynurenate hydroxylase, in liver, 266
Kynurenic acid, excretion of, 264, 266, 276
Kynureninase
in liver, 266
in tryptophan catabolism, in liver, 264, 266

Kynurenine
in liver, 267
in urine, 264–266, 276
Kynurenine aminotransferase, in liver, 151, 266
Kynurenine hydroxylase, activity of, in liver, 267

L

Lactate
accumulation of, in brain, 222
in blood, 254
in bone marrow, 27
formation of
in intestine, 91
in red blood cells, 39
in thymocytes, 64, 65
leakage of, from lens into aqueous humor, 218
in liver, 135
in lymphoid tissue, 50
in muscle, 211, 212
ratio of, to pyruvate, in liver, 135, 139
Lactate dehydrogenase
in body fluids, 268, 276
in intestine, 90
in liver, 129
in lymphoid tissue, 62, 63
in muscle, 212
in serum, 269
in thymus nuclei, 66
in tumors, 243
Lecithin, biosynthesis of, in bone marrow, 15
Lens, 218
Leucine, incorporation of
into cytoplasmic proteins, in liver, 153
into glycogen, in liver, 136
into liver-synthesized plasma proteins, 155
into nucleoproteins
of liver, 153
of thymus, 57
into protein, in brain, 221
into retinal proteins, 218
Life-span
effect of transcapillary migration of red blood cells on, 37
of red blood cell, 30, 36–38
of transfused red blood cell after irradiation, 37
Linoleate
in lecithin, in liver, 119
in skin, 206
Lipase (glycerol-ester hydrolase)
in adipose tissue, 199
in serum, 269
in skin, 206
Lipids
in adrenal cortex, 292
in bone marrow, 14
deposition of, in renal tubules, 216
incorporation
of formate into, in kidney, 216
of precursors into, in liver, 253
metabolism of, role of serum enzymes in, 268
mobilization of, from adipose tissue, 253
in plasma, 253
in skin, 206
in testis, 301
total
in blood, 254, 255
in brain, 222
Lipoamide dehydrogenase (NAD-diaphorase)
in brain, 222
in intestine, 90
in muscle, 212
in skin, 206
Lipocytes, in bone marrow, 13–15, 17
Lipofuscin, in adrenal cortex, 291
Lipogenesis, in bone marrow, 15–17
Lipoproteins
in blood, 253, 254
high and low density, 270
in muscle, 270
Liver
accumulation of deoxycytidine in, 164
acetyl-CoA carboxylase in, 133, 135
alanine aminotransferase in, 150, 159
β-aminoisobutyrate in, 256
aminotransferases in, 147
ammonia in, 146, 147
aspartate aminotransferase in, 150
aspartate carbamoyltransferase in, 151
biosynthesis of cholesterol in, 105, 113–115
of DNA in, 162–164, 170, 171, 176–179, 183–185
of nucleoproteins in, 184
butyryl-CoA dehydrogenase in, 130

Liver (*cont.*)
catabolism of adrenal hormones in, 294
deoxycytidylate aminohydrolase in, 181
DNA nucleotidyltransferase in, 182
DNA polymerase in, 180, 181
fetal, uptake of cortisone by, 294
fructose diphosphate aldolase in, 134
glucokinase in, 117, 127, 130, 140, 156
glucose and glycogen in, after adrenalectomy, 316
glucose-6-phosphatase in, 128
glucose-6-phosphate dehydrogenase in, 129
glucose-6-phosphate isomerase in, 130
L-glutamate dehydrogenase in, 151, 153
glyceraldehydephosphate dehydrogenase in, 129
α-glycerolphosphate dehydrogenase in, 130
glycogen in, 117, 121–125, 288, 289, 316, 318, 320, 321
glycogen phosphorylase in, 133
glycolysis in, 117, 137
hexokinase in, 131
hexose diphosphatase in, 128
histone/DNA ratio in, 61
incorporation
of acetate into DNA in, 169
into RNA in, 174
of adenine into DNA, in, 170
of 5-amino-4-imidazole carboxamide into DNA in, 169
into RNA in, 175
of ATP into nRNA in, 173
of CTP into nRNA in, 173
of dCMP into DNA in, 168
of deoxycytidine into DNA in, 162
of dTMP into DNA in, 168
of dTTP into DNA in, 168
of formaldehyde into DNA in, 167
of formate into DNA in, 167, 170
of glycine into DNA in, 169
of inorganic phosphate
into AMP in, 173
into CMP in, 173
into DNA in, 164, 165, 168, 169
into GMP in, 173
into RNA, in172–174
into UMP in, 173
of orotate into CMP and UMP in, 174
into DNA in, 166
into cRNA and nRNA in, 172–174, 176–178
into mitochondrial RNA of, 174
of thymidine
into DNA of, 162, 166, 167, 169
into mDNA and nDNA of, 167
of tryptophan into NAD in, 266
of uracil into RNA of, 175
of D,L-ureidosuccinate into DNA in, 169
of ureidosuccinate into nRNA, 175
of uridine into microsomal and nRNA in, 175
induction of serine dehydratase in, 161
ketohexokinase in, 131
kynurenate hydroxylase in, 266
kynureninase in, 266
kynurenine aminotransferase in, 151, 266
D-lactate dehydrogenase in, 129
lipogenesis in, after adrenalectomy, 316
malate dehydrogenase in, 129
NAD in, 266
NAD pyrophosphorylase in, 178, 182
ornithine-ketoacid aminotransferase in, 156, 161
oxygen consumption by, 142, 144
phosphofructokinase in, 131
phosphoglucomutase in, 131
6-phosphogluconate dehydrogenase in, 129
phosphoglycerate kinase in, 132
phospholipids in, 109–111, 119
phosphopyruvate carboxylase in, 133
phosphorothioate oxidase in, 160
protein in, 153
pyruvate kinase in, 132
ribonuclease in, 185
RNA nucleotidyltransferase (Mg^{2+}) in, 159, 182
L-serine dehydratases in, 156
thymidine kinase in, 180
thymidine phosphate kinases in, 181
thymidine phosphorylase in, 182
TMP kinase in, 159
TMP synthetase in, 181
transketolase in, 132
triglyceride in, 119
triosephosphate isomerase in, 130
tryptophan oxygenase in, 156–158, 161, 266

Liver (*cont.*)
  tyrosine aminotransferase in, 151, 158, 161
  UDPG glycogen glucosyltransferase in, 133
  UDPG pyrophosphorylase in, 132
Lung, histamine in, 308
Luteinizing hormone, 290
Lymphoblasts, monocytic, biosynthesis of thymidine by, 55
Lymphocyte
  accumulation of ribonucleotides in, 61
  ATP in, 50, 62
  biosynthesis of antibodies by, 52
  defective formation of nuclear ATP in, 54
  DNA metabolism in, 52, 54, 55
  detoxification of hydrogen peroxide in, 68
  effect of inorganic phosphate on breakdown of nuclear structure of, 61
  glutathione in, 66
  glutathione peroxidase in, 68
  glutathione reductase in, 66
  histones in, 61
  K^+ loss from nuclei of, 54
  NAD in, 66
  nuclear ATP formation in, 54
  nuclear composition of, 54, 55
  nuclear phosphorylation in, 62
  nuclei
    ionic composition of, 54, 55
    leakage of catalase from, 68
  nucleic acid extractibility from, 59
  nucleoproteins in, 59–61
  prevention of nuclear structural changes in, by Ca^{2+}, 72
  protection from mitotic arrest by Ca^{2+} in, 72
  rate of DNA biosynthesis in, 53, 54
  release of nuclear histones and globulins to cytoplasm, 60, 61
  SH groups in, 67, 68
  thymidine incorporation into DNA of circulating, 53, 54
  thymidine pool in, 54
Lymphoid tissue
  acid phosphatase in, 68–70
  activation of amino acids in, 56–58
  acyl CoA dehydrogenase, 63
  L-amino acid oxidoreductase in, 63
  aminopeptidase in, 69
  anaerobic glycolysis in, 64
  arylsulfatase in, 68
  ATP in, 50, 62–64, 72
  ATPase in, 62, 63, 68
  autolysis in, 145
  biosynthesis
    of DNA in, 52–55, 56, 58, 62, 73
    of proteins, in 56–58
    of RNA in, 50
  catabolism of RNA in nuclei of, 61
  cathepsins in, 68, 70
  cytochrome oxidase in, 63
  defect of electron transport in, 63
  dehydrogenases of pentose cycle in, 63
  DNA in, 52
  DNA nucleotidyl transferase (polymerase) in, 55
  β-galactosidase in, 69
  β-glucuronidase in, 68, 69
  glycolysis in, 50
  glycylglycine peptidase in, 69
  "homogenization" of cell nuclei, 50
  incorporation
    of adenine into DNA of, 52
    of deoxycytidine into DNA of, 55
    of formate into DNA of, 52
    of glycine into DNA of, 52
    of orotate into nuclear RNA of, 56, 57
    of thymidine into DNA of, 53, 55, 61
  lactate in, 50
  lactate dehydrogenase in, 62, 63
  leakage of NAD from cells of, 65
  loss of glutathione by, 50
  mitochondrial phosphorylation, 50, 62, 63
  NAD in, 50
  NADPH cytochrome c reductase in, 63
  nucleoprotein in
    dissociation of DNA and protein from, 59, 60, 72
    viscosity of, 59, 60
  oxygen consumption by, 50, 63, 72
  P/O ratio decrease in, 143
  peroxide formation in lysosomes of, 70
  phosphoprotein phosphatase in, 69
  proteases in, 69
  radiation damage to template activity of DNA in, 54
  radiation protection in, by adrenaline, 310
  RNA in, 52
  succinate dehydrogenase in, 63
  thymidine kinase in, 55

Lymphoid tissue (*cont.*)
tricarboxylic acid cycle dehydrogenases in, 63
Lymphosarcoma, catabolism of NAD in, 66
Lysine
incorporation of
into nucleoproteins, in liver, 184
into plasma proteins, in liver, 155
into serum proteins, 273
Lysosomes
activation of enzymes in, of tumors, 243
in adrenal cortex, 292
enzymes
in bone marrow, 13
in lymphoid tissue, 50
in small intestine, 82, 90, 91
in liver, liberation of hydrolases from, 145
in lymphoid tissue, peroxide formation in, 70
release of arylsulfatase from, in lymphoid tissue, 70
of DNase II from, in spleen, 70
in thyroid gland, 303

M

Magnesium ion, in serum and urine, 253
Malate dehydrogenase
in blood, 269
in kidney, 215
in liver, 129
Malonyl-CoA, in liver, 133
Mercaptoethylguanidine, radioprotection of red blood cells by, 35
Methionine, free
in biosynthesis of creatine, 262
in connective tissue, 198
conversion to taurine, 258
incorporation of
into bone marrow protein, 21
into α- and β-globulins, 273
into histones, in tumors, 239
into muscle protein, 211
into protein, in brain, 221
into serum proteins, 273
methyl group transfer to deoxyribonucleotide in bone marrow, 21
N-Methylnicotinamide
loss of, through intestinal wall, 95
in urine, 257
Mevalonate
conversion to fatty acids in lipids, in liver, 108
incorporation into cholesterol
in brain, 222
in liver, 105, 114, 115
oxidation to CO_2, in liver, 116
Mevalonolactone, incorporation into cholesterol, in liver, 105, 114
Mineralization, inhibition of, in irradiated bone, 202
Mitochondria
cytochrome c bound by
in Ehrlich AT cells, in tumors, 143
in thymocytes, 143
differences in radiosensitivity of various sites along electron transport chain, in lymphoid tissue, 62
effect of cytochrome c on oxidative phosphorylation in, in lymphoid tissue, 62
electron transport in, in lymphoid cells, 50
loss of K^+ by, in liver, 144
number of, in relation to radiosensitivity of tumors, 236
oxidative phosphorylation
in brain, 221
in lymphoid tissue, 62, 63
oxygen consumption by, in brain, 221
phospholipids in, in liver, 119
Mitotic arrest, in epidermis, 204
Monoamine oxidase
in brain, 223
in liver, 308
in small intestine, 82, 90, 91
Motility, of gastrointestinal tract, 82, 91, 92
Mucoproteins, in urine, 257
Muscle
accumulation of lactate in, 211
aminotransferases in, 147
ATP in, 208, 209, 211, 212
ATPase in, 208, 212
cardiac, Rb^+ uptake by, 211
cholesterol granulomas in, 199
collagen in
degradation of, 213
loss of amino acids from, 198
soluble and insoluble, 197, 198
creatine in, 319
creatinuria onset affected in, by thyroxine, 316, 331
glycogen in, 321

Muscle (*cont.*)
glycolysis in, 208, 209, 211, 212
incorporation of amino acids into protein in, 211
K^+ loss by, 208–211, 213
Na^+ uptake by, 208, 211
oxygen consumption by, 208
phosphoprotein content of, 211
protein composition of, 213
pyruvate kinase in, 208
replacement of soluble protein in, 212
utilization of acetate by, 209, 211
of glucose by, 211
Myelin, formation of, in brain stem, 224
Myeloma, radiosensitive, 260
Myoglobin, incorporation of, 211
Mytomycine, and taurinuria, 260

N

Nerve, efflux of Na^+ from, 225
Neurohormones, 223
Neurohumoral agents, liberation from skin, 204
Neuron
localization of acid phosphatase in, 222
Na^+/K^+ ratio in, 223
Nicotinamide
action of, in prevention of NAD loss by tumor cells, 238, 239, 241, 242
leakage from tumor cells, 244
protection by, of NAD, NADP and nuclear structures in thymocytes, 66
Nicotinamide-adenine dinucleotide (NAD)
biosynthesis of
in liver, 268
in thymus, 66
in tumors, 241
catabolism of, 268
in lymphosarcoma cells, 66
enzymatic degradation of, in tumors, 241
leakage of, from lymphoid tissue cells, 65
in liver, 266, 268
loss of
by thymocytes, 49
by tumor cells, 238, 239, 241, 242
in lymphocytes, 66
in lymphoid tissue, 50
metabolites of, excretion of, 268
ratio of $NAD^+/NADH$ of, in liver, 142
in red blood cells, 39
reduced (NADPH), effect of, on inhibited biosynthesis of testerone and androgen, in homogenates of testis, 300, 301
in thymocytes, 66, 67
Nicotinamide-adenine dinucleotide nucleosidase
in spleen, 66
in tumors, 241
Nicotinamide-adenine dinucleotide pyrophosphorylase
in liver, 178, 182
in thymus, 66
Nicotinamide-adenine dinucleotide (reduced) cytochrome c reductase, in lens, 218
Nicotinamide-adenine dinucleotide phosphate (NADP)
lack of, in kynurenine hydroxylation, 267
in red blood cells, 39
in thymus, 65, 66
in urine, 276
Nicotinamide-adenine dinucleotide phosphate nucleosidase, in spleen, 66
Nicotinamide-adenine dinucleotide phosphate (reduced) dehydrogenase, in lens, 218
Nicotinamide-adenine dinucleotide phosphate (reduced) (NADPH) cytochrome c reductase, in lymphoid tissue, 63
Nitrogen, nonprotein
in blood, 254
metabolism of, 255
Nitrogen balance, 253
Noradrenaline, in brain, 223
Normoblast, in bone marrow, 29
Nucleic acid
biosynthesis of, in testis, 301
extractibility of, from lymphocytes, 59
metabolites of, excretion of, 255, 262
phosphorus in, in bone marrow, 23
in testis, 301
Nucleoprotein(s)
acid-soluble, biosynthesis of, in liver, 184
biosynthesis of, in thymus, incorporation of leucine into, 57
of phenylalanine into, 57
in liver, 183, 184
in lymphocytes, 59–61

Nucleoprotein(s) (*cont.*)
in lymphoid tissue
dissociation of DNA and protein from, 59, 60, 72
viscosity of, 59, 60
primer for RNA polyermase, in tumors, 238
priming activity in the heterochromatin fraction of tumor nuclei in, 238
release from nuclei of lymphoid-tissue cells, 50, 60, 61
residual fraction of, in bone marrow, 22
structure of, in lymphocytes, 59
Nucleosides, of guanine in red blood cells, 39
5′-Nucleotidase
in bone marrow, 28, 29
in skin, 205, 206
in spleen, 58, 68
Nucleotides, biosynthesis of, in testis, 301
Nucleus
ATP formation in, in lymphoid cells, 50, 62–64, 72
composition of
in lymphocytes, 54, 55
in thymus, 55
"homogenization" of, in lymphoid-tissue cell, 50
ionic composition of, in lymphocytes, 54, 55
leakage of catalase from, in lymphocytes, 68
loss of K^+ from lymphocyte, 54
membrane of, aggregation of chromatin in, in lymphoid tissue, 50
phosphate uptake by, in thymus, 55
phosphorylation of adenylate in, in lymphoid tissue cell, 62–64
pycnosis of, in lymphoid tissue, 50, 51
release of histones and globulins from, in lymphoid tissue cells, 50, 60, 61

O

Oleate
incorporation of, into phospholipids
of bone marrow, 15
in liver, 109
in total lipids, of liver, 108
in triglycerides, of bone marrow, 14
Organic acids, in blood, 254
Ornithine, effect of on urea production, in liver, 146, 147
Ornithine-ketoacid aminotransferase, in liver, 156, 161
Orotate
incorporation of
into DNA, in liver, 166
into nuclear RNA, of lymphoid tissue, 56, 57
into RNA, in liver, 172–174, 176–178
into UMP and CMP, in liver, 174
uptake of, by thymus nuclei, 61
Osmotic pressure, in red blood cells, 32, 36
Osteoblast
acid phosphatase in, 202
alkaline phosphatase in, 202, 203
Ovaries, 301–303, 315
Oxalate, in urine, 257
Oxygen consumption
by bone marrow, 13, 27, 29
by brain
adult, 221
newborn, 223
by gastric and intestinal mucosa, 91
by kidney, 214
by liver, 142, 144
by lymphoid tissue, 50, 63, 72
by muscle, 208
regulation of, by thyroid gland, 304, 305
by retina, 217
by skin, 206
by tumors, 239, 240, 242

P

P/O ratio, in lymphoid tissue, 143
Palmitate
in blood, 253
incorporation of
into lipids, in thymus, 58
into phospholipids, in liver, 109
into triglycerides, in liver, 112, 113
into triglycerides and phospholipids, in bone marrow, 15
in total lipids, of liver, 118
in triglycerides, of bone marrow, 14
Pancreas, 91
Pentose cycle (hexose monophosphate shunt), dehydrogenases of, in lymphoid tissue, 63
Peptidases, in kidney, 215

Peptide hydrolases, 269
Permeability
of lens, to Rb^+, Na^+, inorganic phosphate, and I^-, 218
of mitochondria in liver, to NAD, 144
of red blood cells, 31, 32, 35, 36, 38, 40
membrane of, in relation to SH group damage, 36, 38
of reticulocyte, 19
of skin, 204
of synovial membrane, to inorganic sulfate, 199
Peroxidation
of fatty acids in triglycerides of lysosomes, in liver, 144, 145
of lipid
in bone marrow, 17
in red blood cells, 33, 34, 36, 37
Peroxide, formation of, in lysosomes, of lymphoid tissue, 70
Phenylalanine
CO_2 formation from, in liver, 155
incorporation of
into albumin, in whole animal and liver, 273
into globin in reticulocytes, 20
into globulins, in whole animal and liver, 273
into nucleoproteins in thymus, 57
into liver-synthesized plasma proteins, 155
into serum proteins, in liver, 272
pool of, in liver, 273
Phosphatase(s)
acid
activity of, in spleen and thymus, 70
in brain, 222
in epidermis, 205
in intestine, 90
in lens, 218
localization of neurons of, 222
in lymphoid tissue, 68–70
in osteoblasts, 202
in retina, 218
in skin, 206
in tumors, 244
alkaline
in adrenal cortex, 292
in blood, 202
in bone, 200, 202
in bone marrow, 24, 26
in brain, 222
loss of, from renal tubules, 215
in muscle, 212
in osteoblasts, 202, 203
in skin, 206
in spleen, 58
in kidney, 215
in small intestine, 82
Phosphate
inorganic
acceleration of breakdown of nuclear structure of lymphocytes by, 61
incorporation of
into AMP, in liver, 173
into AMP and ATP, in muscle, 208, 209
into ATP, in tumors, 243
into CMP, in liver, 173
into DNA
in bone marrow, 24–26
in brain, 221
in liver, 164, 165, 168, 169
of lymphocytes, 54
in thymus, 55, 56
in tumors, 237
into GMP, in liver, 173
into knee joint, 202
into lysosomal fraction, of liver, 111
into microsomal fraction, of liver, 111
into nuclear RNA, in thymus, 56
into phospholipids
in brain, 222
of liver, 110, 111
in spleen, 58
into phosphoprotein
in brain, 221
in muscle, 211
into red blood cell membrane, 39
into renal lipids, 216
into RNA
in brain, 221
in liver, 172, 173, 174
in tumors, 238
into ribonucleotides, in bone marrow, 24
into sphingomyelin, in liver, 111
into UMP, in liver, 173
in serum, 253
in urine, 253, 276
Phosphatidyl choline (lecithin), in liver, 119

Phosphatidyl ethanolamine (cephalin), in liver, 119
Phosphatidyl glyceride, in liver, 119
Phosphodiesterase, in intestine, 90
Phosphofructokinase
activation of, in tumors, 241
in liver, 131
in thymus, 64
Phosphoglucomutase, in liver, 131
Phosphogluconate dehydrogenase, in liver, 129
after hypophysectomy, 316
Phosphoglycerate kinase, in liver, 132
Phospholipids
biosynthesis of
in bone marrow, 14, 16
in intestine, 91
in blood, 253, 255
in brain, 224
incorporation of inorganic phosphate into
in brain, 222
in liver, 110, 111
in spleen, 58
in liver, 109–111, 119
in mitochondria, of liver, 119
Phosphoprotein
in brain, 221
incorporation of inorganic phosphate into
in brain, 221
in muscle, 211
in muscle, 211
Phosphoprotein phosphatase, in lymphoid tissue, 69
Phosphopyruvate carboxylase, in liver, 133
Phosphorothiate oxidase, in liver, 160
Phosphorylases, in liver, 141
Phosphorylation
in mitochondria of lymphoid tissue, 50
nuclear
of AMP, in tumors, 243
in lymphocytes, 62
in tumors, 242, 243
in nuclei of lymphoid tissue, 50
oxidative
of AMP, in mitochondria, of tumors, 243
in mitochondria
of brain, 221
of liver, 141, 144
of lymphoid organs, effect of added cytochrome c on, 62
of spleen, thymus and lymphocytes, 50, 62, 63
P/O ratio, in hepatic mitochondria, 141–144
Pituitary gland
ACTH in, 289–291, 294, 315, 317, 319
cellular composition of, 289
hormones of, 289–291, 294, 303, 315, 317, 319
Plasma
aldolase in, 213
DNase II in, 269
volume of, 250, 251
Plasmalogen, in blood, 254
Polysaccharides, in blood plasma, 254
Polyuria, 250, 252, 291, 312, 316, 318
Polyvinylpyrrolidine, loss of, into gastrointestinal tract, 273
Pons, oxygen consumption by, 221
Porphobilinogen, incorporation of into heme in bone marrow, 19, 20
Porphobilinogen synthase, in bone marrow, 20
Porphyrins, in urine, 257
Potassium ion
in blood, 211, 253
in bone marrow, 29
in brain, 220
leakage of, from tumor cells, 244
liberation of, in central nervous system death, 253
loss of
from brain, 225
by hepatic mitochondria, 144
from lymphocyte nuclei, 54
by muscle, 208–211, 213
from radiosensitive organs, 253
from red blood cells, 36
in organs, 252, 253
in red blood cells, 31–38, 213
in small intestine and mucosa, 82, 95
in urine, 251, 252
Prealbumin
autoimmune response to, 270
in blood, 254
Progesterone conversion of
to hydroxycortisone, in adrenal gland, 294
to testosterone, in testis, 300
Proline, in urine, 256

Proteases
  in gastrointestinal tract, 82, 90, 91
  in lymphoid tissue, 69
  in skin, 204
Protein
  biosynthesis of
    in bone marrow, 21, 22
    in intestine, 82, 86, 89
    in lymphoid tissue, 56–58
    in serum, 271, 273
    in stomach, 82
  in blood, 254, 268, 271, 276
  in brain stem, 221
  catabolism of
    in liver, 146, 273
    in tissue, 267
    tissue breakdown and amino aciduria, 198
  disappearance of, from serum, 273
  in epidermis, 205, 206
  incorporation of
    acetate into, in brain, 221
    glycine into
      in intestine, 86
      in muscle, 211
    leucine
      in brain, 221
      in retina, 218
    methione, in brain, 221
  labeled
    ^{131}I, in stomach, 273
    ^{131}I and ^{3}H, incorporation of into albumin and globulin, in serum, 272
  leakage of, from intestine, 82, 95
  in lens, 218
  in liver, 153
  metabolism of, in serum, 271
  in muscle, 213
  plasma, biosynthesis of, in liver, 154, 155
  ribosomes and polysomes in biosynthesis of, in liver, 154
  in serum, 95
  soluble, metabolic replacement of, in muscle, 212
Protein-bound iodine, 303, 304
Pseudouridine, in urine, 256, 262, 276
Purine nucleoside phosphorylase
  in brain, 222
  in lens, 218
  in red blood cells, 39, 40
  in spleen, 58
Purines
  in blood, 254
  excretion of, 262, 276
Pyridone, in urine, 257
Pyridoxal phosphate, lack of, in tryptophan catabolism, 265, 266
Pyrophosphatase, inorganic, in skin, 205
Pyrrolcarboxylate, in urine, 198, 257
Pyruvate
  in bone marrow, 27
  leakage of, from tumor cells, 244
  in liver, 135
Pyruvate kinase
  in liver, 132
  in muscle, 208
  in tumors, 241

R

Radiation injury, biochemical diagnosis of, 276
Radiation protection, by adrenaline, in lymphoid tissue, 310
Red blood cell
  acetylcholinesterase in, 38, 40
  acid phosphatase in, 38
  ADP in, 39
  antioxidant activity in, 36, 37
  ATP in, 37, 39
  avian
    5-aminolevulinate synthase in, 20
    glycine uptake by, 20
  Ca^{++} in, 38
  catalase in, 39, 41
  cation transport in, 31–36, 38
    inhibitors of, 32, 33, 35
    temperature effects on, 32–34
  correlation of osmotic fragility with glycolysis, 36
  creatine in, 254, 260
  effect
    of *p*-aminobenzoate on, 35
    of *p*-chloromercuribenzene sulfonate on transport phenomena in, 32, 35
  fragility under thermal, mechanical, or osmotic stress, 36
  glucose transport in, 35
  glucose utilization by, 33–36, 38, 40
  glucose-6-phosphate dehydrogenase in, 38, 40
  glutathione metabolism of, 35, 39–41

Red blood cell (*cont.*)
glutathione reductase in, 40, 41
glycolysis in, 30, 36, 38, 40
osmotic fragility after irradiation in and, 36
glyoxalase in, 39, 41
guanine nucleosides in, 39
Heinz bodies in, 40
hemolysis of, 36, 37
hexokinase in, 38
incorporation of inorganic phosphate into membrane of, 39
iron in, 250
K^+ in, 31–33, 35, 36, 38, 213
lactate formation in, 39
life-span of, 30, 36–38
loss of K^+ from, 36
mechanical fragility of, 36
membrane
radioprotection of, by *p*-aminobenzoate, sucrose, and MEG, 35
electron microscopy of, 34
inactivation of enzymes in, 36, 38
peroxidation of lipids in, 34, 36, 37
proteins in, 34, 35
proteolysis and lipolysis of, 34
radiation injury of, 34–36
sulfhydryl group oxidation in, 34, 35
Na^+ in, 31–33, 36, 38, 213
NAD in, 39
NADP in, 39
osmotic fragility of, 36, 38
osmotic pressure in, 32, 36
ouabain-resistant ATPase in, 33, 39
ouabain-sensitive ATPase in, 33, 39
permeability of, 31, 35, 36
production of, 251
prolytic swelling after irradiation, 36
purine nucleoside phosphorylase in, 39, 40
radiosensitization of, 35
resistance to osmotic stress, 32, 36
thermal fragility of, 36
urate in, 39
Renal function, 214
Respiratory quotient and thyroid gland function, 305
Reticulocytes
biosynthesis of globin from phenylalanine by, 20
permeability of, 19
rate of maturation, 19
release from bone marrow, 19
uptake
of glycine by, 20
of iron by and heme synthesis in, 13, 18–20, 29
Retina
acid phosphatase distribution in, 218
glycogen in, 217, 218
glycolysis in, 217
oxygen consumption by, 217
Ribonuclease
acid
in small intestine, 82, 90
in spleen, 70
in thymus, 68, 70
in tumors, 244
alkaline
in intestine, 90
in skin, 206
in spleen, 70
in thymus, 68
in lymphoid-tissue lysosomes, 50
in spleen, 289, 318
effect of head irradiation on specific activity of, 71
total
in liver, 185
in skin, 205
Ribonucleic acid
biosynthesis of
"albuminoid" (insoluble), in lens, 218
in brain, 221, 223
in lymphoid tissue, 50
in small intestine, 82, 86, 89
in tumors, 238, 239
bone marrow, 23–26, 29, 30
in brain, 220, 221
catabolism of
in bone marrow, 23, 26, 29
in intestine, because of cell death, 87
in lymphoid tissue, 61
in muscle, 213
cytoplasmic (cRNA)
biosynthesis of, in liver, 163, 172–174, 178
catabolism of, in brain, 221
incorporation
of adenine into, in liver, 174
of inorganic phosphate into, in liver, 172–174

Ribonucleic acid—*cont.*
cytoplasmic (cRNA) (*cont.*)
of orotate into, in liver, 172–174, 176, 178
incorporation
of adenine into, in lens, 218
of cytidine into, in brain, 221
of inorganic phosphate into
in brain, 221
in intestine, 86
in lens, 218
microsomal
biosynthesis of, in liver, 172
incorporation
of adenine into, in liver, 174
of inorganic phosphate into, in liver, 172
of uracil into, in liver, 175
of uridine into, in liver, 175
mitochondrial, incorporation
of inorganic phosphate into, in liver, 172
of orotate into, in liver, 174
nuclear
biosynthesis of, in liver, 163, 171–174, 177, 178
incorporation
of adenine into, in liver, 174
of ATP into, in liver, 173
of CTP into, in liver, 173
of inorganic phosphate into, in liver, 172–174
of orotate into, in liver, 172–174, 176–178
of uracil into, in liver, 175
of ureidosuccinate into, in liver, 175
of uridine into, in liver, 175
in thymus, 56
precursors of, in intestine, 89
in skin, 205, 206
soluble, in lens, 218
in thymus nuclei, 56, 57
transfer of, from nucleus to cytoplasm, in brain cells, 221
m-Ribonucleic acid, in liver, 161, 177
Ribonucleic acid nucleotidyltransferase (polymerase)
in adrenal cortex, 293
in liver, 159, 163, 177, 182
nucleoprotein in tumor tissue as primer for, 238
Ribonucleotides, in bone marrow, 22–24
Ribose, oxidation of, to CO_2, in liver, 136

S

Saliva, production of, by salivary gland, 91
Serine
decarboxylation of, in liver, 146
in urine, 256
L-Serine dehydratase, in liver, 156, 161
Serotonin (5-hydroxytryptamine)
in blood, 309
in brain, 308
in intestine, 82, 308, 309, 311
liberation of
catabolism of, in brain, 223
from tryptophan, in liver, 264
in uterus, 308
in spleen, 309
in urine, 264
Serum
acylcholine acylhydrolases in, 269
aminotransferases in, 269
cholesterol in, 254, 255
creatine kinase (phosphokinase) in, 269
glutamate oxalacetate transaminase (SGOT) in, 269
glutamate pyruvate transaminase (SPGT) in, 269
glycerol-3-phosphate (glycerophosphate) dehydrogenase in, 269
isozymes (mice) in, 269
ketose-1-phosphate in, 269
lactate dehydrogenase in, 269
lipase (glycerol-ester hydrolase) in, 269
lipids in, 255
neutral fats in, 255
peptide hydrolases in, 269
phospholipids in, 255
Serum glutamate oxalacetate transaminase (SGOT) (aspartate aminotransferase), 269
Skin
collagen in, 197, 198
DNA in, 206
biosynthesis of, 204
glycogen in, 206
histamine in, 308
permeability of, 204, 313

Skin (*cont.*)
replacement of elastin in, 199
total protein in, 205
transfer of albumin in, 313, 314
Sodium ion
absorption of, in intestine, 316
in bone marrow, 29
in brain, 220
concentration of, in serum, 252
depletion of, 253
efflux from nerve, 225
excretion of, in urine, 251, 252
leakage from intestine, effect of bile on, 98
in organs, 252
reabsorption of, 253
in red blood cells, 31–33, 36, 38, 213
in small intestine and mucosa, 82, 95
uptake of, by muscle, 208, 211
Somatotrophic hormone of pituitary gland, 289, 291
Spermatogenesis, role of pituitary gland in, 291
Spermidine, protection of nuclear structure and delay of interphase death of lymphoid cells by, 72
Sphingomyelin
biosynthesis of, in brain, 224
incorporation of inorganic phosphate into, in liver, 111
Spleen
accumulation of deoxyribonucleotides and deoxyribonucleosides in, 54
acid ribonuclease in, 70
adenosine deaminase (aminohydrolase) in, 58
adenosine triphosphatase in, 58
alkaline phosphatase in, 58
alkaline RNase in, 70
aminotransferases in, 58
catabolism of DNA in, 54
cathepsins in, 70
cysteine (free) in, 58
DNA in, 51
effect of head irradiation on, 71
DNase II in, 60, 70
folic acid in, 58
formation of citrate in, 63
glutathione in, 58
guanase in, 58
histidine (free) in, 58
histone/DNA ratio in, 61
incorporation
of formate into DNA in, 58
of thymidine into DNA of, 53, 55
of valine into nucleoproteins of, 58
loss of K^+ from, 253
NAD nucleosidase in, 66
NADP nucleosidase in, 66
nuclear phosphorylation in, effect of head irradiation on, 71
5′-nucleotidase in, 58, 68
phospholipids in, incorporation of inorganic phosphate into, 58
purine nucleoside phosphorylase in, 58
RNase activity in
effect of head irradiation on, 71
total, 70
serotonin in, 309
taurine (free) in, 58
transfer of DNA liberated from nucleoproteins to microsomes in, 59
tryptophan (free) in, 58
tyrosine (free) in, 58
Squalene, incorporation of, into cholesterol, in liver, 105, 114
Stearate
incorporation of
into phospholipids, in liver, 110, 111
into triglycerides, in bone marrow, 15
in triglycerides, in bone marrow, 16
Steroid hormones, in adrenal gland
biosynthesis of, 294, 299
11-β-hydroxylation of, 294
Stomach
biosynthesis of proteins in, 82
content of, 311
gastric emptying time, 82, 92
histidine decarboxylase in, 90
production of gastric juice by, 91
Succinate dehydrogenase
in bone marrow, 27
in intestine, 90
in lymphoid tissue, 63
in muscle, 212
in tumors, 243
Succinate oxidase
in brain, 222
in muscle, 212
in tumors, 243
Sulfate, inorganic
excretion of, 252, 253, 259

Sulfate (*cont.*)
incorporation of
into acidic glycosaminoglycans and ground substance, 199
into aorta, 199
into chondroblastoma, 202
into components of healing wounds, 198
into glycosaminoglycans
in epiphysis of bone, 202
in knee joint, 199
into ground substance surrounding tumors, 199
into sulfatides, in brain, 222
permeability of knee joint to, 199
uptake of, by bone, 199
Sulfatides, incorporation of inorganic sulfate into, in brain, 222
Sulfation, as detoxification, in liver, 147, 149
Sulfhydryl group
free, in tumors, 236, 243, 244
in lymphocytes, 67, 68
oxidation of
in lymphoid tissue, 50
in red blood cell membrane, 38
protein-bound
in lymphocytes, 70
in tumors, 236, 244
SH/S—S ratio for, in lymphocytes, 72
in tumors, 236

T

Taurine
biosynthesis of, 258
conjugation of, with bile acids, 258, 259
enterohepatic circulation of, 258
excretion of, from lymphatic tissue, 260
free, in spleen, 58, 259, 260
liberated from intestine, 260
in muscle, 259
pools of, 258
in radiosensitive tissue, 259, 260
uptake of, by muscle, 259
in urine, 252, 255, 256, 258, 260, 266, 276
Taurinuria
after pancreatomy, 260
prevention of, with cortisone and mitomycine, 260
after splenectomy, 260
Temperature dependence in red blood cells
of cation transport, 33
of permeability, 32
Tendon, collagen in, 198
Testis
metabolism of testerone in, 300–302, 315
nucleic acid in, 301
sperm cells, Leydig and Sertoli cells in, 299
Testosterone, in testis
accumulation of 20α-dihydroxyprogesterone in, 301
biosynthesis of, 300
conversion of, to estradiol, 302
Thiolester hydrolases, in muscle, 212
Thymidine
biosynthesis of
in bone marrow, 23, 25, 29
in monocytic lymphoblasts, 55
incorporation of
into DNA
in bone marrow, 23, 25, 29
in brain, 221
in intestine, 85–87
in kidney, 215–217
in liver, 162, 166, 167, 169
in lymphoid tissue, 53, 55, 61, 62
in skin, 204
in spleen, 53, 62
in thymus, 53, 57, 62
into mDNA and nDNA, in liver, 167
pool
in lymphocytes, 54
in urine, 256, 262
Thymidine diphosphate, in bone marrow, 13, 23
Thymidine kinase (phosphotransferase)
in kidney, 216, 217
in liver, 177, 179, 180, 183
in lymphoid tissue, 55
Thymidinemonophosphate kinase, in liver, 159, 163, 177, 181
Thymidine phosphate kinases, in liver, 181
Thymidine phosphorylase, in liver, 182
Thymidine triphosphate
in bone marrow, 13, 23
incorporation of, into DNA
in liver, 168
in tumors, 238
rates of biosynthesis of, in intestine, 88
Thymidylate, in bone marrow, 13, 23, 26

Thymidylate kinase, in bone marrow, 23
Thymidylate synthetase, in thymus, 55
Thymocytes
accumulation of fructose-1,6-diphosphate in, 64
anaerobic glycolysis in, 64, 65
cytochrome c binding in, by mitochondria, 143
formation of lactate in, 64, 65
incorporation of precursors into DNA and RNA of, radiosensitivity of, 56
NAD in, 66, 67
nucleoproteins in, reactivity of sulfhydryl groups of, 61
Thymus
accumulation of deoxyribonucleotides and deoxyribonucleosides in, 54
acid RNase in, 68, 70
alkaline RNase in, 68
biosynthesis of NAD in, 66
catabolism
of DNA in, 54
of NAD in, 66, 72
change in nuclear composition of, 55
DNA in, 51
after adrenalectomy, 319
DNase II in, 59
histone/DNA ratio in, 61
incorporation
of acetate into lipids in, 58
of deoxycytidine into DNA of, 56
of inorganic phosphate into nuclear RNA of, 56
of palmitate into lipids in, 58
of thymidine into DNA of, 53, 55, 57, 62
of valine into nucleoproteins of, 57, 58
loss of NAD and NADP by thymocytes, 49
NAD/NADH ratio in, 66
NAD pyrophosphorylase in, 66
NADP in, 65, 66
nuclei
activation of amino acids in, 57, 58
aldolase in, 66
incorporation
of adenine and uracil into RNA of, 57
of leucine into nucleoprotein of, 57
of phenylalanine into nucleoproteins of, 57
of lactate dehydrogenase in, 66
of loss of glycolytic enzymes by, 64, 66
of phosphorylation of AMP in, 64
of RNA in, 56, 57
of triosephosphate dehydrogenase in, 66
of uptake of inorganic phosphate by, 55
phosphofructokinase in, 64
pools of precursors for DNA synthesis in, 54
RNase activity (total) in, 71
thymidylate synthetase in, 55
triglycerides in, 58
Thyroid gland, 291, 303–305, 315, 318, 322
Thyroid-stimulating hormone
of pituitary gland, 289, 291
secretion of, 291
in urine, 291
Thyroxine
action of, 315
effect of, on creatinuria onset, in muscle, 316, 331
Transaminases, in body fluids, 268
Transferrin level, effect of, on serum iron, 253
Transketolase, in liver, 132
Tricarboxylic acid cycle
dehydrogenases, in lymphoid tissue, 63
utilization of intermediates of, in bone marrow, 26
Triglycerides
in adrenal cortex, 293
biosynthesis of
in bone marrow, 14, 15
in intestine, 91
fatty acids in, in bone marrow, 14, 16
in liver, 119
after adrenalectomy, 120
in microsomes of liver, 119, 120
in mitochondria, in liver, 119, 120
in thymus, 58
Triosephosphate dehydrogenase, in thymus nuclei, 66
Triosephosphate isomerase, in liver, 130
Tryptamine, in urine, 265
Tryptophan
in blood, 254
catabolism of, 254
by intestinal bacteria, 92
in liver, 264, 266, 267

Tryptophan (*cont.*)
decarboxylation of, to tryptamine, in liver, 267
free, in spleen, 58
hydroxylation of, in liver, 267
incorporation of
into albumin, in total animal and liver, 273
into globulins, in total animal and liver, 273
into NAD, in liver, 265, 266
in liver, 267
metabolites of, excretion of, 264, 265, 267
pool of, in liver, 273
in urine, 256, 265, 276
Tryptophan hydroxylases, in liver, 308
Tryptophan oxygenase (pyrrolase), in liver, 156–158, 161, 266, 267
Tumors
biosynthesis
of DNA in, 236–238, 242
of proteins in, 237, 239, 242
of RNA in, 238, 239
Crabtree effect in, 239
DNA in, 236, 242
Ehrlich AT cells in, binding of cytochrome c in, by mitochondria, 143
glycolysis of, 237, 239–243
incorporation of thymidine into DNA in, 237, 242
oxygen consumption of, 239, 240, 242
Tyrosine
in blood, 254
free, in spleen, 58
in urine, 256, 276
Tyrosine aminotransferase, in liver, 151, 158, 161

U

Uracil
incorporation of, into RNA
in liver, 175
of thymus nuclei, 57
uptake of, by thymus nuclei, 61
Urate
in red blood cells, 39
in urine, 255, 256, 262
Urea
in blood, 254
catabolism of, 253
in intestine, 93
metabolism of, in liver, 146, 147
in tissues, 253
in urine, 252, 256
Ureidosuccinate, incorporation of, into nRNA, in liver, 175
D,L-Ureidosuccinate, incorporation of, into DNA, in liver, 169
Uridine, incorporation of
into microsomal and nRNA, in liver, 175
into RNA, in bone marrow, 29
Uridine diphosphate glucose, in metabolism of corticosterone, in liver, 294
Uridine diphosphate glucose glycogen glucosyltransferase, in liver, 133
Uridine diphosphate glucose pyrophophorylase, in liver, 132
Uridylate
in bone marrow, 13, 23, 24
incorporation in liver
of inorganic phosphate into, 173
of orotate into, 174
Urine
allantoin in, 256
amino acids in, 255, 256
due to protein catabolism subsequent to tissue breakdown, 198
2-aminoethanol in, 257
β-aminoisobutyrate in, 256
androgens in, 299
ascorbate in, 257
bile acids conjugated with taurine in, 259
biogenic amines in, 255, 307
Ca^{2+} in, 202
corticosterone in, 291
cortisol in, 295
deoxycytidine in, 184, 256
deoxyribonuclease II in, 269
deoxyribose derivatives in, 256
deoxyuridine in, 256
electrolytes in, 251, 253
end products of RNA catabolism in, 213
ethanolamine and serine in, 146
β-glucuronidase in, 269
glucuronides in, 257, 294
glycine in, 256
glycosaminoglycans in, 199
hippuric acid in, 146
hormones in, 255

Urine (*cont.*)
indoxyl sulfate in, 257
K^+ and Na^+ in, in gastrointestinal syndrome, 97
17-ketosteroids in, 295–297
malate dehydrogenase in, 215
N-methylnicotinamide in, 257
mucoproteins in, 257
nucleic acid metabolites in, 255
oxalate in, 257
peptide hydrolases in, 269
porphyrins in, 257
pseudouridine in, 256
pyridone in, 257
pyrrolecarboxylate in, 257
pyrrole-2-carboxylate in, 198
steroid hormones in, 291
thymidine in, 256
tryptophan in, 256
urate in, 255, 256
urea in, 146, 256
vitamin B, and related compounds in, 257
xanthine in, 256
Urocanic acid, in skin, 206

V

Valine, incorporation of, into nucleoproteins
in spleen, 58
of thymus nuclei, 57, 58
Vitamin B
in radiation disease, 267
in urine, 257

W

Water
in brain, 220
incorporation of, into cholesterol, in liver, 113–115
loss of, by small intestine, 82, 97, 98
metabolism of, 250
in urine, 250, 251
Wound, healing, 198

X

Xanthine, in urine, 256, 262
Xanthurenate, in urine, 264–266, 276

Z

Zinc, in serum, 253